材料物理
双语教程

◎密保秀 高志强 编著

人民邮电出版社
北京

图书在版编目（CIP）数据

材料物理双语教程 / 密保秀，高志强编著. -- 北京：人民邮电出版社，2018.11
21世纪高等学校规划教材
ISBN 978-7-115-45788-2

Ⅰ. ①材… Ⅱ. ①密… ②高… Ⅲ. ①材料科学—物理学—高等学校—教材—汉、英 Ⅳ. ①TB303

中国版本图书馆CIP数据核字(2017)第267945号

内 容 提 要

本书深入浅出地讲解了与材料性质相关的物理机制，以及材料性质的应用。主要内容包括：（1）材料的组织结构的基本知识及电子理论基础；（2）材料的各种物理性质。本书语言以英文为主，每一章都有对重点、难点句子的中文翻译。

本书在结合教学大纲和教学实践的基础上，融合了国内外同类教材的特点，并根据现代科学技术的发展增加了新内容。本书不同于传统的材料物理教科书，主要介绍金属及非金属中的无机材料，并加入了有机材料的物理性质，同时将侧重点由传统的力学及热学性质转向了现代的光电磁性质。

本书可以作为材料专业本科生的材料物理教材或者低年级硕士生的参考教材，也可以作为材料科学与工程领域相关科技工作者的参考书。

◆ 编　　著　密保秀　高志强
责任编辑　李　召
责任印制　彭志环

◆ 人民邮电出版社出版发行　　北京市丰台区成寿寺路 11 号
邮编　100164　　电子邮件　315@ptpress.com.cn
网址　http://www.ptpress.com.cn

◆ 开本：787×1092　1/16
印张：24.75　　　　2018 年 11 月第 1 版
字数：593 千字　　　2018 年 11 月河北第 1 次印刷

定价：98.00 元

读者服务热线：(010)81055256　印装质量热线：(010)81055316
反盗版热线：(010)81055315

前 言

Foreword

“材料物理”是高等院校中材料科学与工程专业及其相关专业的基础课程，是学生掌握材料性质的基础。将其设为双语教学，有助于学生在掌握专业知识的同时，拥有国际视野，并与国际高等教育逐步接轨。如果要实施双语教学，双语教材是必不可少的。然而，目前国内还没有出版材料物理的双语教材。

“材料物理”课程在南京邮电大学开设以来，编者经历了从母语到双语教学的全过程。本书是通过教学实践以及与学生的充分沟通后逐步完善而定稿的。此外，编者均毕业于物理与材料科学专业（博士），一直从事与材料相关的科研工作，紧跟该领域的前沿发展，并对材料性质有着深刻的理解，这些为本书的成稿奠定了雄厚的基础。

本书采取双语编排，每个章节包括：（1）正文（英语，局部中文翻译）；（2）本章小结（中文），（3）专业词汇（双语）；（4）思考题（英文）。全书文字深入浅出，非常适合作为学生教材。本书在满足国内教学大纲要求的同时，也为学生今后从事材料相关的科研奠定了基础。同时，书中的英文大部分来自原汁原味的英文原版书籍，并通过作者的整理加工编辑而成，有利于学生专业英语水平的提高。此外，书中含有英文重点、难点句子的中文翻译，可帮助英文相对薄弱的学生更好地理解专业知识，同时提高英文水平。

本课程的教学时数为 32～48。各章的参考教学课时见以下的课时分配表。

章	课程内容	学时
Chapter 1	Introduction	1～2
Chapter 2	Basic Structure and Organization of Atoms in a Material	2～4
Chapter 3	Fundamentals of Electron Theory	3～4
Chapter 4	Solid State Phase Transformation	4～6
Chapter 5	Mechanical Properties	4～6
Chapter 6	Electrical Properties	6～8
Chapter 7	Magnetic Properties	4～6
Chapter 8	Optical Properties	4～6
Chapter 9	Thermal Properties	4～6
课时总计		32～48

本书的编写参考了大量的中文教材和英文原版书籍，在此对这些参考书的作者表示感谢。同时，感谢南京邮电大学提供的江苏省首批赴美双语教学研修的机会，感谢材料院领导的关心与支持。感谢以下基金的支持：江苏省教改项目（2013JSJG225、2015JSJG028）、南京邮电大学教改项目（JG03013JX05）、南京邮电大学卓越教师培育计划（ZYJH201402）、南京邮电大学名师培育项目（2015 年）、江苏先进生物与化学制造协同创新中心、江苏省有机电子与信息显示协同创新中心、江苏高校优势学科建设工程（YX03001）。同时，南京邮电大学材料学院 GM 组的研究生刘杰、赵璐、王东平、陈洁、王超对本书的格式校对及图形绘

制做了大量工作。材料学院的钱洁、张晶晶、程佩珊、朱琦、张慧、徐若尘、邓尚平、姜茜、徐雪琪、王东琳、李锐、张伟、李进翔、宋雨鑫、王子恺对本书的出版也提供了帮助，在此一并表示感谢。

由于作者能力所限，加之双语特点，疏漏和不妥之处在所难免。恳请广大读者在使用过程中给予批评指正，以利于再版之际的修改和完善。与本书相关的讨论与交流，也十分欢迎。联系方式：iambxmi@njupt.edu.cn；iamzqgao@njupt.edu.cn.

密保秀，高志强

2017 年 8 月

目录

Contents

Chapter 1 Introduction

This chapter talks about the course scopes of material physics. After giving the definition and classification of materials, the history of material development has been briefly introduced. The relationships among material science and engineering, material physics, material chemistry, quantum dynamics, solid-state physics are discussed. Meanwhile, the content-emphasis of material physics is pointed out. Finally, the importance for material science is stated.

1.1 Definition of Material

Material is a broad term for the (chemical) substance, or a mixture of substances that constitute a thing. Based on their properties they can be used in structures, machines, devices, commercial products, and systems. There is a myriad of materials around us—they can be found in anything from buildings to spacecraft. In short, materials are the useful matter of the universe, the usefulness of materials is governed by the property of material itself.

The term property describes the behavior of materials when subjected to some external impacts such as force, heating, cooling and electric field etc.[1] For examples, the tensile strength of a metal is a measure of the material's resistance to a pulling force and the electrical conductivity of a material refers to the ability of conducting charge under an electric field.

1 性质一词用来描述材料对某种外界作用下的反应行为，这些作用包括力、加热、冷却以及电场等。

A property may be a constant or may be a function of one or more independent variables, such as temperature, electronic filed etc. Materials properties often vary to some degree according to the direction in the material in which they are measured, a condition referred to as anisotropy. The behavior of material properties that relate to different physical phenomena often can be modeled by the differential constitutive equations. In most case, these behaviors can model as linear in a given operating range and significantly simplify the differential constitutive equations that the property describes.

Some materials properties are used in relevant equations to predict the attributes of a system a priori. For example, if a material of a known specific heat gains or loses a known amount of heat, the temperature change of that material can be determined. Materials properties are most reliably measured by standardized test methods. Many such test methods have been documented by their respective user communities and published through ASTM International.

In general, material properties include but not limit to acoustical properties, atomic properties, chemical properties, electrical properties, environmental properties, magnetic properties, manufacturing properties, mechanical properties, optical properties, radiological properties, thermal properties etc. Our course will deal with the properties related to the physical phenomena such as mechanical and electrical behavior of materials.

1.2 Family of Materials

Metals and nonmetals should be the first of material classification by atomic bonding nature. Then along with civilization and material development, organic materials are isolated from nonmetals' family. Therefore, the nonmetal materials are divided into inorganic materials and organic materials; the latter usually refers to carbon-based materials generally made of covalently bonded carbon, hydrogen, oxygen, and nitrogen, etc, with minor species also containing elements of halogen, sulphur and phosphor.[2] Nevertheless, some carbon-containing compounds are traditionally considered as inorganic, such as carbon monoxide, carbon dioxide, carbonates, cyanides, cyanates, carbides, since their components and properties are very similar to other inorganic materials.

2 因此，非金属材料被分为无机材料和有机材料。有机材料通常是指碳基材料，一般含有通过共价键结合的碳、氢、氧和氮等，偶尔也含有卤素、硫及磷元素。

Sometimes, we also typically identify materials either as structural materials or functional materials, with mechanical or opto-electro-magneto/thermal properties to be considered, respectively. And according to the nature of chemical bonding or components, these functional materials can be further divided into four main families: Metals (e.g., steel), Polymers (e.g., plastics), Ceramics (e.g., porcelain), and Composites (e.g., glass-reinforced plastics). Additionally, considering atom stacking mode, there are crystalline, polycrystalline and amorphous materials; by conductivity, there are insulator, semiconductor, conductor and superconductor; according to space scale, there are bulky, nanomaterials and thin films; in terms of availability, there are natural and artificial materials; etc.

1.3 Brief History of Material Development

Materials played the key role in the history of world civilization and the progress of humankind development. The most important aspect of materials is that they are enabling; materials make things happen.[3] For example, in the history of civilization, natural materials used, such as stone, iron, and bronze, shown an important milestone in mankind's development, even coined the names of historical periods, such as Stone Age, Chalcolithic

3 材料是人类文明进步的关键，材料最重要的是有能力促使事件发生。

Age, Bronze Age and Iron Age. In today's fast-paced world, the discovery of making high quality silicon single crystals and the understanding of their electricity properties open the door of the information age.

In many cases different cultures leave their materials as the only records which anthropologists can use to define the existence of such cultures. The progressive use of more sophisticated materials allows archeologists to characterize and distinguish between peoples. This is partially due to the major material of use in a culture and to its associated benefits and drawbacks. For example, the innovation of smelting and casting metals in the Bronze Age started to change the way that cultures developed and interacted with each other. Starting around 5500 BCE, early smiths began to re-shape native metals of copper and gold - without the use of fire - for tools and weapons. The heating of copper and its shaping with hammers began around 5000 BCE. Melting and casting started around 4000 BCE. Metallurgy had its dawn with the reduction of copper from its ore around 3500 BCE. The first alloy, bronze came into use around 3000 BCE. Iron-working came into prominence from about 1200 BCE.

Prehistorically, before the consummation of material science, human beings had learned to use materials by empirical knowledge, the materials include: (1) organic materials: wood, ivory, bone, fiber, and rubber; (2) inorganic materials: minerals and ceramics including stone, flint, mica, quartz, clay, and diamond; (3) metals and alloys: iron, copper, silver and gold.

Just as shown in Figure 1.1, the accumulation of empirical knowledge pushed human civilization forward, which renders the creation of scientific knowledge that further promote the civilization and give insight to empirical knowledge. Thus, the science and technology on materials were gradually developed, which can further promote the development of new materials and explore their unique applications.[4]

4（通过对材料的使用），人类积累了大量经验知识，这促进了人类文明的进步，也产生了科学知识。科学知识又进一步促进了人类文明的进步，并揭示了经验知识的本质。由此，材料科学与技术也就逐渐形成，并成为新材料发展以及开发其独特应用的基础。

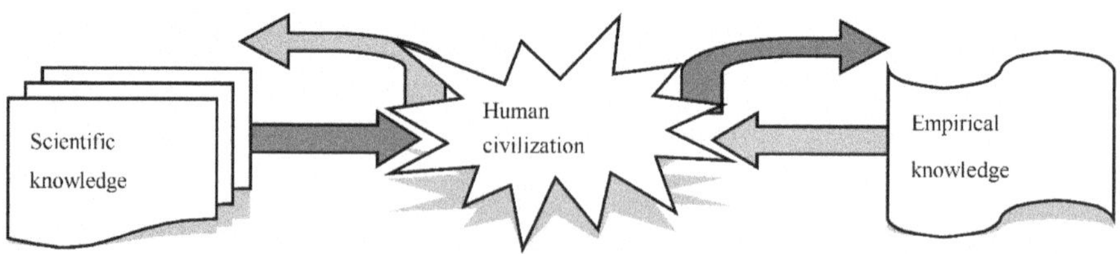

Figure 1.1 Both the empirical and scientific knowledge promote the progress of human civilization

Historically, in the development of material science, the discovery of modern atomic and molecular theory is the fundamental basis of all materials science and a key to understanding material structure and processes, which includes the following milestones:[5]

In 1869, Dmitri Mendeleev published the periodic table of the elements, correlating all atoms in materials.

In 1900, Planck's hypothesis well explained the blackbody radiation by introducing the concept of quantum for radiation.

In 1901, X-ray was found by the German physicist W. K. Roentgen. Since then, the X-ray analysis technology has been used in material science to investigate the construction and structure of materials.

In 1905, Einstein used photon theory explained the photoelectric effect, evidencing the particle nature of light.

In 1911, super conductivity was found by Kamerlingh Onnes in mercury at temperature of 4.2 K, adding a new special member in conducting materials.

In same year 1911, Rutherford introduced his 'classical' atomic model consisting of small nucleus surrounded by electrons, firstly disclosing the core-shell structure of atoms.

In 1924, de Broglie suggested wave particle duality, initializing an approach of quantum dynamic explanation between material macro-properties and microstructure.

In 1928, Bloch's theorem explained the conductivity of metal, which was further developed by Wilson in 1931 to insulator.

In 1960, the practical development of field effect transistor leads to the development of integrated circuit (IC).

In 1960, the first laser (ruby laser) was invented.

It can be said that the science of material began with metals, which was termed as metallurgy. Metals possess excellent mechanical properties such as formability and strength, thus serve as prime materials for structural applications.[6] Additionally, their excellent electrical and thermal conductivity has rendered them indispensable for electrical engineering. Despite its very long tradition, metallurgy is not a classical scientific discipline itself. Up to medieval ages, the knowledge of the extraction and fabrication of metals had been considered as a national secret asset and had been only traded orally from generation to generation.[7] As a matter of fact, metallurgy was a discipline of alchemies in medieval times and comprised a colorful mixture of empirical recipes and superstition. With increasing scientific character of more recent centuries metallurgy became a discipline of chemistry, where it remained even up to now at many universities. The rapid development of the technology to understand the material properties, in particular due to the

5 纵观历史，现代原子及分子物理的发现是材料科学发展的基础，也是理解材料结构及工艺的关键。其中的里程碑包括：

6 可以说材料科学始于金属，即冶金学。金属拥有出色的机械性能，其可成型性及强度都很好，可以用作基础建筑材料。

7 虽然冶金学历史较长，但它本身不是一门经典的学科。直到中世纪，金属冶炼及铸造知识都被认为是国家机密仅仅通过口述方式代代相传。

discovery of radiative beam, such as X-rays, electron beam and ion beam, and their application to material composition and structure, revealed that the properties of metals were not determined by the gross chemical composition, in contrast to common belief at that time. This made metallurgy become a discipline of physical chemistry.[8]

The development of the atomistic foundations of our understanding of mechanical and electronic properties of metallic materials in the frame work of dislocation theory and electron theory of metals finally shifted the focus of metallurgy to physics at the beginning of the 20th century. Eventually it engendered the discipline of metal physics, which has dominantly influenced the science of metals in the past 50 years.[9] In fact, our current understanding of metallic and non-metallic materials on the basis of atomistic models has essentially been developed only in the past 80 years of research in metal physics. The objective of this research has been to describe the properties of a material on the basis of atomistic physical models, which can be formulated in terms of equations of state. This allows for a prediction of materials behavior on a theoretical basis and changes the material research from try and error to the theoretical guiding. This will significantly reduce the cost and time on the experimental investigations and testing of materials behavior.

In the sixties and seventies of the 20th century it became obvious that the urgent demand for high performance materials and competitive mass products had to include the development of non-metallic materials, for instance ceramics for high temperature components and plastics for a weight savings in automobiles and airplanes.[10] Materials research revealed soon, however, that the foundations of physical metallurgy within limits can be readily applied to other materials, in particular to crystalline solids. Crystallography, constitution, diffusion, phase transformations, physical properties, and so on, are the foundations of the understanding of all kinds of engineering materials. Of course, there are also specific differences. For instance, dislocation theory, which is indispensable for an understanding of plastic deformation of metals, is of less importance for brittle ceramics, but it teaches the reasons for the brittleness and, therefore, offers respective for counter actions. For polymers which are usually non-crystalline, an appropriate dislocation concept of their mechanical properties is still too complicated to be useful, and the deformation behavior of plastics is, therefore, currently restricted to phenomenological tribological models.

Material development generated the strong belief that it is possible to

8 由于对材料性能理解技术的迅速发展，尤其是X-射线、电子及离子束的发现及其在材料成分和结构上的应用，人们逐渐意识到金属的性质并不像通常认为的那样，是由其总的化学组成决定的。这使得冶金学转变为物理化学的一个学科。

9 在20世纪初，在位错理论及电子理论框架下的金属材料的力学及电学性质，进一步向原子水平发展，并将冶金学的焦点转移到物理。最终形成了金属物理学科，并在过去的50年里，主导着金属科学的发展。

10 20世纪六七十年代，基于人们对高性能材料及有竞争力的大规模生产的渴望，非金属材料得以发展。例如，陶瓷可用于高温组件，塑料可用于机动车及飞机来减轻重量。

derive a comprehensive description comprising the different classes of materials and that the future world be multi-material.[11] This worldwide trend in the seventies of the past century caused the classical independent disciplines of metallurgy, ceramics, and plastics to merge to a new discipline "material science and engineering", encompassing both the science and engineering aspects of materials, and which has become our modern powerhouse of materials research and development.

11 材料的发展使人们坚信，有可能得到一个包括不同类型材料的综合性描述模型，而且未来世界应该是多种材料的。

1.4 Material Physics and Other Related Science

As shown in Figure 1.2, the development of novel materials and processes requires a deep knowledge of the physical and chemical foundations of materials, especially the knowledge of the relation between microscopic structure and macroscopic properties of materials, which facilitates the systematic tailoring of materials properties. These form the multidisciplinary including material science.[12]

12 新型材料及其工艺的开发需要对材料物理及化学基础、尤其是材料宏观性质与微观结构之间关系的深入理解，以利于对材料性能的系统裁剪。由此形成了包括材料科学在内的多个学科。

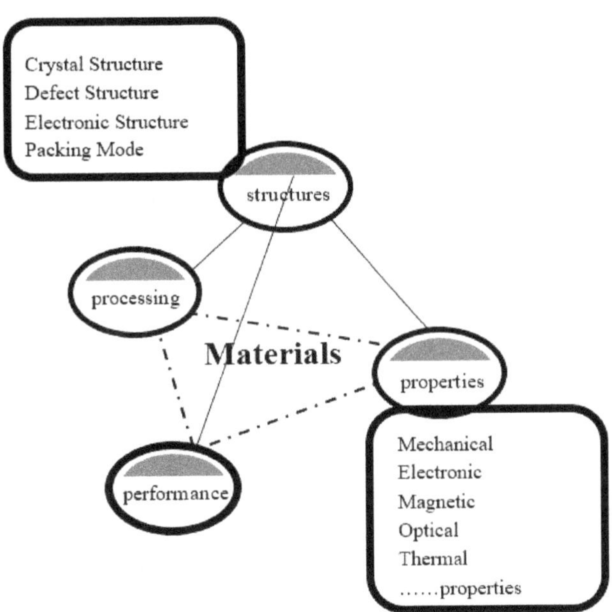

Figure 1.2 Different aspects of materials

- **Material science and engineering**

Materials science and engineering can be thought of as a combination of the sciences of chemistry and physics within a backdrop of engineering. Chemistry helps to define the synthetic pathways, and provides the chemical makeup of a material, as well as its molecular structure. Physics provides an understanding of the ordering (or lack thereof) of atoms and/molecules and electronic structure, and physics also provides the basic principles that enable a description of materials properties. The combined information provided by physics and chemistry about a material leads to the determination and correlation of materials properties with the process used to prepare the material, and with the materials structure and morphology.

The material physics and material chemistry are the roots of material science. The term material science is relatively young and not very precisely defined. Sometimes it is understood as an extension of metallurgy to non-metallic materials. Material engineering in some degree, means, material processing.[13]

13 材料物理及材料化学构成了材料科学的基础。材料科学的概念相对比较新，目前没有精准的定义。有时材料科学被理解为冶金学到非金属材料的延续。材料工程在某种程度上代表材料工艺。

Generally speaking, the “science” focuses on discovering the nature of materials, which in turn leads to theories or descriptions that explain how structure relates to composition, properties, and behaviours. The “engineering,” on the other hand, deals with use of the science in order to develop, prepare, modify, and apply materials to meet specific needs. The field is often considered an engineering science because of its applied nature. Materials science and engineering is interdisciplinary or multidisciplinary, embracing areas such as metallurgy, ceramics, solid-state physics, and polymer chemistry.[14]

14 材料科学与工程是交叉学科，或者说是多学科；它包含冶金学、陶瓷、固体物理及聚合物化学等领域。

- **Material physics**

Material physics is a part of material science; it is also the largest branch of condensed matter physics. The purpose of material physics is to study the physics phenomenon, the effect of microstructure on material properties, as well as the physical mechanisms in materials. Alternatively, it can be said that material physics uses the physics to describe materials in the aspects of force, heat, conductivity, magnetism and light, and they are the key elements of this textbook.

- **Solid-state physics**

Solid-state physics deal with the condensed matter, or solids, through methods such as quantum mechanics, crystallography, electromagnetism, and metallurgy. Solid-state physics studies how the large-scale properties of solid materials result from their atomic-scale structure. Thus, solid-state physics forms the theoretical basis of materials science. It also has direct applications, for example in the technology of transistors and semiconductors.

- **Quantum mechanics**

Quantum mechanics (QM – also known as quantum physics, or

quantum theory) is an important branch of modern physics, which deals with physical phenomena at atomic level where the action is on the order of the Planck constant. It departs from classical mechanics primarily at the quantum realm of atomic and subatomic length scales. QM describes of the dual particle-like and wave-like behaviour and interactions of energy and matter in a mathematical language. It also provides a substantially useful framework for many features of the modern periodic table of elements including the behaviour of atoms during chemical bonding and has played a significant role in the development of many new functional materials.[15]

15 量子力学使用数学语言描述能量和物质的波粒二象性及相互作用，并为现代元素周期表中元素的许多特性提供了有用的框架式解释，包括：描述原子在形成化学键时的行为，这为许多新功能材料的发展起到了重要作用。

1.5 Importance of Material Science

The basis of materials science involves studying the structure of materials, and relating them to their properties. Once a materials scientist knows about this structure-property correlation, he/she can then go on to study the relative performance of a material in a certain application. The major determinants of the structure of a material and thus of its properties are its constituent chemical elements and the way in which it has been processed into its final form. These characteristics, taken together and related through the laws of thermodynamics and kinetics, govern a material's microstructure, and thus its properties.

Materials science underlies all technological advances and an understanding of the basics of material and their application will not only make you a better engineer, but will help you during the design process.[16] In order to be a good designer, you must learn what materials will be appropriate to use in different applications. You need to be capable of choosing the right material for your application based on its properties, and you must recognize how and why these properties might change over time and during processing. Any engineer can look up materials properties in a book or search databases for a material that meets design specifications, but the ability to innovate and to incorporate materials safely in a design is rooted in an understanding of how to manipulate materials properties and functionality through the control of the material's structure and processing techniques.

16 材料科学强调技术的先进性。对材料基础知识及其应用的理解，不仅可以造就更好的工程师，还有助于工艺的设计。

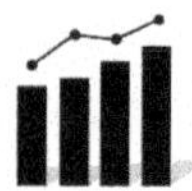

本章小结

1. 内容概要

本章主要介绍材料物理的范围和方向。首先，介绍了材料的定义和分类，简单回顾了材料的发展历程；然后，介绍了材料物理的概念，简单地界定了材料科学与工程、材料物理、材料化学、量子力学、固体物理之间的关系，同时，指出材料物理的内容侧重点。最后，简述了材料科学的重要性。

2. 基本概念

材料、材料物理、材料化学、材料科学、材料工程。

Vocabulary

acid	酸
action	作用
alchemy	炼金术
allowed	考虑，允许
alloy	合金
amorphous	无定型的
artificial material	人工材料
base	基底，碱
behavior	状态，行为
blackbody radiation	黑体辐射
bone	骨架
brief	摘要
carbide	碳化物
carbon	碳
carbonate	碳酸盐
carbonic acid	碳酸
century	世纪
ceramic	陶器
chemical	化学用的
clay	黏土
clue	线索
component	成分

compounds	混合物
concept	概念
conductor	导体
confirm	确定
contain	包含
continuous	连续的
copper	铜
covalent bond	共价键
crystal	结晶
crystallography	晶体学
crystal structure	晶体结构
crystalline	结晶性的，透明的
cyanide	氰化物
defect structure	缺陷结构
diamond	钻石
disclose	揭露
discovery	发现
discrete energy states	离散的能量状态
electric	导电的
electrical	电的
electron beam	电子束
electronic	电子的
electronic structure	电子结构
element	元素
elemental substance	元素物质
energy	能量
fiber	纤维
flint	打火石
fluorescent tube	荧光管
frequency	频率
fullerene	富勒烯
functional materials	功能材料
gold	黄金
govern	支配
halogen	卤素
heat	热，加热

hydrogen	氢
incandescent lamp	白炽灯
inner shall	内层
inorganic	无机的
inorganic material	无机材料
insulator	绝缘体
integrated circuit	回路，电路
interdisciplinary	跨学科的
iron	铁
ivory	象牙
laser	激光
light	光
magnetic	有磁性的
major	专业
material physics	材料物理
material processing	材料加工
mechanism	原理
metal	金属
metallurgy	冶金学
mica	云母
microstructure	显微结构
mineral	矿物
minor	较少的，少数的
molecule	分子
nano material	纳米材料
natural material	天然物质
nitrogen	氮
nonmetal	非金属
opaque	不透明
organic	有机的
organic material	有机材料
outer shall	外层
oxide	氧化物
oxygen	氧
performance	性能，表现
periodic table	周期表

phosphor	磷
photoelectric effect	光电效应
Planck's hypothesis	普朗克假设
property	性质
quantum mechanics	量子力学
quartz	石英
reflect	反射
rubber	橡胶
ruby laser	红宝石激光
salt	盐
semiconductor	半导体
silver	银
similar	相似的
solid state physics	固态物理
species	种类
stone	石头
structural material	结构材料
sulfuric acid	硫酸
sulphur	硫磺
superconductor	超导体
theorem	定理
transistor	电晶体
treat	看作
universe	宇宙
water	水
wood	木材
X-ray diffraction	X 射线衍射

Problems

1. What is the target of material physics?
2. In terms of typical classification, how to classify materials?
3. Please give five material properties.
4. True or false questions.

(1) Inorganic and organic materials are all belong to metal-type material.

(2) All carbon based materials are organic materials.

(3) Material science includes material chemistry and material physics.

(4) In some sense, material engineering can be treated as material processing.

5. Based on conductivity, how many types of materials can be classified? What are they?

6. In the following items, which are materials and which are material properties?

Air, Sound, Light, Copper, Electricity, Triboelectrification, Magnet, Glass, Radio frequency.

Chapter 2 Basic Structure and Organization of Atoms in a Material

Materials science is often described as being comprised of structure-property relationships. In this context structure refers not only to the arrangement of the basic building blocks, or long-range ordering but also to the chemical structure or short-range ordering. Material is made of atoms/molecules. In order to better understand structure-property relationships of materials, two fundamental concepts need to be taken into account. The first is the forces that hold the atoms/molecules together in a matter, in other words, the bonding between atoms/molecules, which can be basically categorized into metallic bonding, ionic bonding, covalence bonding and molecular bonding. The second important aspect is the atoms/molecules ordering within a material. This chapter will discuss these issues.

2.1 Bonding Modes Among Atoms

The structural elements of matter are the atoms, which consist of a nucleus and the atomic shell. The properties of solids are essentially determined by the electron shell structure. According to the Bohr model of an atom, the electrons occupy specific orbitals (Figure 2.1), the configuration of which, i.e. number of electrons and their spatial arrangement, follows the laws of quantum mechanics. The most important electrons for the properties of a solid are the electrons in the outermost orbital, because they determine the interaction with other atoms. The dominant principle of atomic interaction is the tendency of an atom to have its outermost shell filled with eight electrons, i.e. the noble gas configuration. This simple principle is the foundation of chemical bonding. If an atom has already a complete outer shell with eight electrons, like the noble gases, then its tendency to interact with other atoms, i.e. for chemical bonding or even for solidification is very small. All elements which do not have a noble gas configuration have the tendency to accept, to donate, or to share the outermost electrons, also referred to as valence electrons, when in contact with other atoms. From these principles we obtain the fundamental types of atomic bonding. An understanding of this helps categorize materials as metals, nonmetals, organics, or inorganics, etc. It also permits us to draw some general conclusions concerning material properties. There are four types of atomic bonding in materials: metallic bonding, ionic bonding, covalent bonding, and molecular bonding.

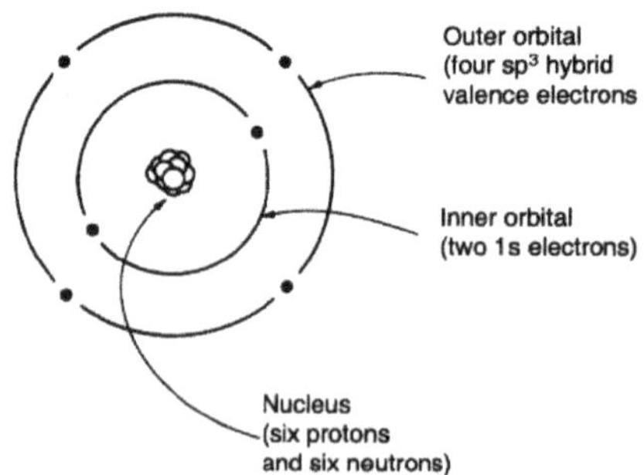

Figure. 2.1 Diagrammatic illustration of the electronic configuration of the ^{12}C atom after Bohr's atomic model.

- **Metallic bond**

As shown in Figure 2.2b, If the number of valence electrons is less than four, a noble gas configuration cannot be established by electron pairs in a three dimensional lattice. In this case the atoms prefer to donate their valence electrons to a common electron gas and form the metallic bond. So that the ionic cores attain the noble gas configuration and the electrons in the electron gas are not associated with a particular atom. The metallic bond is essentially coulombic, but the negatively charged species are the electrons that are delocalized and free to move throughout the solid.[1] This is the most frequent type of bond among the elements, because about three quarters of all natural elements are metals. In the metallic bond the electrons are not localized and practically belong to all atoms at the same time. The weak localization of the electrons in the metallic bond is the reason for its weakness in comparison to the other types of bonds. In turn, this causes the high mobility of defects in metals and, therefore, their excellent formability, which has made metals the dominant structural materials. The electrical conductivity and reflectivity are quite high, and the melting point is high, although it tends to be lower than that of the ionic solids. Metals exhibit a characteristic malleability and ductility.

1 金属键本质上是一种库仑力，但是其负电组分是可以在整个金属固体中自由移动的电子。

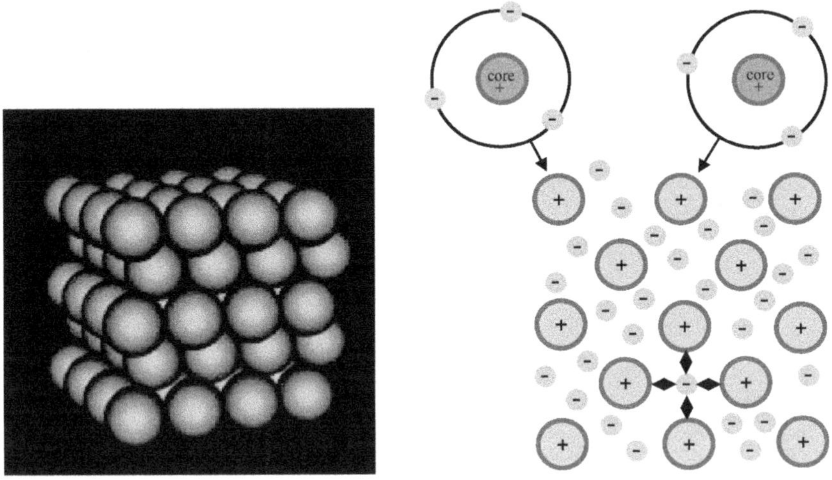

(a) (b)

Figure 2.2 (a) Stacking of metallic atoms. (b) Metallic bonding.

- **Ionic bond**

An ionic bond is built between two unlike atoms with different electronegativity.[2] The number of valence electrons of the partners adds up to eight. The partner with smaller valency donates its valence electrons to the other partner with higher valence. Both elements attain a noble gas configuration, but the atoms lose their charge neutrality. For instance, in sodium chloride when sodium donates its valence electron to chlorine, each atom becomes an ion, coulombic attraction occurs, forming the ionic bond

2 离子键产生于两个不同电负性的不同原子之间。

(Figure 2.3). Ionic solids tend to have relatively high melting points (sodium chloride melts at 801 ℃) and are rather brittle. They are not good electronic conductors at low temperatures and exhibit ionic conductivity at temperature close to and above the melting point.

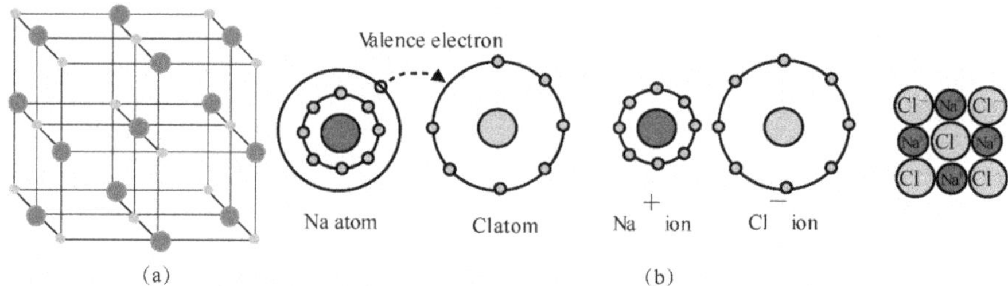

Figure 2.3 (a) Atom stacking in NaCl. (b) Ionic bond of NaCl.

- **Covalent bond**

If the noble gas configuration cannot be established by exchange of electrons since the sum of valence electrons does not add to eight, a stable arrangement of atoms in a molecule can be obtained by the formation of electron pairs. A covalent solid is formed and characterized by bonding that results from the sharing of electrons between neighboring atoms (Figure 2.4); the bonding can have a polar character.[3] Diamond, silicon carbide, quartz, silicon, and germanium are examples of covalently bonded solids. These solids tend to have a low electronic conductivity and high melting point and are likely to be hard and brittle.

3 共价固体由相邻原子之间通过共享电子成键而形成，共价键可以带有极性。

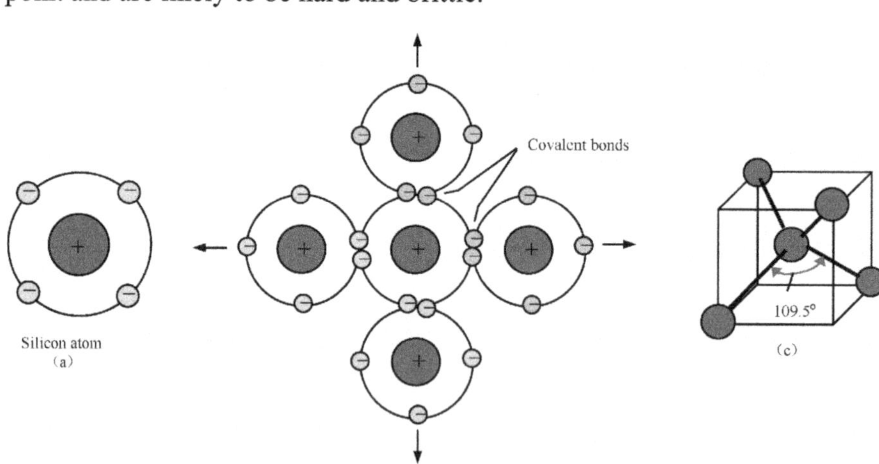

Figure 2.4 (a) Covalent bonding requires that electrons be shared between atoms in such a way that each atom has its outer sp orbitals filled. (b) In silicon, with a valence of four, four covalent bonds must be formed. (c) Covalent bonds are directional. In silicon, a tetrahedral structure is formed with angles of 109.5° required between each covalent bond.

- **Molecular bond**

The previously discussed three bondings are primary bonding, i.e., the atomic (chemical) bonding. In materials of some cases, there is a secondary bonding, namely the so-called Van der Waals bond, which is not associated with the donating-accepting or share of electrons but caused by a dipolar (either an induced or permanent dipolar) interaction of atoms. This dipole interaction is caused by the fact that the center of gravity of charge of the nucleus is not identical with that of the electron shell. This causes a dipole moment of the atoms which invariably generates an attractive interaction with other atomic dipoles. This attractive force is the reason for bonding in noble gas molecules and the interaction of far apart atoms, when no exchange of electrons can take place.

Molecular solids are composed of discrete molecules held together by weak van der Waals forces; though the molecules themselves can consist of atoms held together by covalent bonds. Van der Waals forces between atoms and molecules have their origin in interactions between dipoles that are induced or in some cases interactions between permanent dipoles that are present in certain polar molecules.[4] There are three types of van der Waals interactions, namely **London force, Keesom force** and **Debye force**.

If the interactions are between two dipoles that are induced in atoms or molecules (e.g., carbon tetrachloride), we refer to them as London forces, which are also termed as dispersive force.[5] Figure 2.5 illustrates this type of force.

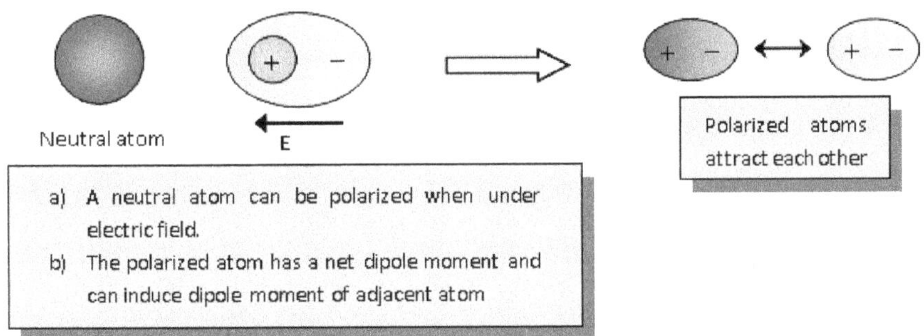

Figure 2.5 Illustration of London forces, a type of a van der Waals force, between atoms.

When an induced dipole (that is, a dipole that is induced in what is otherwise a non-polar atom or molecule) interacts with a molecule that has a permanent dipole moment, we refer to this interaction as a Debye interaction[6] (Figure 2.6).

If the interactions are between molecules that are permanently polarized, it is referred as Keesom interactions. For instance, the positively charged region of one water molecule and the negatively charged regions of a second water molecule render an attractive bond between them[7] (Figure 2.7).

4 诱导偶极子或者某些极性分子中永久偶极子之间的作用力形成了原子或分子之间的范德华力。

5 如果原子或者分子（例如，四氯化碳分子）之间的两个偶极作用是通过诱导产生的，这种相互作用力称为伦敦力，也称为色散力。

6 一个诱导偶极子（非极性原子或分子中被诱导的偶极子）与一个拥有永久偶极子的分子之间的作用力叫作德拜力。

7 极化分子之间的作用力称为基森力。例如，一个水分子带正电的部分和另一个水分子带负电的部分之间的吸引力。

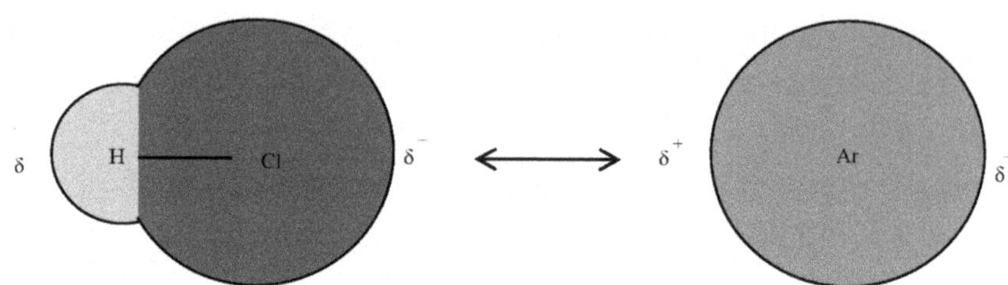

Figure 2.6 Debye interaction between HCl and Ar molecules.

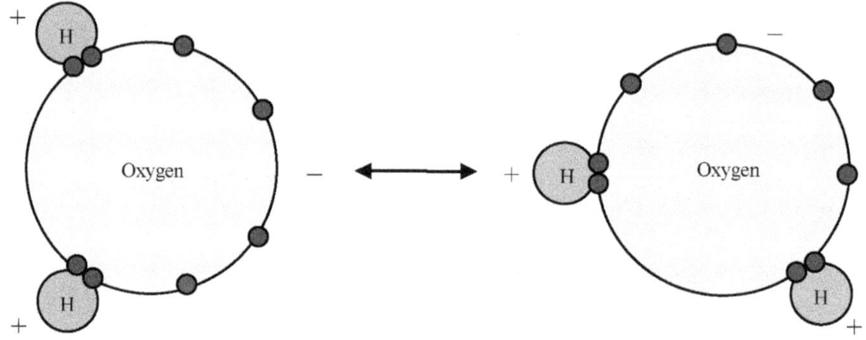

Figure 2.7 Dipole-dipole interaction (Keesom force) between water molecules.

The bonding between molecules that have a permanent dipole moment is often referred to as a hydrogen bond, where hydrogen atoms represent one of the polarized regions. Thus, hydrogen bonding is essentially a Keesom force and is a type of van der Waals force.

Note that Van der Waals bonds are secondary bonds, but the atoms within the molecule or group of atoms are joined by strong covalent or ionic bonds. Although termed "secondary," based on the bond energies, van der Waals forces play a very important role in many material properties.[8] First of all, because of the weak nature of the bonding between molecules in a molecular solid, the energy orbitals between molecules cannot form band structure, but exist isolated, leading to low conductivity. So the molecular solids exhibit narrow band width, wide bandgap and low carrier mobility. It is to be expected that the properties of the individual molecules are retained in the solid state to a far greater extent than in solids with other three types of bonding. Secondly, due to the weak interaction among molecules, molecular solids are soft with low melting points. In addition, low interaction renders loose stack of molecules and large lattice variation. Especially, factors such as temperature, pressure, and external magnetic field can change the space arrangement of molecular solid easily, which results in change of material properties. Specifically, van der Waals forces

8 要注意到范德华力是次级键，但是分子内或者一组原子内的原子之间是通过很强的共价键或离子键结合的。虽然基于键能被称为“次级”，范德华力在很多材料的性质方面起非常重要的作用。

between atoms and molecules play a vital role in determining the surface tension and boiling points of liquids.

In many plastic materials, molecules contain polar parts or side groups (e.g., cotton or cellulose, PVC, Teflon). Van der Waals forces provide an extra binding force between the chains of these polymers. Polymers in which van der Waals forces are stronger tend to be relatively stiffer and exhibit relatively higher glass transition temperatures (T_g). The glass transition temperature is a temperature below which some polymers tend to behave as brittle materials (i.e., they show poor ductility). As a result, polymers with van der Waals bonding (in addition to the covalent bonds in the chains and side groups) are relatively brittle at room temperature (e.g. PVC). In handling such polymers, they need to be "plasticized" by adding other smaller polar molecules that interact with the polar parts of the long polymer chains, thereby lowering the T_g and enhancing flexibility.[9]

9 在加工这样的聚合物时，需要加入一些较小的极性分子将其塑性化。这些极性小分子与聚合物长链中的极性部分相互作用，从而降低聚合物的玻璃化温度，使其柔性增强。

2.2 Crystal Structure

As previously mentioned, materials are made of atoms and primarily based on the nature of atoms, different types of bonding can be found, as discussed in previous section. Besides different atoms and different atomic bonding, atom organization (atom packing) is another key factor influencing material properties.[10] The properties of advanced materials are not so much affected by their overall chemical composition but rather by the specific arrangement of their constituents which we usually can't discriminate with our bare eye. The arrangement of the constituents of a material, i.e. the spatial distribution of elements, phases, orientations, and defects are subsumed under the term microstructure.

10 除了不同原子以及不同的原子键合方式，原子的排列（原子堆叠）是另一个影响材料性质的主要因素。

Atom arrangement in materials can be quite different, as shown in Figure 2.8. According to the space arrangement of atoms in solids, there are:

Single crystalline: atoms are arranged periodically in space, possessing only one long-range regularity.

Polycrystalline: atoms are arranged partially in periodic way, possessing many/several long-range regularity.

Amorphous structure: atoms are randomly distributed in space, possessing short-range order but disordered in a long range.

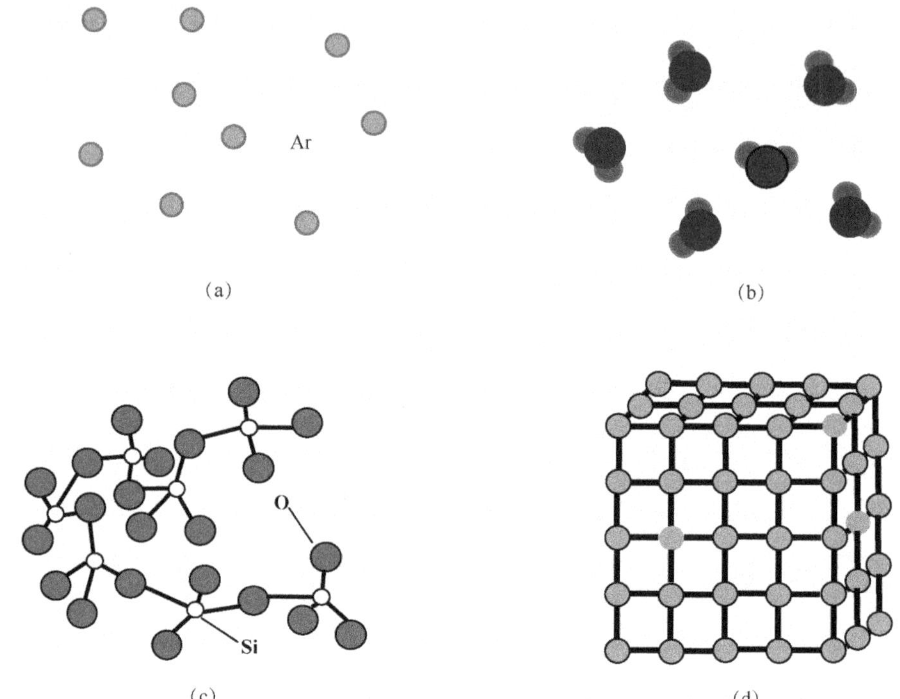

Figure 2.8 Levels of atomic arrangements in materials: (a) disordered and isolated atoms in noble gases, (b) and (c)short-range ordered materials of water vapor and silicate glass, (d) long-rang ordered materials of metals, alloys, many ceramics and some polymers.

Quasicrystalline: having space order feature between crystalline and amorphous solids.

2.2.1 Unit Cells, Space Lattices and Lattice Point

Metallic and ceramic materials are usually crystalline. Also polymers can partially crystallize, a crystalline structure means in a physical sense, a strictly periodic arrangement of atoms. However, long before the atomistic structure of solids was known, crystals of minerals fascinated man and became a subject of scientific interest. The prominent feature of crystalline minerals is their external appearance with planar facets, which are characteristic of each mineral. According to the geometry of the facets, the crystallographers were able to group the crystals in terms of their shapes and symmetries into 32 classes (also referred to as point groups), which could be subdivided into seven crystal systems. These seven crystal systems can be defined by the macroscopic orientation of the crystal surfaces and their lines of intersection in an appropriately chosen crystal coordinate system. If there is no apparent symmetry, the crystal structure is called triclinic, and the angles between the crystal axes and their respective lengths are all different. The highest symmetry is represented by cubic crystals, where all crystal axes are equally long and are arranged under a mutual angle of 90°. The detail information of crystallography can be found in the textbook of solid state physics.

As mentioned, solids are made of atoms that are chemically bound to one another. Most solids, including all metals and most ceramics, have a crystalline structure. The strength and nature of the forces between the atoms of the crystal lattice and also the electron clouds around the nuclei determine the macroscopic properties of the solid, e.g., strength, elasticity and conduction of heat and charge.[11] Taking two carbon materials, graphite and diamond, as an example, the importance of the crystalline structure for the determining of matter behavior can be clearly seen.

11 晶格中的原子成键的本质与强度，以及环绕原子核的电子云，决定了固体的宏观性质，例如，强度、弹性、导热和导电特性。

As shown in Figure 2.9, graphite belongs to hexagonal crystal system. It has a parallelly-layered, planar structure. In each layer, there are six-membered rings of carbon atoms. Within one layer, the carbon atoms are held together by strong covalent bonds. Between the layers, the atoms are held by weak van der Waals bonds. This is the reason for the weakness of graphite towards shearing forces and it can work as a lubricant. Contrary to graphite, diamond is made of tetrahedrally shaped and covalently bound carbon atoms, forming face-centered cubic crystal structure. Due to the strong covalent bonding, as well as the extremely rigid lattice, diamond is one of the hardest elements in nature.

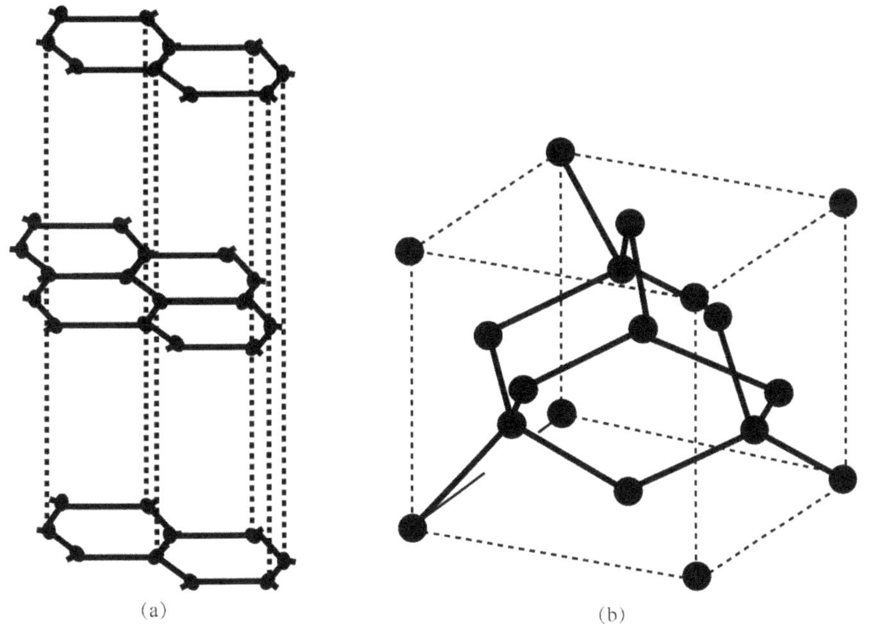

Figure 2.9 Crystalline structure of (a) graphite and (b) diamond.

When atoms are chemically bound to one another they have well-defined equilibrium separations that are determined by the condition that the total

12 当原子之间形成化学键时，它们有明确的平衡距离，这个距离相当于总能量最小时的情形。

energy is minimized.[12] Therefore, in a solid composed of many identical atoms, the minimum energy is obtained only when every atom is in an identical environment. This leads to a three-dimensional periodic arrangement that is known as the crystalline state. A metal would be representative of a crystalline solid (though not perfectly crystalline). The same is true for solids that are composed of more than one type of element. In this case, certain building blocks comprising a few atoms are the periodically repeated units.

Periodicity gives rise to a number of typical properties of solids. Periodicity also simplifies the theoretical understanding and the formal theory of solids enormously. Although a real solid never possesses exact three-dimensional periodicity, one assumes perfect periodicity as a model and deals with the defects in terms of a perturbation. Three-dimensional periodic arrangements of atoms or building blocks are realized in many different ways.

13 如此这样，我们把原子、离子或者分子等效为在空间占据不同距离、不同大小的球，因此需要一个坐标系。

Therefore, in our study of crystalline structures, we will limit ourselves to the orderly arrangement of the atoms in their microscopic world. In this way, we represent atoms, ions, or molecules essentially as spheres of varying sizes occupying points at various distances from each other in space, hence the need for an axis system.[13] Such a system is shown in Figure 2.10.

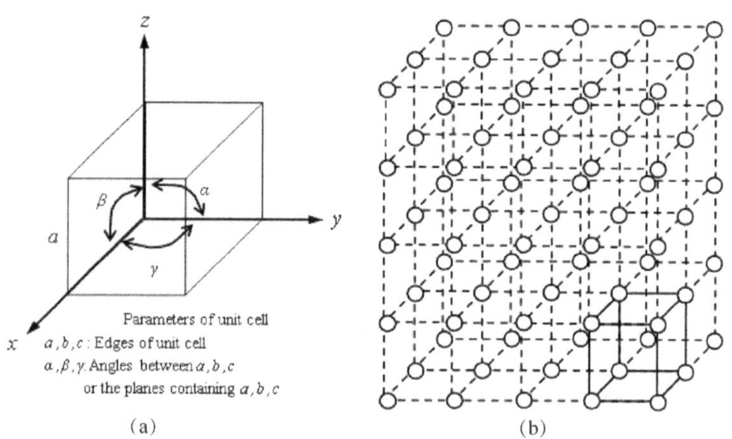

Figure 2.10 (a) A three-dimensional box representing a unit cell. (b) Repeated unit cells in three dimension, leading to a crystal lattice.

Unit cell: The term unit cell is used to describe the basic building block or basic geometric arrangement of atoms in a crystal. You can compare a unit cell to a single brick in a brick wall. The box in Figure 2.10a is a unit cell. The angles between the principal planes are named α, β, and γ, where α is the angle measured in degrees between the x–y and x–z planes; angle β, between x–y and y–z planes; and angle γ, between the x–z and y–z

planes. The sides of the box (unit cell), labeled *a*, *b*, and *c*, are the lattice parameters in the *x*, *y*, and *z* directions, respectively. These distances are also known as intercepts. To describe a particular axes or crystal system adequately, all six dimensions are needed, which are known as **lattice constants**.

Space/crystal lattice and lattice points: If you repeat duplicating the unit cell in all three dimensions, you create a crystalline structure with a definite pattern (Figure 2.10b). This larger pattern of atoms in a single crystal is known as a space lattice or crystal lattice.

A space lattice is defined as a collection of mathematical points, arranged in such a way that each of the points is surrounded by precisely the same configuration, such that the constructing space is divided into small volumes of equal size, with atoms (ions or molecules) located at the intersections of these lines or between the various lines.[14] The view will always be the same, independent of which point we chose as observation site. The points in space lattice are called lattice points. Every lattice point on one side of the observation site always has a corresponding lattice point in an identical position but situated on the opposite side.[15]

Primitive cell: We must keep in mind that the lines and points in a space lattice are only imaginary. The lattice concept is used to show the positions of atoms, molecules, or ions in relation to each other. We must also remember that the actual atoms in solids are located as close to each other as possible, thus attaining the lowest possible energy. Two atoms closest to each other would be represented by two spheres touching each other. The closer the atoms are, the denser the solid. In addition, we also must keep in mind that even if the lattice structure is basically simple, the crystal structure can be very complicated as the unit cell might consist of tens of thousands of atoms.[16] This is the case in proteins and other organic system. If the unit cell only contains one atom/molecule, it is called primitive cell. The number of atoms or molecules per unit cell is given by:

$$N_{\text{unit cell}} = N_{\text{interior}} + \frac{N_{\text{face}}}{2} + \frac{N_{\text{corner}}}{8} \tag{2-1}$$

where N_{interior}, N_{face}, N_{corner} is the number of atoms/molecules inside, on the face, on the corner, respectively, within a unit cell.

Metals are generally made of atoms with a close-packed structure, which means that the number of atoms per unit cell is greater than 1, the number of atoms in a simple crystal structure.[17] The most common close packed crystal structures of metals are given in Table 2.1:

14 空间点阵被定义为有如下排列方式的数学点的集合，每个点都被周围完全相同构象的点围绕，因此所构成的空间就被分成了相同尺寸的小空间，每个原子（离子或分子）都分布在这些线的交点上或者在不同的线之间。

15 观测位置一侧的每一个晶格点，在相反的一侧总是有与之相对应的、完全等同的晶格点。

16 另外，我们也必须铭记的是，即使晶格结构基本上是简单的，晶体结构也可能非常复杂，这是因为一个晶胞可能包含成千上万个原子。

17 金属一般由原子通过密堆积结构组成，这就意味着一个晶胞中的原子数目大于简单晶体结构晶胞中的原子数 1。

Table 2.1 The most common close-packed metal structures.

Name	Structure	No. of atoms per unit cell
BCC	Body-centred cubic	2
FCC	Face-centred cubic	4
HCP	Hexagonal close-packed	2

Organic materials are made of molecules holding together by van der Waals forces. Examples of their molecules per unit cell are given in Figure 2.11.

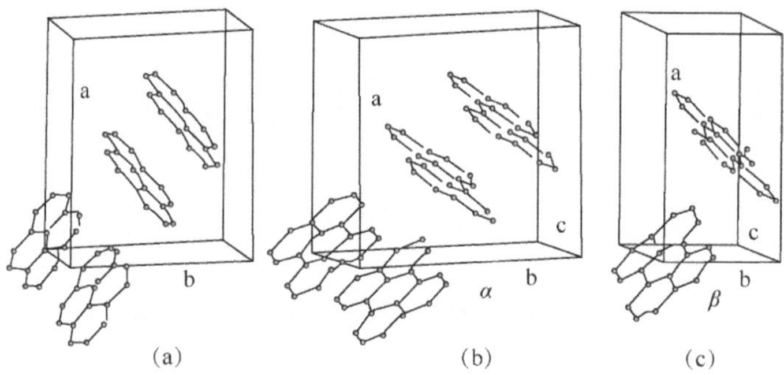

Figure 2.11 (a) and (b): pyrene and α-perylene crystals with 4 numbers of molecules in a unit cell; (c): β-perylene crystal with 2 numbers of molecules in a unit cell. Note: all these organic crystals possess more than 10 atoms in the unit cell.

2.2.2 Crystal Systems

In this section, we will emphasize crystal systems composed of atoms and ions, but the structural particles of crystalline solids can be atoms, ions, or molecules. For example, solid methane, CH_4, a molecular solid, has a face-centered cubic (FCC) structure, which means that there is a CH_4 molecule at each corner and at the center of each face in its unit cell.[18] The forces acting among these structural particles may be metallic bonds, interionic attractions, van der Waals forces, or covalent bonds.

18 属于分子固体的固态甲烷，是面心立方结构（FCC），这意味着在它的晶胞中每个角和每个面的中心都有一个 CH_4 分子。

Ⅰ. Cubic cells

The cubic system includes simple, body-centered and face-entered cubic cells.

The **simple cubic unit cell** (Figure 2.12a) consists of eight atoms located at each corner of the cube. Only polonium is crystallized in the simple cubic lattice. It must be remembered that if you represent these eight atoms with hard rubber balls and arrange them in accordance with this simple cubic unit cell, all eight atoms would be touching each other.[19]

19 必须记住的是，如果用硬的橡胶球代替简单立方晶胞里的八个原子，则这八个原子是紧密挨着的。

Another unit cell (Figure 2.12b) is known as the **body-centered cubic** (BCC). It is similar to the simple cubic unit cell, but contains an additional atom located in the center of the cube.

The third type of unit cell formed from the cubic axis system is the **face-centered cubic** (FCC). One atom at each corner and one in the center of each of the cube faces make up the complement of atoms. There is no atom at the center of the cube (Figure 2.12c).

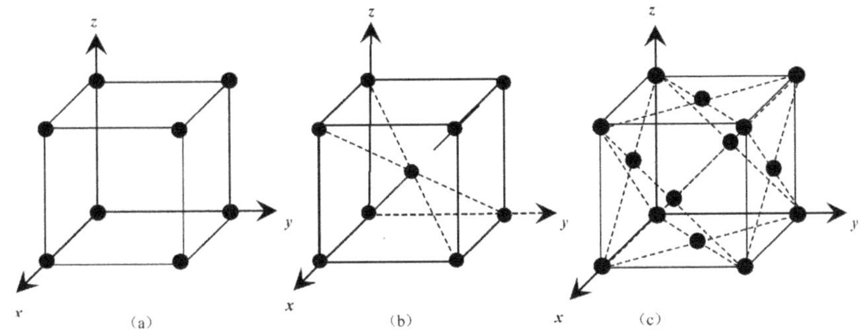

Figure 2.12 (a) A simple cubic unit cell. (b) A body-centered cubic unit cell. (c) A face-centered cubic unit cell.

For both the BCC and FCC structures, the same type of atoms must fill both the center/face-centered locations, as do the cell corners. If, in fact, a different type of atom fills the center position than the cell corners in a body-centered cubic structure, the structure is correctly called a simple cubic (SC) structure.[20] An alloy made up of about 50% copper and 50% zinc forms such a structure.

20 事实上，如果在体心立方结构中占据中心位置的原子与占据角位的原子不同，这个结构应该正确地称为简单立方结构。

Ⅱ. Tetragonal unit cell

The tetragonal crystal system has similar unit cells to the cubic, but the sizes are not equal. As an example, the **body-centered tetragonal** (BCT) crystal lattice unit cell is shown in Figure 2.13. Tin forms a tetragonal unit cell. The tetragonal is similar because the axes are all normal to each other. The difference lies in the length of the intercepts. The x and y intercepts have the same magnitude. The z intercept is larger than the x or y intercept. Martensite, a combination of iron and carbon that is contained in a hard steel, has its atoms of iron and carbon in a tetragonal lattice structure.

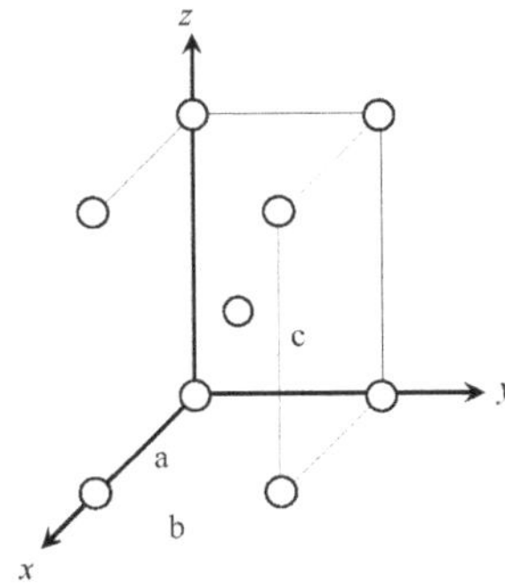

Figure 2.13 A body centered tetragonal crystal lattice unit cell.

Ⅲ. Hexagonal unit cell

The hexagonal crystal system (Figure 2.14a) can best be described using three axes (a_1, a_2, and a_3) in the x–y plane 120° apart and a fourth axis (z) at 90° to the x–y plane. The intercepts along the three axes in the horizontal plane are equal in length ($a = a = a$), but the fourth intercept, labeled c, is of a different length.[21] This unit cell is made up basically of two parallel planes (top and bottom basal) separated by a distance equal to the dimension c. The atoms shown in the figure trace out a right hexagonal prism. Each of these two planes can be divided into six equilateral triangles, with each side equal to the intercept a (Figure 2.14b).

21 水平面上三个轴的截距是相等的（$a=a=a$），但是第四个截距 c 不同。

Atoms of solid materials do not form the purely hexagonal unit cell as in Figure 2.14a because they cannot satisfy equilibrium conditions by being so far apart. In other words, they would be unstable. Consequently, they form the hexagonal unit cell, called close-packed hexagonal (CPH), as shown in Figure 2.14c, with its three mid-plane atoms at a distance of $c/2$. Zinc, titanium, and magnesium form CPH unit cells.[22]

22 因此，它们形成六方晶胞，也称为六方密堆（CPH），如图 2.14c 所示，在它的 $c/2$ 位置会有 3 个位于中间平面的原子，例如，锌、钛和镁。

According to the French crystallographer, A Bravais, in 1848's statement, there are seven types of unit cells (Table 2.2), of which three were mentioned above; and atoms can form 14 patterns in space (Brevais lattices, see Figure 2.15), with which the earth's elements form their particular atomic structures.[23]

23 根据法国晶体学家奥古斯特·布拉维在 1848 年提出的理论，总共有 7 种晶胞（见表 2.2），其中的 3 种上面已经提到。原子在空间能够形成 14 种空间构型（Bravais 点阵，见图 2.15），自然界中的原子基于此形成各自的原子结构。

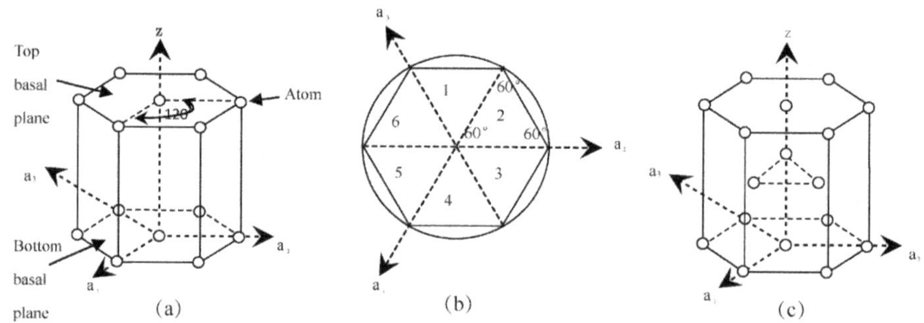

Figure 2.14 A hexagonal crystal lattice unit cell. (b) Top plane of unit cell showing six equilateral triangles. (c) A close-packed hexagonal crystal lattice unit cell.

As early as 1848, long before X-ray methods of crystallography were known, the French crystallographer A. Bravais stated that the atomic arrangements in all crystalline solids can be referred to 14 fundamental crystal classes (Brevais lattices, see Figure 2.15), consisting of four types of space lattice in combination with seven systems of unit cells (Table 2.2), of which three were discussed above. The earth's elements form their particular atomic structures.

Table 2.2 Types of crystal systems with the intercepts and the angles between axes

System	Intercepts	Angles between axes
Cubic	$a=b=c$	$\alpha=\beta=\gamma=90°$
Hexagonal	$a=b\neq c$	$\alpha=\beta=90°,\ \gamma=120°$
Tetragonal	$a=b\neq c$	$\alpha=\beta=\gamma=90°$
Rhombohedral	$a=b=c$	$\alpha=\beta=\gamma\neq 90°$
Orthorhombic	$a\neq b\neq c$	$\alpha=\beta=\gamma=90°$
Monoclinic	$a\neq b\neq c$	$\alpha=\gamma=90°,\ \beta\neq 90°$
Triclinic	$a\neq b\neq c$	$\alpha\neq\beta\neq\gamma\neq 90°$

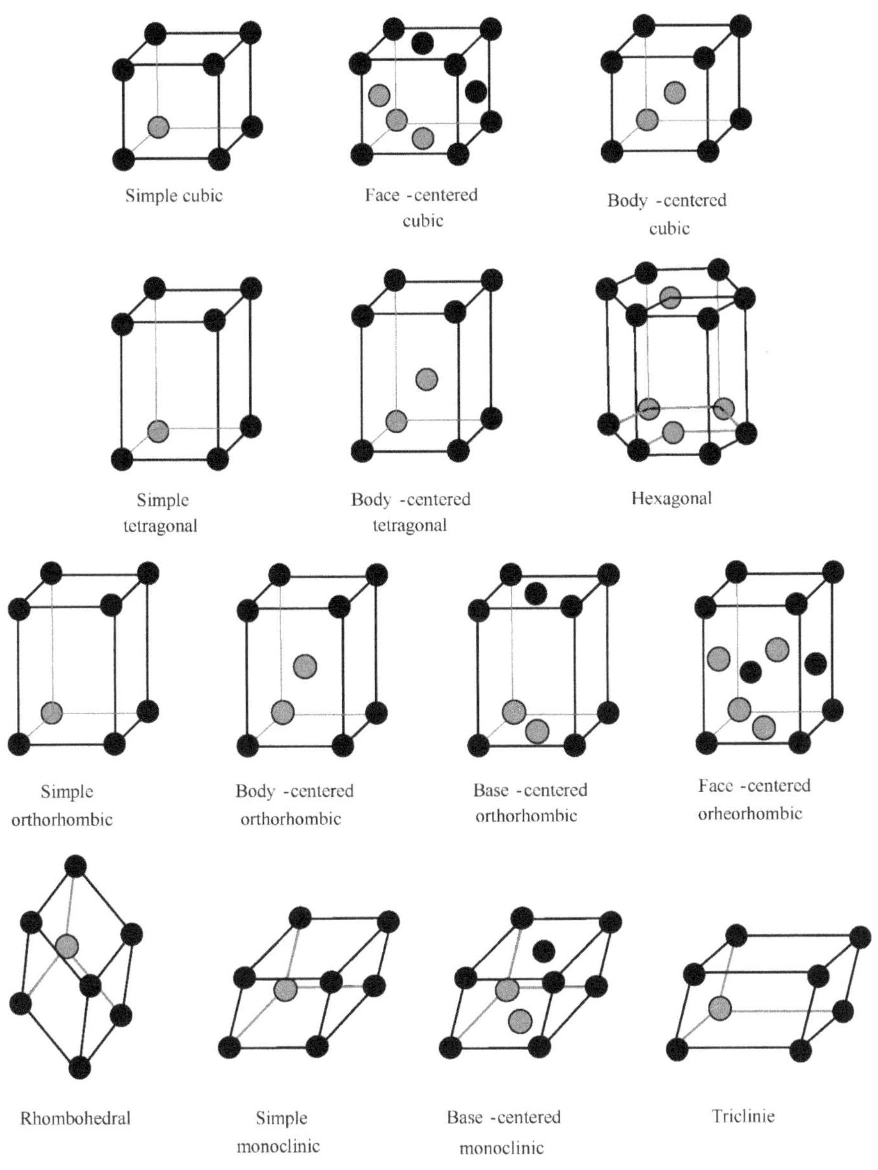

Figure 2.15 Bravais's 14 fundamental types of crystal lattices.

2.2.3 Representative Parameters for Crystal Systems

Ⅰ. Miller indices for atomic planes (lattice planes)

In a crystal system with huge amount of periodically arranged lattice points, there are several different situations (Figure 2.16). For instance, the atom distance in different direction is generally different, also in lattice planes of different orientation, the in between distance is different. This anisotropic material structure leads to anisotropic material properties.[24]

24 例如，不同方向上的原子间距通常是不同的，不同取向的晶面间距也是不同的。这种材料结构的各向异性导致了材料性质的各向异性。

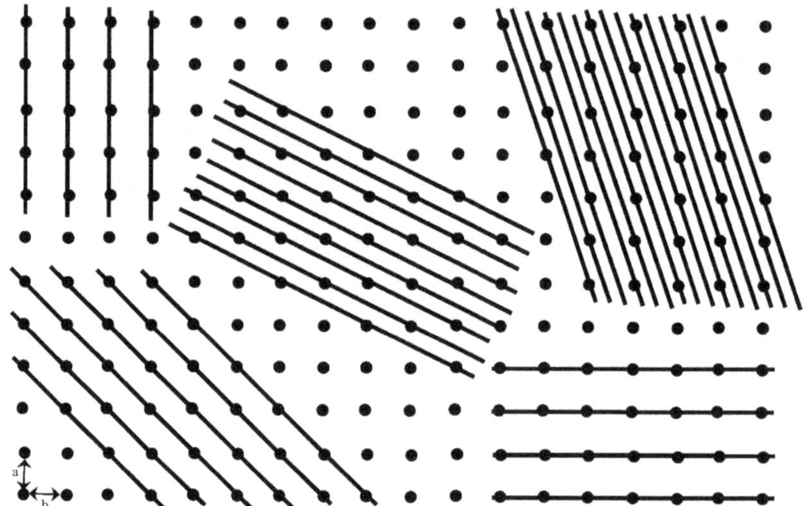

Figure 2.16 Different atomic distances at different directions.

Therefore, it is appropriate to be able to describe the location of atoms in a unit cell as well as the direction of their movement. The sites or locations of atoms and/or points in a unit cell are described by atomic planes in unit-cell dimensions, which are designated by their Miller indices.[25]

25 因此，能够描述原子在晶胞中的位置，以及它们运动的方向是恰当的。在晶胞尺度上，原子以及（或者）格点在晶胞中的位置被描述为原子面，用密勒指数表示。

As shown in Figure 2.17, the Miller indices for atomic plane can be expressed:

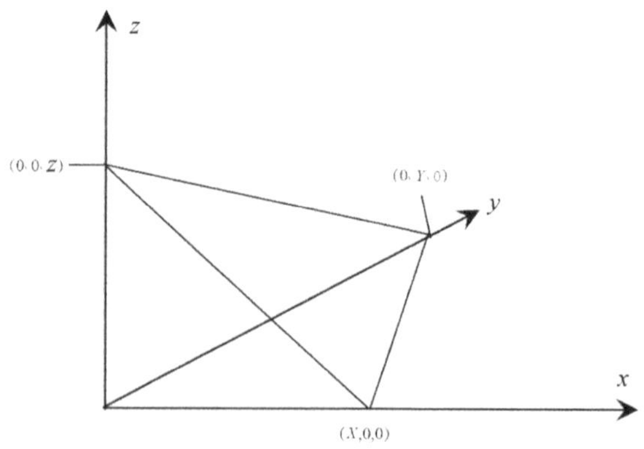

Figure 2.17 The intersections between the coordinate axes and the plane.

$$(hkl) = \left(\frac{D}{X}\frac{D}{Y}\frac{D}{Z}\right) \tag{2-2}$$

where X, Y, Z = coordinates of the intersections between the plane and the coordinate axes; D = the smallest common multiple of X, Y and Z. The miller indices, written within curly brackets {}, mean that all permutations of the hkl indices and combination of signs occur.[26] They define the directions of the normals to the type of planes in question. If one refers to a specific plane with the given directions, ordinary parentheses are used, (hkl). If $X = Y = Z = 1$, the Miller indices of the plane are (111). If the plane is parallel with one of the coordinate axes, the corresponding Miller index is zero as the intersection occurs at infinity. A bar over an index indicates a negative value of the intercept. Some examples of Miller indices for atomic planes are shown in Figure 2.18.

26 这里，X、Y、Z 分别是平面与坐标轴的交点坐标；D 是 X、Y、Z 的最小公倍数。密勒指数写在大括号{}里，意味着 hkl 包括正负号的所有排列组合。

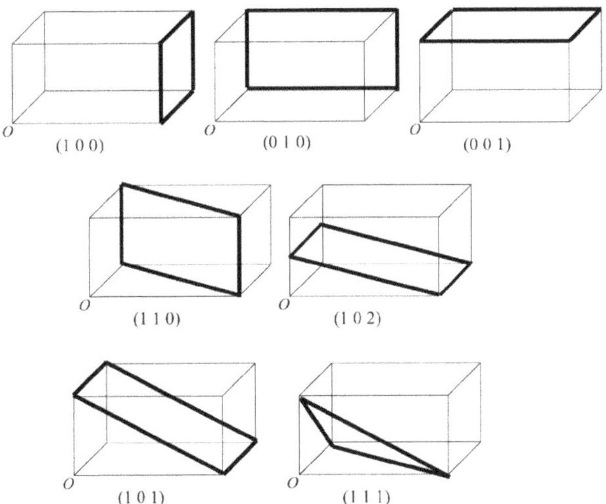

Figure 2.18 Examples of Miller indices for atomic planes (The origin is indicated as a small circle in each unit cell).

Ⅱ. Crystal directions (Lattice directions)

As with Miller indices for specifying atomic planes in a crystal-lattice system, there is Miller indices for directions, which are generally written as <uvw>. This means that all permutations of the uvw indices and combination of signs occur.[27] If one refers to a specific line with the given direction, square brackets are used, [uvw].

27 与晶体点阵体系中特定原子面的密勒指数类似，也有针对方向的密勒指数，一般写作<uvw>。这表示 uvw 包括正负号的所有排列组合。

To calculate the Miller indices for a direction, there are four steps: The first step is to determine the coordinates of two points that lie in the particular line of direction. The first point, sometimes called the "head point," is farthest from the origin. Using the origin as the second point simplifies the procedure. The second step is to subtract the second point from the first point. The third step calls for clearing of any fractions using the smallest common multiple to obtain indices with the lowest integer values. The fourth step is the writing of the results. Negative integer values are indicated by the use of a bar placed over the integer. In more complex crystal systems, the determination of directions requires temporary relocation of the origin to another point in the unit cell to simplify the procedure.[28] If the properties of a crystal measured along two different directions are identical, the two directions are termed equivalent.[29] Equivalent directions are referred to as a family of directions, which can be specified by using angle brackets: <110>. Figure 2.19 are examples of the Miller indices for directions.

28 在更复杂的晶系中，方向的确定需要临时在晶胞中重新定义原点以简化过程。

29 如果沿着两个不同的方向测得的晶体性质是完全相同的，这两个方向就被称为等效的。

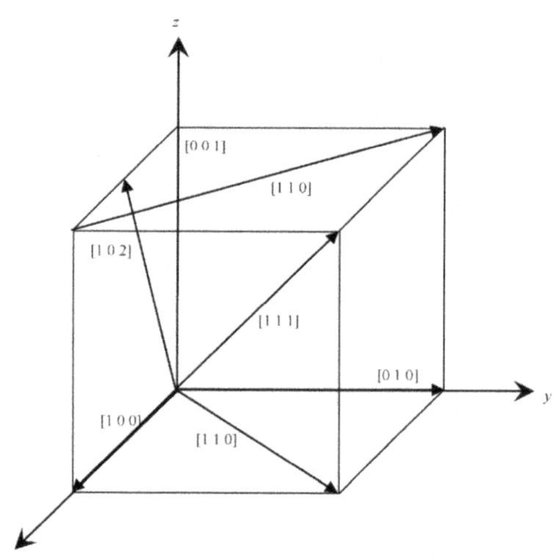

Figure 2.19 Examples of the indices of the directions.

Ⅲ. Coordination number

To describe how many atoms are touching each other in a group of coordinated atoms, the term coordination number (*CN*) is used. The *CN* is the number of neighboring atoms that are directly surrounding it. Note in Figure 2.14c that each upper and lower basal plane of a CPH unit cell contains an atom at its center. Each atom touches six atoms in its own plane, plus three atoms above and below in adjacent planes. Consequently, the *CN* for these atoms would be 12. The number of nearest atoms is dependent on two factors: (1) the type of bonding, and (2) the relative size of the atoms or ions involved. In our discussion of bonding, for example, we learned that valence electrons determine the type of bonding as well as the number of bonds an atom or ion can have. Carbon (C) in group IV has four covalent bonds and therefore a *CN* of 4. The group VII elements, such as chlorine (Cl), form only one bond (*CN* is 1).[30] The relative size of the atoms determines how many neighboring atoms will touch another atom. Ionic

30 例如，在对化学键的讨论中，我们知道价电子决定了原子或者离子形成化学键的类型和数量。第 4 族的碳（C）有 4 个共价键，所以配位数 *CN* 是 4。第 7 族元素，如氯（Cl），只能形成一个化学键（*CN*＝1）。

bonding involves ions of different charges, hence, different sizes. The limiting factor in this case is the ratio of the size (radii) of the combined atoms of ions. The minimum ratios of atomic (ionic) radii produce various *CNs*. During ionization, atoms decrease in size when they change to cations and increase in size as they form anions.

Figure 2.20 represents the five ions occupying one of the six faces of the FCC unit cell for NaCl. The Na ion is just the right size to fit between the Cl ions at the corners of the unit cell. Thus, the ions are closely packed, with each cation separated from other cations by a layer of anions. Each cation and each anion are shared equally by six oppositely charged ions. Therefore, a *CN* of 6 describes this geometric arrangement. Actually, in the case of ionic crystal, the radius ratio between the anion and the cation determines the *CN* value. As the difference between r and R decreases, higher *CNs* are possible. A *CN* of 12 is the maximum, which occurs when the atoms (ions) have the same radius and the ratio becomes 1. In other words, as the r gets smaller than R (radius of surrounding atoms), the fewer neighboring atoms can make contact or touch the smaller atoms.[31] Table 2.3 lists the minimum radii ratios for some common *CNs*.

31 当原子（离子）有相同的半径且它们的比例为 1 时，配位数最大为 12。换句话说，与 R（周围原子的半径）相比，r 越小，原子就越小，相邻原子中与其相接触的数目就越少。

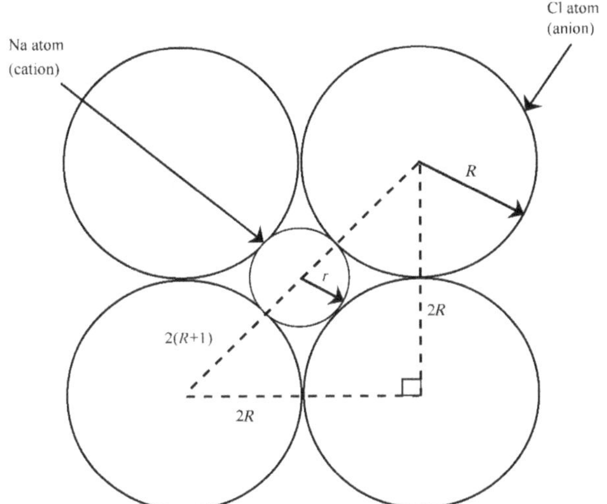

Figure 2.20 A face plane in NaCl crystal, along the axes perpendicular to the page are not shown for clarity.

Table 2.3 Minimum radii ratios for *CNs*

CN	r/R
3	⩾ 0.155
4	⩾0.225
6	⩾0.414
8 (bcc)	⩾0.732
12 (cph or fcc)	1.0

With a *CN* of 12, each atom has contact with 12 other atoms. Each atom in an fcc or cph unit cell meets this description, provided that their radii are of similar size.

Ⅳ. Volume changes and packing factor

In discussing crystal structure changes, we mentioned that every change in atomic structure brings changes in properties of the solid.[32] One of these changes is volume. When pure steel transforms from a BCC to FCC structure at 912°C, it decreases about 1.06% in volume. The explanation for this phenomenon involves the density of atoms in the various unit cells. Density is the ratio of the mass to the volume of a substance, which stays constant provided it is non-allotropic.

32 在讨论晶体结构变化的时候，我们提到了原子结构的每一个变化都会改变固体的性质。

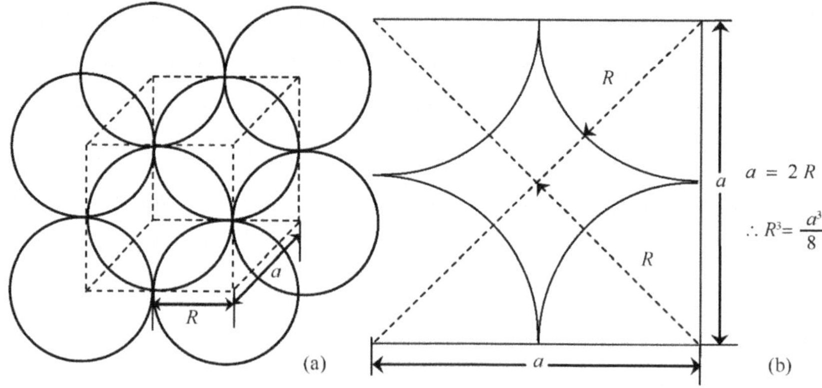

Figure 2.21　(a) A simple cubic cell. (b) (100) face plane of simple cubic unit cell.

The atomic packing factor (*APF*), or packing factor (*PF*), is the ratio of the volume of atoms/molecules present in a crystal (unit cell) to the volume of the unit cell. In calculating the volume of an atom, we assume the atom is spherical. The difference between the *PF* and unity (1) is known as the void fraction, that is, the fraction of void (unoccupied or empty) space in the unit cell.[33]

33 1与*PF*的差值被称为空隙率，即在晶胞中空隙部分（未被占据或空置）的空间。

$$PF = \frac{volume\ of\ atoms\ /\ molecules\ in\ a\ unit\ cell}{volume\ of\ unit\ cell} \tag{2-3}$$

Using the simple cubic crystal unit cell sketched in Figure 2.21a, with atoms of equal radius (*R*) located at each corner, the volume of the cell occupied by the eight atoms is equivalent to one atom. Each corner atom contributes one-eighth of its volume. Therefore, the volume of one atom is $(4/3)\pi R^3$. The (100) face plane of this simple cubic unit cell shows four corner atoms touching each other (see Figure 2.21b). The relationship of the radius *R* of an atom to the lattice parameter or edge length *a* of the unit cell is 2*R*. The volume of the unit cell is a^3. The volume of atoms is $(\pi/6)a^3$.

Thus,

$$PF = \frac{volume\ of\ atoms}{volume\ of\ unit\ cell} = \frac{(\pi / 6)a^3}{a^3} = \frac{\pi}{6} \cong 0.52$$

Solving for the PF in terms of the radius R produces the same results. The void factor is therefore $1 - 0.52 = 0.48$. What the calculation tells us is that only about half (52%) of the space in the simple cubic unit cell is occupied by the atoms. This is too inefficient, so atoms of metals do not crystallize in this structure except Polonium. Remember, the closer the atoms come to each other, the less energy they have and the more stable is their structure.

The BCC unit cell is quite similar to the simple cubic unit cell with the addition of one atom in the very center of the unit cell. Therefore, the bcc unit cell contains the equivalent of two atoms.

For the FCC unit cell, there are four net atoms. Each of the eight corner atoms contributes one-eighth of an atom. Each of the six face atoms contributes one-half. The total is 1 + 3, or 4, atoms. Notice that the atomic radius (R), or a as used in the formula, cancels out in all these calculations, which tells us that the PF is not dependent on the radius of the spheres being packed if all the atoms are of the same size.[34]

The FCC structure has the maximum PF for a pure metal. The CPH structure also has a PF of 0.74. Finally, we should note that the coordination number varies directly with the PF. As an example, the $CN_{\text{bcc}} = 8$ and $PF_{\text{bcc}} = 0.68$; the $CN_{\text{fcc}} = 12$ and $PF_{\text{fcc}} = 0.74$.

In metallic materials, for example, the pure iron will change its structure on heating from BCC to FCC at 910℃. Knowing the PF for both these structures will lead us to the conclusion that iron will contract in volume as it is heated above 910℃. This change in structure forms the basis for the production of steel as well as the heat treatment of steel.

In organic materials, the concept of packing factor is the same as the above description, just that the atoms change to molecules. Since the unit is molecules that are hold together by weak van der Waals force and strong short-range repulsion, the molecules will adopt the densest possible packing with the least possible repulsion.[35] The arrangement of the molecules will be determined by atom-atom potentials. The lattice energy is minimized when the number of van der Waals atom-atom contacts is as large as possible. The values of PF for the aromatic hydrocarbons lie between 0.68 (benzene) and 0.80 (perylene). For comparison: the packing coefficient of ice, which is bound through dipolar forces (hydrogen bonding) is only 0.38.[36]

34 需要注意的是：原子半径或者如公式中的“*a*”，在计算中被消去了，这就说明如果所有原子大小相同的话，*PF* 不取决于所堆叠球的半径。

35 由于晶胞中是分子，其聚集在一起的力是微弱的范德华力和强大的短程斥力，所以分子会采取最低排斥势能下的最密堆积方式。

36 相比较而言，由于受到偶极力（氢键）的作用，冰的堆积系数仅为 0.38。

2.2.4 Crystal structures of Metals and Organic Materials

Ⅰ. Metal

Many metals have a close-packed structure, which means that the number of atoms per unit cell is greater than 1, the number of atoms in a simple crystal structure. The most common close packed crystal structures have already been given in Table 2.1.

- BCC structure

This structure is a simple cubic lattice with an additional atom in the centre of the cube (Figure 2.22). Number of atoms per unit cell = $(8\times1/8) + 1 = 2$; Coordination Number = 8. Examples of metals with a BCC structure are **Cr**, **Ti**, **W**, **Mo**, **V**, and α-**Fe** etc.

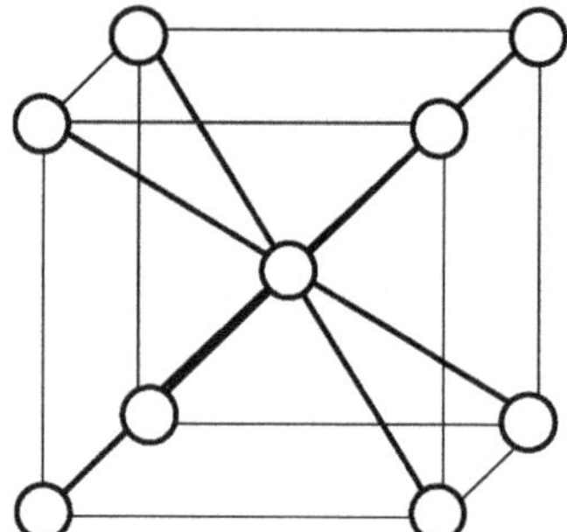

Figure 2.22 BCC structure of metal.

- FCC structure

This structure is a simple cubic lattice with an additional atom in the centre of each of the six faces of the cube. Number of atoms per unit cell = $(8\times1/8) + (6\times1/2) = 4$.

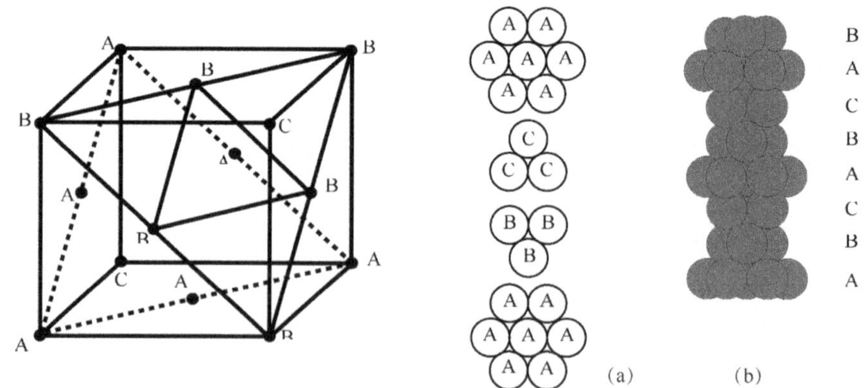

Figure 2.23 FCC structure of metals.

Some {111} places of FCC structure are shown in Figure 2.23(b). As labeled, there are A, B and C planes. The pattern of these {111} planes are illustrated as following: Take the central atom in plane A as sample, besides the 6 peri-surrounding atoms, at top and bottom of plane A, there exist 3 surrounding atoms, respectively. Hence, the coordination number is 12. Examples of metals with a FCC structure are **Al**, **Cu**, **Au**, **Ag**, and γ–**Fe** etc.

- HCP structure

BCC and FCC structures are simple in the sense that atoms are directly placed in the corners of the Bravais point lattice. The hexagonal close-packed structure is more complicated as every lattice corner is occupied by an atom, along with three mid-plane atoms at a distance of *c*/2, like a dumb-bell.[37] All dumb-bells are parallel in space. Number of atoms per unit cell = (8×1/8) + 1 = 2, Number of nearest neighbours = 12. The HCP structure can be regarded as three single unit cells connected rigidly with one another.[38] Figure 2.24 is the unit cell of HCP cell. Examples of metals with a HCP structure are **Mg**, **Zn**, and **Cd** etc.

37 六方密堆积结构更加复杂，因为每个晶格的角上都被一个原子占据，同时在二分之一 *c* 轴处，有 3 个中间平面的原子，像哑铃一样。

38 HCP 结构可以看作由 3 个晶胞彼此通过刚性结合而成。

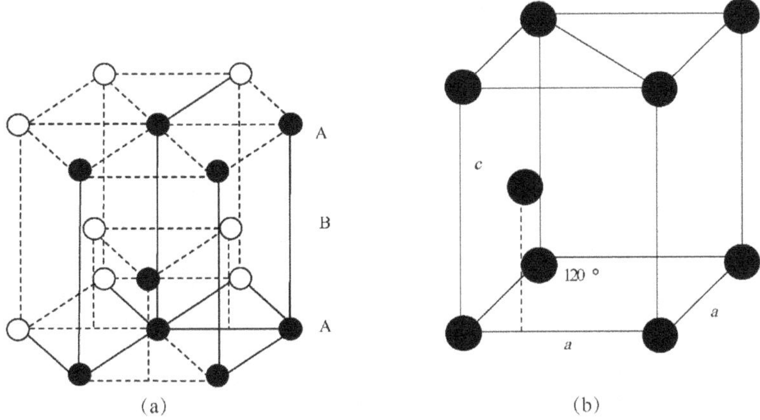

Figure 2.24 (a)Structure of HCP, (b) Single unit cell of HCP structure.

The stacking sequence of the atomic planes in an HCP structure is shown in Figure 2.25.

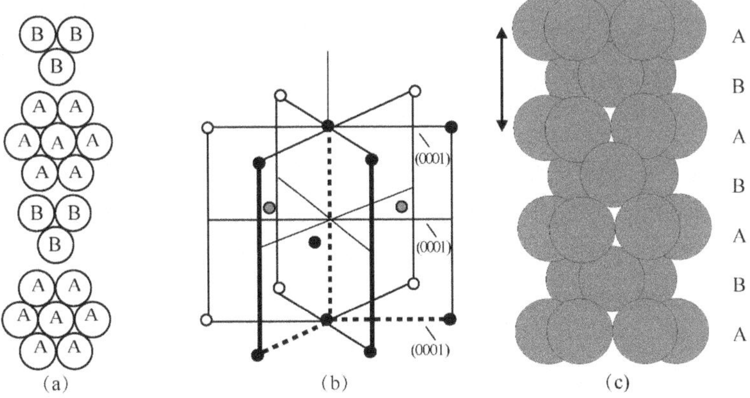

Figure 2.25 Stacking sequence of the atomic planes in an HCP structure.

Ⅱ. Organics

In considering the packing modes within a crystal of organic materials, due to the relatively weak dispersive forces and strong short range repulsion, the trend is to pack as close as possible to maximize the van der Waals force in order to reduce lattice energy.[39] And we also need to remember that even with planar molecules, the molecular surfaces are not structureless. The positions of the atoms correspond to 'hills', the positions in between to valleys in the molecular contour, as shown with anthracene molecule in Figure 2.26.

39 考虑到有机材料中晶体的堆叠方式，由于色散力非常弱，同时短程排斥力非常强，所以堆叠时尽可能紧密从而得到最大的范德华力以减小晶格能。

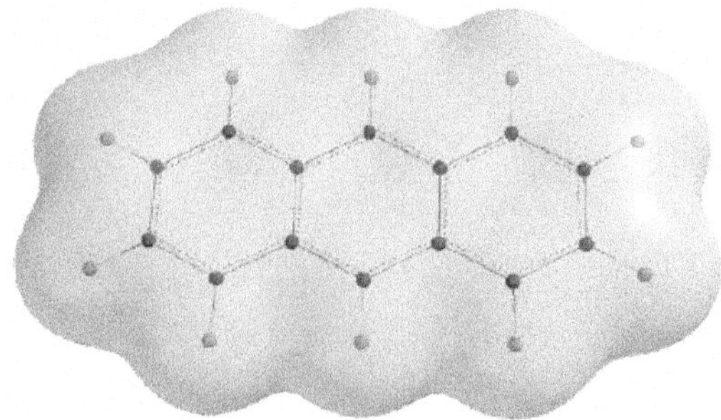

Figure 2.26 The overall distribution of the π electrons in the electronic ground state of the anthracene molecule, $C_{14}H_{10}$.

Therefore, an arrangement in which the hills of one molecule lie above the valleys of the neighboring molecules is generally energetically more favorable for nonpolar molecules than one in which the molecules lie directly above one another.[40] This is most noticeable in the 'herringbone' pattern in many aromatic molecules crystalline (Figure 2.27).

40 因此，对于2个相邻的非极性分子，相比于原子直接相对，原子通过一个分子的峰在另一个分子的谷上的错位排列通常在能量上是有利的。

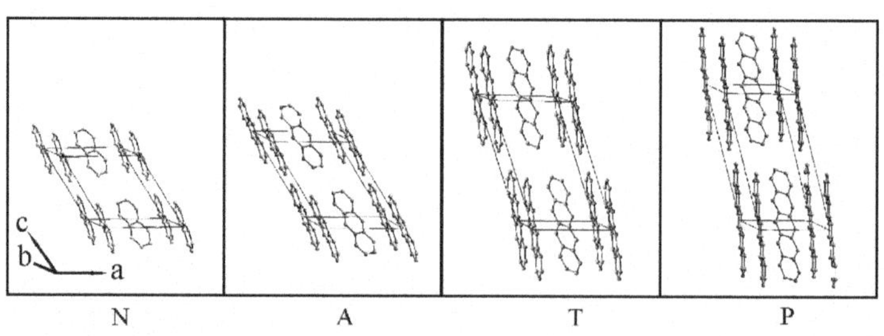

Figure 2.27 The crystal structure of naphthalene (N), anthracene (A), tetracene (T), and pentacene (P). These aromatic compounds crystallise in the herringbone pattern.

In the case of long-chain alkanes, in particular the linear hydrocarbons with n carbon atoms, the preferential arrangements are the CH_2 zig-zag chains of the individual molecules, which are parallel to each other, and thus form a layered structure (Figure 2.28).[41] Such an arrangement guarantees the strongest dispersive interactions in view of the anisotropic polarizability of these molecules.

41 长链烷烃中，特别是有 n 个碳原子的直线型碳氢化合物，最合适的排列是每个分子中的 CH_2 形成锯齿形链，链与链互相平行，形成层状结构（图2.28）。

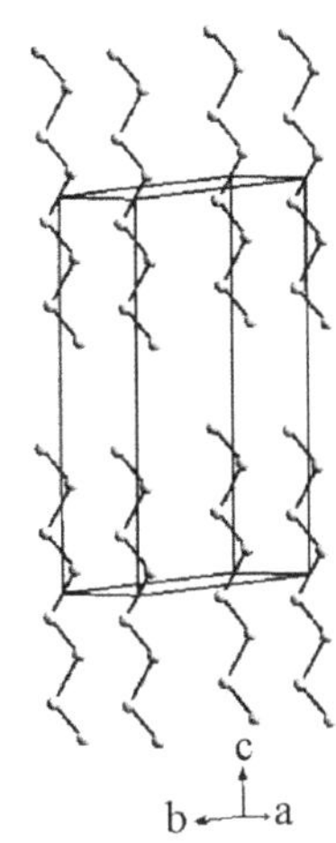

Figure 2.28 Crystal structure of n-octane, $CH_3-(CH_2)_6-CH_3$.

2.3 Crystal Defects

Real crystals are never perfect. They always contain crystal defects, which give a distorted lattice. Even if the fraction is not more than one atom out of place in 10000, this is enough to have a strong influence on many important properties of the material, for example mechanical strength, heat transfer etc.[42]

Of course, this lack of perfection in the microstructure of materials is far from being all bad. If it were not for imperfections of many kinds in solid materials, these solids would not possess the properties that we desire them to have. An example would be the heat-treating process used with high-carbon steel to control the properties of the steel to suit certain technical requirement. Without imperfections in the crystalline arrangement of atoms, these processes would be severely limited in, if not incapable of, changing the structure and hence the properties of steel.[43] The whole semiconductor industry owes much of its existence to the imperfections in bonding arrangements of the atoms' outer shell electrons. Therefore, it is now essential to delve into the several imperfections of a solid's atomic structure so that we can learn how to take advantage of such disorder in the atomic structure.

To classify crystal defects, according to the atomic nature of defects, there are two types: (1) structural defect--No impurity exists and the stoichiometry is perfect, the defects are caused by vacancy, dislocation etc; (2) chemical defect--The defects are originated from impurities, isotopes, or the non-ideal stoichiometry etc.[44] According to the shape of defects and the

42 实际的晶体绝不是完美的，它们总是包含晶体缺陷，使得晶格发生畸变。即使只是 10000 个原子里的一个原子发生畸变，就足以对材料的很多重要性质产生很大的影响，例如，机械强度、导热性等。

43 如果在原子的晶体分布中没有缺陷，不是不可能，在改变钢材的结构乃至性质方面，这些过程（注：指上面的热处理）将会受到严重的局限。

44 基于缺陷的原子特征，晶体缺陷的分类有 2 种：（1）结构缺陷——没有杂质的存在并且化学计量是完美的，这种缺陷是由空位、位错等因素引起的；（2）化学缺陷——造成这种缺陷的原因是杂质、同位素以及不理想的化学计量等。

involved region, all sorts of deviations from the regular crystal pattern can be divided into three main classes (Figure 2.29): (1) The point defects are local. They concern just a few atoms close to the defect; (2) The line defects are more long-range. Even atoms far from the centre of the defect are influenced; (3) The interfacial defects are three-dimensional phenomena.

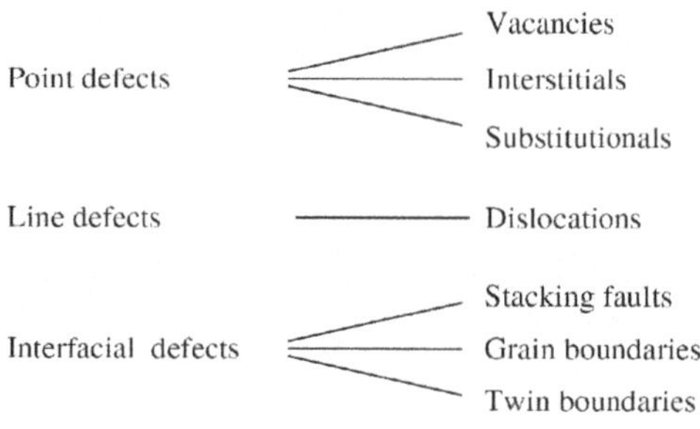

Figure 2.29 Types of defect in crystals

Ⅰ. Point defects

Point defects consist of either a vacant lattice site, a substitutional atom, or an interstitial atom within one or several lattices (Figure 2.30). A substitutional atom is a foreign atom that occupies a lattice point. An interstitial atom can be the same atom as the lattice, or be a foreign atom, who shoves in the interstitial site between or among atoms.[45] By themselves, point defects do not affect strength as much as they affect diffusion (e.g., the migration of atoms).

45 间隙原子可以是与晶格中的点阵原子相同的原子，也可以是外来原子，它们挤在两个或者多个原子的间隙之间。

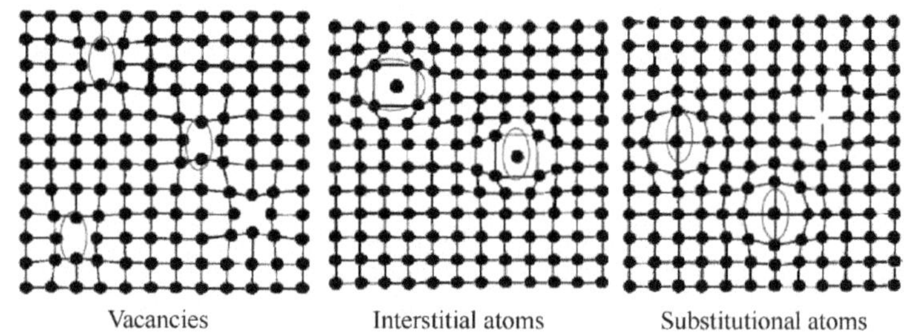

Figure 2.30 Examples of point defects: vacancies, interstitial and substitutional, respectively.

- **Vacancy**

A missing ion or atom in a crystal lattice is called a vacancy. Vacancies are the most important point defects in metals. It is customary to illustrate a vacancy as a square.

The density of the vacancies in a crystal depends strongly on the temperature and increases rapidly with increasing temperature. The upper limit is of magnitude $\geqslant$ 0.1 at-% close to the melting point temperature for most solids. The vacancies cause an expansion of the crystal lattice. This is one of the reasons for thermal expansion of solids. Vacancies disrupt the electronic bonding of the adjacent atoms, which changes the effective radii of these atoms. This weakens the crystal. If sufficient vacancies were produced by heating of a crystalline solid, the crystal structure would lose its long-range symmetry and order, resulting ultimately in porosity or a change to a fluid.[46] It is important to note that these local imperfections in the crystal structure produce a disequilibrium that has a great effect on the important properties of crystalline solids such as density, mechanical strength, diffusion, and electrical conductivity.[47]

46 如果通过加热晶态固体获得足够的空位，晶体会失去其长程有序的对称性和序列性的结构，最终变成多孔结构或者变成流体。

47 需要重点指出的是：晶体结构中这些局部的缺陷造成的不平衡对晶态固体的重要性质有很大的影响，比如密度、机械强度、扩散以及电导率。

There are two type vacancies related point defects, i.e., Frenkel defect and Schottky defect (Figure 2.31). A Frenkel defect is formed by excitation and migration of an atom from its normal site in the lattice with an empty left. The atom becomes an interstitial and formed a vacancy-interstitial pair. A Schottky defect is a vacancy in a crystal lattice where the atom has been removed from its site to the surface of the lattice. The vacancy is not coupled to any interstitial.

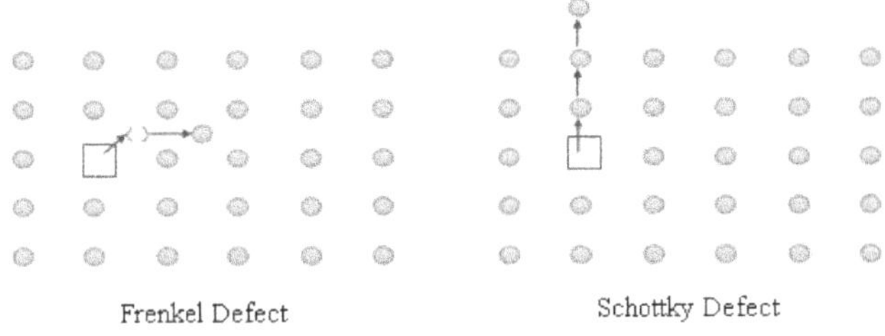

Figure 2.31 Comparison between Frenkel and Schottky vacancies.

- **Substitutional solid solutions**

If the impurity atom occupies the lattice position, a substitutional defect is formed, and the physical properties, such as acoustic, optical, electrical and magnetic properties, can be significantly influenced.[48] For example: introducing "soft additive", e.g., La, Nd, Bi etc, into ferroelectric ceramic $Pb(Zr_xTi_{1-x})O_3$, these atoms can substitute Pb in the lattices, and the dielectric constant can be improved, the mechanic quality factor can be reduced; When "hard additive", such as Fe、Co、Mn etc is added, these atoms can occupy the positions of Zr or Ti in the lattices, the electric quality factor can be significantly improved.

48 如果杂质原子占据了点阵的位置，就形成了取代缺陷。取代缺陷会对材料的物理性质，如声学、光学、电学和磁学等，造成很大的影响。

- **Interstitial solid solutions**

If the impurity atoms take up sites in the lattice structure that are normally unfilled or unoccupied by the pure (solvent) atoms, they form an interstitial solid solution. These normally unfilled voids or vacant spaces are called interstices. In the FCC unit cell, we know there is a relatively large interstice in the center and smaller interstices near each corner atom. It is worthwhile to point out that steel making is made possible because of the formation of an interstitial solid solution. First, we know that iron is allotropic. At temperatures below 912℃, iron is in the BCC form. Above that temperature, the BCC structure changes to FCC to accommodate a higher energy level of the atoms. In the FCC structures, carbon atoms can form in the interstices of the iron unit cell. At temperatures below 912℃, the BCC structure contains no room for the carbon atoms to fit between the iron atoms. This fact forms the basis for many of the heat-treating procedures used to produce a multitude of steels with many different properties required by our technological society.[49]

49 这一点成为许多热处理过程的基础，被用来生产我们科技社会需要的许多不同性质的多种钢材。

Ⅱ. Line defects

Dislocations are the only line defects that exist in crystalline solids. In the strict geometrical sense, they are really cylindrical defects of about five atom spacings in diameter. They thread their way through the crystal in all sorts of directions, not usually as straight lines. The dislocations form during solidification and recrystallization. They can also be generated in significant quantities during plastic deformation. An annealed metal that contains a dislocation density of 10^6 per cm^2 will contain about 10^{12} per cm^2 after the metal has been severely deformed (e.g., by 80% reduction in area). They are one of the more important crystalline imperfections, even though we have known about their existence for only a little more than 50 years. For instance, line defects have a great deal to do with the strength of a solid. An abundance of them will cause a mutual interference in their movement through a crystal, preventing the planes of atoms from slipping, thereby strengthening the material.[50] The presence of a few dislocations increases the ductility of a crystalline solid.

50 大量的（线缺陷）会对它们在晶体中的运动产生相互干扰，阻止原子平面的滑动，从而强化材料。

There are two different configurations of atoms in dislocations, called edge and screw types.

The structure of the edge dislocation, whose symbol is (⊥), is depicted in Figure 2.32. It can be viewed as if a half-plane of atoms has been removed from the lattice and then the neighboring atoms collapsed to fill the planar void.[51] One might think that this would be a planar defect, but not so. Look at the atom positions. At points only a few atom spacing away from the missing half-plane of atoms, all of the other atoms are in near-perfect registry, all around the dislocation core.[52] If one mentally projects a distance of about 10 atom distances, the perfect registry can be visualized. The disregistry resulted from the collapse of the planes.

51 它可以看作是半个原子平面被从晶格中移出，然后相邻的原子倒塌从而填补于平面空位。

52 在距离缺失半原子面仅有几个原子间距的位置处，所有其他原子都处于接近完美的状态。

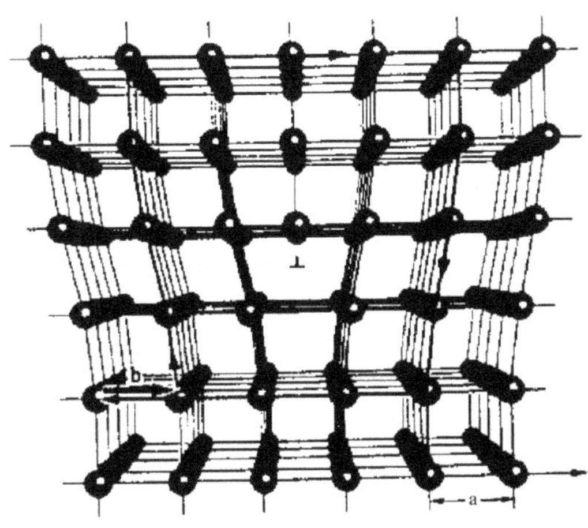

Figure 2.32 Line defects (edge dislocation).

Figure 2.33 illustrates the movement of an edge dislocation. Figure 2.33a contains the line of local disturbance, which may extend to the boundaries of the crystal with higher energy. This region contains the line defect or dislocation. The dislocation plane contains this line. Above this line the atoms are under a compressive stress, while those atoms below the line are experiencing a tensile stress. A force would cause the rows of atoms to move in the direction of the force, each row moving one at a time much like dominoes striking each other (Figure 2.33b). As each row of atoms moves, the next follows in turn until all planes of atoms in the area have been displaced sufficiently to make all planes continuous, as shown in Figure 2.33c. These bonding forces are not as strong as in a perfect crystal lattice, which permits a relatively small shear force to break the bonds, allowing the dislocation to move.[53] The bonds reform after the dislocation passes.

53 这些键不像在完美的晶格中那样强健，破坏这些键只要提供一个相当小的剪切力，从而允许位错的移动。

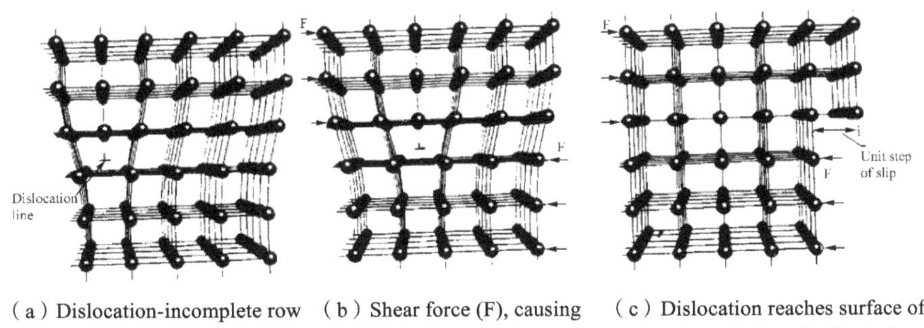

（a）Dislocation-incomplete row of atoms above slip plane.

（b）Shear force (F), causing dislocation to move.

（c）Dislocation reaches surface of crystal, producing plastic strain (deformation).

Figure 2.33 Edge dislocation movement (slip).

A screw dislocation is produced by adapting a shear stress along the line AB (Figure 2.34a) in the absence of atomic bonds within the area

ABCD. The shear stress produces a permanent displacement of the crystal on the side of AB relative to the other. The displacement is one lattice spacing. This dislocation can also be achieved if we use another approach. Let a vector AD, with the end D fixed along the line DC, rotate in the vertical (100) plane in the clockwise direction, seen from D towards C, around the axis DC and simultaneously move forward one lattice spacing, AA' or BB', per revolution around DC.[54] The screw dislocation converts a pile of crystal planes into a single continuous helix (Figure 2.34b). When the helix intersects the surface a step is formed, which cannot be eliminated by adding further atoms. The crystal grows as a never-ending spiral (Figure 2.34c).

54 在轴DC附近由D看向C，让一个终点D固定在DC线上的矢量AD沿垂直平面（100）顺时针转动，同时每旋转一个DC，向前移动一个点阵间距AA'或者BB'。

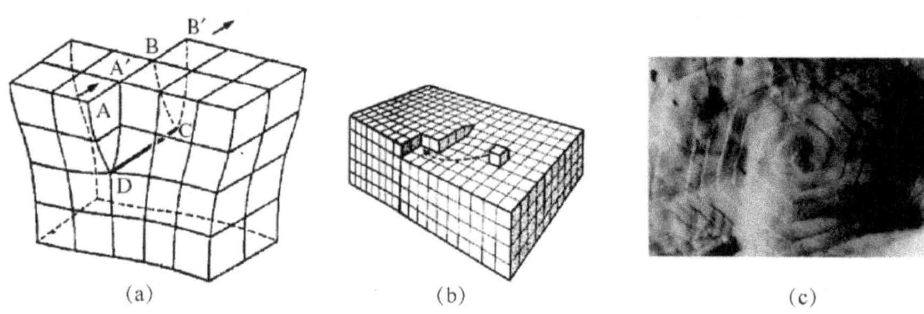

Figure 2.34 A screw dislocation

Ⅲ. Planar defects

Planar defects (interfacial or two-dimensional) are the third type of imperfection and exist in the form of grain boundaries (Figure 2.35) or surface. As each crystal grows, it establishes its own axis system, on which the atoms/ions orient themselves. Adjacent crystals with their differently oriented lattice structures close in on each other. The last atoms to take up position in a crystal find it more difficult to occupy normal lattice sites. Consequently, a transition zone is formed that is not aligned with any of the adjoining crystals. The atoms making up the grain boundary possess greater disorder, and hence greater energy, than their counterparts within the crystals themselves. Furthermore, the atoms are less efficiently packed together. These factors signify that the atoms in the grain boundaries are ready to act as sources of new crystal formation (nucleation sites) once the right conditions are met.[55] Second, they assist in the diffusion of atoms through the solid. Third, they offer resistance to the movement of dislocations and therefore modify the strength and the ability of materials to plastically deform.[56] Fourth, they act as sinks for vacancies. A solid with a large number of individual crystals (also called grains) has more grain

55 这些因素表明在条件成熟时，位于晶界的这些原子会立刻成为新晶体形成的源泉（成核点）。

56 第三，它们阻碍了位错运动，从而改变了材料强度和塑性变形能力。

boundaries than a solid with a lesser number of grains. Fine-grained material will, at normal temperatures, be stronger than coarse-grained material. Figure 2.35 is a two-dimensional sketch showing grain boundaries with different orientations of grains.

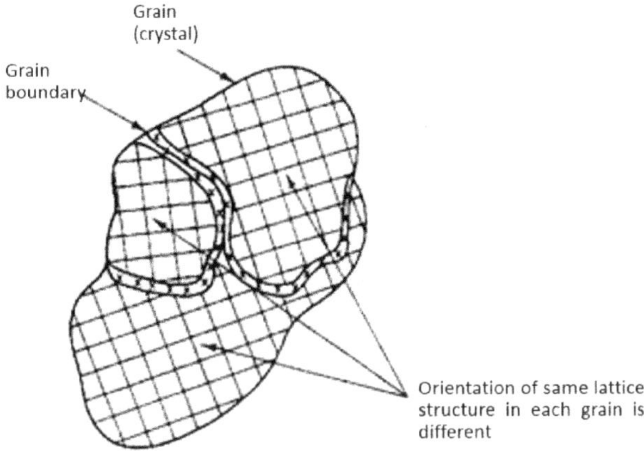

Figure 2.35 Grain and grain boundaries

2.4 Nanocrystalline Structures

A nanocrystalline material is a polycrystalline material with a crystallite size of only a few nanometers. These materials fill the gap between amorphous materials without any long range order and conventional coarse-grained materials. Definitions vary, but nanocrystalline material is commonly defined as a crystallite (grain) size below 100 nm. Grain sizes from 100–500 nm are typically considered "ultrafine" grains.

The shapes of nanocrystal material vary with material system and growth conditions. The spherical nanocrystals of monodisperse semiconductor nanocrystals, especially the sufides and selenides of cadmium, had been successful synthesized and widely used in optoelectronic application. The one dimension nanocrystals of nanotube or nanowire are the common form of carbon nanomaterials.

Nanocrystals of metals, oxides and semiconductors have been studied intensely in the last several years by different chemical and physical methods. In the past decade the realization that the electronic, optical, magnetic and chemical properties of nanocrystals depend on their size has motivated intense research in this area. This interest has resulted in better understanding of the phenomena of quantum confinement, mature synthetic schemes and fabrication of exploratory nanoelectronic devices.

2.5 Amorphous Structures

Any material that shows only a short-range order of atoms or ions is an amorphous material; that is, a non-crystalline one. In general, most materials want to form periodic arrangements since this configuration maximizes the thermodynamic stability of the material.[57] Amorphous materials tend to form when, for one reason or other, the kinetics of the process by which the material was made did not allow for the formation of periodic arrangements. The investigation of amorphous materials is a very active area of research. Despite enormous progress in recent years, our understanding of amorphous materials still remains far from complete. The key problem is the absence of the simplifications associated with periodicity. Nonetheless, from comparison of the properties of materials in a crystalline and an amorphous state we have learned the essential features of the electronic structure, and thereby also macroscopic properties, are determined by short-range order.[58] Thus these properties are similar for solids in the amorphous and crystalline state. Examples of amorphous solids are glasses, ceramics, gels, most polymers, rapidly quenched metals and some thin-film systems deposited on a substrate at low temperatures.

57 通常地，大多数材料希望形成周期性的排列，这样能最大限度地提高材料的热力学稳定性。

58 然而，对比晶态与非晶态材料的性质，我们得到的结论是电子结构的基本特征，乃至材料宏观性质取决于其短程有序性。

- **Glass-ceramics**

By deliberately nucleating ultrafine crystals in amorphous glasses, glass-ceramics can be achieved, which can be made up to about 99.9% crystalline and are quite strong. Some glass-ceramics can be made optically transparent by keeping the size of the crystals extremely small (about 6100 nm). The major advantage of glass-ceramics is that they are shaped using glass-forming techniques, yet they are ultimately transformed into crystalline materials that do not shatter like glass.

- **Amorphous plastics**

Similar to inorganic glasses, many plastics are amorphous. They do contain small portions of material that are crystalline. During processing, relatively large chains of polymer molecules get entangled with each other, like spaghetti. Entangled polymer molecules do not organize themselves into crystalline materials. During processing of polymeric beverage bottles, mechanical stress is applied to the preform of the bottle (e.g., the manufacturing of a standard 2-liter soft drink bottle using polyethylene terephthalate (PET plastic)). This process is known as blow-stretch forming. The radial (blowing) and longitudinal (stretching) stresses during bottle formation actually untangle some of the polymer chains, causing stress-induced crystallization.[59] The formation of crystals adds to the strength of the PET bottles.

- **Metallic glasses**

Metals, as they are found in nature, are crystalline; however, when particular multicomponent alloys are cooled rapidly, amorphous metals may form.[60] Some alloys require cooling rates as high as 10^6 K・s^{-1} in order to form an amorphous (or "glassy") structure. This technique of cooling metals and alloys very fast is known as rapid solidification. Recently, new compositions have been found that require cooling rates on the order of only a few degrees per second. This has enabled the production of so-called "bulk metallic glasses"—metallic glasses with thicknesses or diameters as large as 5 cm.

Before the development of bulk metallic glasses, amorphous metals were produced by a variety of rapid solidification techniques, including a process known as melt spinning. In melt spinning, liquid metal is poured onto chilled rolls that rotate and "spinoff" thin ribbons on the order of 50 microns thick. It is difficult to perform mechanical testing on ribbons; thus, the development of bulk metallic glasses enabled mechanical tests that were not previously possible. Bulk metallic glasses can be produced by several methods.

Since metallic glasses are not crystalline, they do not contain dislocations. Dislocations lead to yield strengths that are lower than those that are theoretically predicted for perfect crystalline materials. The high strengths of metallic glasses are due to the lack of dislocations in the amorphous structure.

Despite their high strengths, typical bulk metallic glasses are brittle. Most metallic glasses exhibit nearly zero plastic strain in tension and only a few percent plastic strain in compression (compared to tens of percent plastic strain for crystalline metals). This lack of ductility is related to a lack of dislocations. Since metallic glasses do not contain dislocations, they do not work harden. Thus, when deformation becomes localized, it intensifies and quickly leads to failure.

To summarize, amorphous materials can be made by restricting the atoms ions from assuming their "regular" periodic positions. This means that amorphous materials do not have a long-range order. This allows us to

59 在瓶子成型时，径向（吹）和纵向（拉伸）应力确实打乱了聚合物的某些链，导致应力诱导结晶。

60 正如自然界中所发现，金属是晶态的。然而，当特定多组分的合金快速冷却时，可以形成非晶金属。

form materials with many different and unusual properties. Many materials labeled as "amorphous" can contain some level of crystallinity.

Since atoms are assembled into non-equilibrium positions, the natural tendency of an amorphous material is to crystallize (i.e., since this leads to a thermodynamically more stable material). This can be done by providing a proper thermal (e.g., a silicate glass), thermal and mechanical (e.g., PET polymer), or electrical (e.g., liquid crystal polymer) driving force into their "regular" and periodic arrangements.[61]

61 通过提供适当的热（例如，硅酸盐玻璃）、热和机械（例如，PET 聚合物）、或者电（例如，液晶聚合物）的驱动力，它们可以转变为有规律且周期性的排列。

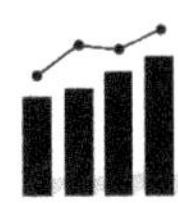

本章小结

1. 内容概要

材料科学通常描述的是组成材料的结构和性质之间的关系。这里结构不仅仅是指组成材料的基本架构或长程的顺序，也指材料的化学结构和短程的顺序。材料是由原子构成的，除了原子本身的特性，原子成键方式以及堆叠方式对材料性质也有重要影响。本章从原子出发，论述了与材料及其性质密切相关的两个基本概念：第一是材料中原子之间的各种成键方式，例如，金属键、离子键、共价键和分子间作用力；第二是材料中原子的堆叠方式，不同的堆叠方式可以形成单晶、多晶、非晶和准晶，从而导致了不同的材料性质。

2. 基本概念

金属键、离子键、共价键、分子间作用力、范德华力、伦敦力、德拜力、基森力、单晶、多晶、准晶、非晶、晶胞、空间点阵、格点、晶格常数、原胞、体心立方、面心立方、六方密堆、简单立方、四方晶格、密勒指数（晶面指数）、晶向指数、配位数、堆积系数、晶格缺陷、点缺陷、线缺陷、面缺陷、结构缺陷、化学缺陷、位错、取代粒子、间隙粒子。

3. 主要公式

（1）晶胞中原子/分子数目：$N_{\text{unit cell}} = N_{\text{interior}} + \frac{N_{\text{face}}}{2} + \frac{N_{\text{corner}}}{8}$

（2）密勒指数：$hkl = \left(\frac{D}{X}\frac{D}{Y}\frac{D}{Z}\right)$

（3）堆积系数：$PF = \frac{volume\ of\ atoms / molecules\ in\ a\ unit\ cell}{volume\ of\ unit\ cell}$

Vocabulary

amorphous	无定形的
atom	原子
atomic level defects	原子水平的缺陷
close-packed (CP) structure	密堆积结构
close-packed direction	密堆积方向
coulombic	库仑的
covalent bonding	共价键
crystal cell	晶胞
crystal lattice	晶格
crystal structure and defect	晶体结构与缺陷
crystal system	晶系
crystalline	晶态

crystalline materials	晶体材料
crystallization	结晶
crystallography	晶体学
cubic	立方
delocalize	离域
density	密度
diffraction	衍射
direction indices	晶向指数
extended defects	扩展缺陷
Frenkel defect	弗仑克尔缺陷
geometry	几何学
glasses	玻璃
grain	晶粒
grain bondaries	晶界
hexagonal	六方
impurities	杂质
indice	指数
infinite	无限的
interfacial defect	界面缺陷
interrupt	阻断
interstitial defect	间隙缺陷
interstitial sites	间隙位置
ionic bonding	离子键
isotropic	各向同性的
kinetic energy	动能
lattice	晶格
lattice constants	晶格常数
lattice direction	晶向
lattice parameters	晶格参数
lattice plane	晶面
lattice point	格点
lattice types	晶格类型
Long-Rang Order (LRO)	长程有序
lubricant	润滑剂
macroscopic properties	宏观性质
melting point	熔点
metallic bonding	金属键
molecular bonding	分子键

monoclinic	单斜
negatively charged	带负电荷的
noble gas	惰性气体
orthorhombic	正交的
overlap	重叠
parabola	抛物线
plane indices (miller indices)	晶面指数
point defects	点缺陷
primitive cell	初级晶胞
primitive lattice	初级点阵
quasicrystalline	准晶体
rhombohedral or trigonal	菱方
Schottky defect	肖特基缺陷
space lattice	空间点阵，晶格
species	物种
strength	强度
substitutional defect	取代型缺陷
tetragonal crystal system	四方晶系
triclinic	三斜
unit cell	晶胞
vacancy	空位
width	宽度

Problems

1. True or false questions.

(1) Metallic and ionic bondings are coulombic in nature.

(2) The melting points of metals are generally high due to the strong metallic force between nucleus and electrons.

(3) Ionic materials are electrical conductive at room temperature.

2. Please schematically draw each of the following site in a crystal.

(1) cubic site (2) octahedral site (3) tetrahedral site (4) rhombohedral site

3. Please write the full name of fcc, bcc and hcp.

4. For crystal of pure metal with fcc, bcc, and hcp, respectively, please tell the (1) coordination number (CN) and (2) number of atoms per unit cell (N_{unit}).

5. Please enumerate all the atomic bonding modes, and state material properties with each of the bonding mode, respectively.

6. Please classify van der Waals force with brief explanation

7. Can van der Waals force exist between atoms, give an example.

8. According to atom stacking, how many types of materials? What are they?

9. What are Frenkel defect and Schottky defect?

10. In the following figure, two planes A and B are shown with their intercepts. Please determine the Miller indices of plane A and plane B.

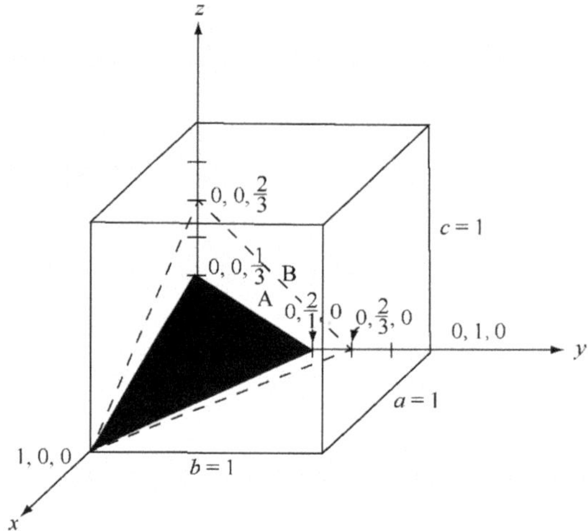

Chapter 3 Fundamentals of Electron Theory

As a basic knowledge of the following chapters, in this chapter, we will discuss the fundamental of electron theory, which is the basis for electrical, optical, magnetic and thermal properties of materials. For example, the electrical properties of solids are mainly determined by the properties of electrons in solid. Protons can usually be relegated to subordinate roles, like ensuring charge neutrality. Neutrons may sometimes need to be considered in the case of some superconducting materials, in which the critical temperature depends on the total mass of the nucleus. But on the whole, the energies and movements of electrons hold the key to the electrical properties of solids, which are the focus of this chapter. Here, after introducing wave particle duality of electron, Schrödinger's equation and the theory of statistics are presented; then free electron theory and band theory based on quantum mechanics will be introduced.

3.1 Introduction

In atomic physics, we learned that the electrons of free atoms occupy fixed and discrete energy levels. The electrons obey the Pauli Exclusion Principle and two electrons, at most, occupy one energy level, or orbital. When *N* atoms come together to form a solid, the energy levels of the electrons change depending on the position of electrons. Considering two atoms approaching each other to form a bond, the orbitals that contain the valence electrons are located (on average) farther from the nucleus than the orbitals that contain the "core" or inner most electrons.[1] At first, the orbitals that contain the valence electrons of one atom thus interact with the orbitals that contain the valence electrons of the other atom. Since the orbitals of these electrons have the same energy when the atoms are in isolation, the orbitals shift in energy or "split" so that the total energy of whole system will be minimum. The electrons of the atoms will occupy these new orbitals by first filling the lowest energy levels. One orbital will be occupied by two electrons because of Pauli Exclusion Principle.

1 考虑两个原子彼此靠近成键时，相比于含有“核心”或最内层电子的轨道，（平均来讲）含有价电子的轨道距离原子核较远。

If we postulate the existence of a certain number of electrons capable of conducting electricity, how to describe and predict their behaviours? First of all, how to represent a conducting electron? In equilibrium, what kind of statistical distribution does these electrons follow? Is it an analogue or the same as the Maxwell-Boltzmann statistics for gases in equilibrium, which depends strongly on the temperature of the system? We must keep in mind that besides electrons, a corresponding amount of positive charge exists in the solid in order to keep electrically neutral to the outside world. Second, in analogy with our picture of gases, we may assume that the electrons bounce around in the interatomic spaces, colliding occasionally with lattice atoms. How do these conditions influence the behaviour of the conducting electrons? In this chapter, a brief picture describing these contents will give out.

3.2 Wave Particle Duality and Quantum Mechanics

According to classical theories, electron is a particle because it has a rest mass of 9.1091×10^{-31} kg and light is a wave because it exhibits diffraction, interference, polarization, reflectance properties and no static mass.[2] However, with the capacity of human cognition than ever, it was found that light also possesses particle behavior, and electron shows wave properties simultaneously. One famous manifestation is the finding of photoelectric effect. In 1905, Einstein published a paper on the photoelectric effect (Figure 3.1), explaining the experimental phenomenon that when a monochromatic light hits a metal surface, if the frequency υ of the light exceeds a particular frequency υ_0, electrons are emitted from the metal surface. The velocities of the electrons can be determined from the deviation in a known magnetic field.[3] Einstein assumed that the incident light behaved as a stream of light quanta or photons, each with energy $h\upsilon$, and not like a wave. He set the minimum energy, required to release an electron from a metal surface, equal to the energy $h\upsilon_0$, which is just able to knock out the electron, i.e. $\phi = h\upsilon_0$. Einstein applied the energy law to the process:

$$h\upsilon = \phi + \frac{m_e v_e^2}{2} = h\upsilon_0 + \frac{m_e v_e^2}{2} \tag{3-1}$$

where v_e is the velocity of the electron and ϕ is called the work function of metal. The last term represents the kinetic energy of the released electron, which equals $h\upsilon - h\upsilon_0$, where $\upsilon > \upsilon_0$. If $\upsilon < \upsilon_0$, no electrons are emitted even if the intensity of the light increases substantially.

This is in excellent agreement with the experimental observations, which are impossible to explain by wave property of light, and demonstrating the particle property of light.[4] It was necessary to accept the duality of electromagnetic radiation, i.e., it has both wave and particle properties.

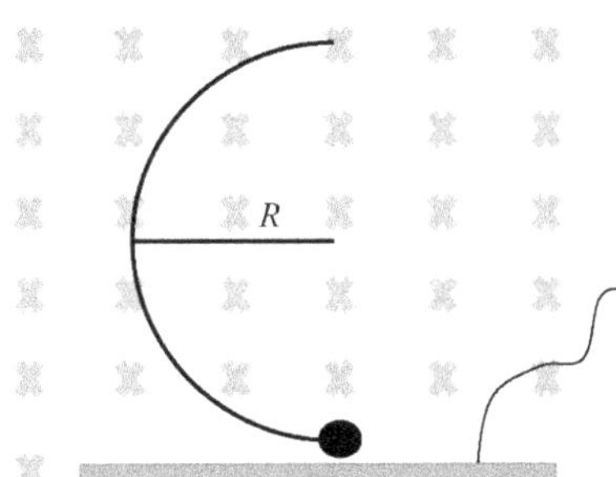

Figure 3.1 Photoelectric effect. The released electron deviates in a magnetic field and its velocity can be calculated from the measured radius R and known strength of the magnetic field B.

Based on the works of Einstein and others, in 1924, Louis de Broglie proposed the matter wave concept. He boldly suggested that all matter has both wave and particle properties. He started from the momentum of a photon:

$$p = \frac{h\upsilon}{c} = \frac{h}{\lambda} \tag{3-2}$$

and assumed that the equation $p = h/\lambda$ is valid for both waves and particles. A particle has a momentum equal to mv, which gives the wavelength of the

2 根据经典理论，电子是一个粒子因为它的静止质量是 9.1091×10^{-31} kg，光是一个波，因为它表现出衍射，干涉，偏振，反射的性质，并且没有静止质量。

3 电子的速度可以通过它对一个已知磁场的偏离来确定。

4 这和实验观测非常吻合，实验中的观测结果是不能用波的性质来解释的，展示了光的粒子特性。

matter wave:

$$\lambda_{\text{deB}} = \frac{h}{mv} \tag{3-3}$$

where v is velocity of the particle, λ_{deB} calls de Broglie wavelength of the particle, which moves with the velocity v, h is Planck's constant, m is the mass of the particle. The expression mv of the momentum is often denoted by p. It is valid in both classical and relativistic mechanics. In the latter case, m represents the relativistic mass of the particle.

Actually, the wave particle duality Equation (3-2) can be deduced by using both particle concept and wave concept.[5]

5 实际上，波粒二象性方程（3-2）可以通过同时使用粒子和波动概念来推导。

de Broglie's hypothesis was confirmed by two famous experiments. The wave character of electrons was demonstrated by Davisson and Germer in US in 1927 and independently by G. P. Thomson (son of J. J. Thomson, who discovered the electron in 1897) in England in 1928 by electron diffraction through a thin metal foil (Figure 3.2).[6] Their experiments showed excellent agreement with Equation (3-3) and the resemblance with X-ray diffraction (Figure 3.3) is striking. The electron distribution behind the foil in Figure 3.2 can be calculated as a diffraction phenomenon if the wave properties of the moving electrons are considered. The result can be verified experimentally on a screen or a photographic plate. Alternatively, the electrons can be regarded as particles and the electron distribution behind the foil can be found by statistical calculations. The electrons can be registered by a mobile particle detector at various points in the plane of the screen. The statistical distribution of a large number of electrons agrees with the diffraction pattern. The probability of finding electrons at the bright areas of the diffraction pattern is high whereas few electrons hit the detector at the dark positions of the diffraction pattern. A beam of electrons shows wave properties during propagation. On interaction with matter, for example a photographic plate or particle detector, the electrons behave like particles.

6 电子的波动特性，由美国的 Davisson 和 Germer 在 1927 年，以及英国的 G. P. Thomson（1897 年发现了电子的 J. J. Thomson 的儿子）在 1928 年，分别独立地通过电子在金属薄片上的衍射实验得到证明（图 3.2）。

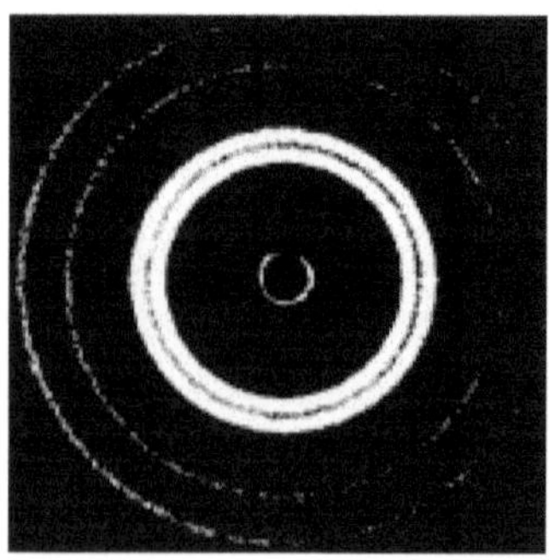

Figure 3.2 Diffraction pattern on passing a beam of electrons through an Al foil. The de Brogile wavelength of the electrons was the same as the wavelength of the X-rays in the experiment.[7] Reproduced from F. Blatt, *Modern Physics*. McGraw-Hill Inc. (1992).

7 电子的德布罗意波长和实验中 X 射线的波长相同。

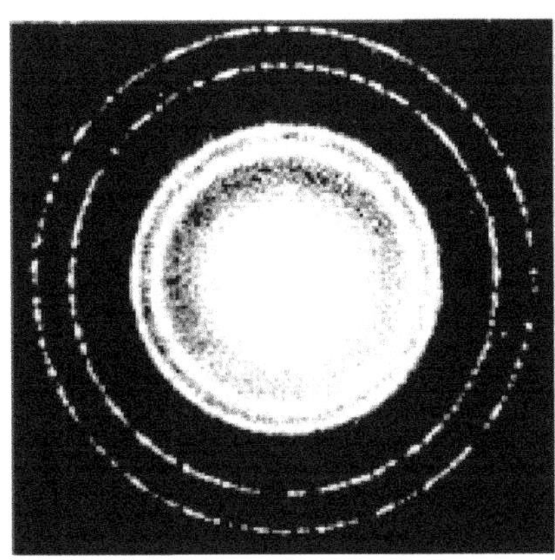

Figure 3.3 Diffraction pattern obtained by passing monochromatic X-rays through the same Al foil as in the experiment show in Figure 3.2. Reproduced from F. Blatt, Modern Physics. McGraw-Hill Inc. (1992).[8]

Table 3.1 shows the comparison between the properties of wave and particles.[9] The discovery of the wave-particle duality led to a fruitful period of physics and is closely associated with the development of two new theories, wave mechanics and quantum mechanics. Schrödinger concentrated on de Broglie's matter waves and developed the theory of wave mechanics. In 1925, he proposed a general wave equation for the determination of the wavelength of the matter wave of a moving particle. This equation is the fundamental in wave mechanics and can be applied to all sorts of waves and particles. Heisenberg concentrated on the particle or quantum aspect and discussed the impossibility of the simultaneous, infinitely accurate determination of both the position and momentum of a moving particle. He published his fundamental uncertainty principle in 1927.

Table 3.1 A comparison between the properties of wave and particles

	Wave	Particles
Energy	$Hf = \hbar\omega$	$1/2mv^2+E_{pot}$
Velocity	$v = \omega/k$, $v_g = d\omega/dk$	v
Wavelength	λ	h/mv

Wave mechanics describes the particle as a matter of wave, whereas quantum mechanics concentrates on the quantization of various physical quantities of particles. It soon became clear that the two theories are two aspects of the same fundamental theory, nowadays generally called quantum mechanics. The two theories proved to be different mathematical formulations of the same physical problem.

Quantum mechanics has been applied to photos, electrons, protons, neutrons, atoms, molecules and even macroscopic particles with great

8 单色 X 射线穿过与图 3.2 实验中相同的铝箔得到的衍射图谱，摘自 F. Blatt，Modern Physics. McGraw-Hill Inc（1992）。

9 表格 3.1 描述了波动特性和粒子特性的比较。

success. Classical mechanics is a special case of quantum mechanics just as geometric optics is a special case of wave optics. Classical mechanics holds for large particles, for example a moving ball, but fails for small particles in high-speed motion. The de Broglie wavelength is very small for a heavy particle and no deviations from classical wave mechanics can be observed. Combining with wave mechanics, modern quantum mechanics uses wave function to describe matter, and introduces quantum feature in micro scale to explain and predict material properties in macro-scale.[10]

10 结合波动力学，现代量子力学用波函数来描述物质，并且在微观上引入了量子特性，来解释和预测宏观上的物质性质。

3.3 Wave Function and Schrödinger Equation

Ⅰ. Wave Function

Matter possesses wave property, thus, a wave function can describe this matter wave. In 1925, Schrödinger proposed the wave function firstly, and established the fundamental equation in quantum mechanics which replaces the equations of classical mechanics for atomic systems.

Taken electrons as an example, electron wave describes the probability of finding the electron within a given volume. At different moment, the probability of finding the electron at different position may be different, thus wave function ($\varPhi$) is position- and time- dependent.[11] That is:

11 在不同时刻、不同位置找到电子的概率或许不同，因此波函数（$\varPhi$）和位置和时间有关。

$$\varPhi = (x, y, z, t) or\ \varPhi = (\vec{r}, t) \tag{3-4}$$

Since the position of an electron is closely related to statistics and probability. In quantum mechanics it is impossible to tell the exact position of an electron. The best one can do is to predict the probability of finding the electron within a given volume. This is an essential feature of quantum mechanics. It is known from elementary wave theory that the square of the amplitude is proportional to the intensity of a wave. In 1926, Born suggested, in analogy with this, that the value of the wave function $\varPhi$ within a volume element dxdydz at position(x, y, z) is related to the probability of finding the particle within the volume element.[12] Therefore, the probability of finding a particle within a volume element dxdydz is

12 1926 年，Born 提出，类比于这个，在位置为(x, y, z)时的体积元素 dxdydz 内波函数$\varPhi$的值与在该体积内找到粒子的概率相关。

$$\mathrm{d}w = C|\varPhi|^2\, \mathrm{d}x\mathrm{d}y\mathrm{d}z = C|\varPhi|^2\, \mathrm{d}\tau \tag{3-5}$$

$$\mathrm{d}\tau = \mathrm{d}x\mathrm{d}y\mathrm{d}z \tag{3-6}$$

In volume V, the probability of finding the particle:

$$w = \int_v \mathrm{d}w = \int_v C|\varPhi|^2\, \mathrm{d}\tau = C\int_v |\varPhi|^2\, \mathrm{d}\tau \tag{3-7}$$

In the whole space, the probability of finding a particle should be 100%. Hence:

$$C = \int_{\infty} |\Phi|^2 \mathrm{d}\tau = 1 \tag{3-8}$$

Let $\psi = \sqrt{C}\,\Phi$, then:

$$\int_{\infty} |\psi|^2 \,\mathrm{d}\tau = 1 \tag{3-9}$$

$\psi(x,y,z)$ is called the normalized wave function, and the left hand of Equation (3-9) represents the probability of finding a particle in the whole space.

In wave theory, there is the definition,

$$|\psi|^2 \,\mathrm{d}x\mathrm{d}y\mathrm{d}z = \psi\psi^* \mathrm{d}x\mathrm{d}y\mathrm{d}x \tag{3-10}$$

where ψ^* is the complex conjugate of ψ. Thus,

$$\int_{\infty} |\psi|^2 \,\mathrm{d}\tau = \int_{\infty} \psi \cdot \psi^* \mathrm{d}\tau = 1 \tag{3-11}$$

The probability concept is a most important interpretation of the wave function. It can be used to find the electron distribution around the nucleus of the atom and to obtain selection rules for electron transition from one orbital to another in an atom or molecule.[13]

13 它可以用来找出电子在原子中围绕原子核的分布情况，并且得出在一个原子或分子中从一个轨道到另一个轨道的电子跃迁规则。

Ⅱ. Schrödinger Equation

In order to determine the stationary energy states as well as the corresponding space shape of the system, we have to set up the wave equation in analogy with the equation of a vibrating string.[14] Since the wave function for micro-particles was firstly proposed by Schrödinger, the matter wave equations are usually termed as Schrödinger equation.

14 为了确定体系的稳态能量，以及相应的空间形状，我们需要建立与弹簧振动类似的波动方程。

Firstly, taking 1D planar wave as an example, for 1D planar wave propagating along x direction with initial phase of zero, the classical wave function can be written as:

$$Y(x,\ t) = A\cos\left[2\pi\left(\frac{x}{\lambda} - vt\right)\right] \tag{3-12}$$

where A is vibrating amplitude, λ is wavelength, v is frequency, and t is time.

Introducing a new variable of the wave vector, $\bar{k}$, with its magnitude representing the wave number within one period ($|\bar{k}| = k = \frac{2\pi}{\lambda}$), the expression (3-12) can be expressed as:

$$Y(x,t) = A\cos[kx - \omega t] \tag{3-13}$$

The ω is angular frequency, and $\omega = 2\pi v$.

In order to facilitate calculation, we can use an exponential function to instead of a cosine function with the time dependence:[15]

15 为了便于计算，我们可以使用一个指数函数来替代与时间有关的余弦函数。

$$Y(x,t)=Ae^{i(kx-\omega t)} \tag{3-14}$$

With wave-particle duality, $\lambda=\dfrac{h}{p}$, and considering $E=hv$, we have:

$$k=\frac{2\pi}{\lambda}=\frac{2\pi}{h/p} \tag{3-15}$$

$$\omega=2\pi v=2\pi\frac{E}{h} \tag{3-16}$$

Then (3-14) can be rewritten to:

$$\psi(x,t)=Ae^{\frac{2\pi i}{h}(px-Et)}=Ae^{\frac{i}{\hbar}(px-Et)}=Ae^{\frac{i}{\hbar}px}e^{-\frac{i}{\hbar}Et} \tag{3-17}$$

In (3-17), the function can be divided into a time independent part ($\varphi=Ae^{\frac{i}{\hbar}px}$) and a time dependent part ($e^{-\frac{i}{\hbar}Et}$). The time-independent part is termed as amplitude function. If a wave function is independent of time, it is called standing wave function. If electron locates in a potential field that is only dependent on the spatial location and independent of time. These electrons will eventually reach steady state, thus the wave function of electron will be only the function of spatial coordinates, the probability of finding the electron in space is:

$$|\psi|^2=\psi\psi^*=\varphi(x)e^{-\frac{i}{\hbar}Et}\varphi(x)e^{\frac{i}{\hbar}Et}=|\varphi(x)|^2 \tag{3-18}$$

The standing wave function $\varphi(x)$ is the solution of the Schödinger equation without time variable. The square of the amplitude of $\varphi(x)$ refers to a time-independent probability of finding electron in a space. [16]

16 驻波函数 $\varphi(x)$ 是与时间无关 Schödinger 方程的解，等于在一个空间中找到电子的概率，$\varphi(x)$ 的平方和时间无关。

Generally, to solve the Schödinger equation, the standing wave function $\varphi(x)$ need to be obtained firstly, then the time dependent part is added :

$$\psi(x,t)=\varphi(x)e^{-\frac{i}{\hbar}Et} \tag{3-19}$$

For electron movement in one dimension with energy of E, the standing wave function $\varphi(x)$ is,

$$\varphi(x)=Ae^{\frac{i}{\hbar}px}=Ae^{\frac{i}{\hbar}\sqrt{2Em}x} \tag{3-20}$$

Applying differential operation, one obtains,

$$\frac{\mathrm{d}^2\varphi}{\mathrm{d}x^2}+\frac{2mE}{\hbar^2}\varphi=0 \tag{3-21}$$

This is one-dimension Schrödinger equation for electron with energy of E.[17] If the electron is not free, but moving in a potential field U, the energy of electron should be:

17 这是能量为 E 的电子的一维薛定谔方程。

$$E=\frac{1}{2}mv^2+U,\quad p^2=m^2v^2=2m(E-U) \tag{3-22}$$

Thus,

$$\varphi(x)=Ae^{\frac{i}{\hbar}px}=Ae^{\frac{i}{\hbar}\sqrt{2m(E-U)}x} \tag{3-23}$$

Similarly, by differential process, one can obtain the one-dimension Schrödinger equation for electron in potential field of U:[18]

$$\frac{\mathrm{d}^2\varphi}{\mathrm{d}x^2}+\frac{2m(E-U)}{\hbar^2}\varphi=0 \tag{3-24}$$

Following the same processes, the three-dimension Schrödinger equation for electron in potential field of U can be expressed as,

$$\frac{\partial^2\varphi}{\partial x^2}+\frac{\partial^2\varphi}{\partial y^2}+\frac{\partial^2\varphi}{\partial z^2}+\frac{2m(E-U)}{\hbar^2}\varphi=0 \tag{3-25}$$

or,

$$\nabla^2\varphi+\frac{2m(E-U)}{\hbar^2}\varphi=0 \tag{3-26}$$

$$\nabla^2=\frac{\partial^2}{\partial x^2}+\frac{\partial^2}{\partial y^2}+\frac{\partial^2}{\partial z^2} \tag{3-27}$$

∇^2 is called Laplacian operator. Under certain situation, solutions of above Schrödinger equation are termed as eigenfunctions, they represent the stationary states of the wave motion, for which the waves do not cancel owing to destructive interference, the corresponding energy is eigenenergy or eigenvalue.[19]

Ⅲ. Heisenberg Uncertainty Principle

Independently of Schrödinger's theory, Heisenberg analysed the possibilities of determining the position and momentum of a moving particle simultaneously. His result, published in 1927, is an essential part of quantum mechanics, by no means in contradiction with wave mechanics. Born's probability interpretation of the wave function represents a link between Schrödinger's wave mechanics and Heisenberg's quantum mechanics. The outlines of his theory will be discussed briefly below.

Consider a particle with mass m moving with the velocity of v. According to de Broglie matter wave concept, its momentum can be written as $mv = h/\lambda_{\mathrm{deB}}$, where λ_{deB} is the de Broglie wavelength of the moving particle. If we want an exact value of the wavelength or the momentum of the particle, then the position of the particle is completely undetermined according to wave mechanics.[20] It is the same everywhere as the de Broglie wave has an infinite extension. Similarly, if we wish to define the position of a particle very accurately, the wave function must differ from zero only at a given position. This can only be achieved by overlapping of many sine waves of all wavelengths from zero up to infinity. This makes the wavelength and hence the momentum completely uncertain. That is, position and momentum cannot be measured exactly simultaneously. This is a consequence of Heisenberg's uncertainty principle, which is closely

18 类似地，通过微分过程，能够得到电子在势场 U 中的一维薛定谔方程。

19 在特定的情况下，以上薛定谔方程的解被称为本征函数，它们代表波动的驻波状态，这种情况下，相消干涉，不会导致波的相互抵消，相应的能量称为本征能量或者本征值。

20 根据波动力学，如果我们想要得到粒子的波长或者动量的精确值，粒子的位置就完全不确定。

related to wave mechanics:

$$\Delta y \Delta p_y \geqslant \frac{h}{4\pi} \tag{3-28}$$

where, h is Planck's constant, Δy is uncertainty in position, Δp_y is uncertainty in momentum.

The Heisenberg uncertainty principle is also valid for another pair of quantities, energy and time:

$$\Delta E \Delta t \geqslant \frac{h}{4\pi} \tag{3-29}$$

Because it is impossible to determine the position and momentum of an electron simultaneously, well-defined electron orbitals are impossible. The only possibility is to calculate the probability of finding an electron within a certain volume element as a function of position.

3.4 Classical and Quantum Statistics

Materials are made from a huge quantity of particles, such as molecules, atoms, electrons, protons, neutron, ions, etc. The macro properties of materials are determined by the average statistical values under equilibrium condition. In physics, there are three kinds of statistical theories. They are Classical Maxwell-Boltzmann statistics, Fermi-Dirac statistics, and Bose-Einstein statistics.

Ⅰ. Maxwell–Boltzmann Statistics

When a large number of distinguishable and non-interacting particles distribute among a number of different available energy states, they obey Maxwell-Boltzmann statistics.[21]

Consider a particle system containing N_0 identical, but distinguishable, and non-interacting particles with a fixed volume V and constant temperature T, their total energy is U. The energy of the particles can be distributed in many different ways. It will be described by the specification of the number of N_i particles with the energy of u_i.[22] The problem is to find the most likely distribution, i.e. to derive N_i as a function of u_i.

Suppose that N_1 particles have the energy u_1, N_2 particles have the energy u_2, … and N_i particles have the energy u_i. This partition N_1, N_2, … and N_i represents a so-called macro state of the system. For the macro state the following sub conditions are valid:

$$\Sigma_i N_i = N_0 \tag{3-30}$$

$$\Sigma_i N_i u_i = U \tag{3-31}$$

21 当大量可区分且无相互作用的粒子分布在若干不同的可用能量状态时，它们遵循 Maxwell-Boltzmann 统计。

22 它将被描述为具有能量 u_i 的粒子数量为 N_i。

Boltzmann assumed that the probability of each partition is proportional to the number of different ways in which the particular partition can be obtained.[23] Permutations of particles within each energy level produce no new distribution and consequently no new macro state. Only if the number of particles within two or more energy levels is changed is a new macro state obtained.

23 Boltzmann 假定，具有一定能量粒子的概率，与产生这部分粒子方式的数量成正比。

Boltzmann deduced the probability for an arbitrary partition, which is proportional to the total number of distinguishable different ways to obtain the partition. The equilibrium of the particle system corresponds to the most probable partition, i.e., the partition which can be obtained in the maximum number of distinguishable different ways.[24] By use of this maximum condition, he found the equilibrium distribution:

24 粒子体系的平衡对应于最可几分布，也就是说，这个分布的粒子具有最大数量的可分辨的分布方式。

$$f(u_i)=\frac{N_i}{N_0}=\frac{e^{-u_i/k_BT}}{e^{-u_1/k_BT}+e^{-u_2/k_BT}+\cdots+e^{-u_i/k_BT}} \tag{3-32}$$

where u_i is the energy of particle i, N_i is the number of particles which have the energy of u_i, N_0 is the total number of particles, k_B is Boltzmann's constant, T is absolute temperature of the system.

We have assumed that all energy levels are equally probable. If this is not the case, we have to introduce the statistical weight g_i of energy level u_i and the Maxwell–Boltzmann distribution law in its general form becomes

$$f(u_i)=\frac{N_i}{N_0}=\frac{g_i e^{-u_i/k_BT}}{g_1e^{-u_1/k_BT}+g_2e^{-u_2/k_BT}+\cdots+g_ie^{-u_i/k_BT}} \tag{3-33}$$

The above expression is normalized equation. Treating other factors as one front factor A, the probability of the system with energy of E can be simplified as,

$$f(E)=Ae^{-E/kT} \tag{3-34}$$

Classical Maxwell-Boltzmann statistics is valid for particles which are identical and distinguishable. That is the structure and components of these particles are identical, and the path of every particle principally can be tracked. These systems include the velocity and energy distribution in a gas, distribution of vacancies in a solid and distribution of atoms and molecules in different energy levels.[25]

25 这些体系包括气体的速度和能量分布、固体中空位分布、以及在不同能级上原子和分子的分布。

For system possessing significant wave property, classical statistics will face discrepancy results, and the following two quantum statistics is needed. In quantum statistics, the particles are indistinguishable, i.e. the number of particles in an energy level is known, however, these particles cannot be identified from each other.

Ⅱ. Fermi–Dirac Statistics

The physicists, Fermi and Dirac, showed that all particles with half integer spins obey the same statistics which named after them. The derivation of Fermi–Dirac statistics is performed in the same way as Maxwell–Boltzmann statistics but with the very important difference that the particles are indistinguishable. The result is that:

$$f(E)=\frac{1}{\exp\left(\frac{E-E_{\mathrm{F}}}{kT}\right)+1} \tag{3-35}$$

where E_{F} is Fermi level.

The Fermi factor $f(E)$ represents the probability that the energy level E will accommodate an electron at temperature T.

Ⅲ. Bose–Einstein Statistics

The Fermi-Dirac statistics is valid for particles which are identical and indistinguishable and belong to such a system that the Pauli Exclusion Principle is valid.[26] However, there are systems that the particles are identical and indistinguishable but do not conform with Pauli Exclusive Principle, such as phonons and photons. In this case, Bose-Einstein statistical distribution is valid:

26 Fermi-Dirac统计适用于那些相同且不可区分的粒子，而且这些粒子遵循泡利不相容原理。

$$f(E)=\frac{1}{\exp\left(\frac{E-\mu}{kT}\right)-1} \tag{3-36}$$

where μ is chemical potential

Figure 3.4 shows a comparison between the three statistics. In Bose-Einstein statistics we have the restriction of $E>\mu$ otherwise $f(E)$ becomes negative. The differences between the three statistics are very large for system with small energy values, which are related to small-size system such as electron.[27] The Maxwell-Boltzmann statistics was initially produced for thermal-statistic usage dealing with relatively larger system of molecules. When comes to electron system for the study of electricity, due to the far more reduced sized compared to molecular system, quantum concept must be included and the Maxwell-Boltzmann statistic becomes invalid. At high energy values they coincide and both the Bose-Einstein and Fermi-Dirac statistics can be approximated by the mathematically simpler Maxwell-Boltzmann statistics.[28] This is because at fixed temperature, the system with high energy corresponds to large-size system, which means that the quantum effect is not such obvious as in system with low energy, and the dominant factor for energy distribution is temperature.

27 对于能量很小的体系这三个统计区别很大，这类体系中粒子的尺寸很小，例如电子。

28 在能量很高的体系，它们是一致的，Bose-Einstein 和 Fermi-Dirac 统计都可以近似看成数学上比较简单的 Maxwell-Boltzmann 统计。

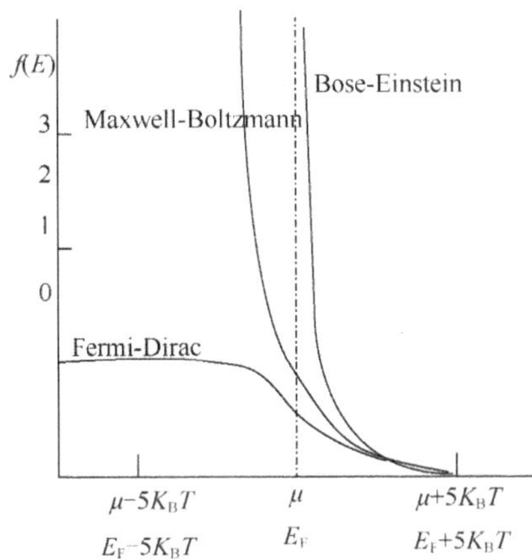

Figure 3.4 Maxwell-Boltzmann, Bose-Einstein and Fermi-Dirac statistics.

After introducing a parameter α, the three equations can be written as one, and we have:

$$f(E)=\frac{1}{\exp\left(\frac{E_i-\mu}{kT}\right)+\alpha} \tag{3-37}$$

(1) $\alpha=1$: Fermi-Dirac statistics, applied to particles with spin quantum of ½, e.g. electron, neutron, proton, etc.

(2) $\alpha=-1$: Bose-Einstein statistics, valid for particles with spin quantum of integer, e.g. photon, phonon, etc.

(3) $\alpha=0$: Maxwell-Boltzmann statistics, applied to particles dominated by particle property.

3.5 Free Electron Theory of Metals

Most metals have a few weakly bound electrons in their outermost electron shells. Since metallic crystals normally have high coordination numbers, there are far from enough valence electrons to form electron pair bonds between all near neighbor atoms in a metal. Instead, all the valence electrons are assumed to be shared by all the metal atoms together. A metallic crystal can be regarded as a three-dimensional array of positive ions firmly held together by the attraction from a common electron cloud, consisting of all the valence electrons in the crystal. The electrons belong to the whole crystal lattice and not to any particular metal ion.

If the metal is exposed to an electric field, the valence electrons of most metals can easily move in a direction opposite to the field and carry the electric current. The electrons also transport momentum and kinetic energy in case of a temperature gradient instead of an electric field across the metal.[29] Hence most metals are good electrical and thermal conductors. The theory suggested above, which has proved to be very successful, is called the free electron model of a metal, which will be discussed below.

29 在金属中存在的是温度梯度而不是电场时，电子也传输动量以及动能。

Ⅰ. Classical Free Electron Postulation –Drude Model

A free metal atom in its ground state has a number of filled electron shells around the nucleus and one or several valence electrons in orbitals of the outermost electron shell. The filled shells are tightly bound to the nucleus whereas the valence electrons are supposed to be free in the sense that they are not bound to any special nucleus in solid. Therefore, these valence electrons can easily move anywhere within the metal volume but not outside. They can be compared with the molecules in an ideal gas, and be termed as ‘electron gas’.[30] Thus the classical statistics was applied to electron gas, i.e. to the valence electron in a metal.

30 它们（电子）可以比作理想气体中的分子，并且被看作是“电子气体”。

In 1900, Drude and Lorentz proposed the Drude model of electrical conduction (also called classical free electron theory), with hypothesis of: (1) When metal atoms aggregate to form solid, the valence electrons break away from their corresponding binding positive cores, moving freely in the metal.[31] These electrons are called free electron; (2) The free electrons are treated as an ideal gas and have no interaction between each other, and their behaviors conform with Maxwell-Boltzmann statistics; (3) The positive cores distribute in the whole volume, neutralizing the negative charges from the free electrons. In the classical free electron theory, the electrons were regarded as free non interacting classical particles trapped in a potential well, which keeps the electrons inside the metal.[32]

31 当金属原子聚集形成固体时，价电子从相应的正离子实中挣脱出来，在金属中自由移动。

32 正电荷离子实分布在整个固体，中和自由电子的负电荷。在经典自由电子理论中，电子被认为是在势阱中自由的、没有相互作用的经典粒子。该势阱将自由电子束缚在金属内。

Classical electron gas theory successfully explained the relationship between electric and thermal conductivity for metals. However, there are some contradictions to experimental results. Such as: (1) it cannot explain the abnormal phenomenon in Hall coefficient; (2) the real mean free path of electron is far more longer than that of evaluated by this model; (3) the experimental specific heat capacity for electron in metal is only one percent of the value obtained from this model; (4) it cannot explain the huge electricity difference among conductor, semiconductor and insulator.

One of the most important reasons for the failure of classical free electron theory is the adopting of Maxwell-Boltzmann statistics. Figure 3.5 shows the Maxwell–Boltzmann energy distribution of particles as a function of particle energy at a temperature of 300 K. It can be concluded that few particles have energies larger than $3k_BT$ and even fewer have energies equal to and above 10 k_BT. Hence we can in practice regard 10 k_BT as an upper limit of the kinetic energy of the particles. Therefore, based on Maxwell–Boltzmann statistics, the energy levels of the valence electrons in metal can be schematically described as Figure 3.6.

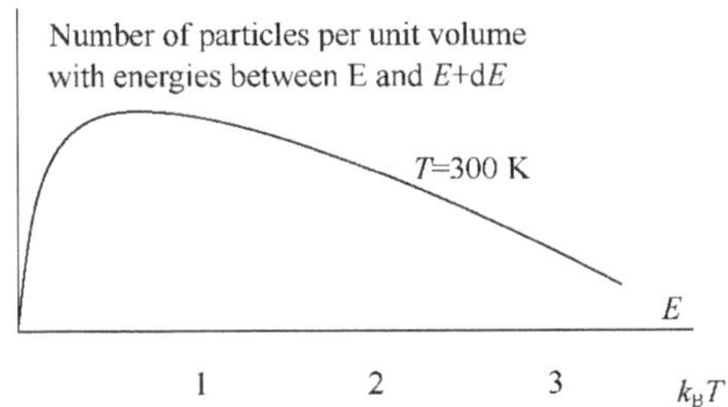

Figure 3.5 Maxwell–Boltzmann distribution of particle energies at room temperature (300 K) and thermal equilibrium.

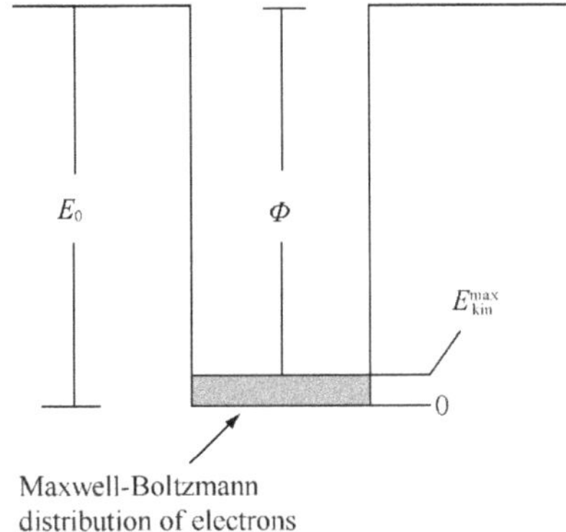

Figure 3.6 Potential well for trapped valence electrons at the surface of a metal. The figure also shows the kinetic energies of the valence electrons provided that the Maxwell-Boltzmann distribution was valid (which it is not). The calculation give $E_{kin}^{max} \approx 10k_BT \approx 0.025$eV

However, the experiments (e.g. photoelectric effect experiment) showed clearly that the valence electrons in metals have much higher kinetic energies than 0.025 eV at room temperature. Besides the invalidity of Maxwell–Boltzmann statistics for electrons, the other reason for the failure of classical free electron theory is that it does not consider the Pauli exclusive rule, which tells that each orbital only can occupy two spin-opposite electrons.[33]

33 除了Maxwell–Boltzmann 统计对于电子不适用之外，经典自由电子理论失败的另外原因是它没有考虑 Pauli 不相容原理，即一个轨道只能被两个自旋相反的电子占据。

Ⅱ. Quantum Free Electron Theory

As shown above strong objections can be raised to the classical model of the electron gas. It is not reasonable to regard the electrons as non-interacting classical particles like the molecules in an ideal gas.

The introduction of quantum mechanics and its successful application to electrons in atoms and molecules raised the idea that quantum mechanics could be applied also to free electrons in a metal.[34] The valence electrons in

34 量子力学的引入和它应用在原子和分子中电子的成功，促使了量子力学应用在金属中自由电子观点的产生。

atoms have energies calculated from quantum mechanics and obey the Pauli exclusive principle. It is highly unlikely that they behave like classical particles as free electrons in a metal.

Sommerfeld replaced the classical model with a model based on quantum mechanics. He published his quantum mechanical model in 1928. The basic hypotheses are: (1) Atoms in metal will lose their valence electrons, leading to positive ions and freely moved valence electrons. (2) Due to the Columbic attraction between the positive cores and the free electrons, the material is neutral and the electrons cannot be scattered by the repulsive interaction between them.[35] (3) These free electrons obey Pauli exclusive principle, and their energies satisfy Fermi-Dirac distribution.

35 由于带正电的原子核和自由电子之间的库仑引力，物质是中性的，电子不能被它们之间的相互排斥作用分散。

Using these ideas to solve the Schrödinger equation for free electron, lots of useful results can be obtained. We will discuss these in detail as following.

- **Schrödinger Equation and Its Solution to Free Electron of Metal**

To study the free electron movement in metal, we need to build Schrödinger equation first. In the classical theory, the interactions between the free electron and the nuclei and among the free electrons were assumed to be zero. Sommerfeld did not neglect the interactions but he made the simplifying assumption that the potential energy of the free electron is constant, independent of position inside the metal (Figure 3.7).[36] As the zero level of the potential energy can be chosen arbitrarily, we choose the value $U(0 < x < L) = 0$.

36 Sommerfeld 没有忽略相互作用，但是他提出了一个简化的假设：自由电子的势能是常数，与其在金属中的位置无关（图 3.7）。

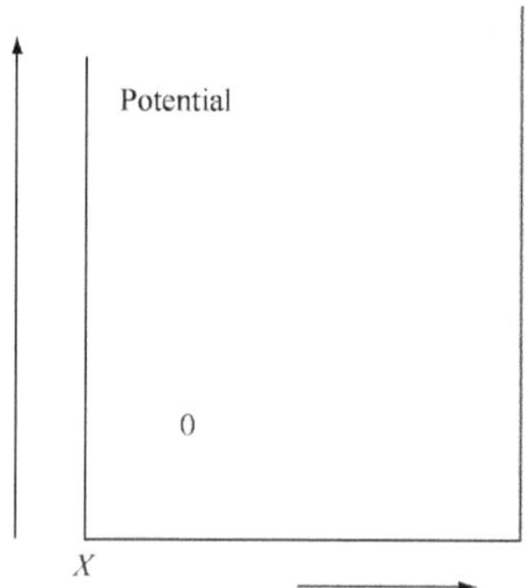

Figure 3.7 Potential well of the free valence electrons in a metal.

Within the well, the one-dimension standing Schrödinger equation should be:

$$\frac{d^2\varphi}{dx^2} + \frac{2mE}{\hbar^2}\varphi = 0 \tag{3-38}$$

And the energy of the electron satisfies:

$$E = \frac{1}{2}mv^2 = \frac{(mv)^2}{2m} = \frac{p^2}{2m} = \frac{h^2}{2m\lambda^2} \tag{3-39}$$

Substitute the E with λ in the above Schrödinger equation:

$$\frac{d^2\varphi}{dx^2} + \frac{4\pi^2}{\lambda^2}\varphi = 0 \tag{3-40}$$

The general solution for above equation is

$$\varphi = A\cos\frac{2\pi}{\lambda}x + B\sin\frac{2\pi}{\lambda}x \tag{3-41}$$

where the A and B are two constants, which will be determined by the boundary conditions.

Electron can be thought as trapped in a one-dimension well and can move back and forth within the well.[37] Within the well ($0 < x < L$) and outside the well, the probability of find electron is 1 and 0, respectively. Therefore, when $x = 0$, $\varphi(0) = 0 = A\cos(0)$, thus $A = 0$. The solution can be simplified to:

$$\varphi = B\sin\frac{2\pi}{\lambda}x = B\cos\left(\frac{\pi}{2} - \frac{2\pi}{\lambda}x\right) = Be^{i\left(\frac{\pi}{2} - \frac{2\pi}{\lambda}x\right)} \tag{3-42}$$

Since the electron cannot move outside the well, the probability of finding electrons inside the well is unity.[38] That is,

$$\int_0^L |\varphi(x)|^2\,dx = \int_0^L \varphi(x)\varphi(x)^*\,dx = B^2L = 1 \tag{3-43}$$

$$\therefore B = \frac{1}{\sqrt{L}}$$

and: $\varphi = \frac{1}{\sqrt{L}}\sin\frac{2\pi}{\lambda}x$

Apply the function to metal wire with length of L: The probability of find electron in the wire at position x is:

$$|\varphi(x)|^2 = \varphi\varphi^* = \left[\frac{1}{\sqrt{L}}e^{-i\left(\frac{\pi}{2} - \frac{2\pi}{\lambda}x\right)}\right]\left[\frac{1}{\sqrt{L}}e^{i\left(\frac{\pi}{2} - \frac{2\pi}{\lambda}x\right)}\right] = \frac{1}{L} \tag{3-44}$$

This means that the probability of finding electron at position x in metal wire of L length is the same, independent of position.

Another boundary condition is when $x = L$, $\varphi(L) = 0$, introduce this to the previous solution, obtain:

$$\frac{1}{\sqrt{L}}\sin\frac{2\pi}{\lambda}L = 0 \tag{3-45}$$

This equation gives the requirement for L, which can be written as,

$$\frac{2\pi}{\lambda}L = n\pi \tag{3-46}$$

where n is an integer, and λ is equal to $\pm 2L/n$, i.e.: $\pm 2L$, $\pm 2L/2$, $\pm 2L/3$, … The positive and negative signs with same absolute value in wavelength can

37 电子被认为是困在一维势阱中，能在势阱中前后运动。

38 因为电子不能移动到势阱外，在势阱中发现电子的几率是一。

39 符号相反、绝对值相同的波长代表相对传播的波函数。

represent for wave functions with opposite propagation directions.[39] The wave function and energy of the electron can be written as:

$$\varphi=\frac{1}{\sqrt{L}}\sin\frac{2\pi}{\lambda}x=\frac{1}{\sqrt{L}}\sin\frac{n\pi}{L}x;\ E=\frac{h^2}{2m\lambda^2}=\frac{h^2}{8mL^2}n^2 \tag{3-47}$$

40 这是能量的量子化条件，它是随着正整数 n 变化的离散值。（见图 3.8）

This is the quantization condition for energies, which are discrete values with variation according to the positive integer n (Figure 3.8).[40]

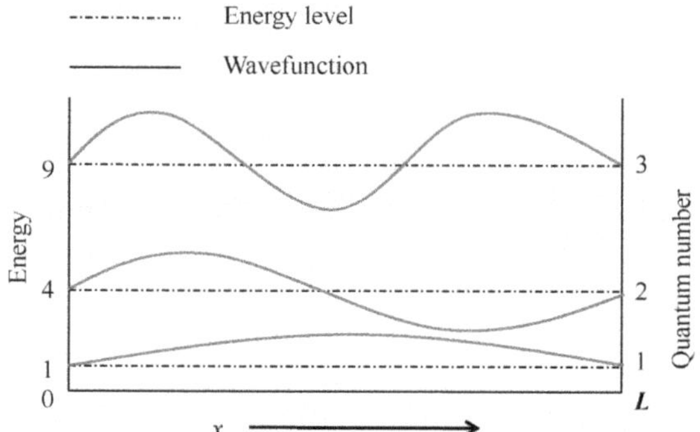

Figure 3.8 Free electron in metal wire with length of L.

Here comes the conclusion: the energy for electrons in metal wire is not continuous but quantized. Since the number of valence electrons in a macroscopic crystal is always very large, of magnitude $10^{28}\ \mathrm{m}^{-3}$, the energies are perceived as continuous.

For three dimensional metals, an electron is bound to move within the space (L_x, L_y, L_z), the standing Schrödinger equation is:

$$\frac{\partial^2\varphi}{\partial x^2}+\frac{\partial^2\varphi}{\partial y^2}+\frac{\partial^2\varphi}{\partial z^2}+\frac{2mE}{\hbar^2}\varphi=0 \tag{3-48}$$

Similar to one-dimension condition, the solution for three-dimension electron is:

$$\begin{aligned}\varphi(x,y,z)&=\varphi(x)\varphi(y)\varphi(z)\\&=\frac{1}{\sqrt{L_xL_yL_z}}\sin\left(\frac{\pi n_x}{L_x}x\right)\sin\left(\frac{\pi n_y}{L_y}y\right)\sin\left(\frac{\pi n_z}{L_z}z\right)\end{aligned} \tag{3-49}$$

where n_x, n_y, n_z is the quantum number in x, y, z direction, respectively (integers). Thus, three independent quantum numbers are needed for three-dimension space.

Introducing wave number k, the solution became:

$$\varphi(x,y,z)=\frac{1}{\sqrt{L_xL_yL_z}}\sin(k_xx)\sin(k_yy)\sin(k_zz) \tag{3-50}$$

In three-dimension metal①, the eigenvalue is found to be:

$$E_x=\frac{h^2}{8mL^2}n_x^2,E_y=\frac{h^2}{8mL^2}n_y^2,E_z=\frac{h^2}{8mL^2}n_z^2 \tag{3-51}$$

① with $L_x=L_y=L_z=L$

$$E = E_x + E_y + E_z = \frac{h^2}{8mL^2}(n_x{}^2 + n_y{}^2 + n_z{}^2) \quad (3\text{-}52)$$

As can be seen that different set of quantum number may yield the same energy (i.e., 112, 121, 211), it called degenerated energy levels. And the same as in 1−D case, the energy states of the 3−D free electrons (E) are quantized. Due to the huge number of electrons in macro-scale materials, E is perceived as continuous.

- ***k* Space**

Consider the solution of Schrödinger equation for a free electron (Equation 3-50), we found that both the eigenfunction and the eigenvalue for energy were expressed in terms of the wave number k.[41] A very fruitful approach for further development of Sommerfield's quantum mechanical model of free electron is to introduce the k space and to discuss the kinetic energy and the energy states of the free electron in this representation.[42]

41 考虑到自由电子薛定谔方程的解（公式 3-50），我们发现能量的本征函数和本征值都能用波数 k 表示。

42 Sommerfield 自由电子模型的一个非常富有成效的发展是 k 空间的引入、以及在这样的表述下对动能和自由电子能量状态的讨论。

We define the wave-vector k of the matter wave as a vector with the magnitude of $2\pi/\lambda$ and a direction of propagation of the matter wave. Corresponding to xyz coordinate system, we can establish coordinate system using unit wave vector $\boldsymbol{k}_x$, $\boldsymbol{k}_y$, $\boldsymbol{k}_z$, forming k space for wave function.

In k space, the wave vector k is represented by a point, i.e. by a vector from the origin to the point k (k_x, k_y, k_z) (see Figure 3.9). The eigenfunction of a free electron can be expressed as:

$$\varphi = Ce^{\pm i\vec{k}\cdot\vec{r}}$$

where both $\vec{k}$ and $\vec{r}$ are vectors:

$$\vec{k} = k_x\vec{x} + k_y\vec{y} + k_z\vec{z}$$

$$\vec{r} = x\vec{x} + y\vec{y} + z\vec{z}$$

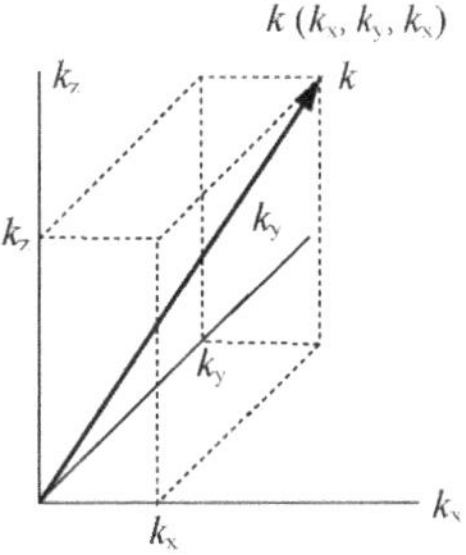

Figure 3.9 A point $\vec{k}$ in the k space.

- **Kinetic Energy of the Free Electron**

Under the free electron theory, the eigenvalue is equal to the kinetic energy of electron as its potential energy is zero.[43] It is possible to use the particle character of the electron to find its kinetic energy:

43 在自由电子理论下，由于电子势能为零，本征值和动能相同。

$$E_{\text{kin}} = \frac{1}{2}mv^2 = \frac{1}{2m}m^2v^2 = \frac{p^2}{2m}$$

According to wave-particle duality of $p = \frac{h}{\lambda}$ and wave vector definition of $k = \frac{2\pi}{\lambda}$

$$E_{\text{kin}} = \frac{p^2}{2m} = \frac{1}{2m}\frac{h^2}{\lambda^2} = \frac{1}{2m}\frac{h^2k^2}{(2\pi)^2} = \frac{\hbar^2}{2m}k^2$$
$$= \frac{\hbar^2}{2m}(k_x^2 + k_y^2 + k_z^2) \quad (3\text{-}53)$$

Equation (3-53) offers two different possibilities to represent the kinetic energy of the free electron as a function of the wave-vector graphically. The first way is to show E_{kin} as a function of the wave number k (Figure 3.10). According to Equation (3-53), the curve in Figure 3.10 is a parabola. It is important to remember that both k and E_{kin} are quantized as discussed previously. This is because the electron does not move in free space but inside the metal. However, huge number of free electrons leads to energy splitting and quasi-continuous. Therefore, the curve is not continuous but consists of a large number of closely situated points, as shown in the enlarged circle of Figure 3.10.[44]

44 因此，曲线不是连续的，而是由大量密集的点组成，就像在图 3.10 中的放大圆圈中所示。

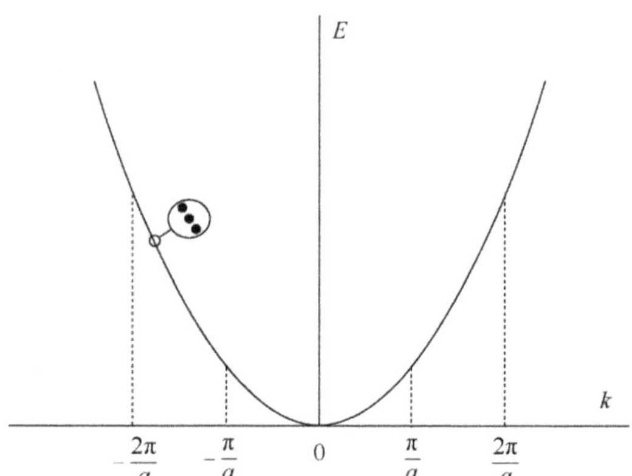

Figure 3.10 Kinetic energy of a free electron in a metal as a function of the wavenumber of the matter wave. a is the lattice constant.

Alternatively, all the allowed values of the k vector can be plotted in k space, i.e. in a three-dimensional k_x, k_y, k_z coordinate system, as shown in Figure 3.11.

Figure 3.11a is a two-dimensional plot. The circle around the points has a radius that represents the same k value which is compatible with Equation (3-53) for a given value of E_{kin}.[45] Each point represents the tip of a k vector, i.e. one of the many possible matter waves inside the metal. Figure 3.11b is an attempt to show the same thing in three dimensions. Each energy

45 围绕点的圆，其半径代表相同的 k 值，和方程（3-53）中给定的 E_{kin} 一致。

level is represented by a sphere, which contains all the tips of the *k* vector on its surface, a sphere that represents a maximum kinetic energy is called a Fermi surface, which represents the maximum energy of non-excited electrons.[46]

46 每一个能级都被一个球面表示，在它的表面上包含了所有 *k* 矢量的尖端，代表最大动能的球面被称为费米面，它代表了非激发电子的最大能量。

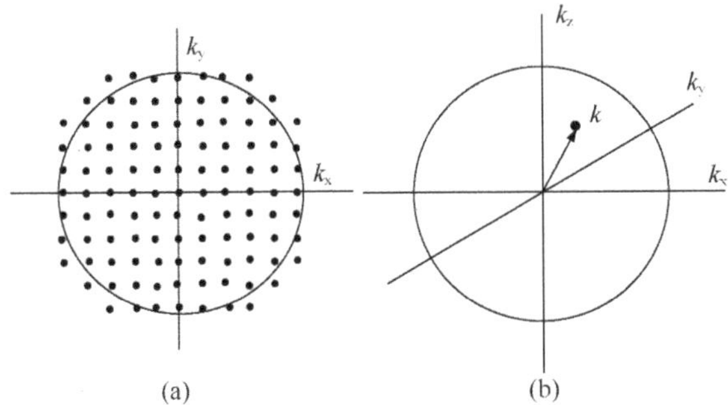

Figure 3.11 (a) Allowed energy levels of a free electron in a metal are represented by points in ***k*** space. (b) Energy levels of a free electron in a metal with constant energy correspond to points in ***k*** space on a spherical surface

- **Born-Karman Model and Density of the State**

Now we introduce Born-Karman model to demonstrate the smallest space size in k space. Previously, in the quantum free electron theory, for the discussion of three-dimension metal, it is supposed that only surface and bulk states are different, meaning that the electrons cannot move out of the solid.[47] However, the process that electrons move back and forth inside the solid by reflection has not been included. Even more, inside a real solid, the potential is not evenly distributed, e.g., there exist periodic changes in single crystal. The interstitial status inside crystal will affect the movement of electrons. Therefore, in order to mathematically represent these situations for free electron, Born-Karman model suggested that a crystal is an unlimited identical system, consisting of innumerable cubes with length of L. Electron enters a cube from one boundary and comes out the cube from the opposite boundary (Figure 3.12).

47 此前，在量子自由电子理论中，对于三维金属的讨论，它假定仅仅表面和内部状态是不同的，这意味着电子不能运动到固体外。

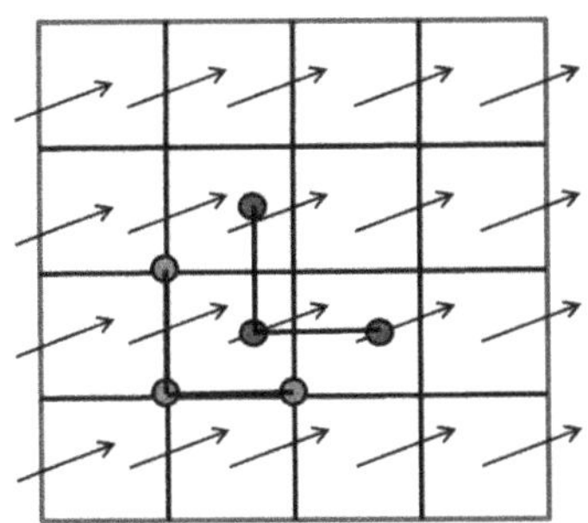

Figure 3.12 Born-Karman model.

Under Born-Karman model, the number of wave functions in each cube is the same, thus the number of free electrons (N) in a unit volume of $V = L^3$, keeps constant.[48] Born-Karman boundary conditions:

48 在Born-Karman模型下，每个立方体中波函数的数目是相同的，因此自由电子的数目（N）在一个单位体积 $V = L^3$ 内，保持不变。

$$\varphi(x,y,z)=\varphi(x+L,y,z)=\varphi(x,y+L,z)=\varphi(x,y,z+L) \tag{3-54}$$

thus,

$$\varphi(x+L)\varphi(y)\varphi(z)=\varphi(x)\varphi(y+L)\varphi(z)=\varphi(x)\varphi(y)\varphi(z+L) \tag{3-55}$$

Because,

$$\varphi(x+L)\varphi(y)\varphi(z)=Ae^{-ik_x(x+L)}e^{-ik_y y_e -ik_z z} \tag{3-56}$$

$$\varphi(x)\varphi(y+L)\varphi(z)=Ae^{-ik_x(x)}e^{-ik_y(y+L)}e^{-ik_z z} \tag{3-57}$$

$$\varphi(x)\varphi(y)\varphi(z+L)=Ae^{-ik_x(x)}e^{-ik_y(y)}e^{-ik_z(z+L)} \tag{3-58}$$

where k_x, k_y, k_z are wave numbers in three dimension of the k-space. Hence, the following equations hold:

$$e^{ik_xL}=e^{ik_yL}=e^{ik_zL}=1 \tag{3-59}$$

Therefore,

$$\cos k_xL=\cos k_yL=\cos k_zL=1 \tag{3-60}$$

$$k_x=\frac{2\pi}{L}n_x \tag{3-61}$$

$$k_y=\frac{2\pi}{L}n_y \tag{3-62}$$

$$k_z=\frac{2\pi}{L}n_z \tag{3-63}$$

where n_x, n_y, n_z are quantum number in x, y, z directions. When $n_x = n_y = n_z = 1$, $k_x = k_y = k_z = 2\pi/L$. $2\pi/L$ is the smallest k value in x, y, z direction, respectively.

As mentioned in free electron theory: (1) There are many electrons in metal, the state of electron can be described by wave function. (2) There are many quantized electron energy states, with the lowest one is the most preferred occupying state. (3) Due to the huge number of electrons in macro-scale materials, the intervals between energy levels are very small and can be considered as semi-continuous.[49] Then the next question is how to describe the state density of these energy levels?

49 由于在宏观尺度上材料中的电子数量庞大，能级之间的间隔很小，可以被视为半连续的。

According to Equation (3-61), (3-62), and (3-63), the smallest k_x, k_y, and k_z values are $2\pi/L$. In k-space, the smallest k_x, k_y, and k_z together determine one electron state, then the smallest electron occupied volume in k space should be:

$$\left(\frac{2\pi}{L}\right)^3=\frac{8\pi^3}{L^3} \tag{3-64}$$

The density of the state $Z(E)$ refers to the number of state (N) per energy (E) unit under unit volume,

$$Z(E)=\frac{\mathrm{d}N}{\mathrm{d}E} \tag{3-65}$$

where dN is the number of energy state between $E \sim (E + \mathrm{d}E)$

Derived from Equation (3-64), the number of electrons under unit volume should be a reciprocal expression of Equation (3-64):[50]

$$\frac{8\pi^3}{L^3}$$

Considering that an electron state can occupy two electrons (spin-paired) and the volume of the spherical ball in k space is: $(4/3)\pi k^3$, the total number of electrons with energies not greater than E (the corresponding wave vector is k) should be:[51]

$$N(E) = 2 \times \frac{4\pi}{3}k^3 / \frac{8\pi^3}{L^3} = \frac{L^3}{3\pi^2}k^3 \tag{3-66}$$

Combining Equation (3-66) and $E = \frac{\hbar^2}{2m}k^2$:

$$N(E) = \frac{L^3}{3\pi^2}k^3 = \frac{L^3}{3\pi^2}\left(\frac{2mE}{\hbar^2}\right)^{\frac{3}{2}} \tag{3-67}$$

Finally, the density of state can be obtained by differential Equation (3-67):

$$Z(E) = \frac{\mathrm{d}N}{\mathrm{d}E} = \frac{L^3}{2\pi^2}\left(\frac{2m}{\hbar^2}\right)E^{\frac{1}{2}} = C\sqrt{E} \tag{3-68}$$

Similarly, we can calculate the state density of free electron in two- and one-dimensions. It is verified that the state density of free electron in two-dimension is constant, while that of one-dimension is reciprocal to the squareroot of energy.[52] Figure 3.13 schematically show the energy state density with the energy in one-, two-, and three- dimensions

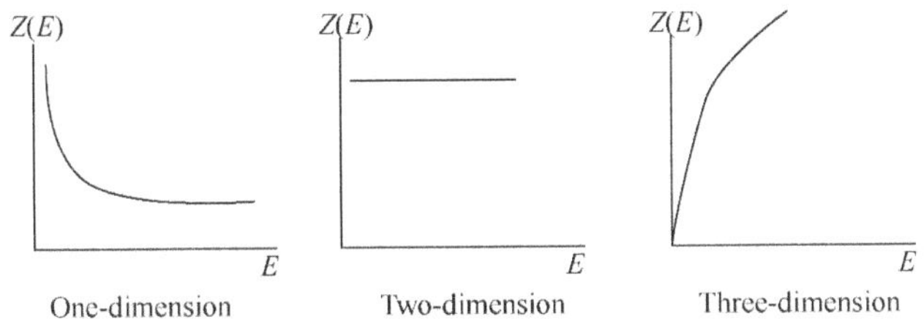

Figure3.13 The state density of free electron in one-, two-, and three- dimensions.

- **Eergy Distribution of Free Electrons in a Metal: Fermi-Dirac Distribution**

Previously, based on quantum free electron theory, we got the wave function, eigenvalue, and density of state. The results show that the energy of free electron is quantized, and the interval of quantized energies is very small because of the huge amount of electron. Thus, the energy spectrum of free electron can be perceived as continuous. The question raised here is how could electron occupy these energy levels?

The free electrons in a metal do not belong to any particular metal ions.

50 通过方程（3-64）推导，单位体积中电子数量应该是（3-64）式的倒数。

51 考虑到一个电子态中可以占据两个电子（自旋相反），在 k 空间中球的体积是（4/3）πk^3，总能量不大于 E 的电子数（相应的波矢为 k）应该是：

52 通过验证表明，二维自由电子的态密度是恒定的，而一维的态密度与能量平方根成反比。

53 因此，整个金属是一个有大量自由电子和大量不同能量状态的系统。

54 因此，价电子不仅在每一个最低的能态上，也被迫在较高的能态上。

Consequently, the whole metal represents one system with a large number of free electrons and a large number of different energy states.[53] Each electron has one eigenfunction and it is associated eigenvalue to the electron pool. For example, 1 mol of a metal with one valence electron. The number of valence electrons in the metal is equal to $N_A = 6.02\times10^{23}$. These free electrons must obey the Pauli Exclusion principle. Hence the valence electrons are forced to be located not only in each of the lowest energy states, but also in higher energy states.[54] The upper occupied energy limit E_F is called the Fermi level, which represents the energy of the most energetic valence electrons in metal at $T = 0$ K.

As we known, electrons are Fermions, and the probability that the energy levels of free electron will accommodate an electron at temperature T can be expressed by Fermi factor shown in Equation (3-35).

$$f(E) = \frac{1}{\exp\left(\frac{E - E_F}{kT}\right) + 1}$$

At absolute temperature of $T = 0$ K: we can get:

$$f(E) = \frac{1}{\exp\left(\frac{E - E_F}{kT}\right) + 1} = \frac{1}{e^{-\infty} + 1} = 1 \text{ for } E < E_F \tag{3-69}$$

$$f(E) = \frac{1}{\exp\left(\frac{E - E_F}{kT}\right) + 1} = \frac{1}{e^{+\infty} + 1} = 0 \text{ for } E > E_F \tag{3-70}$$

This means that at $T = 0$ K, all energy states $\leqslant E_F$ will be occupied and all energy states $\geqslant E_F$ will be empty. The Fermi level is the border between occupied and unoccupied energy levels at $T = 0$ K (Figure 3.14).

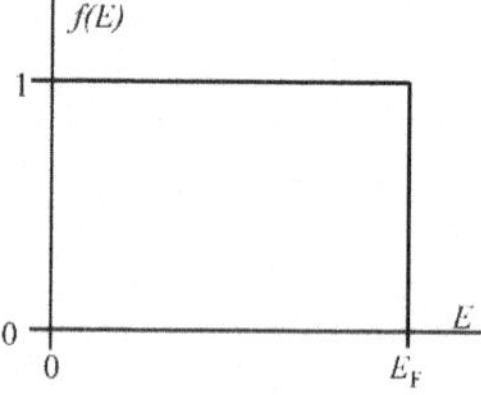

Figure 3.14 The Fermi factor $f(E)$ as a function of energy at $T = 0$ K.

At temperatures $T > 0$ K, according to Equation (3-35):

If $E = E_F$, $f(E) = \frac{1}{\exp\left(\frac{E - E_F}{kT}\right) + 1} = \frac{1}{e^0 + 1} = 0.5$

If $E < E_F$,

$$E \ll E_F:\ f(E) = \frac{1}{\exp\left(\frac{E - E_F}{kT}\right) + 1} \sim \frac{1}{e^{-\infty} + 1} = 1$$

$$E_F - E \leqslant kT:\ \exp\left(\frac{E-E_F}{kT}\right) > 0,\ 0.5 < f(E) < 1$$

If $E > E_F$,

$$E >> E_F:\ f(E) = \frac{1}{\exp\left(\frac{E-E_F}{kT}\right)+1} \sim \frac{1}{e^{+\infty}+1} = 0$$

$$E - E_F < kT:\ \exp\left(\frac{E-E_F}{kT}\right) > 1,\ 0 < f(E) < 0.5$$

The schematic view of Fermi factors at $T > 0$ is shown in Figure 3.15a. At $T > 0$, electrons with kinetic energies $E > E_F - kT$ may be thermally excited up to available energy levels $> E_F$, with probability of $0 < f(E) < 0.5$.[55] The sharp discontinuity between occupied and empty energy levels at $E = E_F$ ($T = 0$) is smoothed out at higher temperatures. In the case of free electron in metal, it means that some of the most energetic electrons become excited up to empty sites above the Fermi level and leave vacant sites below the Fermi level, as shown in Figure 3.15b.[56]

55 在 $T > 0$ 时，动能为 $E > E_F - kT$ 的电子能够被热激发到大于 E_F 的可用能级，概率为 $0 < f(E) < 0.5$。

56 对于金属中自由电子，这意味着一些能量最高的电子被激发到费米能级上的空能级上，在费米能级下留下空位，如图 3.15b 所示。

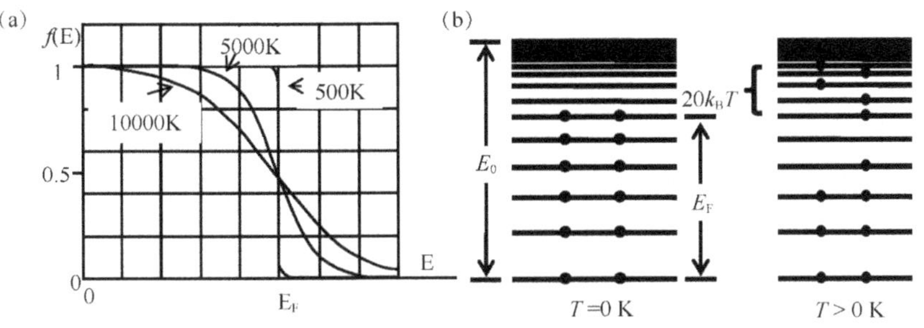

Figure 3.15 (a) The Fermi factor as a function of energy for different temperature T. (b) Distribution of electrons in electron energy states in the vicinity of the Fermi level.

- **Values of Fermi Level**

Using quantum free electron theory, we can determine the Fermi level of the metal:

$$T = 0:\quad E_F^0 = \frac{h^2}{2m^*}\left(\frac{3n}{8\pi}\right)^{\frac{2}{3}} \tag{3-71}$$

$$T \neq 0:\quad E_F = E_F^0\left[1 - \frac{5}{12}\pi^2\left(\frac{kT}{E_F^0}\right)^2\right] \tag{3-72}$$

where n is the number of valence electrons per unit volume, m^* is the effective mass of electron in the metal. The Fermi level or Fermi energy is a function of the number of free electrons per unit volume in the metal.[57] E_F is slightly smaller than E_F^0. Since E_F is generally $>> kT$, the difference between E_F and E_F^0 is lower than 10^{-5}, E_F is usually treated as constant under different T. Table 3.2 shows the values of Fermi energy of some

57 费米能级或费米能量是金属中单位体积内自由电子数目的函数。

common metals.

Table 3.2 Fermi energies of some common metals

Metal	Fermi energy (eV)	Valence
Na	3.2	1
K	2.1	1
Cu	7.0	1
Ag	5.5	1
Au	5.5	1
Mg	7.1	2
Al	11.6	3

- **Average Energy of an Electron**

We can also estimate the average energy of an electron in a metal:

$$T=0:\quad \overline{E}=\frac{3}{5}E_{\mathrm{F}}^{0} \tag{3-73}$$

$$T\neq 0:\quad \overline{E}=\frac{3}{5}E_{\mathrm{F}}^{0}\left[1+\frac{5}{12}\pi^{2}\left(\frac{kT}{E_{\mathrm{F}}^{0}}\right)^{2}\right] \tag{3-74}$$

The average energy of an electron in metal is related to its Fermi energy, and when T is not zero, the average energy of an electron in the metal is slightly higher than $T=0$.[58]

58 金属中的电子的平均能量与费米能量有关，当 T 不等于 0 时，金属中电子的平均能量略高于 T=0 时的电子平均能量。

- **Significances of Quantum Free Electron Theory**

Quantum free electron theory shows great significance in material science:

(1) It successfully explained the heat capacity of metal: In classical thermal dynamic, heating the gas molecules, the thermal energy will be evenly distributed to all particles. Thus all free electrons in metal involve in the thermal conductivity, yielding the several-hundred larger value of heat capacity. In quantum mechanics, heating metal, only small part of electrons whose energies are around $E_{\mathrm{F}}-kT$ can absorb thermal energy, and then jump to the levels above E_{F}. The specific heat capacity is only related to the amount of free electron above the Fermi level, leading to much accurate result of heat capacity.[59]

59 比热容只与在费米能级以上的电子数量有关，从而得到更准确的热容值。

(2) It explained the opaque feature of metal: The energy states of the valence electrons are so closely spaced that they are practically continuous and form an energy band. Valence electrons in low-lying energy states inside the metal can easily absorb photons of arbitrary energies and be excited to higher empty energy levels in the energy band above the upper

limit of filled energy levels. If the bound valence electrons with high kinetic energies are excited sufficiently they may escape from the surface of the metal.[60] In this case the energy of the electrons is no longer quantized, because arbitrary values of the kinetic energy are allowed outside the metal. Hence photons of all energies can be absorbed. Consequently, all wavelengths of visible light become absorbed and no wavelengths can transmit and result in the opaque of metal.

60 如果高能量的束缚价电子被足够的能量激发，它们可能从金属表面逃逸。

3.6 Band Theory of Solids

The quantum free electron model is generally accepted in solid-state physics, which could successfully explain the specific heat capacity of metal and had made a great progress compared to classical free electron theory. However, objections were raised to the assumption that the potential energy inside a metal is constant. And this quantum free electron model cannot explain many other properties of metal such as the binding energy in crystal and conductivity of conductor, semiconductor and insulator.

3.6.1 Schrödinger Equation and Its Solution Under near Free Electron Approximation

In free electron model, it is proposed that the electron is in the potential well. Within the well, the electron potential is zero, and outside the well, electron potential is infinite. Under real circumstances, we can’t neglect the effect of positive charged atomic cores and other electrons. Treating the nucleus as still, the movement of electrons should be described as the movement of electron under the potential of the nucleus and other electrons. In a crystal, due to the periodical arrangement of the atoms, the potential of the electrons should be varied with the same periodicity as the crystal lattice. This led to the development of the band theory of solids.

Single Electron Approximation: Supposing the nucleus is still and each electron is moving under potential of the nucleus and other electrons (Figure 3.16). Hence all electrons are identical, the multiple electrons problem can simplify by solving a single electron equation, this is called single electron approximation.

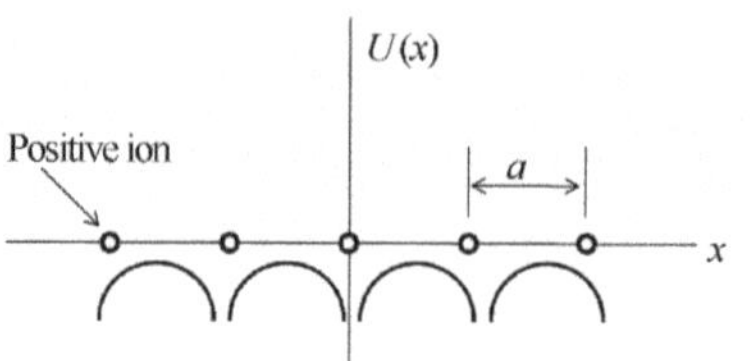

Figure 3.16 Potential of an electron varying periodically with distance.

Nearly Free Electron Approximation: (1) Suppose the crystal is perfect and infinite, without taking account of the surface effect; (2) Neglect the influence from thermal movement of positive ion; (3) Suppose each electron moves independently under the potential of positive atomic core and with neglecting the interaction between electrons; and (4) Treat the potential of the electron as periodical variation, which is very small and can be handled as micro-disturbance.[61]

61 将电子势能按照周期性变化考虑；这一变化是非常小的，可以用微扰来处理。

Therefore, we have

$$U(x+na)=U(x) \tag{3-75}$$

where a is the periodic distance, n is an integer.

Generally, introduce periodical disturbing potential:

$$U(x+na)=U_0+\sum_{\mathrm{n}} U_{\mathrm{n}} e^{i\pi nx/a} \tag{3-76}$$

Combining one-dimension Schrödinger equation gives:

$$\frac{\mathrm{d}^2\varphi}{\mathrm{d}x^2}+\frac{2m(E-U(X))}{\hbar^2}\varphi=0 \tag{3-77}$$

Bloch proposed the solution of the above equation as:

$$\varphi(x)=e^{ikx}f(x) \tag{3-78}$$

$\varphi(x)$ is a planar matter wave modulated by the **Bloch function** of $f(x)$:

$$f(x)=f(x+na)$$

x is the distance from the origin position.

The eigenvalues that correspond to the eigenfunctions of the above Schrödinger equation are complicated functions of k and depend of the geometry of the crystal lattice.[62]

62 相对应于上述薛定谔方程的本征函数的本征值，是 k 的复杂函数，它依赖于晶格的几何形状。

3.6.2 Bandgap

Solving the above equation, we can have the E-k relationship of free electron in crystal lattice, which is the so called Kronig-Penney energy curve.

As shown in Figure 3.17, compared with the free electron model (Figure 3.10), instead of a quasi-continuous curve (consisting of closely spaced points with parabola shape), the discontinuities appear at k values given by $k=\pm n\pi/a$.[63] At k values far from the discontinuity points the

63 如图 3.17 所示，与自由电子模型（图 3.10）相比，它不是一个准连续曲线（由抛物线状的紧密的点组成），不连续的情形出现在 $k=\pm n\pi/a$ 处。

energy is nearly the same as for free electron model and the electrons can move freely through the crystal lattice. For small values of the potential energy, the discontinuous curve is comparatively close to a parabola; at higher energies, due to the discontinuities, energy bands are formed as shown at the right hand of Figure 3.17.[64]

64 势能较小时，不连续的曲线比较近似于抛物线，能量增加时，由于不连续性，形成了如图 3.17 右侧所示的能带。

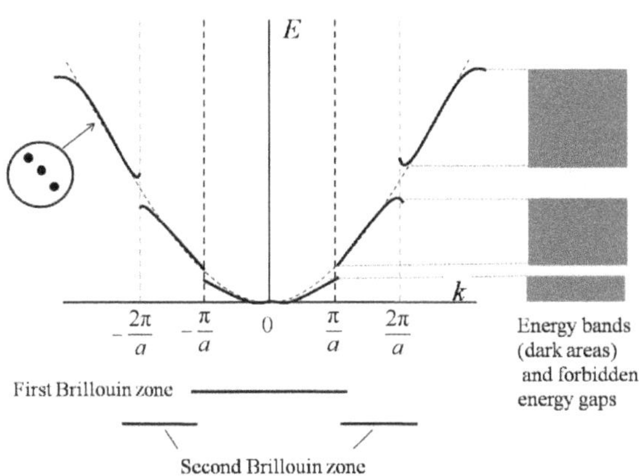

Figure 3.17 Energy levels in a linear periodic lattice as a function of the wave number k. a is the lattice constant. (Derived based on Nearly Free Electron Approximation).

Explanation of the Bandgap by Bragg's Law:

According to Bragg's law, if a crystal surface is exposed to a beam of parallel mono-energetic X-ray with a known angle of incidence (Figure 3.18), when $n\lambda = 2d\sin\theta$, the reflected light can be strengthened (d is distance between the atomic planes, θ is the angle between the incident ray and the atomic plane, λ is the wavelength of the X-ray, n is an integer).[65]

65 根据 Bragg 定律，如果晶体表面受到单能量且已知入射角的平行 X 射线(图 3.18）照射，当 $n\lambda = 2d\sin\theta$时，反射光得到加强（d 是原子面之间的距离，θ是入射光线和原子平面之间的角度，λ是单能 X 射线的波长，n 是一个整数）。

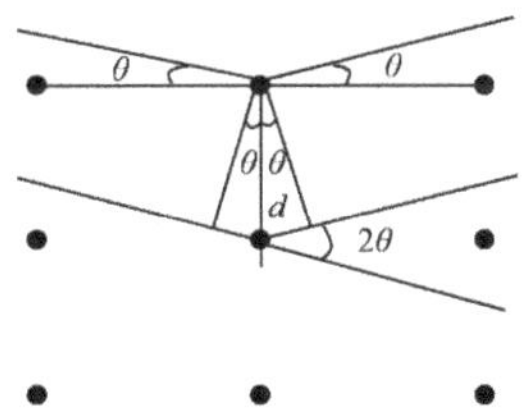

Figure 3.18 A mono-energetic X-ray irradiates a crystal surface with incident angle of θ.

Thus for normal incidence, the reflection light satisfies: $n\lambda = 2d$. In this condition:

$$k = \frac{2\pi}{\lambda} = \frac{2\pi}{2d/n} = \frac{n\pi}{d} \tag{3-79}$$

Electron movement can be treated as a matter wave with wavelength shorter than X-ray, transporting of electron also satisfies Bragg's law. That is for an electron transporting perpendicular to the atomic plan, when the wave-vector $k = n\pi/d$, the wave can be strengthened.

The strengthened electron wave is the linear combination of the incident wave and the reflection wave:

$$\psi_1(x) = Ae^{ikx} + Ae^{-ikx} = 2A\cos kx \tag{3-80}$$

$$\psi_2(x) = Ae^{ikx} - Ae^{-ikx} = 2iA\sin kx \tag{3-81}$$

Thus, the probability of finding an electron can be described as Figure 3.19:

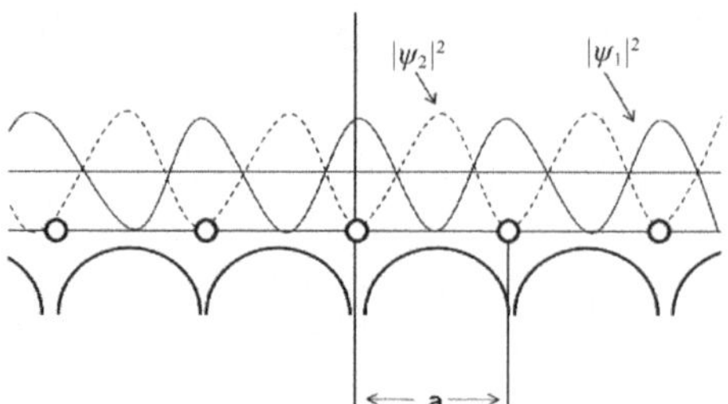

Figure 3.19 The probability of finding an electron in a periodic crystal.

At points of positive ions, $|\psi_1|^2$ is maximum, where the potential is the lowest and the energy of electron is lower than that of a free electron.[66] At peak position of the potential energy, $|\psi_2|^2$ has maximum value, the energy of electron is higher than that of a free electron.

66 在正离子实的位置，$|\psi_1|^2$ 是最大的，在这个位置时势能最小，电子能量比自由电子能量低。

Conclusion:

(1) When the electron matter wave satisfies Bragg's law, the energy of the electron will split into two, corresponding to ψ_1 and ψ_2 with energy of E_1 and E_2. In between E_1 and E_2, no electron can be found.[67]

67 当电子的物质波满足 Bragg 定律时，电子能量会分裂成两个，对应于 ψ_1 和 ψ_2，能量分别为 E_1 和 E_2。在 E_1 和 E_2 之间，没有电子存在。

(2) When the electron matter wave is far from the Bragg's condition, the phases of the reflection waves changed serially, thus eliminating with each other. The total reflection effect is zero. These electrons are transparent to the crystal and behave like free electrons.

3.6.3 Interpretation of Conductivity

- Band of Fully Filled

This kind of band is called filled band. When external electric field applied, electrons in the filled band move oppositely (both k and $-k$ direction) with same velocity, thus no net charge flow (current) occurs.[68]

68 施加外加电场时，电子在满带中以相同的速率相对运动（沿 k 和 $-k$ 方向），因此没有净电荷流动。

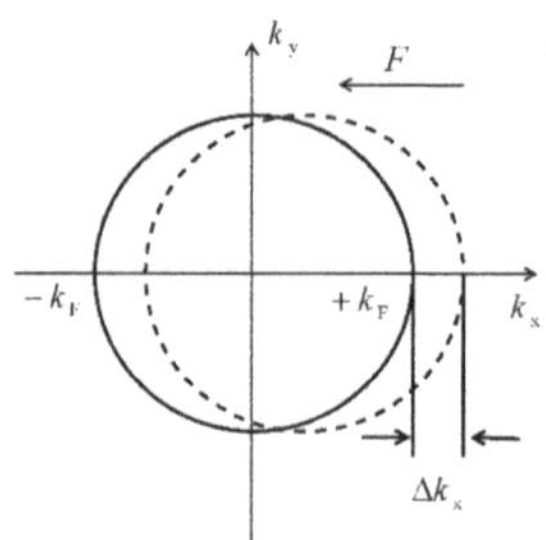

Figure 3.20 Fermi surface and its shift under electric field of F.

- Band of Half Filled:

If the three-dimension Brillouin zone is partially filled, the Fermi surface can be treated as a ball. Applying electric field F, each electron is accelerated by force of eF, i.e. the Fermi ball moved Δk_x oppositely to the electric field as shown in Figure 3.20. Thus, electrons with wave-vector close to k_F, moved along the k_x direction, leading to current.

- Conductor, Semiconductor and Insulator

The classification of solids can base on their electrical conductivities. A material is characterized as an insulator if its resistivity greater than 10^9 Ωm. The electric current through a solid is transported by the valence electrons.[69] Hence the conductivity of a solid is closely related to the widths and positions of the energy bands of the solid.

69 通过固体的电流由价电子输运。

(a) Conductor. As shown in Figure 3.21, in mono valent atoms the valence band is only partially filled. Other metals, the conduction band is close to the valence band or overlaps the valence band.[70] Hence the electrons can move rather freely within the metal.

70 其他金属，导带接近于价带或者与价带重叠。

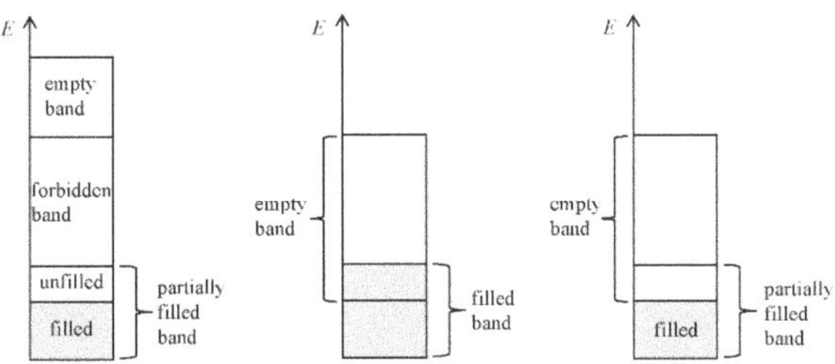

Figure 3.21 Energy levels in conductors.

(b) Semiconductor. In a semiconductor the valence band is filled. The energy gap is of the magnitude ⩽ 2 eV. Thermal excitation of electrons across the energy gap is possible (Figure 3.22a).

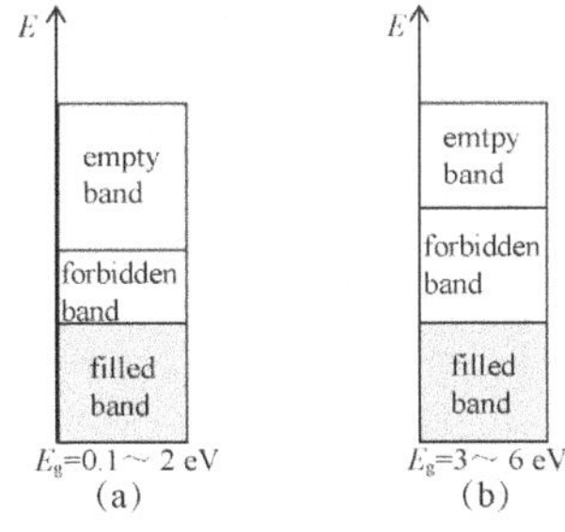

Figure 3.22 Energy levels for (a) semiconductor and (b) insulator.

(c) Insulator. In an insulator the valence band is filled. The energy gap between the conduction band and the valence band is very high (3-6 eV). Very few electrons have enough thermal energies to reach the conduction band (Figure 3.22b).

3.6.4 Brillouin Zone Under Near Free Electron Approximate

The k zones of allowed energies in Kronig-Penney energy curve are also called Brillouin zones. The mean points for Brillouin zones are summarized as following: (1) The Brillouin zones form the energy band for electron. (2) The widths of the energy bands increase with increasing energy. The stronger the electron is bound to the lattice ions, the narrower will be the widths of the energy bands.[71] (3) The number of energy states in each Brillouin zone is equal to the total number (N_{total}) of atoms of the crystal. (4) Each Brillouin zone can accommodate totally a maximum of $2N$ electrons.

71 电子与晶格离子结合得越强，能带宽度越窄。

- **One-Dimension Brillouin Zone**

From Figure 3.17, the width of the first and second one-dimension Brillouin zone are:

$$-\frac{\pi}{a} to \frac{\pi}{a} \text{ and,}$$

$$-\frac{2\pi}{a} to -\frac{\pi}{a} plus -\frac{\pi}{a} to \frac{2\pi}{a}$$

Consider a linear crystal of length L which consists of N atoms, the number of energy states in the first one-dimension Brillouin zone should be:[72]

$$\frac{width\_of\_Brillouin-zone}{the\_size\_of\_one\_energy\_stage\_in\_k-space} = \frac{2\pi/a}{2\pi/L} = \frac{L}{a} = N \quad (3\text{-}82)$$

72 考虑一个由 N 个原子组成、长度为 L 的线性晶体，第一个一维 Brillouin 区域的能量状态应数目应该是：

where $2\pi/L$ is the size of one energy state in one dimension k space. Therefore, a maximum of $2N$ electrons can accommodate in one-dimension Brillouin zone.

- **Two-Dimension Brillouin Zone**

The shape of two-dimension Brillouin zone is complicated and related to crystal structure (Figure 3.23).[73]

73 二维 Brillouin 区域的形状是复杂的，与晶体结构有关。（图 3.23）。

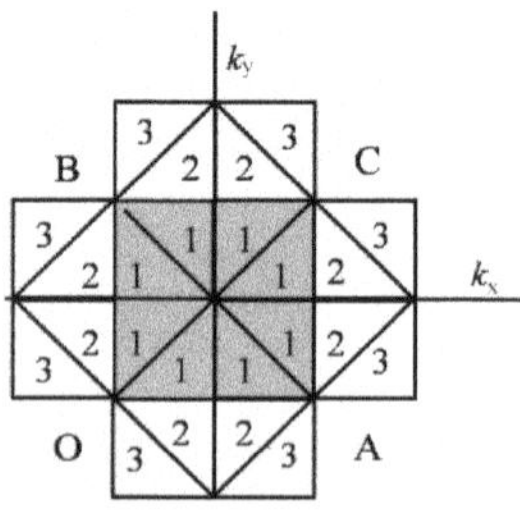

Figure 3.23 First, second and third Brillouin zones of a two-dimensional crystal lattice with simple cubic structure.

All the Brillouin zones have equal areas, which means the number of energy states are the same. Consider a two-dimension metal with length of L and lattice constant of a, which holds $N = L/a$,[74] the number of atoms is N^2,

74 所有 Brillouin 都有相同面积，这意味着能量状态的数目是相同的。考虑一个长度为 L 的二维金属，其晶格常数为 a，则 $N = L/a$ 成立：

and each energy state takes up space of $(2\pi/L)^2$.

The area of a two-dimension Brillouin zone is $(2\pi/a)^2$, thus, the number of energy state can be expressed as:

$$\frac{area\_of\_Brillouin-zone}{area\_of\_one\_energy\_state}=\frac{(2\pi/a)^2}{(2\pi/L)^2}=\left(\frac{L}{a}\right)^2=N^2 \tag{3-83}$$

- **Brillouin Zone in Three-Dimension**

The shape of three dimension Brillouinzone is even more complicated than the two-dimension ones. Example is given in Figure 3.24.[75]

75 三维 Brillouin 区的形状比二维的更复杂。如图 3.24 给的例子所示。

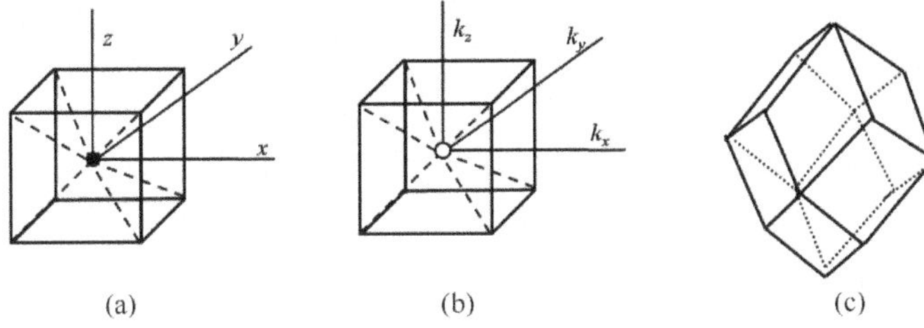

Figure 3.24 (a) A simple cubic lattice in r space. Lattice constant is a. (b) First Brillouin zone of a simple cubic crystal. (c) Outer surface of the second Brillouin zone of a simple cubic crystal.

3.6.5 State Density under Band Theory

In band theory, due to the introduction of periodic potential from nuclei in crystal, the state density changes compared to that of free electron theory (Figure 3.25). [76]

76 在带理论中，由于晶体中原子周期性势场的引入，相比于自由电子理论，态密度发生改变（图 3.25）。

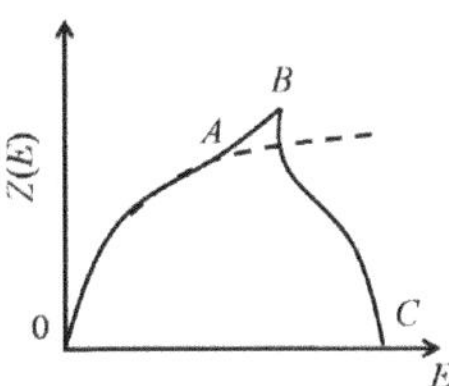

Figure 3.25 The density of state under free electron theory (dash line) and near free electron approximation (solid line).

At low energy part, both curves satisfy the parabola (OA part). When electron energy close to the boundary of the Brillouin zone, in the case of band theory, the increase of the density state is rather abrupt, not along the parabola curve anymore (AB part). The BC part of Figure 3.25 shows the corner states of Brillouim zone, which are rather few hence becomes filled fairly quickly and drops rapidly to zero.

If the energy gap between the first and second Brillouin zones is small, some of the energy states of the second Brillouin zone are lower than the energy states of the first. In this case, electrons are located in the second band before the first energy band is completely filled and the curves of the first and the second bands overlap (Figure 3.26).[77] In metals, the overlapping of energy band is very common.

77 这种情况下，在第一个能带被完全填充之前，电子就占据了第二个能带，第一和第二个能带的曲线相互重叠（图 3.26）。

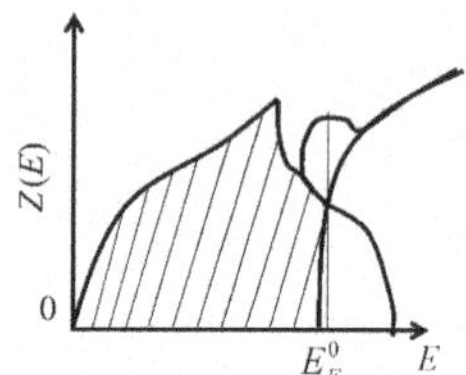

Figure 3.26 Density of state with overlapping in the first and second Brillouin zones.

3.6.6 Effective Electron Mass as a Function of *k*

It has been known for a long time that an electron has a well-defined mass, and when accelerated by an electric field it obeys Newtonian mechanics. What happens when the electron to be accelerated inside a crystal? How will it react to an electric field? In fact, the mass of an electron in a crystal appears, in general, different from the free electron mass, it is usually referred to as the effective mass (expressed as m^*).

In 1-D case, the effective mass of a valence electron in a metal depends on its energy, i.e. m^* must be a function of the wavevector k of the matter wave of the electron:[78]

$$m^* = \frac{\hbar^2}{\frac{d^2E}{dk^2}} \tag{3-84}$$

78 在一维情况下，金属中价电子的有效质量取决于它的能量，也就是说，m^* 必须是电子物质波波矢 k 的函数。

Figure 3.27 exhibits the relationship between the effective mass and its energy. If an electron, initially at rest at $k = 0$, with mass of m, is accelerated by an electric field, it will move to higher values of k and will become heavier and heavier, reaching infinity at the points of reflection ($d^2E/dk^2 = 0$ at $k = n<a$). For even higher values of k the effective mass becomes negative, heralding the advent of a new particle, the hole, which we have casually met from time to time and shall often meet in the rest of this course. The practical effect of this is that electrons with wave numbers within the intervals $-\pi/a < k < -\pi/2a$ and $\pi/2a < k < \pi/a$ behave like positively charged particles when they are accelerated in an electric field, otherwise like negative particles.

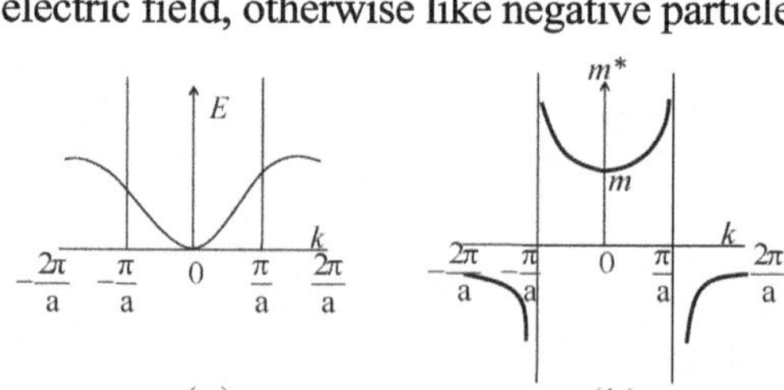

Figure 3.27 (a) Kinetic energy of a valence electron within the first Brillouin zone (one-dimension case). (b) The effective mass of a valence electron as a function of k.

In the case of three-dimension Brillouin zones, m^* is far more complicated. In general case $1/m^*$ is a tensor with nine components.

本章小结

1. 内容概要

本章作为后面章节的基础知识，我们讨论了电子理论的基础知识，该理论是材料的电、光、磁和热学性质的基础。例如，固体的电学性质主要由固体中的电子性质决定。质子通常被看作是起次要作用的因素，来确保电中性。中子一般在超导材料中由于临界温度依赖于原子核的总质量而被考虑。但是总体上来说，影响固体电学性质的关键因素是电子的能量和运动，这是本章的焦点。本章在介绍了电子的波粒二象性后，介绍薛定谔方程以及统计理论、基于量子力学的自由电子理论和能带理论。

2. 基本概念

波粒二象性、量子力学、光电效应、德布罗意波、海森堡不确定性原理、麦克斯维尔—玻尔兹曼统计、费米—狄拉克统计、波色—爱因斯坦统计、经典自由电子理论、自由电子理论、薛定谔方程、k 空间、Born-Karman 模型、费米-狄拉克分布、费米能级、布里渊分区、能带理论、驻波、测不准原理、泡利不相容原理、自旋量子数、霍尔系数。

3. 主要公式

（1）爱因斯坦光电效应方程：$h\upsilon=\phi+\dfrac{m_e v_e^2}{2}=h\upsilon_0+\dfrac{m_e v_e^2}{2}$

（2）波粒二象性方程：$p=\dfrac{h\upsilon}{c}=\dfrac{h}{\lambda}$

（3）物质波：$\lambda_{\mathrm{deB}}=\dfrac{h}{m\upsilon}$

（4）相对论能量：$E=m\upsilon^2$

（5）动量：$p=m\upsilon=\sqrt{Em}$

（6）光能：$E=h\upsilon=h\,(c/\lambda)$

（7）波函数：$\varPhi=(x,y,z,t)\ or\ \varPhi=\left(\vec{r},t\right)$

（8）共轭波函数：$|\psi|^2\mathrm{d}x\mathrm{d}y\mathrm{d}z=\psi\psi^*\mathrm{d}x\mathrm{d}y\mathrm{d}z$

（9）简谐振动波函数：$Y(x,t)=A\cos\left[2\pi\left(\dfrac{x}{\lambda}-vt\right)\right], Y(x,t)=Ae^{i(kx-\omega t)}$

（10）波数：$k=\dfrac{2\pi}{\lambda}=\dfrac{2\pi}{h/p}$

（11）角频率：$\varpi=2\pi\upsilon=2\pi\dfrac{E}{h}$

（12）薛定谔方程：$\nabla^2\varphi+\dfrac{2m(E-U)}{\hbar^2}\varphi=0,\ \nabla^2\dfrac{\partial^2}{\partial x^2}+\dfrac{\partial^2}{\partial y^2}+\dfrac{\partial^2}{\partial z^2}$，

（13）海森堡不确定原理：$\Delta y\Delta p_y\geqslant\dfrac{h}{4\pi}$，$\Delta E\Delta t\geqslant\dfrac{h}{4\pi}$

（14）麦克斯韦—玻尔兹曼统计：$f(E)=Ae^{-E/kT}$

（15）费米—狄拉克统计：$f(E)=\dfrac{1}{\exp\left(\dfrac{E-E_{\mathrm{F}}}{kT}\right)+1}$

（16）波色—爱因斯坦统计：$f(E)=\dfrac{1}{\exp\left(\dfrac{E-\mu}{kT}\right)-1}$

（17）统一的统计公式：$f(E)=\dfrac{1}{\exp\left(\dfrac{E_{\mathrm{i}}-\mu}{kT}\right)+\alpha}$

（18）自由电子的平面波函数：$\varphi=Ce^{\pm \mathrm{i}\vec{k}\cdot\vec{r}}$

（19）Born-Karman 模型

波函数满足条件：$\varphi(x,y,z)=\varphi(x+L,y,z)=\varphi(x,y+L,z)=\varphi(x,y,z+L)$

最小占用空间：$\left(\dfrac{2\pi}{L}\right)^3=\dfrac{8\pi^3}{L^3}$

（20）态密度：$Z(E)=\dfrac{\mathrm{d}N}{\mathrm{d}E}$

（21）一维布里渊区能态的数量：$\dfrac{width\_of\_Birllouin-zone}{the\_size\_of\_one\_energy\_state\_in\_k-space}=\dfrac{2\pi/a}{2\pi/L}$

$=\dfrac{L}{a}=N$

（22）二维布里渊区能态的数量：$\dfrac{area\_of\_Brillouin-zone}{arez\_of\_energy\_state}=\dfrac{(2\pi/a)^2}{(2\pi/L)^2}=\left(\dfrac{L}{a}\right)^2=N^2$

（23）有效质量：$m^*=\dfrac{\hbar^2}{\dfrac{\mathrm{d}^2E}{\mathrm{d}k^2}}$

Vocabulary

amplitude	振幅
amplitude function	振幅函数
angular momentum	角动量
assumption	假设
atom	原子
band theory	能带理论
binding energy	结合能
bragg's law	布拉格定律
bright field	亮场
Brillouin zone	布里渊区
classical and quantum statics	传统及量子统计
classical atomic model	经典原子模型

complicate	复杂
component	组分
concept	定义，概念
conform with	一致
constant	常数
cosine	余弦
dark area	暗区域
deduce	推导
deviation	背离
diffraction	衍射
discharge	放电
distinguish	区分
distinguishable	可区分的
distribution	分布
duration	持续时间
effective mass	有效质量
eigenfunction	本征函数
empirical equation	经验公式
energy	能量
exponential	指数的
expression	表示
facilitate	使容易
filled band	充满带
frequency	频率
hypothesis	假设
interference	干涉
interpretation	解释
ion	离子
Laplacian operator	拉普拉斯算子
mean free path	平均自由程
occupy	占据
orbital	轨道的
parabola	抛物线
photoelectric effect	光电效应
planar wave	平面波
possess	拥有
postulate	假定，假说
potential field	势场

probability	概率
propagate	传播
proportional to	成比例，均衡的
quantity	数量
quantum mechanics	量子力学
quantum statistics	量子统计
rest mass	静止质量
Schrodinger equation	薛定谔方程
second differential form	二次微分形式
solutions to Schrodinger equation	薛定谔方程的解
spectrum	光谱
standing wave function	驻波
tensor	张量
three-dimensional	三维
time-independent	与时间无关
track	轨道
transition	跃迁
uncertainty principle	测不准原理
valid	有效的
wave function	波函数
wave particle duality	波粒二象性
wavelength	波长
wave number	波数
wave vector	波矢

Problems

1. Calculate the maximum wavelength of light that is able to release photoelectrons from a sodium electrode if the work function of sodium is 2.3 eV. If photons of wavelength 200 nm hit the electrode. Calculate the maximum kinetic energy of the released photoelectrons.

2. An X-ray tube has a constant voltage. The most energetic radiation emitted by the tube has the wavelength 0.300 nm. Calculate the smallest de Broglie wavelength of the electrons in the tube. The relativistic effects on the electrons can be neglected.

3. The lowest energy bands in a solid are very narrow because there is hardly any overlap of the wave functions. In general, the higher up the band is, the wider it becomes. Assuming that the collision time is about the same for the valence and conduction bands, which would you expect to have higher mobility, an electron or a hole?

4. True or false questions.

(1) Energy distribution of free electrons in a metal obeys Fermi-Dirac distribution

(2) According to the Pauli principle, each energy state can only accommodate two electrons with opposite spins.

(3) The most energetic valence electrons in the metal at $T = 0$ K locate at Fermi level.

(4) The Fermi level in a metal is related to the density of free electron in metal. And it is usually treated as constant at different temperature.

(5) According to quantum free electron theory, the distribution of free electron agrees with Maxwell-Boltzmann statistics.

(6) Although there are many electron energy levels in metals, electrons generally occupy the energy levels below Fermi level, and only electrons near Fermi level can be excited to higher energy levels by thermal activation.

(7) Maxwell-Boltzmann statistics is not valid to free electron in materials.

(8) Bose-Einstein statistics belongs to quantum statistics, applying to system that does not obey Pauli exclusive rule.

5. In metal, the probability that the energy levels of free electron will accommodate an electron at temperature T can be expressed as,

$$f(E) = \frac{1}{\exp\left(\frac{E - E_F}{kT}\right) + 1}$$

Using this equation, please explain occupation situation of the energy levels of free electrons, (a) at $T = 0$K, (b) $T > 0$K.

6. What is the probability of finding an electron with the energy at the bottom of the conduction band in intrinsic silicon at 300K? (The bandgap of Silicon is $E_G = 1.14$ eV)

7. Considering two Brillouin zones, generally, the band width is wider in the higher energy band, please do the true or false questions.

(1) A wider band width means energy dispersion in the Brillouin zone is larger, and electrons there tends to possess higher delocalization ability.

(2) The number energy states in different Brillouin zone is different.

(3) Electrons in an energy band with wider width will be more mobile than those with narrow band width.

8. Please explain the meaning of $N(E)$~ E curve in the following figure. Note: the solid line is the density of state under different energy, the dash line is the profile of a parabola.

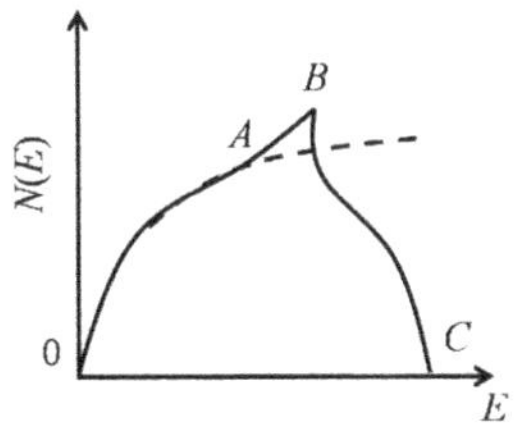

9. Describe the density of state and the K space.

10. Describe the hypothesis and conclusions of the band theory.

11. Please describe the main points of classical free electron postulation.

12. Please points out the main hypothesis of quantum free electron theory.

13. Please give the differences between the hypothesis of classical free electron theory and the quantum free electron theory.

14. Based on the Schrödinger equation of $\frac{d^2\varphi}{dx^2}+\frac{2mE}{\hbar^2}\varphi=0$, and using the free electron model of one dimension metal with length of L, neglecting the lattice effect, solve the equation (i.e., give expression of φ) and based on this wave function, (1) calculate the probability of finding electron at any position within the L length, (2) determine the electron energy. What can be inferred from this energy equation?

15. Tell the limitation of the hypothesis of free electron theory.

16. Please deduce expression of density of state for 2-dimensional free electron employing k space.

17. Heat capacity of metal can be explained by classical free electron theory and quantum free electron theory. Please tell which one is better, and explain both of the explanations.

18. Using free electron theory, explain the opaque feature of metal.

19. The lowest energy bands in a solid are very narrow because there is hardly any overlap of the wave functions. In general, the higher up the band is, the wider it becomes. Assuming that the collision time is about the same for the valence and conduction bands, which would you expect to have higher mobility, an electron or a hole?

20. Classify metals, semiconductors, and insulators on the basis of conductivity.

Chapter 4 Solid State Phase Transformation

In this chapter, phase transformations in solid state will be discussed. Firstly, some special terms for solid state transformation, such as phase interface, misfit of lattice, interfacial energy, orientation relationship, habit plane, and strain energy etc., are explained. Meanwhile, corresponding mechanisms related to the terms are discussed. Then the types of solid state transformation are given in the aspects of thermal dynamics and diffusion. Finally, several typical solid state phase transformations are introduced, including polycrystalline transformation, eutectoid transformation, martensitic transformation, and glass transition. Particularly, the ways that the transformation mechanism affects the microstructure and the properties are also covered.

4.1 Concepts

Materials with different atoms exhibit different properties. Even with the same atoms, material properties can be different, depending on the microstructure. The term of phase is a word that describes some aspect of microstructure.[1] A phase can be defined as any portion, including the whole, of a system which are identical in chemical composition and physical state and might be separated from the rest of the system by a distinct interface called the phase boundary. For example, water has three phases—liquid water, solid ice, and steam. A phase has the following characteristics:

- with the same structure or atomic arrangement inside;
- roughly the same composition and properties throughout;
- a definite interface between the phase and any surrounding or adjoining phases.

1 具有不同原子的材料表现出不同的性质。即使是相同的原子，材料的性能也可能不同，这取决于微观结构。相，这个术语就描述了微观结构的一些方面。

The phases of materials are related to several factors, such as composition, temperature, pressure, etc. In order to better understand these factors on material phase, phase diagram was proposed. A phase diagram typically shows phases that are expected to be present in a system of either pure material or a system with a set of elements, under thermodynamic equilibrium conditions.[2] Sometimes metastable phases may also be shown. From the phase diagram, we can predict how a material will solidify under equilibrium conditions. We can also predict what phases will be expected to be thermodynamically stable and in what concentrations such phases should be present.

2 经典地，相图能够展示出一个含有单组分或者多组分的体系在热力学平衡条件下预期存在的相。

Generally, there are three primary phases in a pure substance, i.e. solid, liquid and vapor. As shown in Figure 4.1a, these phases might be changed mutually via temperature and/or pressure varied. Figure 4.1b shows the phase diagram of a pure substance. The three areas represent solid phase, liquid phase and vapor phase. The curve AB is called the sublimation curve. At the pressures and temperatures along the curve, solid phase and vapor phase coexist in equilibrium with each other, otherwise either as a solid or a vapor. Along the sublimation curve, solid phase can be transformed into vapor or vice versa. The point B is called the triple point. It represents the only temperature and pressure where solid phase, liquid phase and vapor phase can coexist at equilibrium. It is an important point of a pure matter. The triple point of water (273.16 K) is used for definition of the thermodynamic temperature scale. The steep curve BC is the melting curve. It represents the temperatures and pressures where the solid and liquid phases are in equilibrium with each other. It indicates a very high pressure different is required to vary the melting point of the pure matter. Normally the slope of the curve is positive but some substances, for example water, have a negative derivative. In the latter cases the melting point decreases with increasing pressure.[3] The curve BD calls the evaporation curve. At the pressures and temperatures along the curve, liquid phase and vapor phase coexist in equilibrium. The slope of the curve is not as steep as that of the melting curve. For this reason, the boiling point of the liquid varies strongly with the external pressure. At low pressures, for example at high altitudes, water boils at temperatures considerably below 100 ℃. At point D in Figure 4.1b, the evaporation curve stops abruptly. Point D corresponds to the critical point, where there is no difference between the liquid and the vapor. Above the critical temperature, the vapor cannot be compressed to a liquid, even at extremely high pressures.

3 通常来说，这些曲线的斜率是正的，但是有些物质，例如水是负导数。在后一种情况中，物质的熔点随着压力的增加而减小。

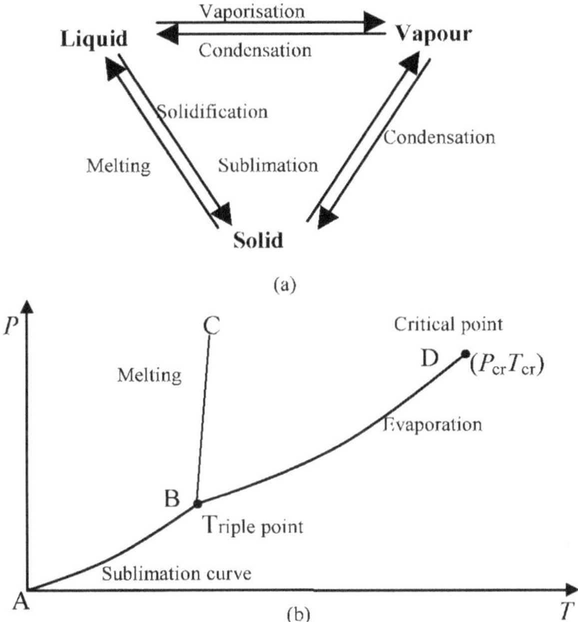

Figure 4.1 (a) Terms in the transformation of three primary phase in a pure substance. (b) Schematic unary phase diagram under pressure and temperature. The term critical point refers to the state, above which distinct liquid and gas phases do not exist.

To describe completely the conditions of equilibrium for a system with more than one component, three externally controlled variables of temperature, pressure, and composition must be specified. Normally, phase diagrams record the data when the pressure is held constant under normal atmospheric conditions.[4] Consequently, phase diagrams especially are graphical representations of a material system under varying conditions of temperature and composition. From this temperature-composition phase diagram,

4 通常来说，相图记录的是正常大气条件下，压力一定时的数据。

5 从温度-组分相图中，我们能够预测材料在平衡条件下如何固化。

we can predict how a material will solidify under equilibrium conditions.[5] We can also predict what phases will be expected to be thermodynamically stable and in what concentrations such phases should be present.

Figure 4.2 is the phase diagram of a two-components system with Ni-Cu. Vertical line X-Y represents our alloy. At point 1 the alloy is at temperature T_1 in the liquid phase. The compositions of liquid are 60% Ni and 40% Cu. Upon slow cooling, point 2 is reached at a temperature of T_2. Our alloy is now entering the two-phase region. Part of our alloy is liquid and part has formed a solid solution. The solid solution is designated α solid solution. To determine the composition of these two phases at any point, such as at 3, a horizontal temperature line called a tie-line is drawn through point 3. This tie-line will cross both the liquidus and solidus curves at points labeled L_3 and S_3, respectively. From both these points, drop vertical lines to the abscissa (horizontal axis) to determine the composition of these two phases at the temperature T_3. The liquid phase formed so far has a composition of 54% Cu and 46% Ni, determined by proceeding from point 3 along the tie-line to the liquidus L_3, then vertically down to the horizontal axis at the point labeled 46% Ni. Similarly, the composition of the solid phase is found to be 70% Ni-30% Cu. This time we proceeded along the tie-line to the solidus at S_3 and vertically downward to the horizontal axis to read 70%. Note that the tie-line drawn across the two-phase region ties in or connects the adjacent phases. In this instance, the solid phase is on the right and the liquid phase is on the left of the two-phase region. This is an excellent way of identifying any two-phase region. As the alloy continues to cool to temperature T_4, observe, using the same procedure, that both phases continue to increase in the percentage of copper. Upon reaching the solidus at temperature T_4, the last liquid, very rich in Cu, solidifies at the grain boundaries. Through diffusion of the atoms, all the solid solution will be at an overall composition of 60% Ni-40% Cu.

6 材料从旧相到新相的转变过程称为相变。固态相变指的是在固态下发生的相变。

7 固态相变不容易观测，大多数依赖于电子显微镜和X射线衍射设备，有时也可以通过颜色变化来判断。

The process that material changes from the old phase to a new one is termed as phase transformation. Solid-state phase transformation refers to a change in phase that occurs in the solid state.[6] Different from the primary phase transformation among solid, liquid and vapor, solid state has its own phase transformation among different crystal structure and/or different chemical composition, as well as different grain size, which can cause the material property changes, such as magnetism, optical property or conductivity, etc. It is not easy to verify solid state phase transformations, most of them depend on electron microscopes, X-ray diffraction, sometime these can also be noticed by color change.[7]

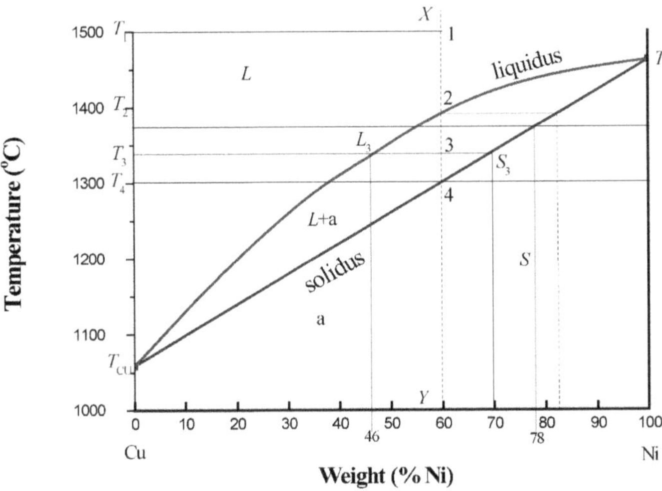

Figure 4.2 Cu-Ni phase diagram. Alloy X-Y is 60% Ni- 40% Cu.

Here are examples about solid state phase transformation.

(1) Preparation of ceramics: it has been known that the temperature of the kiln must be raised especially slowly when passing through the range between 550 to 600 ℃. Otherwise, the ceramic product would be self-cracked in that range, even if they have been dried enough before they are put into the kiln. It is because that due to the large volume expansion associated with the crystal structure change of silica (SiO_2) particles, usually contained in an appreciable amount in clay, from α-quartz to β-quartz at 573 ℃.[8] This was probably the first knowledge on the solid phase transformation of materials obtained by the people in industry. Cooling down the kiln must also be done carefully, but in this case, the reverse transition to α-quartz doesn't always take place.

(2) Nuclear fission reactions of ^{235}U: During the nuclear fission reactions of ^{235}U in α-uranium, the material gets a self-irradiation effect by the produced high energy fission fragments and fast neutrons and makes a large dimensional change or cracking. Therefore, to use α-uranium as a fuel element, as it is, is not practical as the structure must be transformed to another crystal structure by alloying or making an oxide. Today, the uranium fuel elements are usually bars or plates with FCC structure, being alloyed with aluminum or niobium, or oxide pellets.

(3) Non-equilibrium multi-component phases in steels: In steels, there are various non-equilibrium, multi-component phases of Fe and Fe-C composites, as shown in Figure 4.3.

These phases are all formed from the transformation or decomposition of Austenite as the steel is cooled below the eutectoid temperature.

8 那是因为在陶土中通常含有相当数量的 SiO_2，在 573℃时它将发生由α石英到β石英的相转变，同时伴随巨大的体积膨胀。

Austenite is a solid solution of Fe and C with face-centered cubic (FCC) structure. It is termed as γ-Fe; ferrite refers to the almost pure iron with carbon less than 0.03% (0.006%) at high temperature with a body centered cubic (BCC) structure. It is termed as α-Fe; Martensite is formed in carbon steels by the rapid cooling (quenching) of Austenite at such a high rate that carbon atoms do not have time to diffuse out of the crystal structure in large enough quantities to form cementite (Fe_3C).[9] As a result, the face-centered cubic Austenite transforms to a highly strained body-centered tetragonal form of ferrite that is supersaturated with carbon; cementite, also known as iron carbide, is a chemical compound of iron and carbon, with the formula Fe_3C (or Fe_2C:Fe). It has an orthorhombic crystal structure.

9 奥氏体是铁碳固溶体，具有面心立方结构（FCC）。它被称为γ铁；铁素体是指高温下含有少于0.03%（0.006%）碳的纯铁，具有体心立方结构（BCC）。它被称为α铁；马氏体是碳钢，通过将奥氏体快速地冷却（淬火）而得到的，在这么快的速率下，碳原子没有时间去扩散到晶体结构之外来形成渗碳体（Fe_3C）。

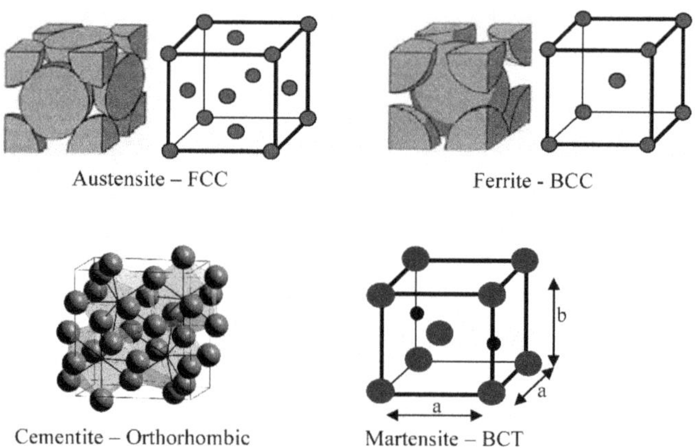

Figure 4.3 Various non-equilibrium, multi-component phases of Fe and Fe-C composites.

Besides the phases shown in Figure 4.3, another important form of steel is pearlite, which is a two-phased, lamellar (or layered) structure composed of alternating layers of alpha-ferrite (88 wt%) and cementite (12 wt%) that occurs in some steels and cast irons.

Many well-known phase transformation of steels are listed in Figure 4.4.

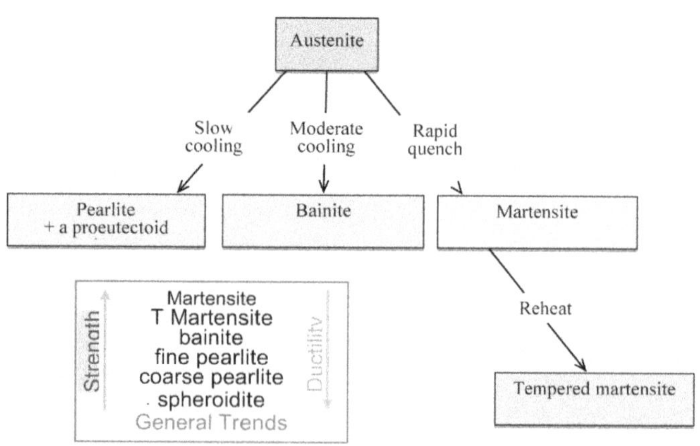

Figure 4.4 Possible phase transformations in steel.

4.2 Features of Solid State Phase Transformation

In solid state phase transformation, both the old phase and the new phase are solids, no other state involved.[10] Therefore, it exhibits many different features compared to the transformations of solid-liquid, gas-liquid, and solid-gas.

10 在固态相变中，旧相和新相都是固态，其他状态不涉及。

4.2.1 General Procedure

A solid state phase transformation generally includes the following steps:

Seeds: Most phase transformations begin with the formation of numerous small particles of the new phase that increase in size until the transformation is complete.

Nucleation: Nucleation is the process whereby nuclei (seeds) act as templates for crystal growth. There are two types of nucleation: one is homogeneous nucleation, which refers to the process that the nuclei form uniformly throughout the parent phase; the other is heterogeneous nucleation, which refers to the formation of nuclei at structural in-homogeneities (container surfaces, impurities, grain boundaries, dislocations etc.).

Growth: Refers to the process that crystal grows from the nucleation site and along the direction of the nucleation, leading to the formation of a new phase.

4.2.2 Phase Interface

During phase transformation, interfaces between the old and the new phases, or between two new phases always present, which is generally termed as phase interface. Different from the interface between the substrate and the thin film which is prepared by gas or liquid deposition with physical adsorption mechanism, in which the interaction is van der Waals force, the phase interfaces are produced between the new and the old phases, in which chemical bonding generally dominates.[11] Even both the new phase and the parent phase are crystal, there are still three types of interface: coherent interface, semi-coherent interface and non-coherent interface. Phase interface can be described by two parameters:

11 通过气相或者液相沉积方法获得的基底和薄膜之间的界面，与新相和旧相之间的相界面是不同的。前者为物理吸附机制，其相互作用力为范德华力；后者通常是化学键占主导。

Misfit of Lattice (m): This refers the relative atomic-distance difference between two adjacent phases, which gives the degree of mismatch between the old and the new phases, it can be described as m = absolute (a-b)/a, where a is the lattice constant of old phase.

Interfacial Energy: The interfacial energy in solid phase interface originates from the energy increase due to the irregular atom stacking, and the chemical potential changes due to the chemical bonding or composition differences between the old and new phases.[12]

12 固相界面的界面能产生于原子不规则堆积所导致的能量增加，以及由于新相和旧相之间化学键或者组分差异造成的化学势变化。

It was commonly thought that: $m < 0.05$, coherent interface is formed, the interfacial energy is about 0.1 J/m^2; $0.05 < m < 0.25$, semi-coherent interface can be formed, the interfacial energy will be smaller than 0.5 J/m^2; when m > 0.25, incoherent interface will be formed, the interfacial energy is about 1.0 J/m^2.

4.2.3 Orientation Relationship

When a new phase is produced from the old phase, if the orientation of the new phase is random, most of the phase interfaces will be incoherent, leading to increase of interfacial energy.[13] In order to reduce interfacial energy, there exists preferred relationship between the old and the new phases, which is termed as orientation relationship. Generally, the crystal planes of the old phase and the new phase tend to parallel in such that both of them possess dense atoms and low indices, as well as low degree of mismatch. For instance, in an allotropic transformation of,

13 当一个新相从旧相中产生时，如果新相的取向是随机的，那么大多数相界面将会是非共格的，从而导致界面能的增加。

$$\alpha-\text{Co}\left(\text{hcp, A3}\right)\xrightarrow{>450℃}\text{Co (fcc, Al)} \tag{4-1}$$

The orientation relationship will be:

$$\{111\}_{A_1}//\{0001\}_{A_3}$$
$$\{110\}_{A_1}//<11\bar{2}0>_{A_3} \tag{4-2}$$

4.2.4 Habit Plane

In solid state phase transformation, the new phase generally grows from specified surface of the old phase, which is called habit plane.[14] For example, a hypoeutectoid steel (carbon < 0.77%) presents as solid solution of Austenite at high temperature. When it is cooled at certain cooling rate, the proeutectoid phase of ferrite appears. Continuing to lower the temperature, eutectoid transformation occurs with simultaneous appearance of the ferrite and cementite, which is called perlite. Before the eutectoid transformation, in the proeutectoid transformation, the parallel needle shape ferrite always grows along the {111} plane of the parent austenite. Thus the {111} plane of the parent austenite is called the habit

14 在固态相变中，新相通常会在旧相的特定表面上生长，这个特定的面就是惯习面。

plane of the proeutectoid ferrite. These proeutectoid ferrites together with its parent Austenite phases are called Widmanstatten as shown in Figure 4.5.

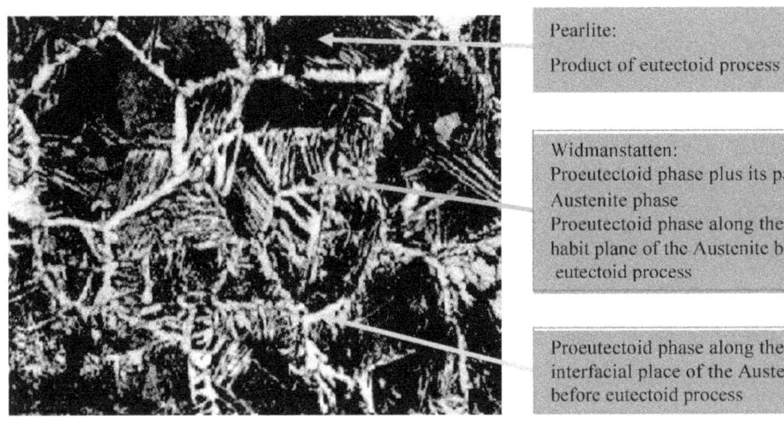

Figure 4.5 Widmanstatten structure in steel.

4.2.5 Strain Energy

When a new phase forms, the specific heat capacities of the new and old phases are different, for e.g. the new phase does not occupy the same volume that is displaced. This may result in the volume expansion or shrinking. Due to the constraint from the mother phase, the new phase cannot expand or shrink freely, thus additional energy is required to accommodate the new phase in the matrix. This energy calls strain energy.

The strain energy refers to the energy required to permit a new phase to fit into the surrounding matrix during the nucleation and growth of the new phase.[15]

15 应变能是指在新相的成核和长大过程中，将新相放置到母相周围所需要的能量。

In order for a new phase of a precipitate to form from a solid matrix of phase, both nucleation and growth must occur. The total change in free energy required for nucleation of a spherical solid precipitate of radius ***r*** from the matrix is

$$\Delta G=\frac{4}{3}\pi r^3\Delta G_{\mathrm{v}}(\alpha\to\beta)+4r^3\sigma_{\alpha\beta}+\frac{4}{3}\pi r^3\varepsilon \tag{4-3}$$

There three parts in the above equation: (1): the free energy change per unit volume from α to β; (2): the energy change needed to create the unit area of the interface between α and β; (3): the strain energy per unit when creating β phase from α phase. As it is, strain energy is related to the volume of the new phase; interface energy is related to the surface area of the new phase.

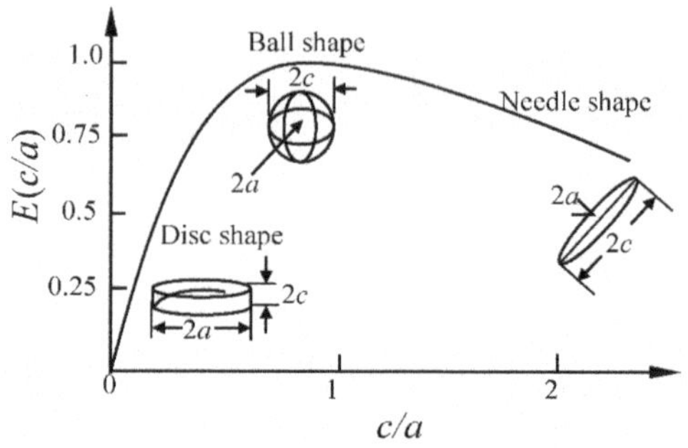

Figure 4.6 The relative strain energies vs the shape of the new phase (c/a).

When a new phase forms, the total interface energy and the strain energy are all related to the shape of the new phase. As illustrated in Figure 4.6, new phase with disc shape will exhibit the smallest strain energy; that with needle shape produces increased strain energy compared to the disc shape; when the new phase present in ball shape, the stain energy is the largest. Thus, under some condition, we can predict the shape of the new phase:

(1) If the new phase is coherent with the parent phase, the interface energy is small; the strain energy is dominant factor. In order to reduce strain energy, the crystal usually adopts disc or needle structure.[16]

16 如果新相与母相共格，界面能就较小，应变能为主导因子。为了减少应变能，晶体通常采用圆盘或针形结构。

(2) In non-coherent phase transformation, the interface energy is large, and strain energy is negligible. In order to reduce interface energy, small surface area is preferred, thus ball structure is adopted.[17]

17 在非共格相转变过程中，界面能大，应变能就可以忽略不计。为了降低界面能，小的表面积是优选的，因此采用球结构。

4.2.6 Influence of Defects

In solid state phase transformation, if the parent phase is crystal, besides the point defects at equilibrium condition, other crystal defects inevitably influence the formation of the new phase. New phase usually begins nucleation at defect points, where the fluctuations of energy, components, and structure are abundant, thus are suitable for nucleation. Because the sites of defect possess higher energy, it can provide driving force for the formation of the new phase; additionally, the irregular arrangement at the defect site cannot be only beneficial to atom diffusion, but may also bearing similar structure to the new phase.

4.2.7 Atom Diffusion

If the components are different between the old and the new phases, diffusion must exist during the phase transformation.[18] One must note that the diffusion rates for liquid and solid species are quite different, e.g. liquid: ~10^{-7} cm^2/S; Solid: ~10^{-7}–10^{-8} cm^2/d. This means that the diffusions

18 如果旧相和新相的组分是不同的，那么在相变过程中一定存在扩散。

of solid atoms are very slow, hence the solid state phase transformation seldom occurs in equilibrium condition, it usually need undercooling. For example, if the eutectoid steel is cooled under very slow rate, atom diffusion occurs and perlite will form; if it is cooled under fast rate, no diffusion occurs and Martensite will form.

4.3 Classification of Solid State Phase Transformation

There are many types of solid state phase transformations, such as: allotropic transformation, precipitation, eutectoid, peritectoid, ordering, magnetic transformation, superconducting transformation, etc. These classifications are complicate and just based on the phenomena during the phase transformation. Here we will discuss two common classifications of solid phase transformation, one is based on thermodynamics, the other is based on atomic diffusion.

4.3.1 Classification Based on Thermodynamics

Based on thermodynamics, solid state phase transformation can be classified as first order phase transformation and second order phase transformation.

In the first order phase transformation, the chemical potentials of new and old phases are equal, but their first order partial derivatives are different.[19] That is the entropy and volume change discontinuously, accompanying by thermodynamic transition and a sudden change of volume at constant temperature. That is:

$$\mu^{\alpha}=\mu^{\beta},\ \left(\frac{\partial\mu^{\alpha}}{\partial T}\right)_{\mathrm{P}}\neq\left(\frac{\partial\mu^{\beta}}{\partial T}\right)_{\mathrm{P}},\ \left(\frac{\partial\mu^{\alpha}}{\partial P}\right)_{\mathrm{T}}\neq\left(\frac{\partial\mu^{\beta}}{\partial P}\right)_{\mathrm{T}} \tag{4-4}$$

Because

$$\left(\frac{\partial\mu}{\partial P}\right)_{\mathrm{T}}=V,\ \left(\frac{\partial\mu}{\partial T}\right)_{\mathrm{P}}=-S \tag{4-5}$$

so:

$$V^{\alpha}\neq V^{\beta}$$
$$S^{\alpha}\neq S^{\beta} \tag{4-6}$$

Here, μ is chemical potential, α and β are old and new phases, T is temperature, P is pressure, V is volume, and S is entropy.

The chemical potential keeps constant during the first order phase

19 在一阶相变过程中，新相和旧相的化学势是相等的，但它们的一阶偏微分是不同的。

transformation means that there is no new component created in the new phase. But due to the phase changes, the volume and the entropy change.

In the second order phase transformation, the chemical potentials and their first order partial derivatives of new and old phases are all equal, but their second order partial derivatives are different. That is:

$$\mu^\alpha=\mu^\beta,\ \left(\frac{\partial\mu^\alpha}{\partial T}\right)_P=\left(\frac{\partial\mu^\beta}{\partial T}\right)_P,\ \left(\frac{\partial\mu^\alpha}{\partial P}\right)_T=\left(\frac{\partial\mu^\beta}{\partial P}\right)_T \tag{4-7}$$

$$\left(\frac{\partial^2\mu^\alpha}{\partial T^2}\right)_P\neq\left(\frac{\partial^2\mu^\beta}{\partial T^2}\right)_P,\ \left(\frac{\partial^2\mu^\alpha}{\partial P^2}\right)_T\neq\left(\frac{\partial^2\mu^\beta}{\partial P^2}\right)_T,\ \frac{\partial^2\mu^\alpha}{\partial T\cdot\partial P}\neq\frac{\partial^2\mu^\beta}{\partial T\cdot\partial P} \tag{4-8}$$

Because

$$\left(\frac{\partial^2\mu}{\partial T^2}\right)_P=-\left(\frac{\partial S}{\partial T}\right)_P=-\frac{C_P}{T},\ \left(\frac{\partial^2\mu}{\partial P^2}\right)_T=\left(\frac{\partial V}{\partial P}\right)_T=V\cdot K \tag{4-9}$$

where $K=\frac{1}{V}\left(\frac{\partial V}{\partial P}\right)_T$ is compression coefficient.

$\frac{\partial^2\mu}{\partial T\cdot\partial P}=\left(\frac{\partial V}{\partial T}\right)_P=V\cdot\alpha,\ \alpha=\frac{1}{V}\left(\frac{\partial V}{\partial T}\right)_P$, is the coefficient of thermal expansion.

For second order phase transformation, $C_P^\alpha\neq C_P^\beta, k^\alpha\neq k^\beta, \alpha^\alpha\neq\alpha^\beta$

In second order phase transformation, beside the chemical potential keeps no change as in first order phase transformation, the volume and entropy of old and new phases also keep constant.[20] But the heat capacity, the compression coefficient, and thermal expansion coefficient of new and old phases changes discontinuously.

20 在二阶相变过程中，除了与一阶相变过程中化学势保持不变一样外，新旧相的体积和熵也保持不变。

As shown in Figure 4.7, the first (left) and the second (right) order phase transformations are compared, it can be seen that in first order phase transformation, only at extrema points (left figure), the composition for the new and old phases is the same. In second order phase transformation, at equilibrium point, the old phase and the new phase always have the same composition.

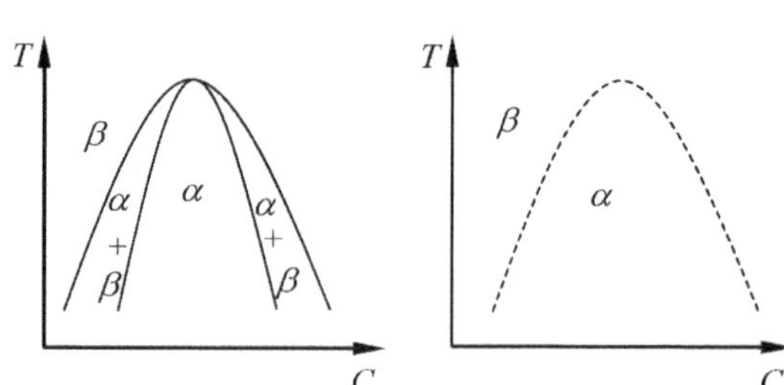

Figure 4.7 Phase diagrams for the first (left) and the second (right) order phase transformation.

In the first order phase transformation, although the elements keep the same before and after phase transformation (chemical potential keeps constant), the composition of the old and new phases changes, leading to the volume and entropy change. In the second order phase transformation, not only the elements but also the compositions keep constant before and after phase transformation, hence the chemical potential, the volume, and the entropy of new and old phases do not change.

4.3.2 Classification Based on Atomic Diffusion

According to atomic diffusion, there are diffusional, diffusionless, and semi-diffusional phase transformations. The diffusion means that during the phase transformation, diffusion of atoms occurs, and the process of phase transformation is controlled by the atomic diffusion.[21] The second term means that no diffusion process occurs in phase transformation. The last one means that the phase transformation is partially accompanied by a diffusion process. Table 4.1 lists some solid state phase transformation processes and their classification based on the thermodynamics and the atomic diffusion phenomenon.

21 扩散是指在相变过程中，发生了原子扩散，并且相变过程是由原子扩散控制的。

Table 4.1 Some solid state phase transformation processes

Process	Classification	
Allotropic transformation	First order	Diffusional
Precipitation	First order	Diffusional
Eutectoid	First order	Diffusional
Peritectoid	First order	Diffusional
Martensite transformation	First order	Diffusionless
Bainite transformation	First order	Semi-diffusional
Spinodal decomposition	First order	Diffusional
Ordering transformation	First/second order	Diffusional
Magnetic transformation	Second order	Diffusionless
Super-conduction transformation	Second order	Diffusionless

4.4 Polycrystalline Transformation

Materials, consisting of either one component or several components, can possess different properties depending on its microstructure.[22] The change of the microstructure which can be termed as polycrystalline transformation. Polycrystalline transformations result in changes in properties of materials and form the basis for the heat treatment of steels and many other alloys.

22 由一个或多个组份组成的材料，可能具有不同的性质，这取决于它的微观结构。

Materials, presenting polycrystalline transformation, include allotropic

23 具有一个以上晶体结构的材料被称为同素异形的或多晶型的。

material, alloy solid solution, one component ceramics, ceramic solid solution etc. Materials that can have more than one crystal structure are called allotropic or polymorphic.[23] The term allotropy is normally reserved for this behavior in pure elements, while the term polymorphism is used for compounds.

Some metals, such as iron, have more than one crystal structure and can have polycrystalline transformation. Figure 4.8 illustrates different phases of pure iron under different temperature and/or pressure.

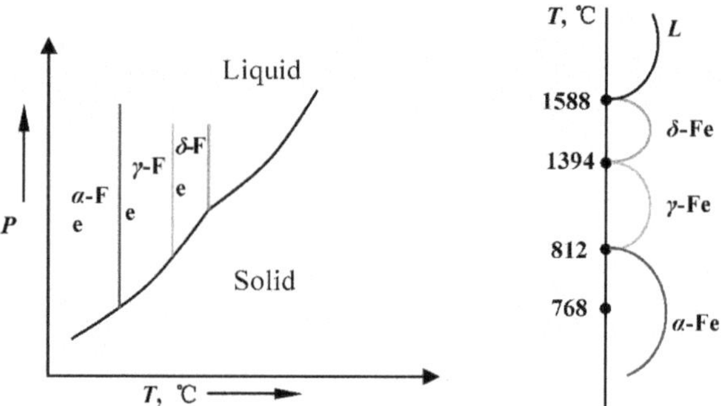

(a) phase changes with both T and P. (b) phase changes only with T.

Figure 4.8 The phase diagram of pure Fe.

Many ceramic materials, such as silica (SiO_2) and zirconia (ZrO_2), are polymorphic, and might have polycrystalline transformation.

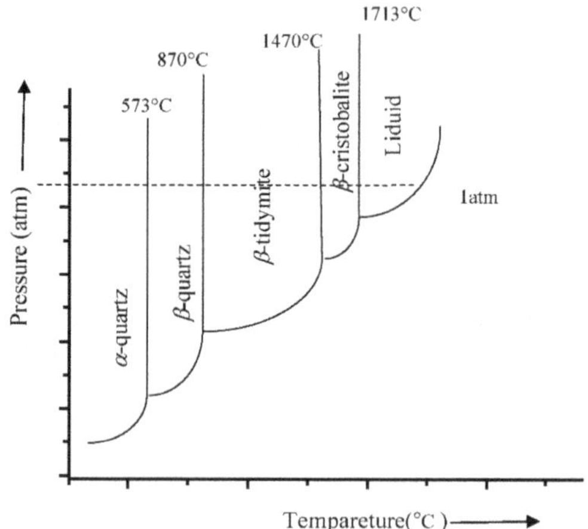

Figure 4.9 Phase diagram of Silica. The dashed line shows one atmosphere pressure.

A volume change may accompany the transformation during heating or cooling; if not properly controlled, this volume change causes the brittle ceramic material to crack and fail.

The change of free energy is the driving force in polycrystalline transformation:

$$G = H - TS$$

$$\mathrm{d}G = V\mathrm{d}P - S\mathrm{d}T \tag{4-10}$$

Under constant pressure: $\mathrm{d}G = -S\mathrm{d}T$, hence:

$$\frac{\mathrm{d}G}{\mathrm{d}T} = -S < 0 \tag{4-11}$$

As can be seen from figure 4.10, for α to β transformation, $S_\alpha \neq S_\beta$, those of G-T curves for α and β phases meet at T_0, with $\Delta G = 0$. So when $T = T_0$, no phase transformation occurs; when $T > T_0$, β phase transform to α phase, when $T < T_0$, α phase change to β phase.

In polycrystalline transformation, the nucleation generally occurs at interfacial sites.[24] These places have the following features: (1) Energy fluctuating: interface generally possesses high energy, which can be used to overcome the resistance for phase transformation. By nucleation at this place, the high energy can be eliminated with disappearing of the interface. (2) Structure fluctuating: irregular structure usually causes the rearrangement of atoms, which is favorite to create a new phase. (3) Components fluctuating: when components are different between the old and the new phases, the aggregation of the solute atoms will tend to form new phase. The growth in polycrystalline transformation refers to the enlargement of the new phase. Atoms from the parent phase move to the nucleus sites of the new phase via self-diffusion, then they are rearranged according to the ordering and stretching pattern in the new phase.[25] With the growth of the new phases from many nuclei sites, the dispersive small new phases connect with each other, leading to the disappearance of the old phase and the completion of the phase transformation.

24 在多晶型转变过程中，成核一般发生在界面处。

25 母相中的原子通过自扩散移动到新相的成核位置，然后它们会根据新相的顺序和延展模式被重新排列。

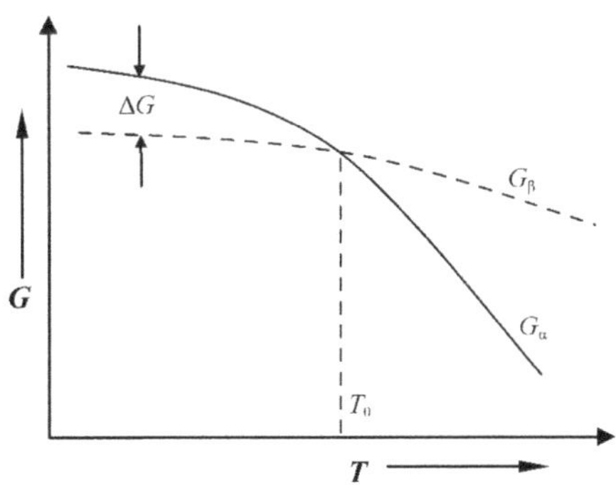

Figure 4.10 The free energy of the α and β phases with temperature.

Polycrystalline phase transformation belongs to diffusional phase transformation, diffusing atoms must be detached from their original locations (perhaps at lattice points in a solid solution), move through the surrounding material to the nucleus, and be incorporated into the crystal structure of the new phase. In some cases, the diffusing atoms might be so tightly bonded within the old phase that the detachment process limits the rate of growth. In other cases, attaching the diffusing atoms to the new phase—perhaps because of the lattice strain—limits growth. Sometimes, this result leads to the formation of new phase that has a special relationship to the matrix structure that minimizes the strain. In most cases, however, the controlling factor is the diffusion step.

4.5 Eutectoid Transformation

Eutectoid transformation is referred to the phase transformation of the simultaneous appearing of two new solid phases from the old solid parent phase: $\alpha \to \beta + \gamma$. It is a kind of diffusion type transformation.

There is a counterpart concept, as eutectic transformation, which is referred to the simultaneous coming out of two solid phases from the liquid parent phase.

4.5.1 Thermodynamics of Eutectoid Transformation

According to phase rule:

$$f = C - Ph + 2 \tag{4-12}$$

where f is the freedom of the system, C is the number of members, Ph is the number of phases.

For the condition of fixed pressure:

$$f = C - Ph + 1 \tag{4-13}$$

In eutectoid phase transformation, three phases coexist, i.e., the parent phase, the two new phases that simultaneously appear, hence $Ph = 3$.

In a system with two members ($C = 2$), the freedom of system at the fixed pressure condition:

$$f = 2 - 3 + 1 = 0 \tag{4-14}$$

The zero freedom means that in eutectoid transformation with three coexisting phases under a fixed pressure, the temperature and each composition of the three phases are fixed.[26]

26 零自由度意味着压力一定时的共析转变中，三相共存时，温度和三个相中每个组分的含量是固定的。

In the two components system (A and B) at the temperature of eutectoid transformation (T_c), the free energy versus composition (G-C) curves of three phases (α, β, γ) can be expressed as Figure 4.11.

At temperature T_C, there is a common tangent for the three curves, with tangent points of x_γ, x_α, and x_β, which are the lowest points in each curve. Eutectoid transformation satisfies the common tangent rule, that is: In the

G-C curves, at a fixed composition in the composition range of x_γ to x_β, the corresponding common tangent value is the free energy of the system consisting of α, β, γ phases and it is not greater than that for α, β, γ phases at the same composition when these phases exist alone. Hence, α, β, γ phases are coexisting in the composition range of x_γ to x_β. The phase diagram of a two components system capable of the eutectoid transformation can be schematically expressed as Figure 4.12 .

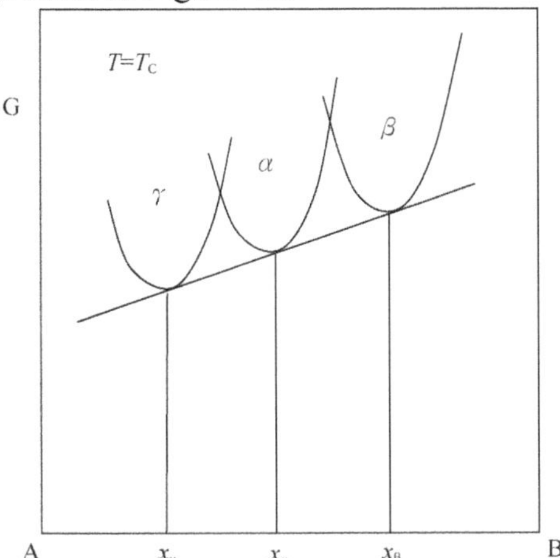

Figure 4.11 The free energy-component diagram under the eutectoid transformation temperature (T_C).

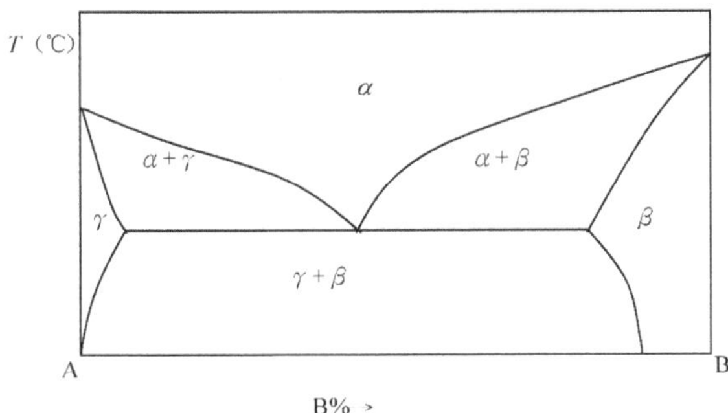

Figure 4.12 Schematically phase diagram of a two components system.

The eutectoid transformation occurs at fixed temperture, thus it is a horizatal line in the phase diagram, the composition of phase α, β, γ is x_γ, x_α, and x_β, respectively. For eutectoid transformation, since it occurred under T_C, and the composition of three phases are not equal to the common tangent values at the beginning, thus eutectoid transformation is diffusional type.

4.5.2 Processes in Eutectoid Transformation

Similar to the polycrystalline transformation, the eutectoid transformation also involves nucleation and phase growth. But system with different composition may produce eutectoid reaction in different ways.[27] In the case

27 类似于多晶形转变，共析转变也包括成核和相生长。但具有不同成分的系统可能通过不同的途径发生共析反应。

of steel, there are three situations:

(1) In hypo-eutectoid steel with carbon content less than 0.77%, before eutectoid transformation, there will be pro-eutectoid α phase in the interface.

(2) In hyper-eutectoid steel with carbon content large than 0.77%, there is pro-eutectoid cementite (Fe_3C) phase in the interface.

(3) In eutectoid steel with carbon content equal 0.77%, the Fe_3C is generally considered as pro-eutectoid phase.

The phase diagrams of eutectoid steel, hypo eutectoid steel and hyper eutectoid steels are summarized in the following three figures (Figure 4.13, Figure4.14, Figure4.15).

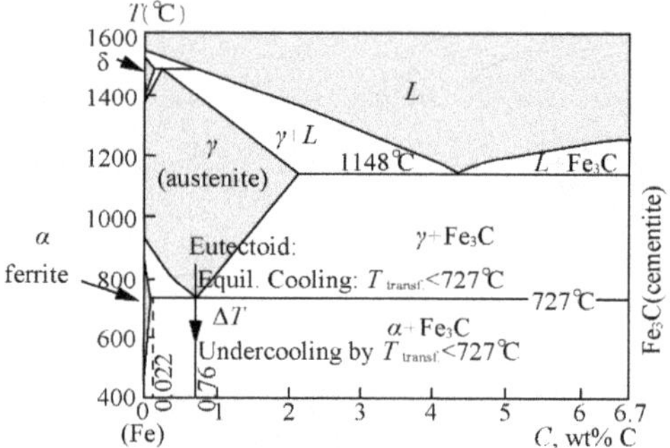

Figure 4.13 Transformations of eutectoid steel.

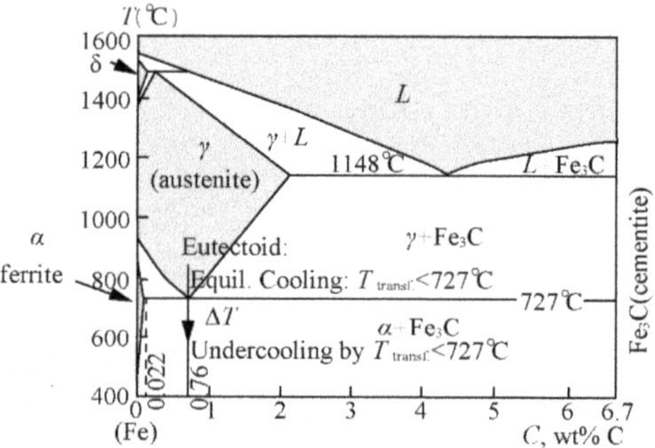

Figure 4.14 Transformations of hypo eutectoid steel.

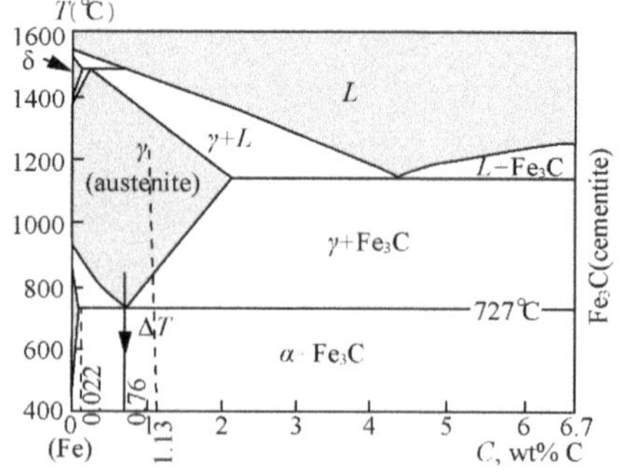

Figure 4.15 Transformations of hyper eutectoid steel.

The nucleation in eutectoid transformation is demonstrated here using by eutectoid steel as example, as shown in Figure 4.16. Firstly, at the interface of γ_1 and γ_2, nucleation of Fe_3C occurs at certain phase direction for the purpose of reducing the energy of nucleation, i.e. $(010)_C//(110)\gamma$. Then in the vicinity of Fe_3C, the concentration of C decreases due to the above Fe_3C formation, thus forming the α nucleus at certain orientation to Fe_3C and γ.These kinds of nucleation repeated along the outside of γ_2 phase to a large extent, along with some non-coherent interfaces produced.

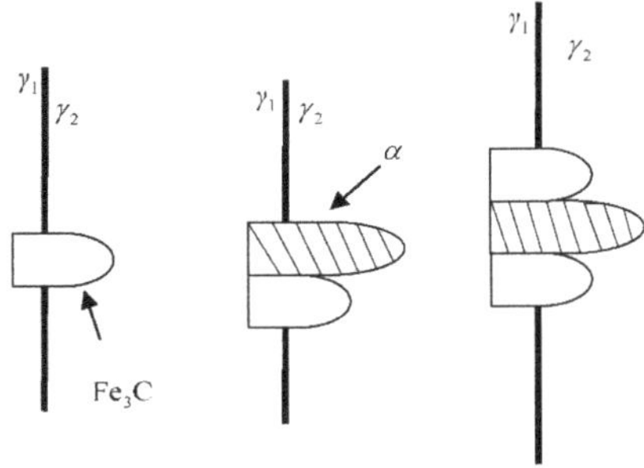

Figure 4.16 Nucleation of eutectoid steel.

After nucleation, growing starts, which will happen in both outside and the forward directions, accompanying with the transverse and the longitudinal diffusion processes.[28] Branches of growth may also happen if condition is preferred. As such, perlite microstructure is formed (Figure 4.17).

28 在成核之后，晶体开始向外向前生长，并且伴随着横向和纵向扩散。

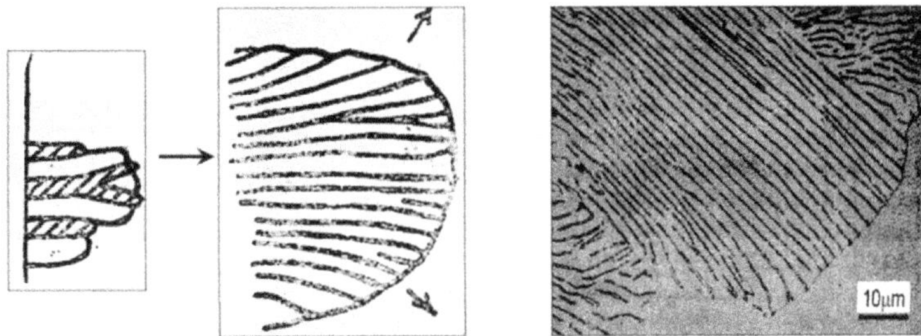

Figure 4.17 Growth of eutectoid steel and the microstructure of the perlite after eutectoid growth.

The evolution of the microstructure of hypo-eutectoid and hyper-eutectoid steels during cooling are summarised in the following Figure 4.18.

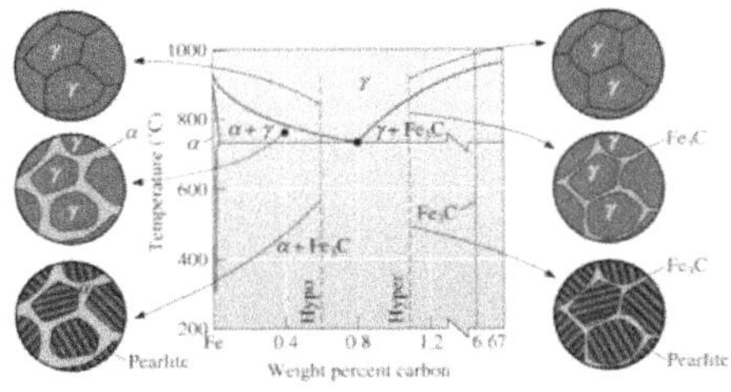

Figure 4.18 Comparison of the eutectoid transformation between hypo-eutectoid steel and hyper-eutectoid steel, in relationship to the $Fe\text{-}Fe_3C$ phase diagram.

4.5.3 Dynamic Characteristics: TTT Curve

Temperature-time-transformation curves are often used to describe the dynamic characteristics of the diffusional phase transformation, such as eutectoid reaction[29], as shown in Figure 4.19. This figure shows at a fixed temperature, the time needed for some degree of phase transformation, e.g, 0%, 50% and 100% completion. At high T, since the temperature gradient (the degree of super cooling) is small, hence transformation is slow; at low T, atom diffusion is slow, also leading to slow transformation. At the middle part of temperature, proper cooling rate with proper diffusion rate results in the fastest transformation, resulting in the nose of the TTT curve.

29 温度-时间变化曲线经常用来描述扩散型相变的动力学特性，如共析反应。

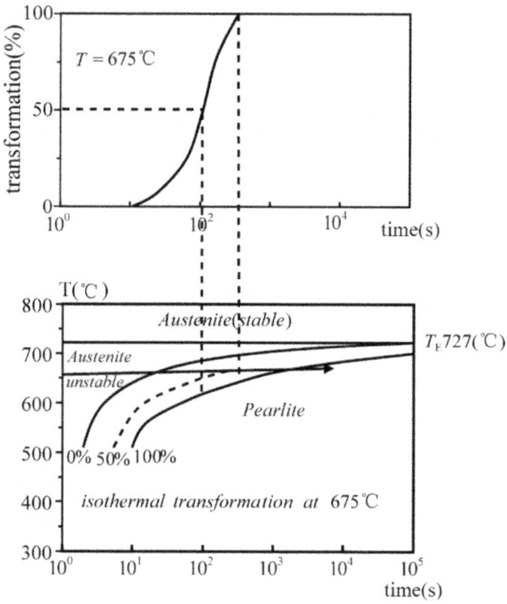

Figure 4.19 The isothermal transformation diagram (TTT curve, temperature-time-transformation diagram) of eutectoid steel.

4.6 Austenite–Martensite Transformation

4.6.1 Historical Remark

Austenite-Martensitic transformation was firstly found in the quenching process of the Austenite steel under 200°C. In the ancient Greeks, the structural transformation in steel has been used to harden sword by rapid cooling. The underlying physics of it was first seriously explored by Martens in the later of 19th century, and because of his work, the diffusionless structural phase transformations in steel and other materials have become known as martensitic transformations, which is not only limited in steel but also in other metals, alloys, ceramics, and even in solid nitrogen.

The martensitic transformation is a first order diffusionless structural transformation mainly by shear deformation of successive atomic planes, which starts at M_s temperature and propagates in the parent phase, producing a peculiar surface relief pattern known as the crystallographic Martensite structure, as shown in Figure 4.20. The transformation doesn't finish until a lower temperature M_f is reached.

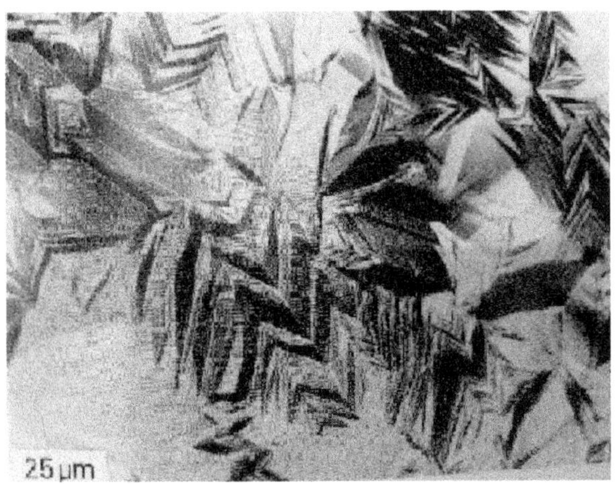

Figure 4.20 The Martensite structure in steel.

Usually, most of the structural phase transformation of metals were not desirable in daily and industrial use. A well-known exception is the martensitic transformation of steel which was an excellent method of strengthening steel, especially for swords and spears, used more than one thousand years ago. This old technique in steel work was suddenly found to be useful again in a new material, that is the shape memory alloys.

4.6.2 Features of Martensitic Transformation

Ⅰ. Coherent Interface with Embossed Surface

One of important feature of Martensite transformation is the appearance of the embossed surface, which means that after surface polishing to produce Martensite transformation, there are local emboss or concavity appeared [30] (Figure 4.21). In this process, the mother phase and the Martensite phase are coherent. This means, if a pre-scratch was made, after polishing it will become a continuous dogleg curve as shown in Figure 4.22a, but not Figure 4.22 b and c.

30 马氏体转变的一个重要特征是表面的凸起，这意味着表面抛光产生马氏体转变后，局部有凸起或凹陷的出现。

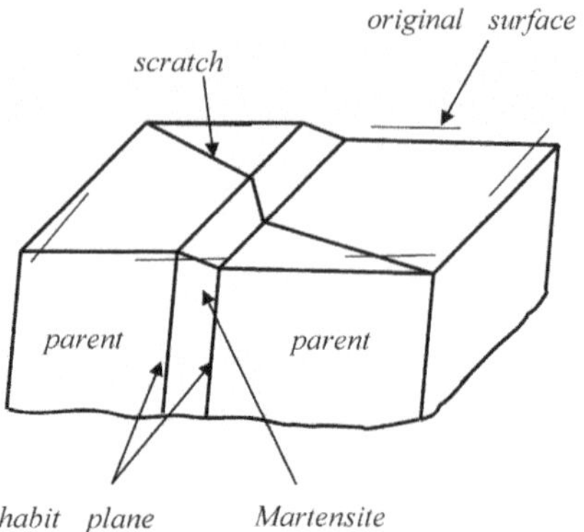

Figure 4.21 The emboss or concavity appeared in Martensite transformation.

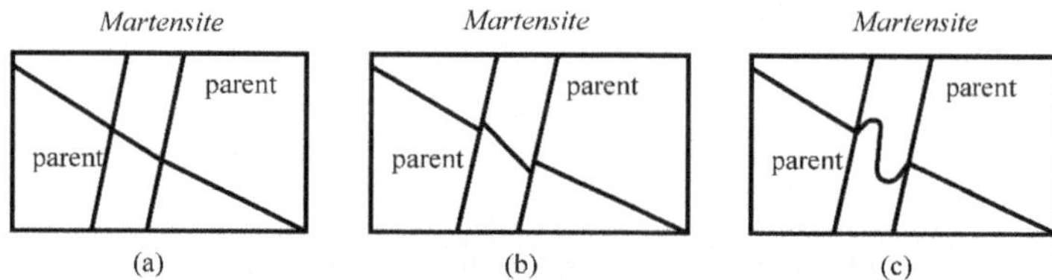

Figure 4.22 The patterns of the scratch line after polishing, (a) coherent at interface, (b) non-coherent with broken line at interface, (c) twist at interface. In Martensite transformation, the situation is as (a).

Ⅱ. Fixed Orientation Relationship and Fixed Habit Plane

Since uniform shears only slightly shift the atoms, generally resulting in coherent phase transformation, thus the old and the new phases keep fixed orientation, and have habit plane, i.e. the plane that is not twisted, and the size, shape and phase direction are kept unchanged.[31] For a same components system, when component concentration is different, the orientation relationship can be different. For example, in carbon steel, the orientation relationship in low and high carbon containing steels is different:

$$(111)_\gamma // (101)_\alpha, [1\bar{1}0]_\gamma // (11\bar{1})_\alpha (C < 1.4\%)$$

$$(111)_\gamma // (101)_\alpha, [1\bar{2}1]_\gamma // (10\bar{1})_\alpha (C > 1.4\%)$$

γ and α refer to mother phase and Martensite phase, respectively.

Ⅲ. Diffusionless

Martensite phase transformation generally cannot be completely finished, parent phase (Austenite) generally always exists.[32] The regions of the Austenite which have transformed to Martensite are lenticular in shape and may easily be recognized by etching, or from the distortion they produce on the polished surface of the alloy. These relief effects indicate that the Martensite needles have been formed not with the aid of atomic diffusion but by a shear process, since if atomic mobility were allowed the large strain energy associated with the transformed volume would then be largely avoided. The lenticular shape of a Martensite needle is a direct

31 由于均匀的剪切力仅仅会产生微小的原子移位，通常导致共格的相转变。因此，新旧相保持固定的取向，并且存在惯习面，也就是说平面是不扭曲的，并且大小，形状和相方向都保持不变。

32 马氏体相变一般不能发生完全，母相（奥氏体）一般总会存在。

consequence of the stresses produced in the surrounding matrix by the shear mechanism of the transformation and is exactly analogous to the similar effect found in mechanical twinning. The strain energy associated with Martensite is tolerated because the growth of such sheared regions does not depend on diffusion, and since the regions are coherent with the matrix they are able to spread at great speed through the crystal. The large free energy change associated with the rapid formation of the new phase outweighs the strain energy, so that there is a net lowering of free energy.[33] The fast transition rate also indicates its diffusionless transformation feature. For instance, in Fe-Ni-C alloy, between temperature of −20 to −195 ℃, Martensitic transition can be finished within 0.05−0.5 ms. Furthermore, the low temperature transformation feature also proves its diffusionless nature.

Ⅳ. Many Crystal Defects

Shear deformation is usually accompanied with slide and twinning, hence Martensite transformation creates a high density of crystal defects, such as dislocations, twins and staggered layers.[34]

Ⅴ. Reversible

Some of the transformation is reversible. For example, cooling the Austenite steel, Martensite transformation occurs; Martensite will change back to Austenite during heating.

4.6.3 Dynamic Process

One of the most distinctive features of the Martensite transformation is that in most systems Martensite is formed only when the specimen is cooling, and that the rate of Martensite formation is negligible if cooling is stopped.[35] For this reason, the reaction is often referred to as an athermal transformation, and the percentage of Austenite transformed to Martensite is indicated on the TTT curve by a series of horizontal lines. The transformation begins at a temperature M_s, which is not dependent on cooling rate, but is dependent on prior thermal and mechanical history, and on composition.

Typically, the Martensite transformation can be classified into three types with features shown in Figure 4.23.

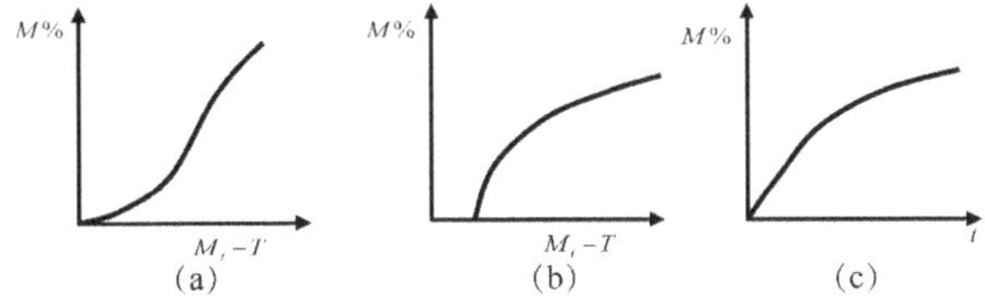

Figure 4.23 (a) Type 1: temperature dependent transformation, (b) Type 2: outbreak transformation, and (c) Type 3: isothermal transformation.

33 马氏体相变过程中的应变能是可以承受的，因为这样的剪切区域不依赖于扩散，而且由于这一区域的马氏体相与母相是共格的，因此它们能够在晶体中迅速扩大。新相的形成很快，其自由能变化大于应变能，导致该过程总的自由能降低。

34 剪切形变通常伴随着滑动和双晶生成，因此，马氏体相变产生高密度的晶体缺陷，如位错，孪晶和交错层。

35 马氏体转变的最明显的特征之一是，在大多数系统中，马氏体只能在样品冷却时产生，如果停止冷却马氏体形成的速率可以忽略不计。

In the type 1 and type 2 transformations, the content of Martensite increases with decreasing of temperature; For type 3 transformation, the content of Martensite will increase with the time.

Depending on the nucleation and growth, the Martensite transformation can be classified into the following four types:

(1) Temperature-varied nucleation and instant isothermal growth (e.g. carbon steel, Figure 4.24)

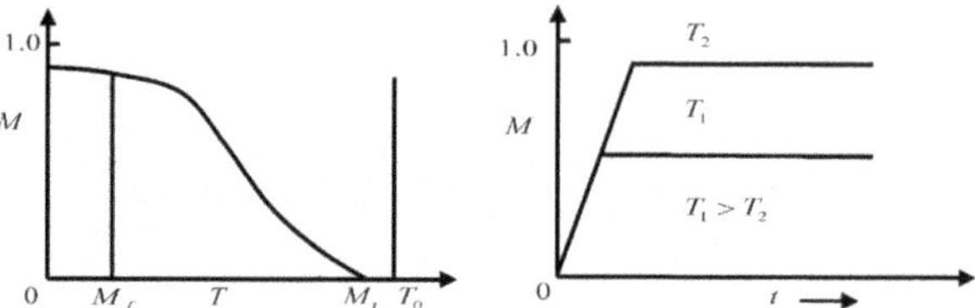

Figure 4.24 Martensite transformation with temperature-varied nucleation and instant isothermal growth.

Below the temperature of M_s, Martensite can be formed immediately. A fix temperature results in a fixed amount of Martensite, and lower temperature yields more Martensite and the size of nucleation is small. In this case, after the instant formation of nucleation, the grow rate can be as high as 2000 m/s and the completion time in the order of $(0.5\sim5)\times10^{-7}$ s. In this case, coherence with the parent phase is damaged.

(2) Outbreak Nucleation, Temperature-varied Growth (e.g. Fe-Ni-C alloy, chromium steel and manganese steel, Figure 4.25)

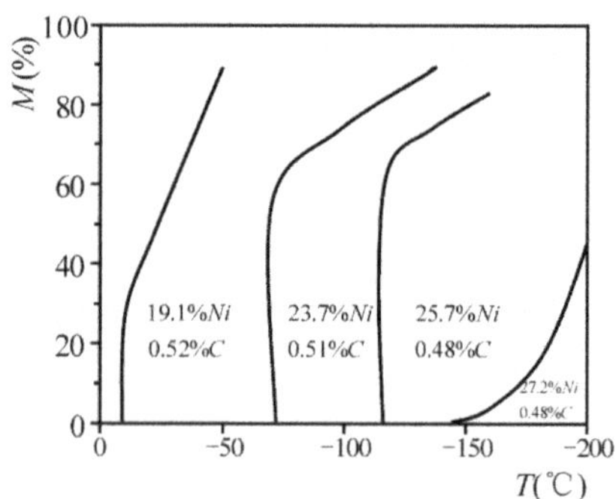

Figure 4.25 Martensite transition curve of Fe-C-Ni alloy.

In some Fe-C-Ni alloy with $M_s < 0$, under M_s, a large scale of nucleation occurs simultaneously, then growing by cooling. This can happen within 10^{-3} to 10^{-4} s. During the outbreak, the sound of formation can be heard.

(3) Temperature-varied Nucleation and Growth (e.g. Au-Cd alloy, Cu-Al alloy)

In this case, below M_s, nucleation occurs, and then growing instantly.

Meanwhile, the coherent relationship between the Martensite and the parent phase does not damage. The Martensite phase can be elongated, thickened, and new nucleation can be formed with reducing temperature.[36]

36 通过降温，马氏体相可以加长、加厚、以及形成新核。

(4) Isothermal Martensite

The Martensite transformations in Fe-Ni-Mn, Cr-steel, MnCu-steel, and high speed steel show isothermal dynamic feature. Their TTT curves also belong to C curve (see Figure 4.26). Although the part of Martensite increases with time, due to the incompleteness feature, the time for 100% Martensite can't be reached.

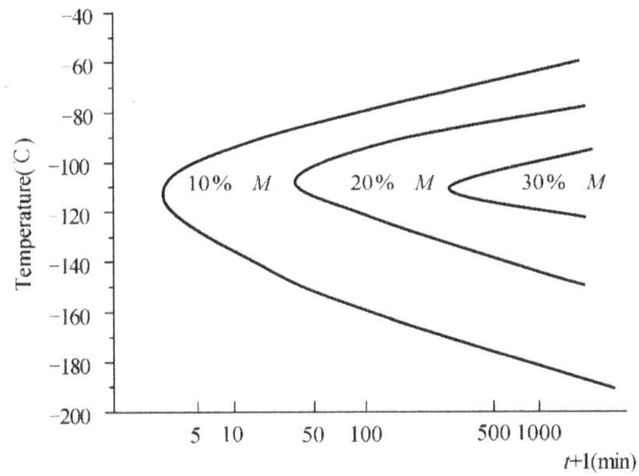

Figure 4.26 The C curve for the Martensitic transformation of Fe-23.2%Ni-3.62% Mn alloy.

The rate of isothermal Martensite transformation is not a constant, as shown in Figure 4.27. It is generally thought that the formation of isothermal Martensite has some degree of self-catalyst: one piece of Martensite can causes several nucleation of Martensite. The rate of nucleation increases with time. After reaching to some point that the Austenite parent phase reduces some amounts and separated by the new Martensite phase, the nucleation rate of Martensite decrease. In the meantime, below the temperature of M_s, new nucleation occurs with time and the percentage of Martensite also increases.

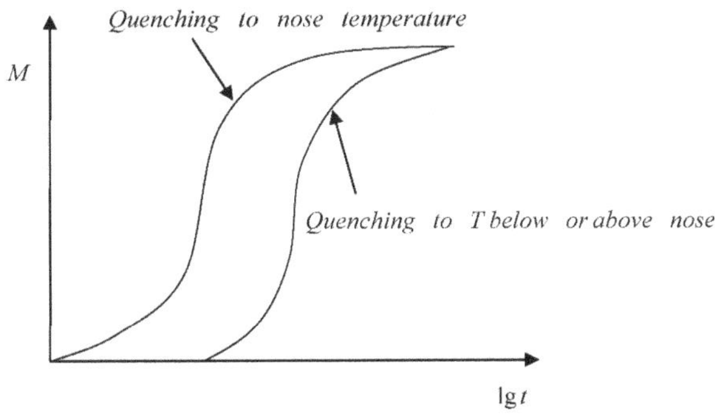

Figure 4.27 Isothermal Martensitic transformation vs time.

4.6.4 Shapes of Martensite

37 依赖于合金体系，马氏体的形状是不同的。

Depended on alloy system, the shape of Martensite is different.[37] The most common two types are lath Martensite and plate Martensite.

(1) Lath Martensite

As shown in Figure 4.28a, in low carbon steel, the lath Martensite is a typical form, with habit plane of $(111)_\gamma$ or $(225)_\gamma$. In this case, the Austenite grain usually changes to several lath Martensite groups. Each group may include laths with width of 0.15 to 0.20 μm. Within the lath Martensite, there are substructures (Figure 4.28b), which generally originated from high density atom dislocation in the lath structure. The density of this substructure is usually as high as $0.3\sim0.9\times10^{12}/\text{cm}^2$.

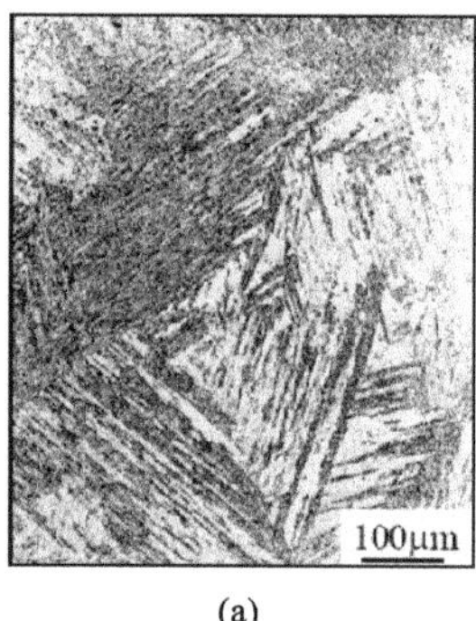

(a)

(b)

Figure 4.28 (a) Lath Martensite in a carbon steel with 0.03%C-2%Mn. (b) Substructure in lath Martensite.

(2) Plate Martensite

38 在板状马氏体生长过程中，第一片通常穿透奥氏体晶粒，然后其他马氏体的增长空间将被第一片马氏体限制，从而形成板状结构。板状马氏体的亚结构为孪晶。

In high carbon steel, plate Martensite usually exists, with habit plane of $(225)_\gamma$ or $(259)_\gamma$. In the growth of plate Martensite, the first piece usually penetrates through the Austine grain, then the growth of other Martensite will be spatially limited by the first one, forming plate structure. The substructure of plate Martensite is twin crystalline.[38] Therefore, plate Martensite is also called twin Martensite (Figure 4.29).

(a)

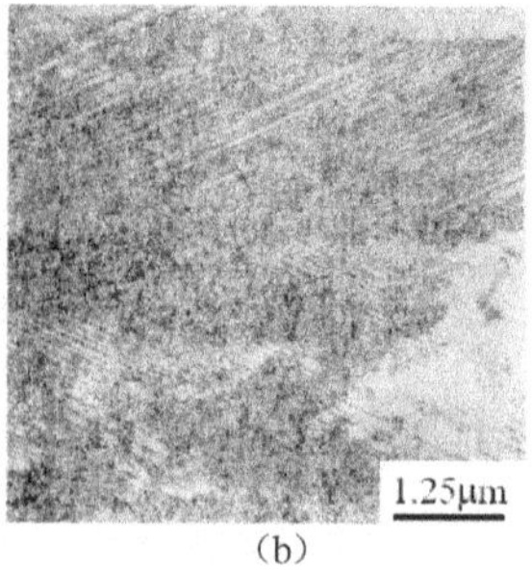

(b)

Figure 4.29 (a) A carbon steel with 1.2%C, exhibiting plate Martensite. (b) Substructure of Plate Martensite: Twin

4.6.5 The Mechanism of Martensite Formation

The Martensite transformation is diffusionless, and therefore Martensite forms without any interchange with neighboring atoms. Accordingly, the observed orientation relationships are a direct consequence of the atom movements during the transformation. The first suggestion of a possible transformation mechanism was made by Bain in 1934. He suggested that, since Austenite may be regarded as a body-centered tetragonal structure of axial ratio $\sqrt{2}$, the transformation merely involves a compression of the *c*-axis of the Austenite unit cell and expansion of the *α*-axis.[39] The interstitially dissolved carbon atoms prevent the axial ratio from going completely to unity and, depending on composition, the *c/a* ratio will be between 1.08 and 1.0. Clearly, such a mechanism can only give rise to three Martensite orientations whereas, in practice, 24 result. To account for this, Kurdjumov and Sachs proposed that the transformation takes place not by one shear process but by a sequence of two shears, first along the elements $(111)_\gamma<112>_\gamma$, and then a minor shear along the elements $(112)_\alpha<111>_\alpha$; these elements are the twinning elements of the FCC and BCC lattice respectively. This mechanism predicts the correct orientation relations, but not the correct habit characteristics or relief effects. Accordingly, Greninger and Troiano in 1941 proposed a different two-stage transformation, consisting of an initial shear on the irrational habit plane which produces the relief effects, together with a second shear along the twinning elements of the Martensite lattice. If slight adjustments in spacing are then allowed, the mechanism can account for the relief effects, habit plane, the orientation relationship and the change of structure.

39 他提出：由于奥氏体可以被认为是一种轴比为 $\sqrt{2}$ 的体心四方结构，该转变仅仅涉及奥氏体 *c*-轴的压缩和 *a*-轴的扩展。

Ⅰ. Driving Force for Martensite Transition

The main driving force usually comes from the difference of free energies of two states in a transformation. As shown in Figure 4.30, the Austine and Martinsite phases have the same free energy at temperature T_0. When $T < T_0$, $\Delta G_{A\to M} = G_M - G_A < 0$, Martensite phase is preferred; and when $T > T_0$, $\Delta G_{M\to A} = G_A - G_M < 0$, Austine phase is preferred. Usually, M_s and A_s represent the starting temperatures for Martensite and Austine, respectively; and M_f, the highest temperature that can form Mantensite; A_f, the lowest temperature that can form Austenite.

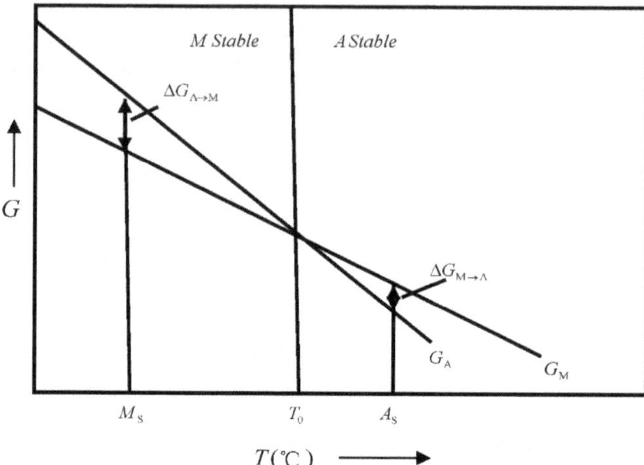

Figure 4.30 Free energy – temperature curve of Austine and Martensite.

Ⅱ. Resistance for Martensite Transformation

Factors that prevent Martensite transformation from Austenite parent phase include interfacial energy and volume aberration energy, etc. In a solid, the interfacial energy, which is caused by the lattice-structure difference between two phases, always hampers the transformation from an old to a new phase.[40] The volume aberration energy includes the shear energy needed for changing the crystal structure, the energy needed for the elastic deformation in adjacent to the Martensite, and the energy stored inside the Martensite, including the dislocation, the twin etc. Volume aberration energy can be very large. Because the resistance for Martensitic transformation is large, a big driving force such as high degree of cooling is generally needed. In this case, M_s is generally required to be low. For instance, in eutectoid carbon steel, the equilibrium temperature for stable eutectoid carbon steel is 727 ℃ (A_d). However, the M_s is 230 ℃, far lower than A_d.

40 在固体中，两个相的晶格的结构差异引起的界面能，总是阻碍从旧相到新相的转变。

Ⅲ. Deformation Induced Martensite Transformation

Above M_s, Martensite can also be produced when elastic deformation occurs in Austenite, leading to the rise of the starting temperature for Matensite transition (M_f). When T is higher than M_f, elastic deformation cannot induce Martensitic transition. The reason for increasing the Martensitic transformation temperature by elastic deformation as follow: elastic deformation can offer additional mechanical driving force for Martensitic transition. As shown in Figure 4.31, at M_s temperature, the chemical driving force is just enough to render the Martensitic transition. At $M_s < T_1 < T_0$, the chemical driving force is mn, not large enough to drive the reaction. If mechanical driving force pm is provided, reaction can take place.

If suitable deformation is realized, it is expected that all the driving force can be offered by this way, and the Martensitic transition temperature can be improved to T_0. However, this is not reached yet.

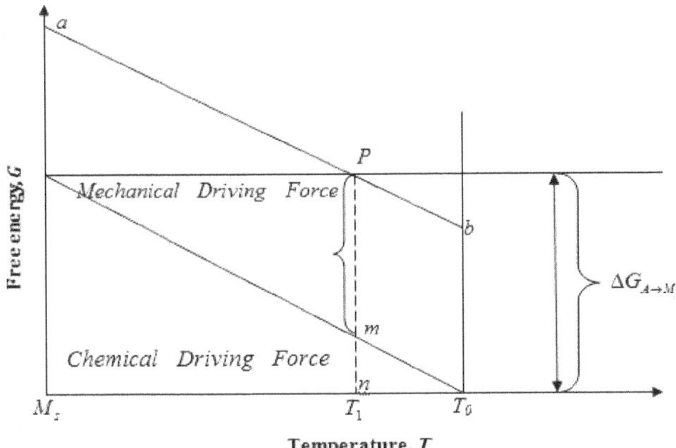

Figure 4.31 Deformation induces the thermodynamic condition of Martensitic transition.

4.6.6 Thermoelastic Martensite and Shape Memory Effect

Thermoelastic Martensite refers to the Martensite that always keeps coherent relationship with parent phase, and the amount of it can reduce with high temperature and increase with low temperature[41] (Figure 4.32). Non thermoelastic Martensite refers to the Martensite that grows to the maximum size instantly, leading to the plastic deformation of the parent phase thus damaging the coherent relationship with its parent phase, therefore it is irreversible (Figure 4.33).

41 热弹性马氏体是指始终与母相保持共格的马氏体，并且其数量在高温时减少、在低温时增加。

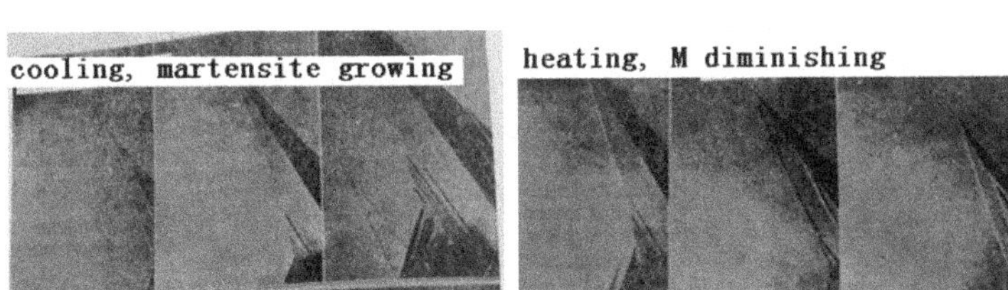

Figure 4.32 Thermoelastic martensite.

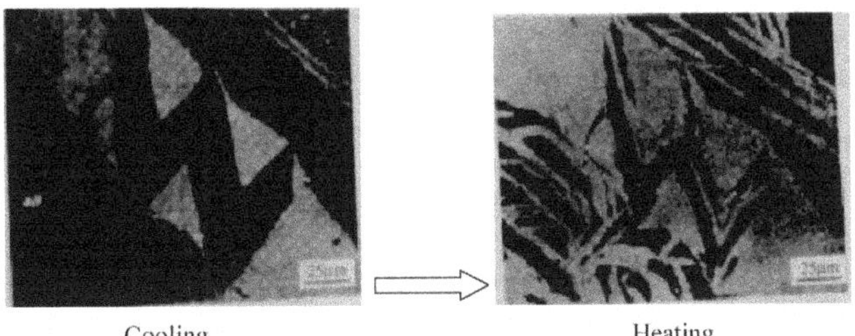

Figure 4.33 Non-thermoelastic Martensite.

One of major difference between the nonthermoelastic and thermoelastic Martensite phases is the resistance - temperature curve during the forward and backward temperature scanning. In the former case, there is a large hysteresis loop; while the later shows small hysteresis loop as shown in Figure 4.34.

The large hysteresis in the phase transformation of nonthermoelastic Martinestie in the Fe-Ni system arises from a strong resistance against the nucleation and/or propagation of the Martensite phase in the parent phase, Austenite, in cooling, and vice versa, mainly due to the damage of coherent relationship in the formation of nonelastic Martinesite in which relatively large energy is required.

Alloys with thermoelastic Martensite transformation exhibit shape memory effect. For example, most of the β-phase alloys, which have the BCC or the CsCl type structure, show small hysteresis in Martensitic transformation and the shape memory effect.

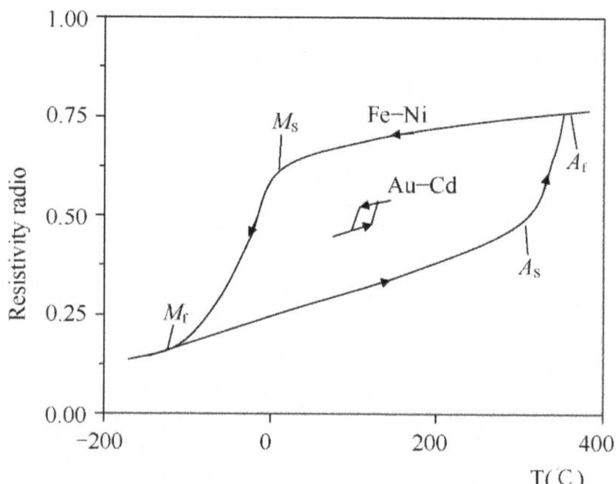

M_s: Martensite starts from austenite under cooling.
M_f: Martensite finishes.
A_s: Austnite conversely starts under heating
A_f: autenite finishes.

Figure 4.34 The electrical resistivity versus temperature curves in Fe-Ni and Au-Cd alloys.

Figure 4.35 shows the mechanism of shape memory microscopically. In the parent phase of Austenite (Figure 4.35a), when temperature decrease to M_s, thermoelastic Martinesite starts to form, and stop to appear when reaching temperature of M_f. Although no macro-changes can be observed, there are microstructure changes inside: from Austine to Martinesite, with shuffled crystal structure and twin boundary (Figure 4.35b). Exerting force to it, the twin boundary motion can easily be done by reshuffling the Martensite structure, yielding deformation. If increasing temperature to A_s, reverse process occurs: Martinesite changes to Austenite structure with the disappearance of the deformation, and leading to the recovery of the original shape.

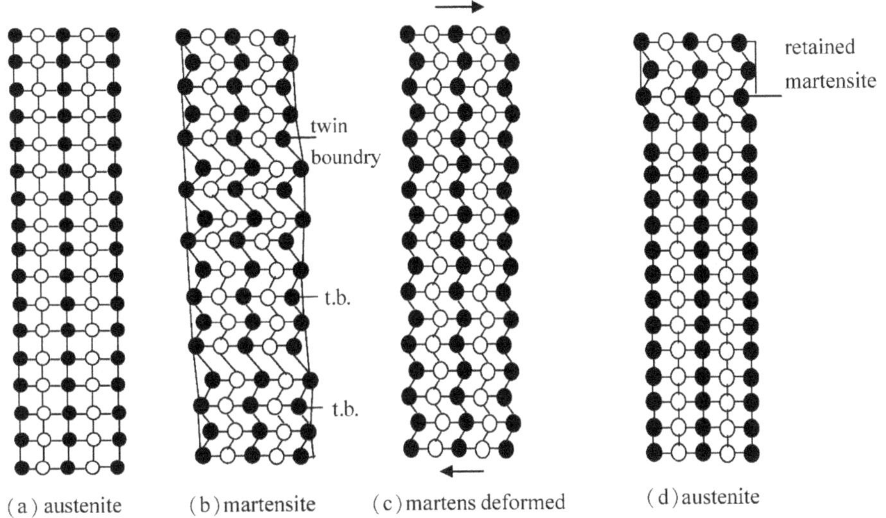

Figure 4.35 A microscopic picture of shape memory.

It is now clear that, in order to realize the shape memory effect, the temperature range of transformation, i.e., between A_f, A_s, M_s, M_f shall be small, twinning must be easier than the motion of dislocation in mechanical deformation, the motion of twin boundary to switch from one variant to another must be easy, and the reverse transformation must follow the unique reversible path. Technologically, there are other requirements to develop the useful shape memory alloys: For instance, they must be perfectly shape restoring, fatigue resistant, capable of large mechanical deformation, desirable transformation and working temperature, etc. Sometimes it is further required to have the two-way memory paths, which means that under the cyclic temperature change the shape must change in cycle between the high temperature form and the low temperature form.[42]

42 有时它还需要有双向的存储路径，这意味着在高温形式和低温形式之间，当温度循环变化时，形状也必须循环地改变。

4.7 Glass Transition

The glass transition for short is the reversible transition in amorphous materials (or in amorphous regions within semicrystalline materials) from a hard and relatively brittle 'glassy' state into a molten or rubber-like state, as the temperature is increased.[43] An amorphous solid that exhibits a glass transition is called a glass (amorphous state). The reverse transition, achieved by supercooling a viscous liquid into the

43 简而言之，玻璃转换过程是指，随着温度的升高，非晶材料（或在半结晶材料中的非晶区域）从较硬且较脆的“玻璃”状态转变为熔融或类似于橡胶状态的一种可逆转变。

glass state, is called vitrification.

It is commonly observed that when a liquid is cooled to below the melting point, T_m, at a constant pressure it crystallizes through a process of nucleation and growth. If the cooling rate is fast enough, the liquid will supercool within a temperature range where the processes of atomic redistribution that characterize crystallization will take place ever more slowly.[44] In the end, below the glass transition temperature T_g, the macroscopic effects of atomic redistributions can no longer be seen and the liquid looks frozen, just like a disordered solid off thermodynamic equilibrium, namely a glass. Conversely, if the transformation path is reversed, the glass is progressively heated until, at T_g, it melts, just as happens with a crystal when it is brought up to the T_m temperature.

44 如果冷却速度足够快的话，液体会在一定温度范围内会产生过冷现象。在此，用以表征结晶的原子重新分布会发生得更慢。

Many physical properties remain almost constant, or show slight changes when a crystal liquefies. For example, the distance between first neighbor atoms hardly changes, the thermal expansion coefficient increases, but still has around the same value as that of the crystal, the heat capacity changes slightly and the volume increases, typically between 2% and 5% for most materials.[45] What immediately makes a solid stand out from a liquid is, from a macroscopic viewpoint, its ability or not to resist shear stresses, namely the fluidity.[46] Thus we can easily find T_g within the narrow temperature range where the measured shear viscosity η reaches the typical value of 10^{12} Pas. This viscosity rises rapidly as the system approaches the transition from the high temperature side.

45 例如，第一个相邻原子之间的距离几乎没有变化；热膨胀系数增加，但仍会与晶体的热膨胀系数相仿；热容量略有变化且体积增加，通常大多数材料在 2%～5%。

46 从宏观上看，能够立刻从液体区分出固体的性质是它的流动性，即对抗剪切应力的能力。

The changes in η as the temperature varies are schematically shown in Figure 4.36 for a metallic system or an ionic compound. The parallel branches (1) and (2) correspond to different cooling rates, where (2) is less than (1), and clearly demonstrate that hysteresis occurs. Typically, $\Delta T_g/T_g$ can vary about 10% of T_g when we vary the cooling rate by several orders of magnitude. Conventionally, T_g is defined as the temperature where, during heating tests carried out at a rate of 0.167 $K \cdot s^{-1}$.[47]

47 传统上，T_g 被定义为当加热速率为 0.167 $K \cdot s^{-1}$ 时的(转变）温度。

The fall in T_g as cooling rate falls confirms that a structural relaxation time, τ, which is representative of microscopic relaxation, depends on temperature. The inverse of τ indicates the rate at which the atomic structure of the system reacts and adapts to the externally change in temperature. We should note that for a liquid-glass transition, within a typical interval of 200 K, τ varies with continuity from 10^{-13} to 10^{17} s with values of around 10^3 s at T_g.

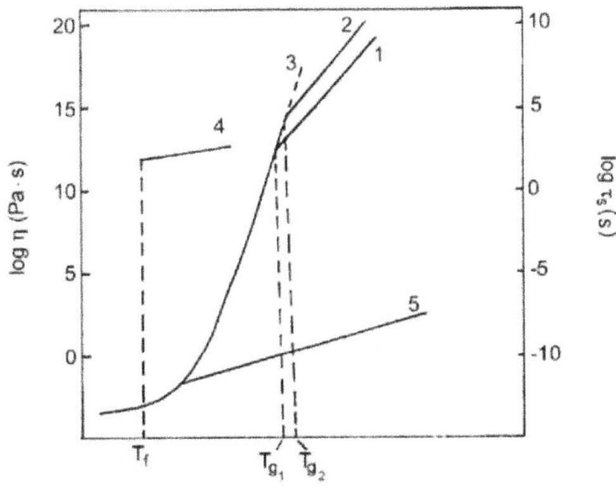

Figure 4.36 Viscosity trend as a function of the temperature in a metallic system in the solid-liquid transition region. Curves 1, 2: glassy system, obtained with various cooling rates (2 < 1); curve 3: ideal glass; curve 4: liquid-crystal transition, T_f is the frozen temperature; curve 5: computer simulated quenching process from the liquid view of glass transition.

The importance of kinetic factors in the glass transition emerges from the fact that branches (1) and (2) tend to converge on the ideal glass curve, as schematically shown in (3), provided we wait long enough, and even if this trend is never completed.

On the other hand, the dependence of T_g on cooling rate, though still not very large, is in any case too marked to indicate that, in the glass transition, kinetic factors alone are the only cause of the change in T_g as the cooling rate changes.[48]

When the temperature is below T_g the configurations of the system do not change appreciably; as such the two states (1) and (2) are defined as isoconfigurational. The schematic curves (4) and (5) represent respectively crystallisation and a computer simulation of a rapid quenching process. The viscosity trend shown in the figure is observed in glassy systems where the bonds are non-directional (metals, ionic solids or van der Waals solids) and cannot maintain a given medium range order as the temperature varies.[49]

48 另一方面，T_g对冷却速率的依赖虽然不是非常大，但在任何情况下，这种依赖关系都非常显著地表明：在玻璃化转变过程中动力学因素是唯一引起 T_g随着冷却速率改变的原因。

49 图中展现的黏滞力变化趋势可在成键无方向性的玻璃态体系（金属、离子固体或范东华力固体）中观察到。这些系统在温度变化时不能保持中程有序。

本章小结

1. 内容概要

本章将会讨论固态变的相关知识。首先，我们将会解释一些关于固态相变的术语，例如，相界面、错配度、界面能、位向关系、惯习面以及应变能等，同时，讨论了与这些术语相关的固体相变机制；然后，基于热力学和扩散对固体相变进行了分类讨论；最后，介绍了几种典型的固态相变，包括多晶型转变、共析转变、马氏体转变以及玻璃化转变，尤其是，固体相变对材料微观结构和性质的影响。

2. 基本概念

相、固态相变、相界面、错配度、界面能、位向关系、惯习面、应变能、同素异构转变、多晶型性转变、共析转变、马氏体转变、玻璃化转变、自由能、自由度

3. 主要公式

（1）固态相变的自由能：$\Delta G = \frac{4}{3}\pi r^3 \Delta G_{v(\alpha\to\beta)} + 4r^3\sigma_{\alpha\beta} + \frac{4}{3}\pi r^3 \varepsilon$

（2）体积 V 和熵 S：$\left(\frac{\partial\mu}{\partial P}\right)_T = V$，$\left(\frac{\partial\mu}{\partial T}\right)_P = -S$

（3）压缩系数 K：$K = \frac{1}{V}\left(\frac{\partial V}{\partial P}\right)_T$

（4）热膨胀系数 α：$\frac{\partial^2\mu}{\partial T\cdot\partial P} = \left(\frac{\partial V}{\partial T}\right)_P = V\cdot\alpha, \alpha = \frac{1}{V}\left(\frac{\partial V}{\partial T}\right)_P$

（5）相律：$f = C - Ph + 2$

Vocabulary

aberration	畸变
acicular	针状的
allotropic transformation	同素异构转变
athermal transformation	非热相变
Austenite	奥氏体
Bainite	贝氏体
cementite	渗碳体
ceramic	陶瓷
coherent interface	共格界面
cold working	冷形变，冷加工
compression coefficient	压缩系数
crystal plane	晶面

diffusion	扩散
diffusionless	非扩散
displacive transformation	推移性相变
ductile	柔软的，易延展的
emboss	浮雕，凸起
enthalpy	焓
entropy	熵
eutectic	共晶
eutectoid	共析
fcc (face centred cubic)	面心立方
ferrite	铁素体
first order phase transformation	一级相变
first-order partial differential equation	一阶偏微分方程
fusion	熔化
habit plane	惯习面
hcp（hexagonal close-packed）	六方最紧密堆积
heat capacity	热容
incoherent interface	非共格界面
interfacial energy	界面能
lath	板条
longitudinal	纵向的
magnetic transformation	磁性转变
Martensite	马氏体
misfit of phases	晶相错配度
niobium	铌
nucleation	成核
nucleus, nuclei	原子核
ordering	有序化转变
orientation relationship	位相关系
peritectoid	包析
pearlite	珠光体
perlite	珍珠岩
phase diagram	相图
phase interface	相界面
phase transformation	相变
polycrystalline transformation	多晶形性转变
precipitation	脱溶/析出
primary phase	领先相，初晶相

proeutectoid phase	先共析相
semicoherent interface	半共格界面
spinodal decomposition/transformation	调幅分解
strain energy	应变能
strain hardening	应变硬化
superconducting transformation	超导转变
supercooling	过冷
thermal expansion coefficient	热膨胀系数
transverse	横断面
triple point	三相点
undercooling	过冷
volume	体积
Widmanstatten	魏氏体

Problems

1. What is phase transformation and solid state phase transformation?

2. Using the following figure, determine the degrees of freedom in a Cu-40% Ni alloy at temperature of (a) above 1280℃, (b) between 1280 to 1240℃, and (c) below 1240 ℃ (assuming constant pressure)

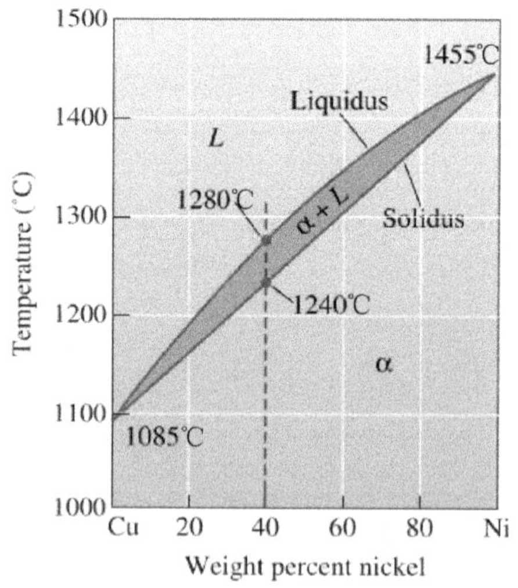

3. Determine the composition of each phase in a Cu-40% Ni alloy at 1300℃, 1270℃, 1250℃, and 1200℃ (see figure)

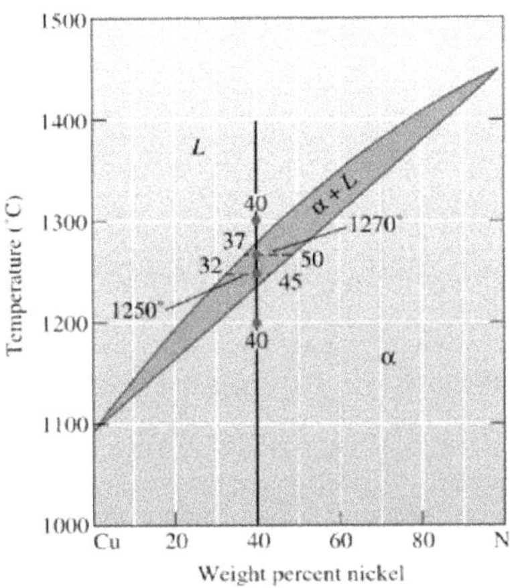

4. Based on the following figure, please state that among disc, ball and needle shapes for new phase in phase transformation process, which type of new phase presents the maximum strain energy?

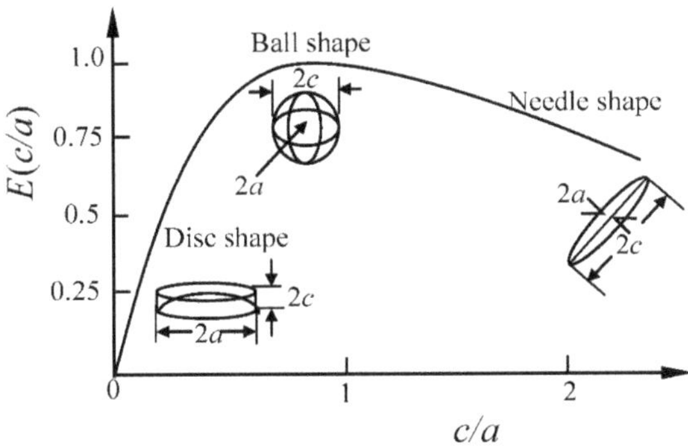

5. Besides the free energy changes, please state the other two factors that account for the driving force of solid state phase transformation. Then please explain why: 1) in coherent phase transformation, the crystal usually adopts disc or needle structure, and 2) in non-coherent phase transformation, ball structure is generally preferred.

6. Please pick out the true statements among the following items:

(1) New phase usually makes nucleation at defect points.

(2) Solid state phase transformation is easier to identify.

(3) Precipitation is a second order solid phase transformation.

(4) Phase diagram reflects phases under thermodynamic equilibrium conditions.

(5) In solid state phase transformation, the new phase generally grows along habit plane.

(6) Polycrystalline transformation is a first order phase transformation and it is diffusional.

(7) In second order phase transformation, there exist volume and entropy changes.

(8) In eutectoid reaction:

① With carbon content < 0.77% (hypoeutectoid steel), there is proeutectoid α phase in the interface.

② With C > 0.77% (hypereutectoid steel), there is proeutectoid Fe_3C phase in the interface.

③ With C = 0.77% (eutectoid steel), the Fe_3C is generally considered as proeutectoid phase.

(9) Eutectoid transformation is a kind of diffusion type transformation.

7. Please briefly describe the general steps in solid state phase transformation.

8. Considering a polycrystalline phase transformation from α to β phase, and their free energy curves are as following figure. Please answer:

When temperature is greater than T_0, can the above phase transformation occur? Why?

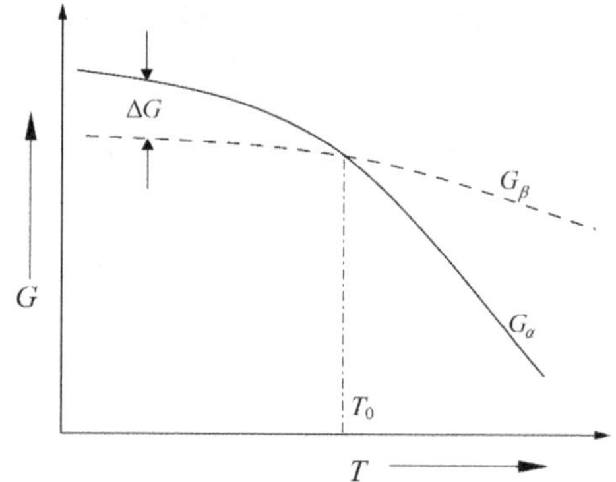

9. Please deduce the freedom of a eutectoid transformation for a system with two components at fixed pressure. And what does this value mean?

10. Please explain the formation procedure of pearlite from a eutectoid steel.

11. According to the free energy diagram of a system with two components (shown in the figure), please schematically draw the phase diagram for this system (Temperature versus component concentration, treat pressure as constant).

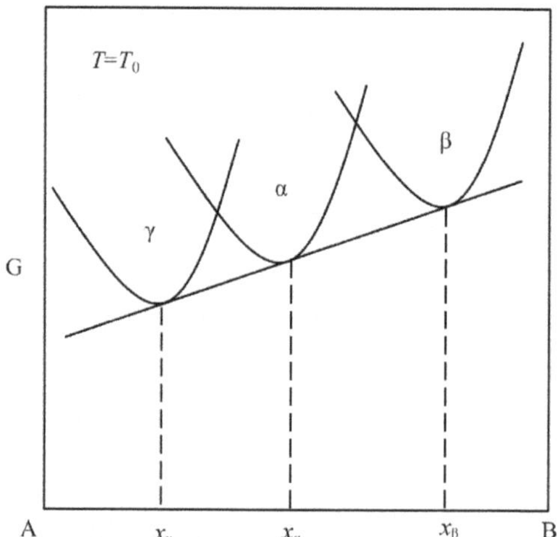

Chapter 5
Mechanical Properties

This chapter deals with the mechanical properties of materials, i.e., the responses of a material upon exerting a force, which include deformation, fracture and fatigue. Firstly, basic terms/processes, such as stress, strain, Young's modulus, yield strength, tensile strength, elongation, hardness, elastic/plastic deformation, and viscoelasticity, are explained and discussed. Then focuses are given to the deformation procedure and the corresponding mechanism. Some testing methods will also be discussed.

5.1 Significance of Material Mechanical Properties

Mechanical property is one of the important characteristics of materials. Different materials possess different mechanical properties and can be used in various ways. In many of emerging technologies, the primary emphasis is on the mechanical properties of the materials. For example, in aircraft industry, aluminum alloys or carbon-fiber-reinforced composites are widely used for aircraft out-shell components because of their light weight, strong, and being able to withstand cyclized mechanical loading for a long and predictable period.[1] Steels used in the construction such as buildings and bridges must have adequate strength so that these architectures can be built without compromising safety. The plastics used for making pipes, valves, flooring, and the like also must have adequate mechanical strength. Materials such as pyrolytic graphite or cobalt chromium tungsten alloys, used for prosthetic heart valves, must not fail. Similarly, the performance of baseballs, cricket bats, tennis rackets, golf clubs, skis, and other sports equipment depends not only on the strength and weight of the materials used, but also on their ability to perform under "impact" loading.

1 例如，在飞机制造工业，铝合金或者碳纤维增强复合材料被广泛应用于飞行器外壳部件，缘于其质量轻巧、强度大、以及可以预期的长时间循环负载。

In many other applications, the mechanical properties of the material also play an important role, even though the primary function is electrical, magnetic, optical, or biological. For example, an optical fiber must have a certain level of strength to withstand the stresses encountered during the installation and daily operation. A biocompatible titanium alloy used for a bone implant must have enough strength and toughness to survive in the human body for many years without failure. A scratch-resistant coating on optical lenses must resist mechanical abrasion. An aluminum alloy or a glass-ceramic substrate used as a base for building magnetic hard drives must have sufficient mechanical strength so that it will not break or crack during operation that requires rotation at very high speeds. Similarly, electronic packages used to hold semiconductor chips and the thin-film structures created on the semiconductor chip must be able to withstand stresses encountered in various applications, as well as those encountered during the heating and cooling of electronic devices.[2] Float glass used in automotive and building applications must have sufficient strength and shatter resistance. Many components designed from plastics, metals, and ceramics must not only have adequate toughness and strength at room

2 类似地，电子器件中的半导体芯片及其薄膜结构在各种应用过程中，必须能够承受一定应力，同时它们必须经受得住加热或冷却过程。

temperature but also at relatively high and low temperatures. For load-bearing applications, engineered materials are selected by matching their mechanical properties to the design specifications and service conditions of the component.[3] The first step in the selection process requires an analysis of the material's application to determine its most important characteristics. Should it be strong, stiff, or ductile? Will it be subjected to an application involving high stress or sudden intense force, high stress at elevated temperature, cyclic stresses, and/or corrosive or abrasive conditions? Once we know the requirements, we can make a preliminary selection of the appropriate material using various databases.[4]

3 在选择用于承受负载的工程材料时，需要将设计规格与材料力学性匹配，并考虑材料实际工作环境。

4 材料是否强韧、坚硬，亦或具有延展性？材料是否会受制于应用环境，例如，在高应力、强冲量、升温时产生的高应力、循环应力、或/及腐蚀磨损的环境？一旦我们知道材料所需的性质，就能初步从各种数据库中选出合适的材料。

5.2 Basic Concepts in Mechanics

Ⅰ. Stress versus strain

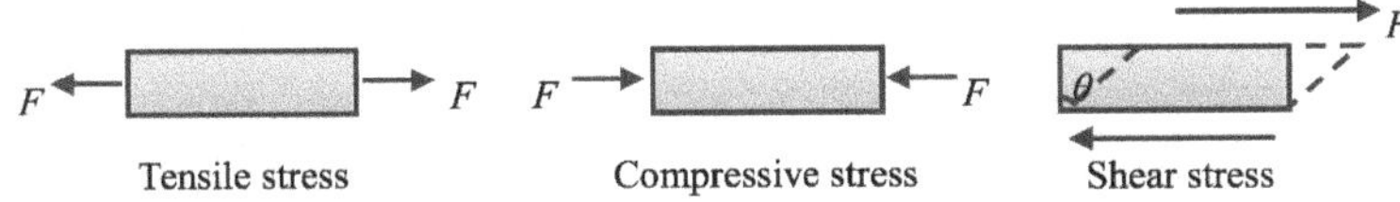

Figure 5.1 Tensile, compressive, and shear stresses.

In general, there are three types of forces or "stresses" that are encountered in dealing with mechanical properties of materials. Tensile, compressive, and shear stresses are illustrated in Figure 5.1. Tensile and compressive stresses are normal stresses. A normal stress arises when the applied force acts perpendicular to the area of interest. Tension causes elongation in the direction of the applied force, whereas compression causes shortening. A shear stress arises when the applied force acts in a direction parallel to the area of interest.

Tensile experiment is the most traditional method to test the mechanical property of material. Figure 5.2 illustrates a general setup. The load necessary to produce a given elongation is monitored as the specimen is pulled in tension at a constant rate. Specimens are machined such that the cross-sectional area in this region is uniform and smaller than at the ends gripped by the testing machine. This smallest area region, referred to as the gage length, experiences the largest stress concentration so that any significant deformation at higher stresses is localized there.[5] A load-versus-elongation curve (Figure 5.2b) is the immediate result of such a test. A more general statement about material characteristics is obtained by normalizing the data of Figure 5.2b for geometry, yielding the concepts of stress and strain, as shown in the Figure.[6]

5 这一最小截面的区域，被称为计量长度，该区域承受着最强的应力，因此在较高应力下发生的任何显著形变都将位于该位置。

6 有关材料性质更通用的描述，可通过将图 5.2b 中的几何形状归一化，得到图中所示的应力和应变的概念。

If F is loaded on the sample with an original (zero-stress) cross-sectional area, A_0, then the stress (σ) (also called engineering stress) is:

$$\sigma = \frac{F}{A_0} \tag{5-1}$$

Strain (ε) is the normalized elongation length. Its expression is as following:

$$\varepsilon = \frac{l - l_0}{l_0} = \frac{\Delta l}{l_0} \tag{5-2}$$

where l is the gage length at a given load, and l_0 is the original (zero-stress) length. Stress is typically expressed in psi (pounds per square inch) or Pa (Pascal). Strain has no dimensions and is often expressed as in/in. or cm/cm.

7 虽然这里我们着重强调了材料在拉伸情况下的行为，但图 5.2 所示的测量设备，可以通过施加相反的力进行常规的压力测试。

Although we are concentrating on the behavior of materials under tensile loads, the testing apparatus illustrated in Figure 5.2 is routinely used in a reversed mode to produce a compressive test.[7] For a compressive force (F), the compressive stress (P) and the produced compressive strain ($\varDelta$) can be expressed as:

$$p = \frac{F}{A_0}$$

$$\varDelta = \frac{V_0 - V_t}{V_0} \tag{5-3}$$

where V_t is volume at a given load (F), and V_0 is the original volume.

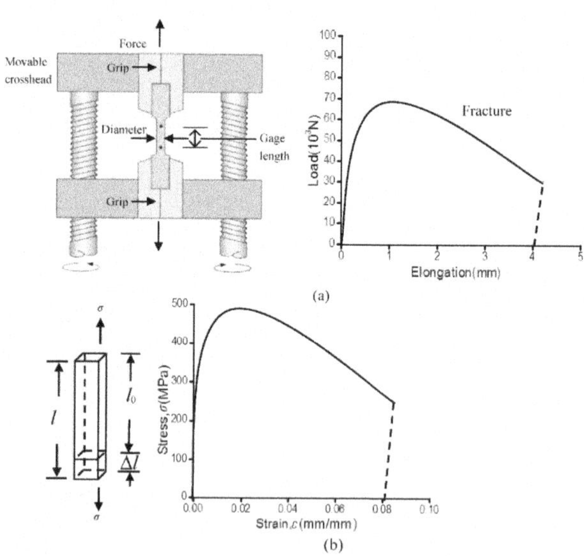

Figure 5.2 (a) Tensile experiment and the immediate results of load-versus-elongation curve (The specimen was aluminum). (b) Stress-versus-strain curve obtained by normalizing the data from a tensile test for specimen geometry.

Similar to above pull/compressive test, a shear stress is defined as (Figure 5.3):

$$\tau = \frac{F_s}{A_s} \tag{5-4}$$

where F_s is the load on the sample and A_s is the area of the sample parallel (rather than perpendicular) to the applied load.

The shear stress (τ), produces an angular displacement (θ) with the shear strain (γ), being defined as:

$$\gamma = \tan\theta \tag{5-5}$$

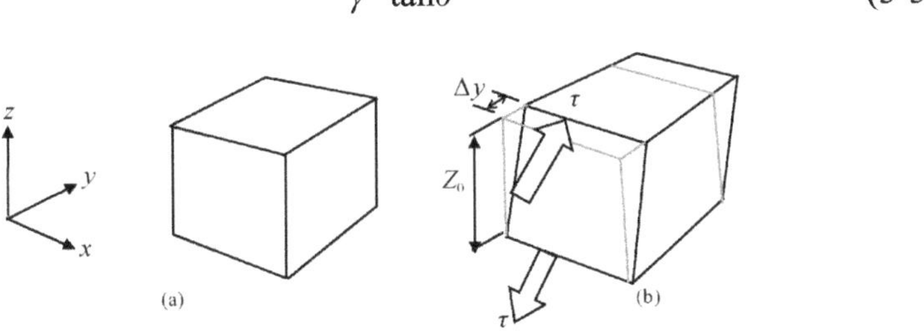

Figure 5.3 Elastic deformation under a shear load. (a) Unloaded and (b) Loaded.

Figure 5.4 shows qualitatively the stress-strain curves for a typical (a) metal, (b) thermoplastic materials (polymers/plastics), (c) elastomer, and (d) ceramic (or glass) under relatively small strain rates.[8] The scales in this figure are qualitative and varied for each material. In practice, the actual magnitude of stresses and strains will be very different. The temperature of the plastic material is assumed to be above its glass-transition temperature (T_g). The temperature of the metal is assumed to be room temperature. Metallic and thermoplastic materials show an initial elastic region followed by a non-linear plastic region. A separate curve for elastomers (e.g., rubber or silicones, Figure 5.4c) is also included since the behavior of these materials is different from other polymeric materials.[9] For elastomers, a large portion of the deformation is elastic and nonlinear. On the other hand, ceramics and glasses show only a linear elastic region and almost no plastic deformation at room temperature.[10] The different mechanical behaviors of the stress-strain response, originated from the material structure and properties, will be discussed in the following sections.[11]

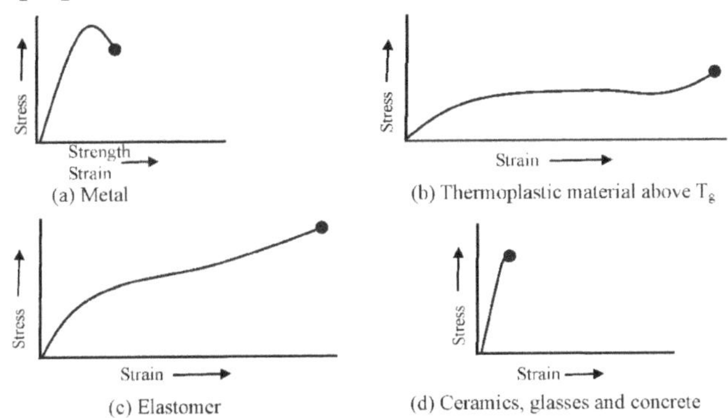

Figure 5.4 Tensile stress–strain curves for different materials. Note that these are qualitative. The magnitudes of the stresses and strains should not be compared.

8 图 5.4 定性展示了（a）金属，（b）热塑材料（聚合物/塑料），（c）弹性体，（d）陶瓷（或者玻璃）在较小应变率下的应力-应变曲线。

9 图中还包括了一个单独的弹性体曲线（如橡皮或者硅胶，见图 5.4c），因为它们与其他聚合物材料的力学性质不同。

10 另一方面，陶瓷和玻璃在室温下仅仅表现出线性弹性区域，几乎没有塑性形变（区域）。

11 缘于不同的材料结构和材料性质的不同应力-应变反应，将在之后的章节中讨论。

Figure 5.5 shows the various response ways of some material types under a force. As shown in Figure 5.5, what we should keep in mind now is that the responses in different type materials can be quite different under the same force conditions.

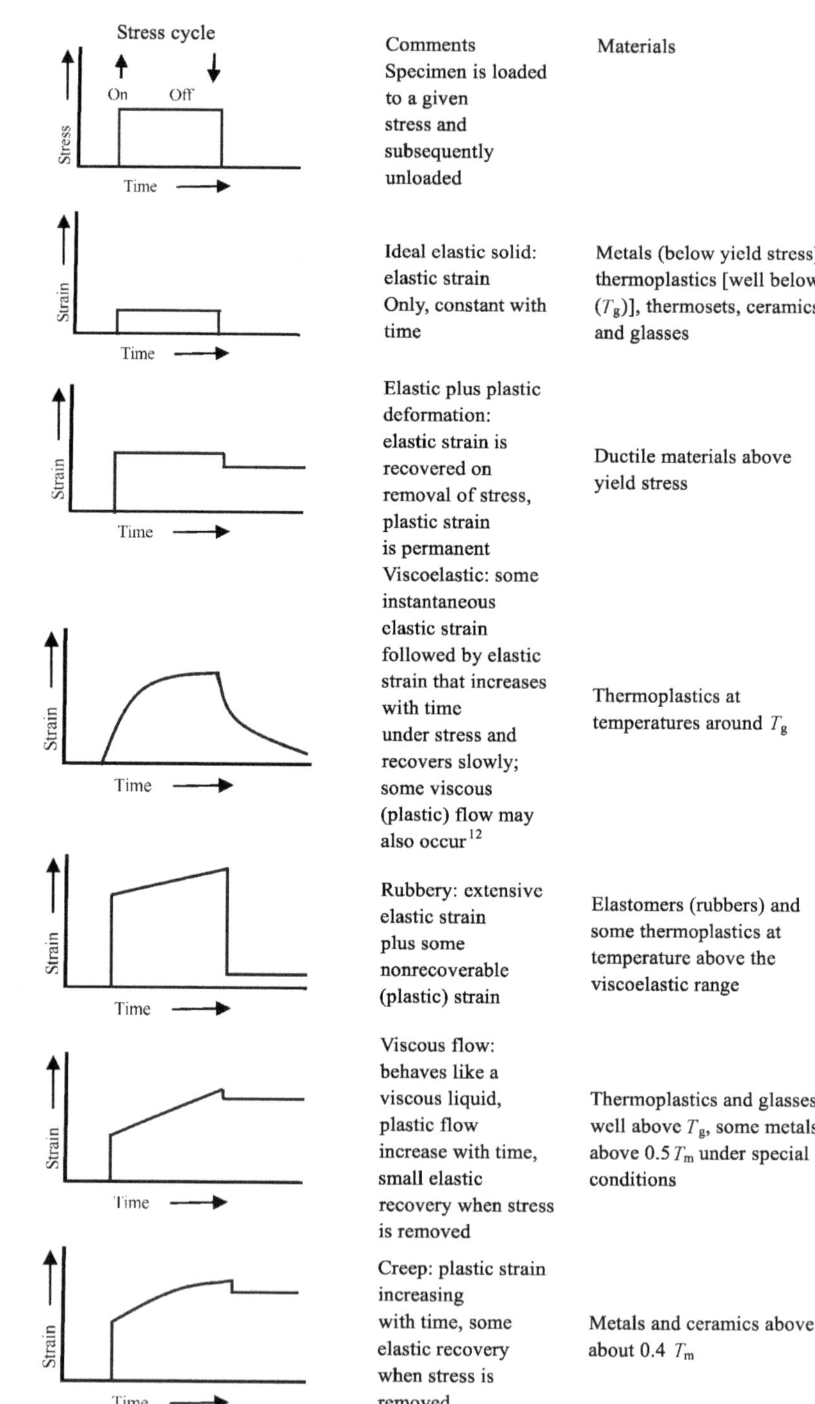

Figure 5.5 Various types of strain response to an imposed stress where T_g is glass transition temperature, and melting point T_m.

12 黏弹性体：黏弹性体在发生一些瞬时弹性应变之后会发生随着时间增加并且恢复缓慢的弹性应变，也会发生一些黏性（塑性）应变。

Ⅱ. Modulus of Elasticity

The slope of the stress-strain curve in the elastic region is defined as the modulus of elasticity, E, also known as Young's modulus. The linearity of the stress-strain plot in the elastic region is a graphical statement of Hooke's law:

$$\sigma = E\varepsilon \tag{5-6}$$

The modulus, E, is a highly practical piece of information. It represents the stiffness of the material (i.e., its resistance to elastic strain), which manifests itself as the amount of deformation in normal use below the yield strength and the springiness of the material during forming.[13]

The shear stress (τ) produces an angular displacement (θ) with the shear strain, γ ($\gamma = \tan\theta$), being defined as

$$\tau = G\gamma \tag{5-7}$$

where $G = \dfrac{\tau}{\gamma}$ is the shear modulus (i.e., the elastic modulus for shear force).

Similarly, for compressive stress (p), in the elastic deformation region, we have the compression modulus (B),

$$B = \frac{p}{(V_0 - V_t)/V_0} \tag{5-8}$$

V_0 and V_t are the initial and the final volumes, respectively.

Ⅲ. Yield Strength

It is often difficult to specify precisely the point at which the stress-strain curve deviates from linearity and enters the plastic region.[14] The usual convention is to define a yield strength. The yield strength (Y.S.) means the intersection of the deformation curve with a straight line parallel to the elastic portion and offset 0.2% on the strain axis (Figure 5.6).[15] The yield strength represents the stress necessary to generate this small amount (0.2%) of permanent deformation. In other words, we can say that when a stress with a value of Y.S. is loaded to a specimen, the produced strain is 0.2% permanent deformation plus the elastic recovery region (Figure 5.6); and upon the removal of loading, there will be a permanent elongation of 0.2%.

Compared with the elastic modulus, E, the yield strength has major practical significance. It shows the resistance of the materials to permanent deformation and indicates the ease with which a metal material can be formed by rolling or drawing operation.[16] Figure 5.7 gives some yield strength of various materials

13 （弹性模量）代表材料的强韧度（即阻碍弹性形变的程度），它显示了材料在低于屈服强度时正常应用中可以到达的形变程度，以及锻造时材料的弹性。

14 在应力-应变曲线上，准确地找出偏离线性区域、进入塑性区域的点很难。

15 屈服强度（Y.S）指的是在在应变轴上数值为 0.2%处做平行于弹性部分的直线，与应变-应力曲线相交点的应力坐标。

16 它（屈服强度）表示了材料对永久形变的抵抗能力，以及金属材料通过滚动或拉伸工序进行加工的难易程度。

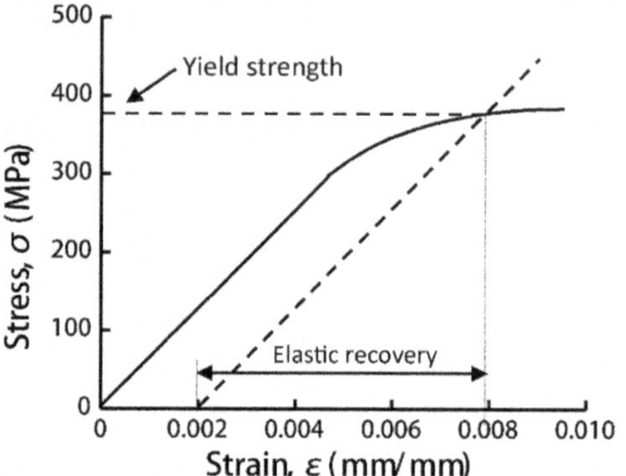

Figure 5.6 Schematic view for yield strength.

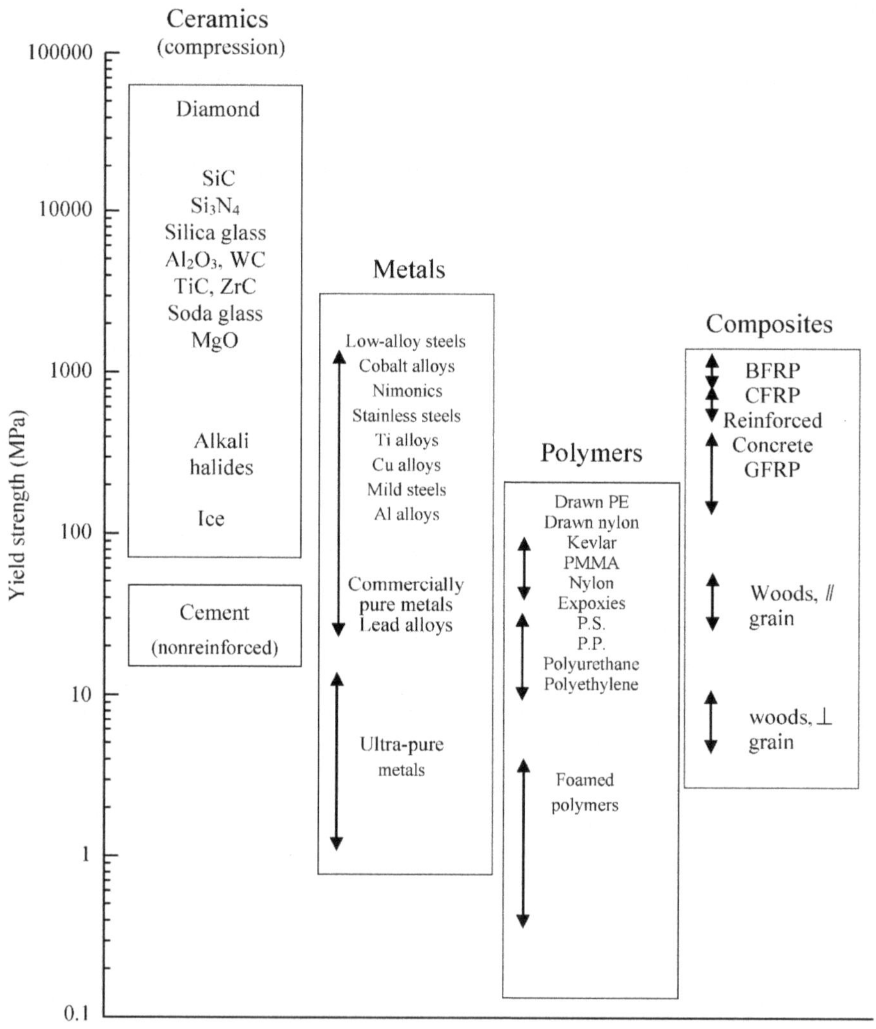

Yield strength (MPa)
100000
10000
1000
100
10
1
0.1

Ceramics (compression)
Diamond
SiC
Si_3N_4
Silica glass
Al_2O_3, WC
TiC, ZrC
Soda glass
MgO
Alkali halides
Ice
Cement (nonreinforced)

Metals
Low-alloy steels
Cobalt alloys
Nimonics
Stainless steels
Ti alloys
Cu alloys
Mild steels
Al alloys
Commercially pure metals
Lead alloys
Ultra-pure metals

Polymers
Drawn PE
Drawn nylon
Kevlar
PMMA
Nylon
Expoxies
P.S.
P.P.
Polyurethane
Polyethylene
Foamed polymers

Composites
BFRP
CFRP
Reinforced Concrete
GFRP
Woods, // grain
woods, ⊥ grain

Figure 5.7 Typical yield strength values for different engineering materials. Note that values shown are in MPa and psi. (Reprinted from Engineering Materials I, Second Edition, M. F. Ashby and D. R. H. Jones, 1996, Fig. 8-12, p. 85. Copyright @ 1996 Butterworth-Heinemann.)

Ⅳ. Tensile Strength (*T.S.*)

As the plastic deformation continues at stresses above the yield strength, the engineering stress continues to rise toward a maximum (Figure 5.8).[17] This maximum stress is termed as the ultimate tensile strength, or simply the tensile strength. It might appear that plastic deformation beyond *T.S.* softens the material because the engineering stress falls.

17 随着塑性形变在应力超过屈服点后的增加，工程应力一直增加直到最大值（图 5.8）。

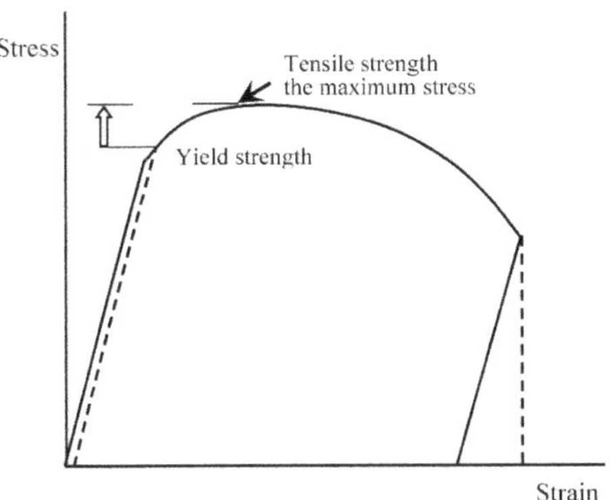

Figure 5.8 Schematic drawing for tensile point with comparison to yield strength.

Ⅴ. Engineering Stress/Strain and True Stress/Strain

Engineering stress and strain are defined relative to original sample dimensions. However, at the ultimate tensile strength, the sample begins to neck down within the gauge length, which leads to a reduction of the sample area.[18] The true stress and true strain are defined as:

18 但是，处于极限拉伸强度时，样品开始在测量长度区域内收缩，导致了样品横截面积的减少。

$$\sigma_T = \frac{F}{A_{actual}} \tag{5-9}$$

$$\varepsilon_T = \int_{l_0}^{l} \frac{dl}{l} = \ln\left(\frac{l}{l_0}\right)$$

where A_{actual} is the area after force exertion. l is the instantaneous sample length, and l_0 is the initial length. Generally, the true stress continues to rise to the point of fracture (Figure 5.9)

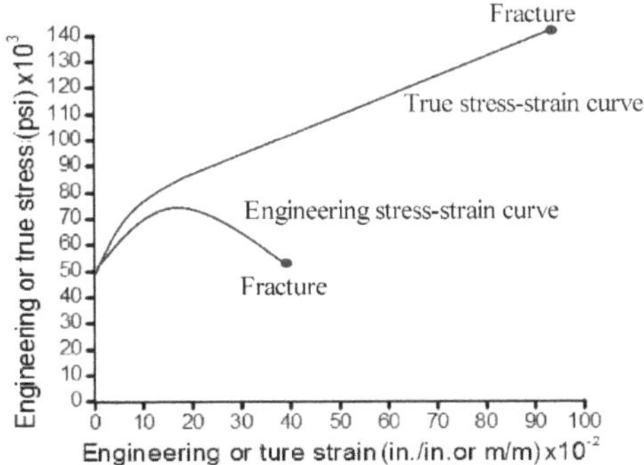

Figure 5.9 Illustration of the differences between true and engineering stress/strain.

Ⅵ. Strain-Hardening

Exerting a stress between the region of *Y.S.* and *T.S* in the stress-strain curve, there will be a permanent deformation after force release (Figure 5.10a); and in the next cycle of force loading, increased stress is needed for producing another plastic deformation (Figure 5.10b), which means the material is strengthened.[19] More cycles of such force treatment yield more strengthened materials (Figure 5.10c). This phenomenon is referred as strain hardening, which is also known as work hardening or cold working[20]. The strain hardening of the metal increases the hardness, yield strength, and tensile strength. Strain hardening is an important factor in shaping metals by cold working.

19 施加这样一个应力，其大小在应力-应变曲线的屈服度和拉伸强度之间，撤销应力后会造成永久形变（图 5.10a）；在下一个循环负载时，必须提高应力才能产生新的塑性形变（图 5.10b），这意味着材料被强化了。

20 多次进行这类力的处理，能强化材料（图 5.10c），这个现象被称为应变强化，也称工程硬化或冷处理。

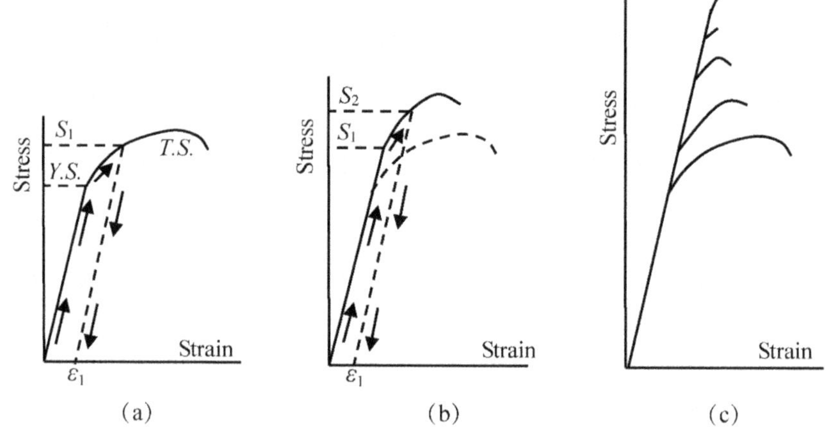

Figure 5.10 (a) The permanent deformation after force release; (b) The increased stress needed for producing another plastic deformation in the next cycle of force loading; (c) More cycles of such force treatment yield more strengthened materials. Notes: The curves are engineering strain-stress curves.

For many metals and alloys, the region between the onset of plastic deformation (corresponding to the *Y.S.*) and the onset of necking (corresponding to the *T.S*) in the true stress (σ_T) versus true strain curve (see Figure 5.11) can be approximated by:[21]

$$\sigma_T = K\varepsilon_T^n \tag{5-10}$$

The *n* of the above equation is termed as strain-hardening exponent. For low-carbon steels used to form complex shapes, the value of *n* will normally be approximately 0.22. Higher values, up to 0.26, indicate an improved ability to be deformed during the shaping process without excess thinning or fracture of the piece.[22]

21 对于许多金属和合金，真实应力-真实应变曲线上，在塑性形变起始点（对应于 *Y.S.*和开始变细点（对应于 *T.S.*）(图 5.11)之间可通过以下公式估算：

22 较高的该值，上限到 0.26，说明材料提高的加工能力，不会过多地变薄或者断裂。

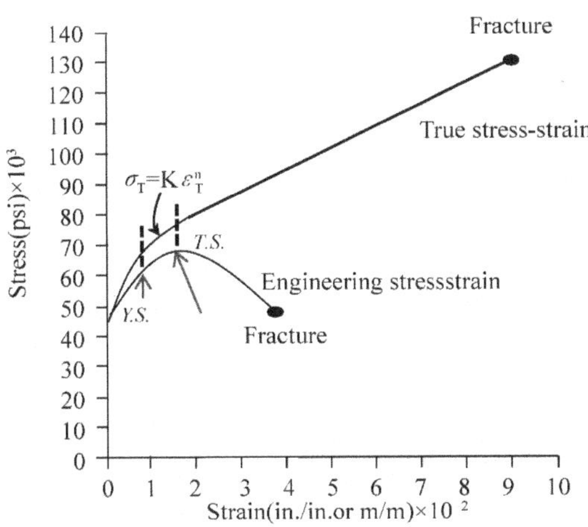

Figure 5.11 The region that satisfies the strain hardening exponential equation in the true stress-strain curve.

Ⅶ. Ductility (Elongation Percentage)

The engineering stress at failure point is sometimes lower than *T.S.* and occasionally even lower than *Y.S.* Unfortunately, the complexity of the final stages of neck down causes the value of the failure stress to vary substantially from specimen to specimen.[23] More useful is the strain at failure. Therefore, ductility is frequently quantified as the percent elongation at failure.

23 遗憾的是，由于颈化最后阶段的复杂性，导致从一个样品到另一个样品的断裂应力都有很大的差异。

$$\delta = \frac{l_x - l_0}{l_0} \times 100\% \tag{5-11}$$

where l_0 and l_x is the original length and length of the specimen at failure (see Figure 5.12).

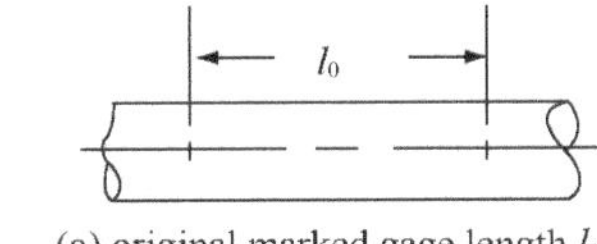

(a) original marked gage length l_0

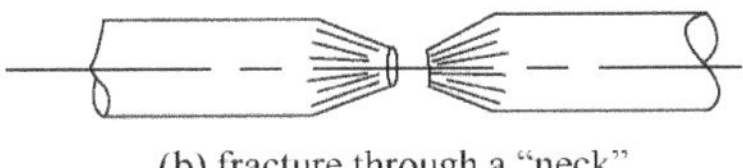

(b) fracture through a "neck"

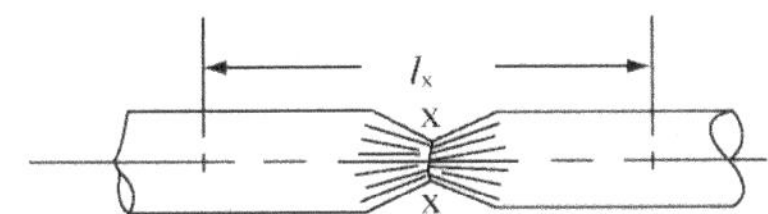

(c) broken parts fitted together to find final gage length l_x

Figure 5.12 Ductility of materials.

Ductility indicates the general ability of a material to be plastically deformed, i.e., the permanent deformation with the value of strain at failure subtracting the recovered elastic strain at failure ($l_x = \varepsilon_{failure} - l_{elastic\ recovery}$).[24]

24 延展率表示材料进行塑性形变的基本能力，即永久性形变，其值等于在断裂处的应变值减去断裂时恢复的弹性应变值（$l_x = \varepsilon_{failure} - l_{elastic\ recovery}$）。

Practical implications of this ability include formability during fabrication and relief of locally high stresses at crack tips during structural loading. It should also be noted that the value of percent elongation at failure is afunction of the gage length used. Tabulated values are frequently specified for a gage length of 2 inch.

Ⅷ. Toughness

Since integrated σ–ε data are not routinely available, we shall be required to monitor the relative magnitudes of strength (*Y.S.* and *T.S.*) and ductility (percent elongation at fracture). The term toughness is used to describe the combination of strength and ductility. It is conveniently defined as the total area under the stress-strain curve. Figure 5.13 shows the meaning of toughness via stress-strain curves. In ductile metallic materials (curve 1 in Figure 5.13), the engineering stress-strain curve typically goes through a maximum; this maximum stress is the tensile strength of the material. Failure occurs at a lower engineering stress after necking has reduced the cross-sectional area supporting the load. In this case, materials possess high toughness.[25] In more brittle materials, failure occurs at the maximum load, where the tensile strength and breaking strength are the same (curve 2). In this case, the toughness is low. In other materials, such as elastomers (e.g., rubber or silicones), which exhibit low tensile strength but large ductility, the toughness is also low (curve 3).

25 在延展金属材料中（图 5.13 中的曲线 1），工程应力-应变曲线通常通过一个最高点；这个最大的应力就是材料的拉伸强度。断裂发生在颈化导致的支撑负载的横截面变细过程之后的较小工程应力下。这种情况材料就有高度的韧性。

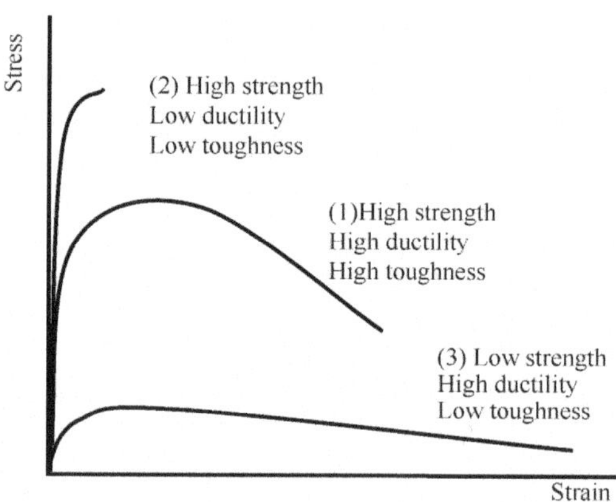

Figure 5.13 The total area under the stress-strain curve is termed as toughness.

Ⅸ. Upper and Lower Yield Points

As shown in the Figure 5.14, for certain alloys (especially low-carbon steels), the stress-strain curve for tensile test exhibits distinct breaks. The break from the elastic region at a yield point is termed an upper yield point. The distinctive ripple pattern following the yield point is associated with nonhomogeneous deformation that begins at a point of stress concentration.[26]

A lower yield point is defined at the end of the ripple pattern and at the onset of general plastic deformation.

26 在屈服点之后的特色波纹图案，与始于应力集中处的各向异性形变相关。

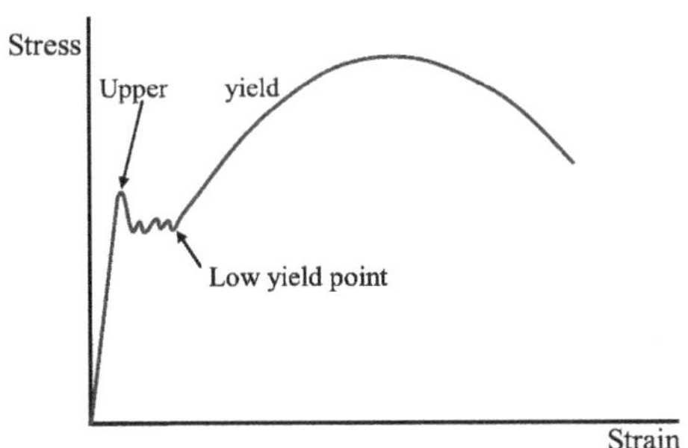

Figure 5.14 Stress-strain curve with upper and lower yield points.

Ⅹ. Flexural Stress

In many brittle materials, the normal tensile test cannot easily be performed because of the presence of flaws at the surface. Often, just placing a brittle material in the grips of the tensile testing machine causes cracking. These materials may be tested using the bend test. Figure 5.15a shows the experimental setup for a three-point bend test. By applying the load at three points and causing bending, a tensile force acts on the material opposite the midpoint. Fracture begins at this location.[27] The flexural stress in three-point bend test is:

27 在三点施加力并造成弯曲，对抗于中点，产生了拉力。断裂始于该点。

$$\sigma_{\text{bend}} = \frac{3FL}{2wh^2} \tag{5-12}$$

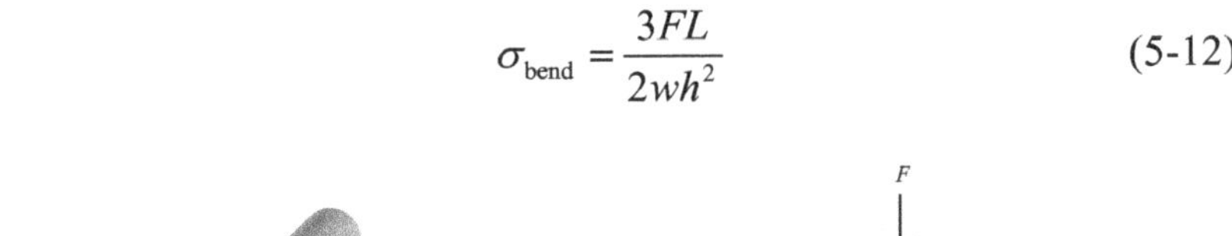

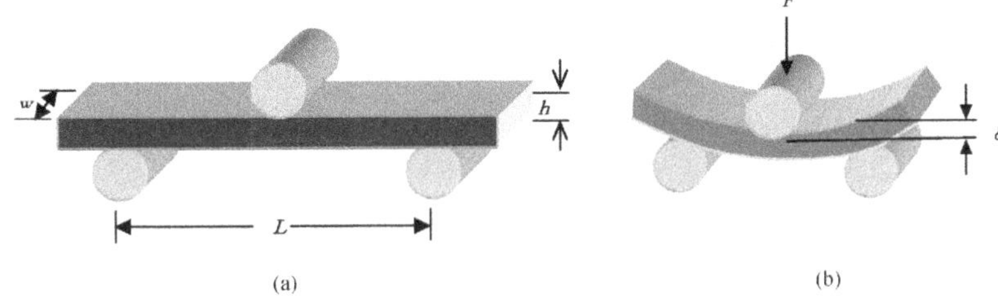

Figure 5.15 (a) The bend test often used for measuring the strength of brittle materials, and (b) the deflection obtained by bending.

where F is the bend load, L is the distance between the two outer points, w is the width of the specimen, and h is the height of the specimen. The flexural stress has units of stress. The results of the bend test are similar to the stress-strain curves; however, the stress is plotted versus deflection (δ) rather than versus strain (Figure 5.15b and Figure 5.16).

The flexural strength (i.e., the maximum strength), or modulus of rupture (MOR), describes the material's strength and can be expressed in three-point bend test as:

$$MOR = \frac{3F_f L}{2wh^2} \tag{5-13}$$

where F_f is the fracture load.

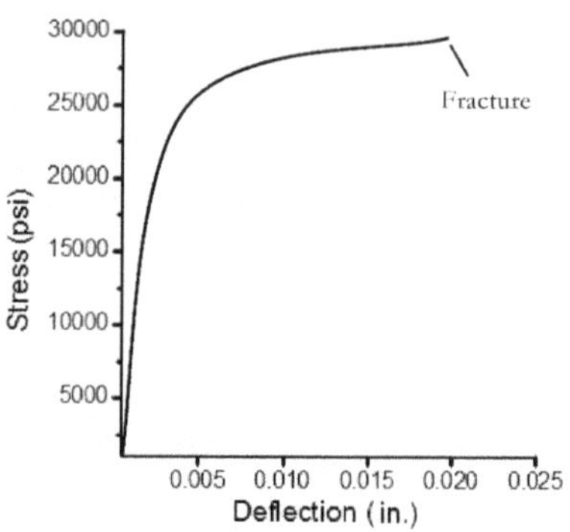

Figure 5.16 Stress-deflection curve for an MgO ceramic obtained from a bend test.

The modulus of elasticity in bending, or the flexural modulus (E_{bend}), is calculated as:

$$E_{bend} = \frac{L^3 F}{4wh^3 \delta} \tag{5-14}$$

where δ is the deflection of the beam when a force F is applied; L, h and w are length, height and width of the specimen, respectively.

Ⅺ. Hardness

The hardness test measures the resistance to penetration of the surface of a material by a hard object. Hardness as a term is not defined precisely. Hardness, depending upon the context, can represent resistance to scratching or indentation and a qualitative measure of the strength of the material.[28] A variety of hardness tests have been devised, but the most commonly used are the Brinell test and the Rockwell test. Different indenters used in these tests are shown in Figure 5.17.

28 基于不同种类，硬度能够表示材料对划刮或者压嵌的抵抗能力，以及定性地描述材料的强度。

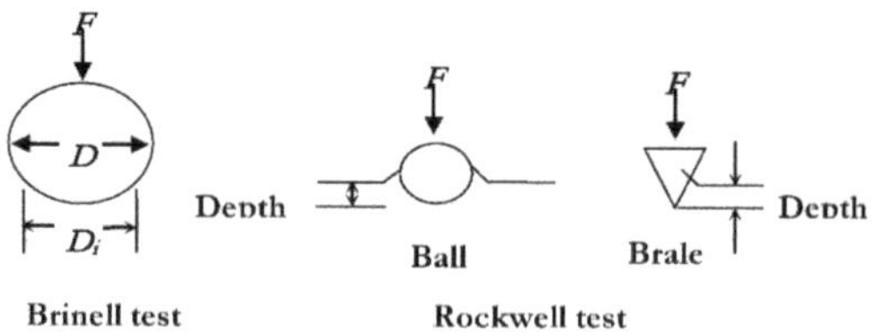

Figure 5.17 Indenters for the Brinell and Rockwell hardness tests.

In the Brinell hardness test, a hard steel sphere (usually 10 mm in diameter) is forced into the surface of the material.[29] The diameter of the impression, typically 2 to 6 mm, is measured and the Brinell hardness number (abbreviated as HB or BHN) is calculated from the following equation:

29 在Brinell硬度测试中，一个坚硬钢球（通常直径为 10 mm）被压入材料表面。

$$HB = \frac{2F}{\pi D[D - \sqrt{D^2 - D_i^2}]} \tag{5-15}$$

where F is the applied load in kilograms, D is the diameter of the indenter in millimeters, and D_i is the diameter of the impression in millimeters. The Brinell hardness has units of $kg \cdot mm^{-2}$.

The Rockwell hardness test uses a small-diameter steel ball for soft materials and a diamond cone, or *Brale*, for harder materials. The depth of penetration of the indenter is automatically measured by the testing machine and converted to a Rockwell hardness number (*HR*). Since an optical measurement of the indentation dimensions is not needed, the Rockwell test tends to be more popular than the Brinell test.[30] Several variations of the Rockwell test are used, including those described in Table 5.1.

30 因为不需对嵌入尺度做光学测量，Rockwell 法的应用倾向于比 Brinell 法更广泛。

Table 5.1 Comparison of typical hardness tests

Test	Indenter	Load	Application
Brinell	10-mm ball	3000 kg	Cast iron and steel
Brinell	10-mm ball	500 kg	Nonferrous alloys
Rockwell A	Brate	60 kg	Very hard materials
Rockwell B	1/16-in. ball	100 kg	Brass, low-strength steel
Rockwell C	Brate	150 kg	High-strength steel
Rockwell D	Brate	100 kg	High-strength steel
Rockwell E	1/8-in. ball	100 kg	Very soft materials
Rockwell F	1/16-in. ball	60 kg	Aluminum, soft materials
Vickers	Diamond square pyramid	10 kg	All materials
Knoop	Diamond elongated pyramid	500 g	All materials

A Rockwell C (*HRC*) test is used for hard steels, whereas a Rockwell F (*HRF*) test might be selected for aluminum. Rockwell tests provide a hardness number that has no units. Hardness numbers are used primarily as a qualitative basis for comparison of materials, specifications for manufacture and heat treatment, quality control, and correlation with other properties of materials.[31] For example, Brinell hardness is related to the tensile strength of steel by the approximation:

31 硬度值主要用作材料比较、加工及热处理规格、质量控制、以及其他相关材料性质的定性基础。

$$\text{Tensile strength (psi)} = 500\ HB \tag{5-16}$$

where HB has units of $kg \cdot mm^{-2}$.

Hardness correlates well with wear resistance. A separate test is available for measuring the wear resistance. A material used in crushing or grinding of ores should be very hard to ensure that the material is not eroded or abraded by the hard feed materials. Similarly, gear teeth in the transmission or the drive system of a vehicle should be hard enough that the teeth do not wear out. Typically, we find that polymer materials are

32 通常地，我们发现高分子材料都异常地柔软，金属和合金的硬度适中，陶瓷最硬。

33 微观硬度测试用于表面硬度比内部硬度大的材料、不同区域硬度不同的材料、或者宏观不平整的样品。

exceptionally soft, metals and alloys have intermediate hardness, and ceramics are exceptionally hard.[32] We use materials such as tungsten carbide-cobalt composite (WC-Co), known as "carbide," for cutting tool applications. We also use microcrystalline diamond or diamond like carbon (DLC) materials for cutting tools and other applications. The Knoop hardness (*HK*) test is a micro-hardness test, forming such small indentations that a microscope is required to obtain the measurement. In these tests, the load applied is less than 2 N. The Vickers test, which uses a diamond pyramid indenter, can be conducted either as a macro or micro-hardness test. Micro-hardness tests are suitable for materials that may have a surface that has a higher hardness than the bulk, materials in which different areas show different levels of hardness, or samples that are not macroscopically flat.[33]

5.3 Elastic Deformation

As stated in the previous section, elastic deformation refers to temporary deformation (the initial linear region in the stress-strain curve) that can be fully recovered when the load is removed, and it satisfies Hooke's law.

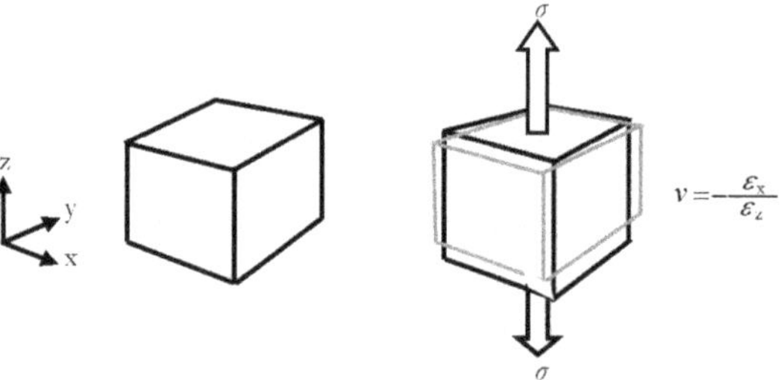

Figure 5.18 The Poisson's ratio (ν) characterizes the contraction perpendicular to the extension caused by a tensile stress.

Figure 5.18 illustrates another important feature of elastic deformation, namely, a contraction perpendicular to the extension caused by a tensile stress. This effect is characterized by the Poisson's ratio, ν,

$$\nu=-\frac{\varepsilon_x}{\varepsilon_z} \tag{5-17}$$

where the strains in the x and z directions are defined as shown in Figure 5.18. (There is a corresponding expansion perpendicular to the compression caused by a compressive stress.) Although the Poisson's ratio does not appear directly on the stress versus strain curve, it is, along with the elastic modulus, the most fundamental description of the elastic behavior of materials.[34] The values of Poisson's ratio generally fall within the relatively narrow band of 0.26 to 0.35.

The compression modulus (B), the shear modulus (G), and the elastic modulus (E) are related, for small strain, by Poisson's ratio; namely,

$$E=2G(1+v)=3B(1-2v) \tag{5-18}$$

Figure 5.19 shows that the fundamental mechanism of elastic deformation, i.e., the stretching of atomic bonds. The fractional deformation of the material in the initial elastic region is small so that, on the atomic scale, we are dealing only with the portion of the force-atom separation curve in the immediate vicinity of the equilibrium atom separation distance (a_0 corresponding to F = 0).[35] The nearly straight-line plot of F versus a across the a axis implies that similar elastic behavior will be observed in a compressive, or push, test as well as in tension test.

34 虽然泊松比并不直接体现在应力-应变曲线中，但它与弹性模量一样，是对工程材料最基础的描述。

35 在初始的弹性区域材料形变非常小，在原子尺度上，对于原子间距和相互作用力关系曲线，我们只考虑在平衡原子间距附近的部分（a_0 对应于 $F=0$）。

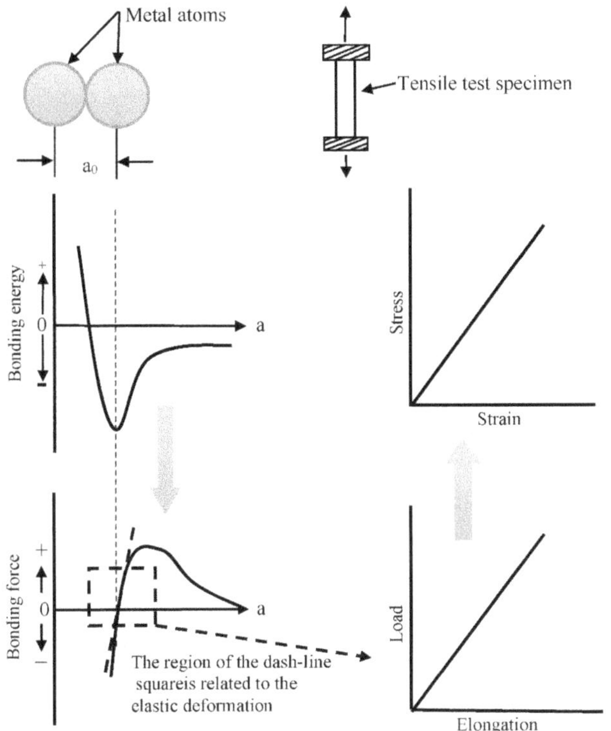

Figure 5.19 Relationship of elastic deformation to the stretching of atomic bonds.

5.4 Plastic Deformation

Plastic deformation is a permanent deformation that can't be recovered completely when the load is free, although a small elastic component is recovered.[36] The fundamental mechanism of plastic deformation is the

36 塑性形变是在撤销力后无法恢复的永久形变，当然其中的一小部分弹性组份是可以恢复的。

distortion and reformation of atomic bonds. The plastic (permanent) deformation of crystalline solids is difficult without dislocations, the linear defects. This deformation would occur by sliding one plane of atoms over an adjacent plane, as shown in Figure 5.20.

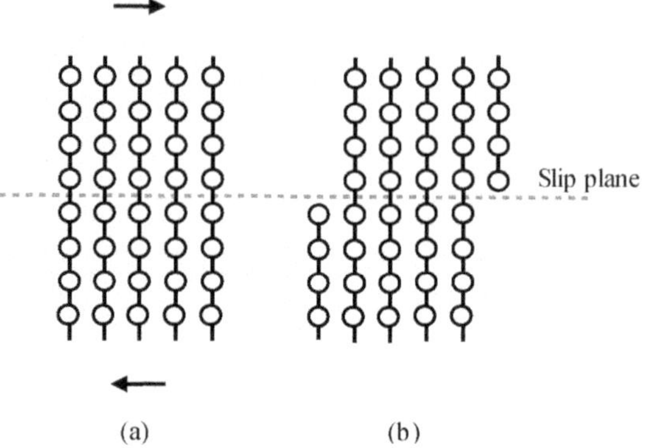

Figure 5.20 Sliding of one plane of atoms past an adjacent one, (a) and (b) is the situation before and after sliding, respectively. This high-stress process is necessary to plastically (permanently) deform a perfect crystal.

The shear stress associated with this sliding action can be calculated with knowledge of the periodic bonding forces along the slip plane. The result obtained by Frenkel was that the theoretical critical shear stress is roughly one order of magnitude less than the bulk shear modulus, *G*, for the materials. For a typical metal such as copper, the theoretical critical shear stress represents a value well over 1,000 MPa. The actual stress necessary to plastically deform a sample of pure copper (i.e., slide atomic planes past each other) is at least an order of magnitude less than this value. Our daily experience with metallic alloys (e.g., opening aluminum cans or bending automobile fenders) represents deformations generally requiring stress levels of only a few hundred MPa.[37] What is the basis of the mechanical deformation of metals, which requires only a fraction of the theoretical strength? The answer is the dislocation.

37 在我们日常生活中，所遇到的金属合金形变（例如，打开铝制易拉罐或弯折汽车挡泥板），通常只需几百兆帕的应力。

Figure 5.21 illustrates the role of a dislocation playing in the shear of a crystal along a slip plane. The key point to observe is that only a relatively small shear force needs to operate in the immediate vicinity of the dislocation in order to produce a step-by-step shear that eventually yields the same overall deformation as the high-stress mechanism of Figure 5.20.[38] A perspective view of a shear mechanism involving a more general, mixed dislocation is given in Figure 5.22.

38 观察的关键在于，仅仅需要一个较小的剪切力就可以操控紧邻位错的位置，从而一步一步地形成剪切应变，最后产生与图 5.20 所表示的高应力机制一样的总形变。

The resistance force to atom dislocation originating from lattice is called P-N force (Peierls-Nabarro force, $\tau_{P\text{-}N}$), which is illustrated in Figure 5.23 and can be expressed as:

$$\tau_{\text{P-N}} = \frac{2G}{1-v}\exp[\frac{-2\pi a}{(1-v)b}] = \frac{2G}{(1-v)}\exp(\frac{-2\pi W}{b}) \tag{5-19}$$

where a is the distance between two sliding planes, b is the atomic distance along the sliding direction, G is shear elastic modulus, v is Poisson's ratio, W is dislocation width.

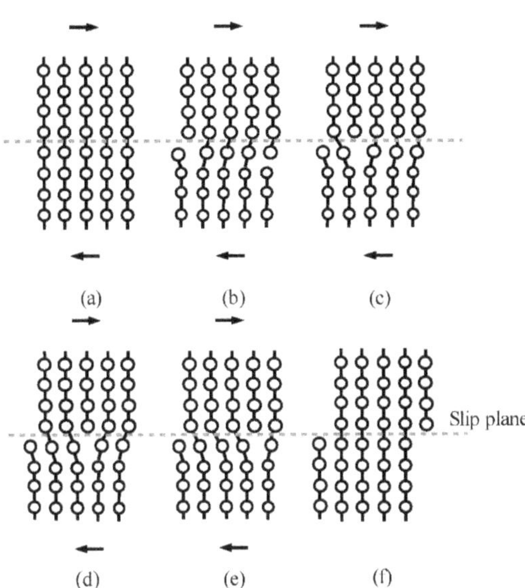

Figure 5.21 A low-stress alternative for plastically deforming a crystal involves the motion of a dislocation along a slip plane.

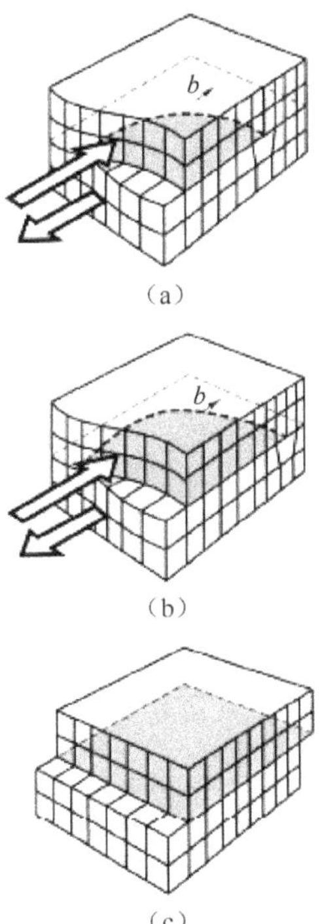

Figure 5.22 Schematic illustration of the motion of a dislocation under the influence of a shear stress. The net effect is an increment of plastic (permanent) deformation.

The dislocation width can be expressed as,

$$W = \frac{a}{1-\nu} \tag{5-20}$$

Larger W means a smaller dislocation resistance ($\tau_{P\text{-}N}$). Metals possess large dislocation width (W is large), resulting in small yield strength, thus more ductile. Ceramics possess small dislocation width, thus lattice dislocation resistance is large, leading to brittle feature.

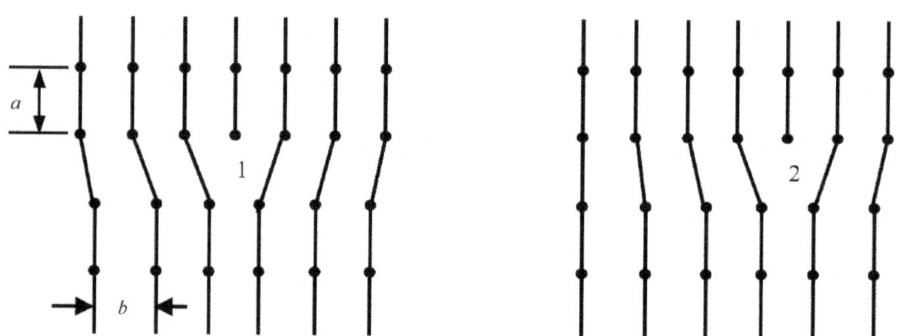

Figure 5.23 Atom dislocation in crystal lattice

Reflecting on Figure 5.21, we can appreciate that the stepwise slip mechanism would tend to become more difficult as the individual atomic step distances are increased.[39] As a result, slip is more difficult on a low-atomic-density plane than on a high-atomic-density plane. Figure 5.24 shows these differences schematically. In general, the micromechanical mechanism of slip-dislocation motion will occur in high atomic-density planes and in high atomic-density directions.[40] A combination of families of crystallographic planes and directions corresponding to dislocation motion is referred to as a slip system. In Figure 5.25, we can label the slip systems in (a) FCC aluminum and (b) HCP magnesium. Aluminum and its alloys are characteristically ductile (deformable) due to the large number (12) of high-density plane-direction combinations.[41] Magnesium and its alloys are typically brittle (fracturing with little deformation) due to the smaller number (3) of such combinations. Table 5.2 summarizes the major slip systems in typical metal structures.

39 回顾图 5.21，我们可以理解逐步滑移机制将随着原子距离的增大而变得比较困难。

40 通常，在微观力学滑移一位错机制上运动将发生在高原子密度晶面和高原子密度晶向上。

41 铝及其合金具有可延展性（可塑性）特征，这是由于密堆积晶面与晶向的组合数量很多(12 个）滑移系统。

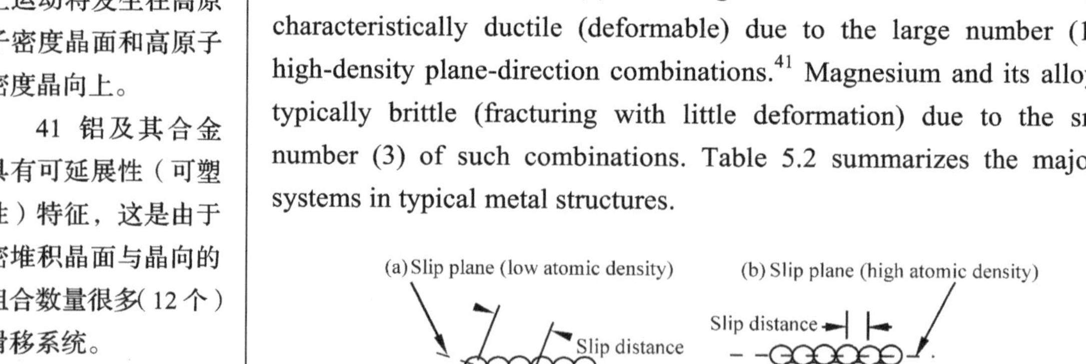

Figure 5.24 Dislocation slip is more difficult along (a) a low-atomic-density plane than along (b) a high-atomic-density plane.

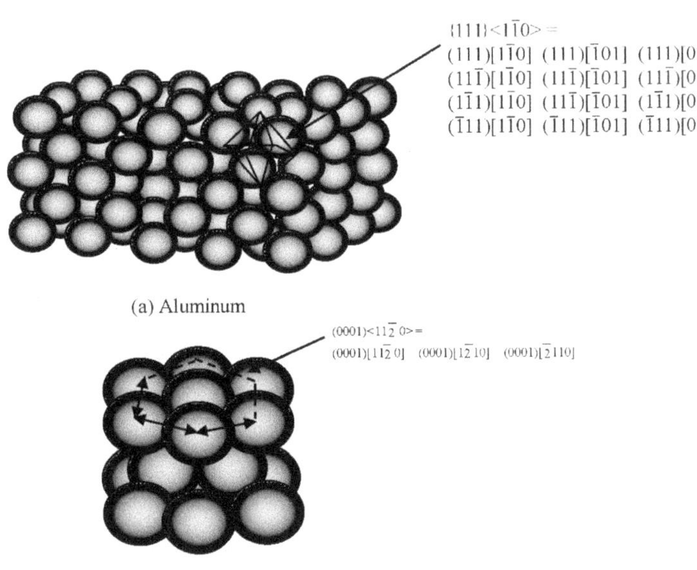

(a) Aluminum

(b) Magnesium

Figure 5.25 Slip systems for (a) FCC aluminum and (b) HCP magnesium.

Table 5.2 Major slip systems in the common metal structures.

Crystal structure	Slip Plane	Slip direction	Number of Slip systems	Unit-cell Geometry	Examples
BCC	{110}	$\langle \bar{1}11 \rangle$	6×2 = 12		α-Fe, Mo, W
FCC	{111}	$\langle 1\bar{1}0 \rangle$	4×3 = 12		Al, Cu, γ-Fe, Ni
HCP	(0001)	$\langle 11\bar{2}0 \rangle$	1×3 = 3		Cd, Mg, α-Ti, Zn

Several basic concepts of the mechanical behavior of crystalline materials relate directly to simple models of dislocation motion and plastic deformations. Here are examples:[42]

42 一些晶体材料力学行为的基本概念与位错和塑性形变的简单模型直接相关。以下是例子：

Ⅰ. Cold Working

The cold working of metals involves deliberate deformation of the metal at relatively low temperatures and forming a stronger product. An

43 冷加工的一个重要特性在于该金属的加工难度将随着形变的增大而增加。

important feature of cold working is that the metal becomes more difficult to deform as the extent of deformation increases.[43] The basic micromechanical reason for this is that a dislocation hinders the motion of another dislocation. This slip mechanism of Figure 5.21 proceeds most smoothly when the slip plane is free of obstructions. Cold working generates dislocations that serve as such obstacles. In fact, cold working generates so many dislocations that the configuration is referred to as a "forest of dislocations" (Figure 5.26).

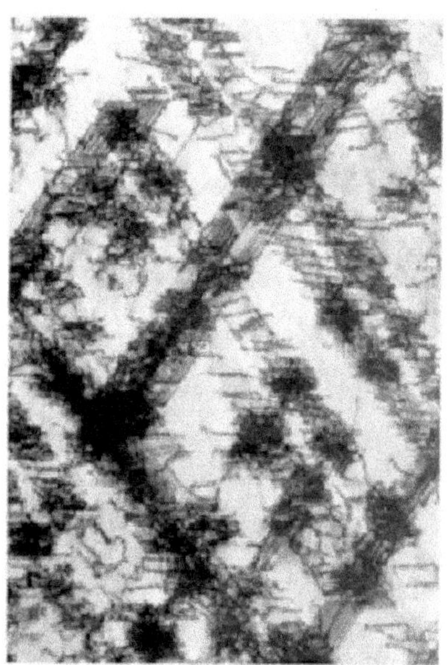

Figure 5.26 Forest of dislocations in a stainless steel as seen by a transmission electron microscope

Ⅱ. Hardening by Fine Grains

Grain boundary prevents the moving of dislocation, leading to accumulation of dislocation at boundary sites. There is the Hall-Petch equation for grain size dependent yield strength:

44 从公式(5-21)中可以看出，小尺寸颗粒（d 很小）导致很大的屈服强度（σ_s），因此材料坚硬。

$$\sigma_s=\sigma_0+Kd^{-\frac{1}{2}} \tag{5-21}$$

where σ_s is *Y. S.*, d is grain radius, σ_0 and K are constants. From Equation 5-21, it can be seen that, small grain size (d is small), leads to large yield strength (σ_s), hence material is hard.[44] This is one of the two considerations for fine grain hardening. On the other hand, small grain size presents large area of grain boundary, which causes dislocation accumulation. Thus the dislocation resistance is large, and material is hard. As shown in Figure 5.27, there is an accumulation of dislocations clogs, which prevents further dislocation and renders material hardening.[45]

45 如图 5.27 所示，位错阻塞的存在阻止了进一步的位错，因而材料得到硬化。

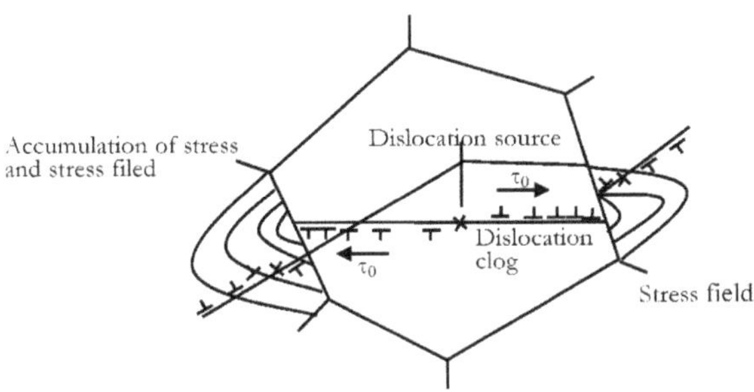

Figure 5.27 Prevention of dislocation by grain boundary.

Ⅲ. Solute Hardening

Small solute above the edge dislocation or large solute under the edge dislocation can reduce lattice distortion, leading to the reduction of dislocation restriction.[46] This can increase the yield point (i.e., the elastic deformation increased). However, when dislocation departs from the solute, system energy generally increases, restricting plastic deformation.[47] This scheme is called string model for dislocation (Figure 5.28).

46 在刃型位错上方的小溶质，或者下方的大溶质能够减少晶格扭曲，降低位错限制。

47 但是，当位错离开溶质时，系统能量通常会增加，限制塑性形变。

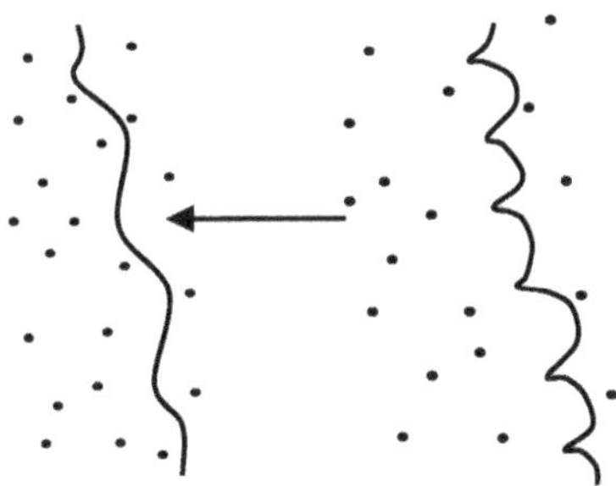

Figure 5.28 String model for dislocation.

The string model thought that due to the small shear force at solute site, dislocation tends to be close to the solute, forming string line. In this case the solid solutes serve as pinning sites. This string line, on one hand can reduce dislocation force by its formation, which can increase the yielding point; on the other hand, forms obstacle for further dislocation, giving hardening effect.[48] When the string bending is larger enough, system energy increases, leading to the departure of the dislocation from the solute. The disadvantage of this solute hardening lies in that while increase hardness by increase yield strength, the ductility decrease, due to the non-compatibility of the solute to the host.

48 这个弦线，一方面,它的形成可以降低位错力,导致屈服点的提高；另一方面，也是进一步位错的阻力，造成硬化效应。

Explanation of Yield Phenomenon in Metal: According to the string model for dislocation, the initial movement of dislocation in metals is limited by solute atoms, i.e. pinning effect by solute atoms. Thus, higher stress is needed for dislocation to escape from pinning and start to slip. This stress is corresponding to the upper yield point. Once the dislocation is apart

49 一旦位错离开丁扎位置，位错阻力将被减小，滑动可以在较小的阻力下继续。

from pinning, the dislocation resistance will reduce, and the slip can be continued under smaller resistance.[49] The stress, which is responsible for the beginning of the continuous plastic deformation, is termed as lower yield point.

Ⅳ. Second Phase (Particle) Hardening

In high-temperature designs, interfaces between dissimilar metals, which forms the second phase, is a common concern for the purpose of hardening. When a dislocation circumvents or over-cuts a second phase particle, obstacle is produced. These are explained in the following two ways:

Circumvent Scheme (Orowan Mechanism):

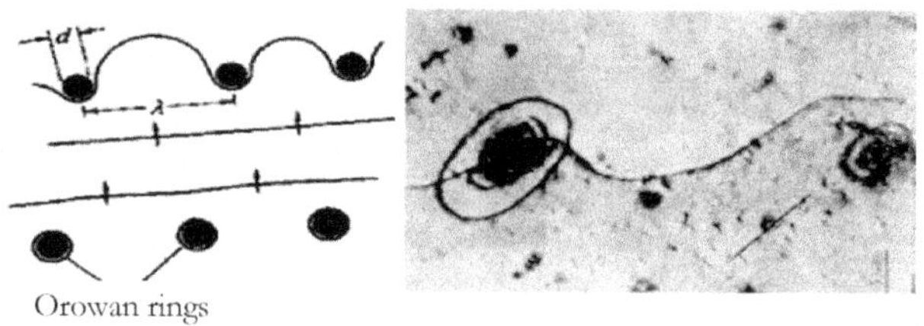

Figure 5.29 Explanation of second phase hardening by Orowan mechanism.

50 当位错遇到这样的第二相粒子时，即很难改变它的形状（意味着第二相更硬），位错将会在第二相粒子周围弯曲，因此阻碍了位错的继续。

When the dislocation meets a second phase of such that it is difficult to change its shape (means: the second phase is much harder), the dislocation will bend around the second phase, leading to resistance.[50] This kind of second phase will eventually results in Orowan ring (Figure 5.29).

Over–Cut Scheme:

51 切过机制认为具有很强滑移力的位错将会切过第二相粒子，形成更多的晶界。

Second phase hardening can also be explained by over-cut scheme, which has been observed under microscope, as shown in Figure 5.30. Over-cut scheme thought that the dislocation with more slip strength will over cut a second phase, leading to more grain boundaries.[51] The increased grain boundaries by cutting, cause further obstacle for dislocation, which results in material hardening.

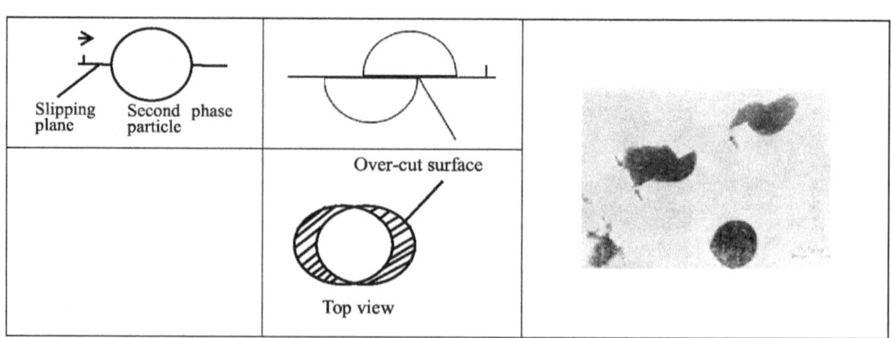

Figure 5.30 Over-cut scheme for second phase hardening

Second phase hardening is one of the methods to strengthen metal/alloy materials. For example, by adding fine particles of second phase, material strengthening can be realized. However, it is difficult to achieve homogeneous second phase by this dispersion strengthening. More common method is precipitation. Figure 5.31a shows a two composition alloy with solubility curve of MN, and the vertical dash line represents the sample of the alloy. Heating the alloy to the solid solution state (α region), there will be an isothermal dissolving process, leading to the solid solution of α phase.[52] Then decrease temperature, due to the decrease of solubility at low temperature, the excessive B component will be precipitated from the α phase, forming homogeneously dispersed second phase.[53] However, due to the coarse size of the second phase, the hardening effect is not obvious. In real application (see Figure 5.31b), after formation of solid solution, supper cooling is exerted to yield supersaturated solid solution, which is called solution treatment. After that, the alloy can be either kept at a relatively low heating temperature for a short time (artificial aging) or at room temperature for a long time (natural aging) to give fine particles of the second phase. By this, good precipitation hardening can be achieved.

52 加热合金使其达到固溶相（α 区域），将会发生等温溶解过程，形成 α 固溶相。

53 之后降低温度，由于低温下溶解性的降低，过多的 B 组分会从 α 相中沉淀，形成分散均匀的第二相。

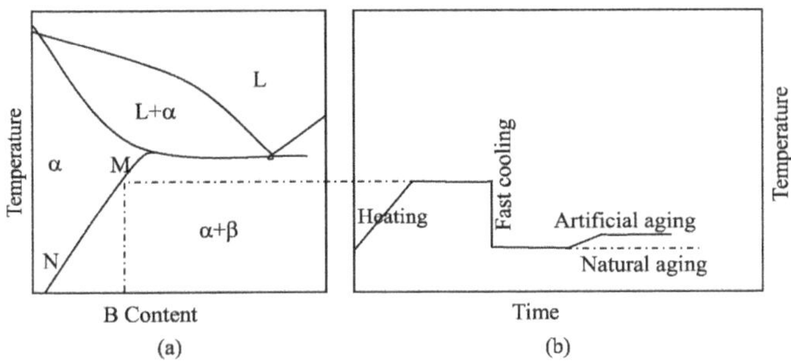

Figure 5.31 Process of second phase hardening by solid solution treatment.

V. Brittleness of Ceramics (e.g., Al_2O_3)

One additional basic concept of mechanical behavior is that the more complex crystal structures correspond to relatively brittle materials. Relatively large Burgers vector combined with the difficulty in creating obstacle-free slip planes creates a limited opportunity for dislocation motion.[54] An additional consideration adding to the brittleness of them is that many slip systems are not possible, owing to the charged state of the ions. The sliding of ions with like charges past one another can result in high columbic repulsive forces. As a result, even ceramic compounds with relatively simple crystal structures exhibit significant dislocation mobility only at a relatively high temperature.

54 相对大的 Burgers 矢量和产生无阻力滑移面的困难，限制了发生位错的机会。

Ⅵ. Softening at High Temperature

Obstacles to dislocation motion harden metals, but high temperatures can help to overcome these obstacles and thereby soften the metals. The micromechanical mechanism here is rather straightforward. At sufficiently high temperatures, atomic diffusion is sufficiently great to allow highly stress crystal grains produced by cold working to be restructured into more nearly perfect crystalline structures.[55] The dislocation density is dramatically reduced with increasing temperature, which permits the relatively simple deformation mechanism to occur free of the forest of dislocations. At this point, we have seen an important blending of the concepts of solid-state diffusion and mechanical deformation. In each case, a useful rule of thumb will apply: The temperature at which atomic mobility is sufficient to affect mechanical properties is approximately one-third to one-half times the absolute melting point, T_m.

55 在足够高的温度下，足够强的原子扩散能够使通过冷加工产生的高应力晶粒重新构筑，达到接近完美的晶体结构。

5.5 Creep and Stress Relaxation

Creep can be defined as plastic (permanent) deformation occurring at high temperature under a constant load over a long time period.

If a structural material is used at elevated temperature, the tensile test alone cannot predict its mechanical behaviors. Because at high temperature (T greater than one-third to one-half times the melting point on an absolute temperature scale), the strain with a loading below its yield point increases with time.[56] This behavior is called creep. Figure 5.32 shows a typical creep test design, and Figure 5.33 shows a typical creep curve in which the strain, ε, gradually increases with time after the initial elastic loading.

56 由于高温下（T高于三分之一或者一半以绝对温度表示的熔点），在低于屈服点的负载下的应变将随着时间的增大而增大。

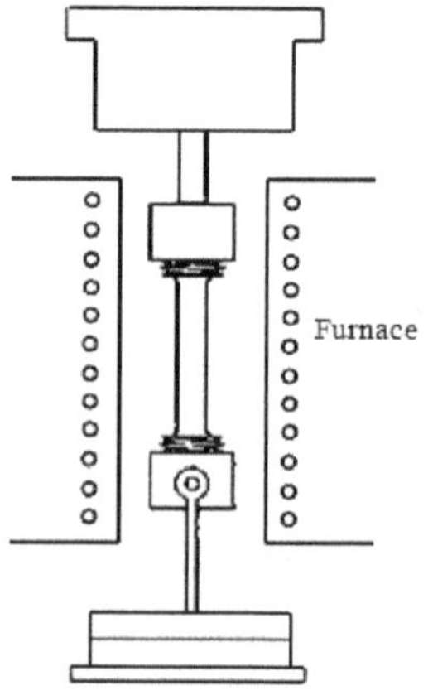

Figure 5.32 Typical creep test setup.

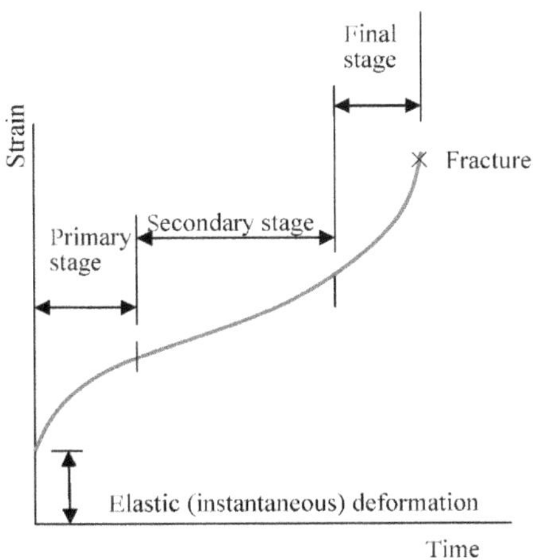

Figure 5.33 Creep curve: plastic strain versus time for a material stressed at high temperatures (above about one-half the absolute melting point).

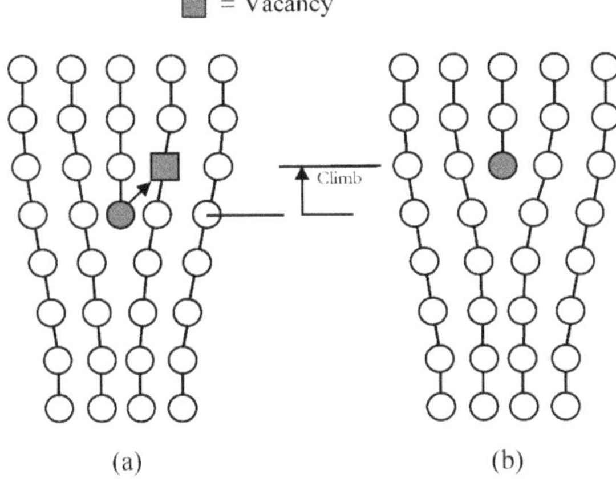

Figure 5.34 Mechanism of dislocation climb. Obviously, many adjacent atom movements are required to produce climb of an entire dislocation line.

As shown in Figure 5.33, there are three stages of creep deformation. The primary stage is characterized by a decreasing strain rate (slope of the ε vs. t curve). The relatively rapid increase in length induced during this early time period is the direct result of enhanced deformation mechanisms. This enhanced deformation comes from thermally activated atom mobility, giving dislocations additional slip planes in which to move (Figure 5.34).[57] The secondary stage of creep deformation is characterized by straight-line, constant-strain-rate data. In this region, the increased ease of slip due to high-temperature mobility is balanced by increasing resistance to slip due to the buildup of dislocations and other microstructural barriers.[58] In the final tertiary stage, strain rate increases due to an increase in true stress. This increase results from cross-sectional area reduction due to necking or internal cracking. In some cases, fracture occurs in the secondary stage, thus eliminating this final stage.

57 这类增强的形变来自于热激活的原子迁移，它使得位错拥有更多的可以在其中移动的滑移平面（图 5.34）。

58 在这区域内，通过高温迁移增加的滑移容易特性，被通过内建位错和其他微观结构性障碍导致的滑动阻力所平衡。

Three existing schemes are related to the mechanism for creep deformation: One is dislocation creep, which occurs at high temperature. Due to the assistant of thermal energy, dislocation can slip and climb under relatively small stresses, leading to strain increase. The second scheme is diffusion creep (Nabarro-Herring creep), which is related to a vacancy diffusion within a grain at high temperature and leading to atom deformation. The third creep scheme is grain boundary creep (Coble creep), which means under relatively low temperature and stress, the diffusion of vacancies along the grain boundary induces a deformation.[59]

59 第三个蠕变机制是晶界蠕变（Coble 蠕变）。这是指在相对低的温度和应力下，空位沿着晶界的扩散导致了形变。

Creep is probably more important in ceramics than in metals because high temperature applications are so widespread. The role of diffusion mechanisms in the creep of ceramics is more complex than in the case of metals because diffusion, in general, is more complex in ceramics. The requirement of charge neutrality and different diffusivities for cations and anions contribute to this complexity. As a result, grain boundaries frequently play a dominant role in the creep of ceramics. Sliding of adjacent grains along these boundaries provides for microstructural rearrangement during creep deformation. In some relatively impure refractory ceramics, a substantial layer of glassy phase may be present at the grain boundaries. In that case, creep can again occur by the mechanism of grain-boundary sliding due to the viscous deformation of the glassy phase. This “easy” sliding mechanism is generally undesirable due to the resulting weakness at high temperatures.

For metals working at high temperature, creep deformation must be considered for safety reason. For example, boiler need to work under high temperature and high pressure, the production of creep will lead to wall thinning, which may induce security problem even break down or explosion.[60] Creep in components of engine, aerospace, and chemistry industry etc., must be paid more attention.

60 例如，锅炉需要在高温和高压下工作，蠕变的产生将会导致炉壁减薄，这可能会导致安全问题甚至损坏或爆炸。

Stress relaxation refers to these phenomena that involve decreasing stress with time under constant strain, contrasting with the creep deformation which involves increasing strain with time under constant stress. The mechanism of stress relaxation is viscous flow (i.e., molecules gradually sliding past each other over an extended period of time). Viscous flow converts some of the fixed elastic strain into non-recoverable plastic deformation. A familiar example is the rubber band, under stress for a long period of time, which does not snap back to its original size upon stress removal. Stress relaxation is also a safety issue in structural components. For example, for a bolt used to fasten two components, the stress produced by a fixed strain plays the role of fastening. However, at high temperature, if stress relaxation occurs, the fixed position cannot produce enough stress, the components will be loose.[61]

61 但是，在高温下，如果发生应力松弛，固定的位置不能产生足够的应力，部件将不牢固。

5.6 Fracture and Fatigue

Ⅰ. Fracture Mechanics

If the load of stress beyond the strength of a material, fracture will occur. Crack or other small flaws tend to induce fracture. The term "flaw" refers to such features as small pores (holes), inclusions, or micro-cracks; but does not refer to atomic level defects such as vacancies or dislocations.[62] For mechanical applications, it is generally desired to know the maximum stress that a material can withstand if it contains flaws of a certain size and geometry. There are two types of fractures:

Brittle Fracture means any crack or imperfection limits the ability of a ceramic to withstand a tensile stress. This is because a crack (sometimes called a Griffith flaw) concentrates and magnifies the applied stress.[63]

Ductile Fracture normally occurs in a transgranular manner (through the grains) in metals that have good ductility and toughness.[64] Often, a considerable amount of deformation—including necking—is observed in the failed component. The deformation occurs before the final fracture. Ductile fractures are usually caused by simple overloads, or by applying too high a stress to the material. Deformation by slip also contributes to the ductile fracture of a metal.

62 如果应力超出了材料的强度，将产生断裂。裂纹或者其他小瑕疵容易引起断裂。术语“瑕疵”指的是这样的特征：小孔（洞）、内嵌物、或者微裂纹；却不是指原子水平上的缺陷，如空位或者位错。

63 脆性断裂指的是裂纹或者缺陷对陶瓷抵抗拉伸能力的限制。这是由于裂纹（有时称为格里菲斯裂纹）的集中和对所施加应力的放大作用。

64 韧性断裂通常在具有良好延展性和韧性的金属中以穿晶方式（通过晶粒）发生。

Ⅱ. Fatigue

Fatigue is the lowering of strength or failure of a material due to repetitive stress which may be above or below the yield strength.[65] It is a common phenomenon in load-bearing components in cars and airplanes, turbine blades, springs, crankshafts and other machinery, biomedical implants, and consumer products, such as shoes, that are subjected constantly to repetitive stresses in the form of tension, compression, bending, vibration, thermal expansion and contraction, or other stresses. These stresses are often below the yield strength of the material; however, when the stress occurs a sufficient number of times, it causes failure by fatigue! Quite a large fraction of components found in an automobile junkyard belongs to those that failed by fatigue. The possibility of a fatigue failure is the main reason why aircraft components have a finite life.

65 疲劳指的是重复性施加大于或者小于屈服强度的应力，使得材料强度降低或者材料崩溃。

66 通常，成核位置位于应力最大的表面或者附近，包括表面缺陷如刮擦或凹痕、由于不好的设计或者制造导致的尖角、内嵌物、晶界、或者位错集中。

67 聚合物疲劳机制与金属材料不同。当对聚合物施加重复应力时，在裂痕尖端会产生可观的热量，疲劳与另外一个称为蠕变的机制共同影响聚合物的行为。

68 疲劳崩溃通常很容易识别。断裂表面-尤其是在（断裂）开始的附近，通常是光滑的。当起始裂纹大小增加时表面粗糙度增加，并且在最后的裂纹拓展过程中可能是纤维状的。

Fatigue failures typically occur in three stages. First, a tiny crack initiates or nucleates often at a time well after loading begins. Normally, nucleation sites are located at or near the surface, where the stress is at a maximum, and include surface defects such as scratches or pits, sharp corners due to poor design or manufacture, inclusions, grain boundaries, or dislocation concentrations.[66] Next, the crack gradually propagates as the load continues to cycle. Finally, a sudden fracture of the material occurs when the remaining cross-section of the material is too small to support the applied load. Thus, components fail by fatigue because even though the overall applied stress may remain below the yield stress, at a local length scale, the stress intensity exceeds the tensile strength. For fatigue to occur, at least part of the stress in the material has to be tensile. In practice, the fatigue of metallic and Polymeric materials is normally concerned.

The mechanism of fatigue in polymers is different to that in metallic materials. In polymers, as the materials are subjected to repetitive stresses, considerable heating can occur near the crack tips and the interrelationships between fatigue and another mechanism, known as creep, affect the overall behavior.[67]

Fatigue failures are often easy to identify. The fracture surface—particularly near the origin—is typically smooth. The surface becomes rougher as the original crack increases in size and may be fibrous during final crack propagation.[68] Microscopic and macroscopic examinations reveal a fracture surface including a beach mark pattern and striations (Figure 5.35). Beach or clamshell marks are normally formed when the load is changed during service or when the loading is intermittent, perhaps permitting time for oxidation inside the crack. Striations, which are on a much finer scale, show the position of the crack tip after each cycle. Beach marks always suggest a fatigue failure, but—unfortunately—the absence of beach marks does not rule out fatigue failure.

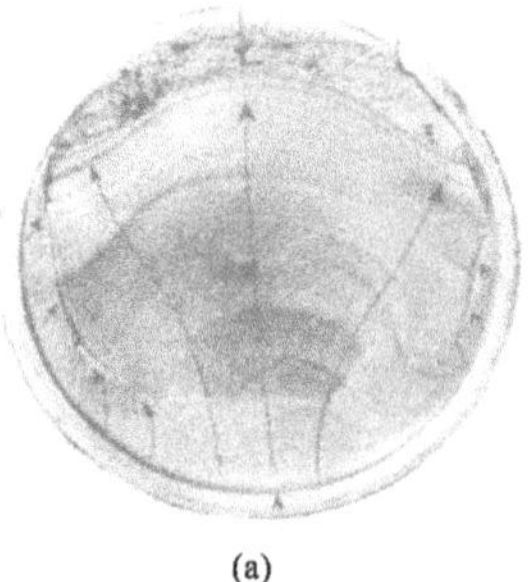

(a)

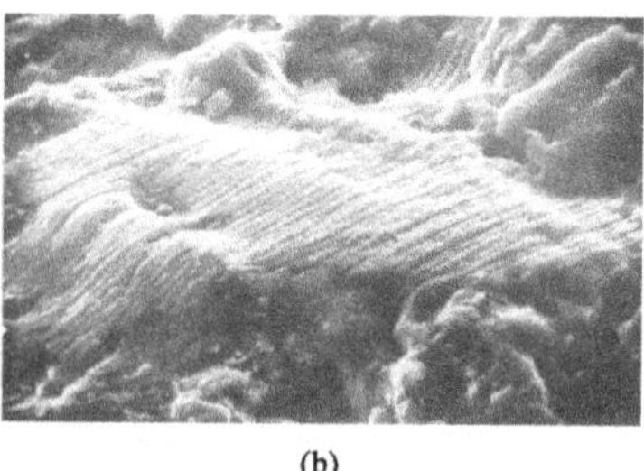

(b)

Figure 5.35　Fatigue fracture surface. (a) At low magnifications, the beach mark pattern indicates fatigue as the fracture mechanism. (b) At very high magnifications, closely spaced striations formed during fatigue are observed. (Reprinted from The Science and Engineering of Materials)

本章小结

1. 内容概要

本章阐述了材料的力学性质，即材料受力后的变化，包括形变、断裂、疲劳等。首先，解释和讨论了描述材料力学性质的基本术语及过程，如：应力、应变、杨氏模量、屈服强度、拉伸强度、延长率、硬度、弹性/塑性形变、以及黏弹性等；然后，重点阐述了形变过程和相应的机理，同时也介绍了材料力学性质的一些常用测试方法。

2. 基本概念

张力、压力、剪切力、塑性形变、弹性形变、扭转角、剪切应变、黏性形变、玻璃化转变温度、杨氏模量、屈服强度、屈服点、压缩模量、抗拉强度、工程应变、真实应变、应变硬化、延长率、泊松比、韧性、弯曲应力、断裂模量、弯曲模量、硬度、耐磨、冷处理、溶质硬化、二相硬化、蠕变、应力松弛。

3. 主要公式

（1）工程应力：$\sigma = \dfrac{F}{A_0}$

（2）工程应变：$\varepsilon = \dfrac{l - l_0}{l_0} = \dfrac{\Delta l}{l_0}$

（3）压缩应力：$p = \dfrac{F}{A_0}$

（4）压缩应变：$\varDelta = \dfrac{V_0 - V_t}{V_0}$

（5）切应力：$\tau = \dfrac{F_s}{A_s}$

（6）切应变：$\gamma=\tan\theta$

（7）弹性模量：$E = \dfrac{\sigma}{\varepsilon}$

（8）剪切模量：$G = \dfrac{\tau}{\gamma}$

（9）压缩模量：$B = \dfrac{p}{(V_0 - V_t)/V_0}$

（10）真实应力：$\sigma_T = \dfrac{F}{A_{actual}}$

（11）真实应变：$\varepsilon_T = \int_{l_0}^{l} \dfrac{dl}{l} = \ln\left(\dfrac{l}{l_0}\right)$

（12）应变指数 n：$\sigma_T = K\varepsilon_T^n$

（13）延展率：$\delta = \dfrac{l_x - l_0}{l_0} \times 100\%$

（14）弯曲应力：$\sigma_{\text{bend}}=\dfrac{3FL}{2wh^2}$

（15）断裂模量：$MOR=\dfrac{3F_{\text{f}}L}{2wh^2}$

（16）弯曲模量：$E_{\text{bend}}=\dfrac{L^3F}{4wh^3\delta}$

（17）布氏硬度：$HB=\dfrac{2F}{\pi D\left[D-\sqrt{D^2-D_{\text{i}}^2}\right]}$

（18）泊松比：$\upsilon=-\dfrac{\varepsilon_{\text{x}}}{\varepsilon_{\text{z}}}$

（19）不同模量之间关系：$E=2G(1+\upsilon)=3B(1-2\upsilon)$

（20）派-纳力（Peierls-Nabarro force）：$\tau_{\text{P-N}}=\dfrac{2G}{1-\upsilon}\exp[\dfrac{-2\pi a}{(1-\nu)b}]=\dfrac{2G}{(1-\nu)}\exp(\dfrac{-2\pi W}{b})$

（21）位错宽度：$W=\dfrac{a}{1-\nu}$

（22）霍尔—佩奇关系（Hall-Petch equation）：$\sigma_{\text{s}}=\sigma_0+Kd^{-\frac{1}{2}}$

Vocabulary

damping	阻尼
deflection	偏斜，偏度，偏转角
dislocation	位错
dislocation width	位错宽度
dispersion strengthening	弥散强化
ductility (elongation percentage)	延长率
edge dislocation	刃型位错
elastic Deformation	弹性形变
elastomer	弹性体
engineering strain	工程应变
engineering stress	工程应力
flexural stress	抗弯强度，弯曲
gage length	标距长度
glass transition temperature	玻璃转变温度
hardness	硬度
Hooke’s law	胡克定律
hysteresis	假设
indentation	凹入
internal friction	内耗

intersection	交点
lattice distortion	晶格畸变
lower yield point	下屈服点
magneto-elasticity	磁弹性
melting range	熔化范围
microhardness	显微硬度
modulus (moduli)	模量
modulus of elasticity, E	弹性模量
modulus of rigidity	刚性模量
modulus of rupture	断裂模量
Moh's hardness	莫氏硬度
natural aging	自然时效
nonlinear	非线性
pinning site	钉扎位置
plastic deformation	塑性形变
Poisson’s ratio	泊松比
precipitation strengthening	析出（沉淀）强化
refractory material	耐火材料
relaxation time	弛豫时间
residual stress	剩余应力
residual torsion angle	残余扭转角
ripple	波纹
Rockwell hardness	洛氏硬度
screw dislocation	螺型位错
shear	剪切
shear strain	剪切应变
shear strength	抗剪强度
shear stress	剪切应力
slip system	滑移系统
softening point	软化点，软化温度
softening temperature	软化温度
solution hardening	溶质硬化
stiffness	顽固，僵直，硬
strain hardening	应变硬化
strain-hardening exponent	应变硬化指数
stress versus strain	应力与应变
string model	弦模型
tempered glass	回火玻璃
tensile	张力的，拉长的
tensile strength, T.S.	拉伸强度

thinning	变薄
torque moment	扭矩
torsion angle	扭转角
torsion shear strain	扭转切应变
toughness	韧性
transgranula	穿晶
true stress	真实应力
upper yield point	上屈服强度
viscous deformation	黏性形变
yield point	屈服点
yield strength for torsion	扭转屈服强度
yield strength, Y.S.	屈服强度
Yong's modulus	杨氏模量

Problems

1. Define “engineering stress” and “engineering strain”, “true stress” and “true strain”.

2. According to the direction of forces, please classify the stress.

3. If a tensile, shear and compressive stress produce a linear, angular and volume displacement of $(l-l_0)$, θ, (V_0-V_t), respectively, what is the corresponding strain?

4. Define “modulus of elasticity”.

5. True or false questions:

(1) The yield strength shows the resistance of the materials to permanent deformation.

(2) The plastic deformation is a kind of temporary deformation.

(3) The yield stress of ceramics is generally higher than that of metals.

(4) The maximum strength in the stress-strain curve is the yield strength.

(5) Ductility indicates the general ability of a material to be plastically deformed.

(6) Plastic deformation is generally realized by atomic plane slipping.

(7) Metals possess large dislocation width, resulting in small yield strength, thus more ductile. While ceramics possess small dislocation width, thus lattice resistance is large, leading to brittle feature.

(8) Material with more slip systems exhibits better ductility.

6. What is strain hardening?

7. Please state the micromechanism for elastic deformation.

8. What is the fundamental mechanism of plastic deformation?

9. What is meant by the term “stress relaxation?”

10. The aluminum alloy with initial length of 2.000 in. has a final length after failure of 2.195 in. Calculate the ductility of this alloy.

11. Using string model to explain the upper and lower yield behavior in metal.

Chapter 6
Electrical Properties

This chapter deals with electrical properties of materials, including conductors, semiconductors, superconductors, as well as the dielectric and thermoelectricity. Some applications based on them are also introduced. In particular, corresponding properties of organic materials are discussed.

6.1 Basic Concepts for Electricity

6.1.1 Ohm's Law

The electrical properties of matter can be described in two ways, either by resistivity or by conductivity.

If a voltage U is applied to a homogeneous solid with length of l in a closed circuit (Figure 6.1), an electric field of

$$F = \frac{U}{l} \tag{6-1}$$

is present at every point in the solid and electric current I occurs in the circuit. In most cases, such as metal, the electricity produced due to this electric field conforms with Ohm's law, which states that the current through a conductor between two points is directly proportional to the potential difference across the two points,[1]

$$I = \frac{1}{R}U \tag{6-2}$$

where I is the current (amperes or amps, A), and R is the resistance (ohms, Ω). The resistance (R) of a resistor is a characteristic of the size, shape, and properties of the resistor, can be described as:

$$R = \rho\frac{l}{A} = \frac{i}{\sigma A} \tag{6-3}$$

where A is the cross-sectional area (cm^2) of the resistor, ρ is the electrical resistivity (Ohm·cm or Ω·cm), and σ, which is the reciprocal of ρ, is the electrical conductivity ($ohm^{-1}\cdot cm^{-1}$). The magnitude of the resistance depends upon the dimensions of the resistor, i.e., proportional to the length and reciprocal to the cross-sectional area of the resistor[2]. The resistivity or conductivity does not depend on the dimensions of the material. Thus, resistivity or conductivity can be compared among different materials. For example, silver is a better conductor than copper because of silver's higher conductivity. Resistivity is a microstructure-sensitive property, similar to yield strength in mechanic property.The resistivity of pure copper is much less than that of commercially high purity copper due to scattering electrons of impurities, which contribute to raising resistivity. Similarly, the resistivity of annealed high purity copper is slightly lower than that of cold-worked high purity copper because of the scattering effect associated with dislocations[3]. In electrical energy transportation, minimizing power losses is very important, not only to conserve energy, but also to minimize thermal effect. The electrical power P (in watts, W) lost when a current flows through a resistance is given by

1 在大多数情况下，如在金属中，电场产生的电流符合欧姆定律，即通过导体的电流与加在导体两端的电压成正比。

2 电阻的大小取决于电阻器的尺寸，电阻值正比于电阻器的长度，反比于电阻器的横截面积。

3 纯铜的电阻率比商业上高纯度的铜小很多，这是因为杂质产生的电子散射导致电阻率的增加。相似地，由于位错引起的散射效应，退火高纯铜的电阻率比冷加工高纯铜稍微低一些。

$$P=UI=I^2R \tag{6-4}$$

A high resistance R results in larger power losses. These electrical losses are known as Joule heating losses.

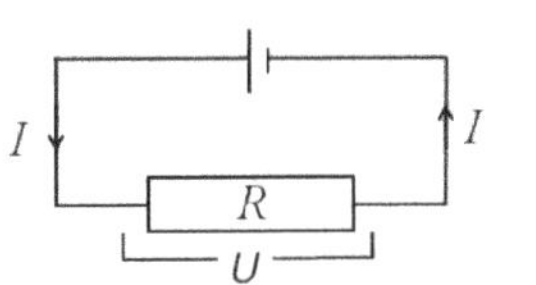

Figure 6.1 Simple electrical circuit.

6.1.2 Conductivity

Conductivity and resistivity are the macroscopic parameters for judging material electricity. In the micro point of view, electrical conductivity refers to the degree of carrier drift under electric field, which can be expressed by carrier density and mobility.[4] Combining Ohm's law (Equation 6-2) and the definition for resistivity (Equation 6-3), one has

$$I=\frac{1}{R}U=\frac{1}{\left(\frac{l}{A}\right)\frac{1}{\sigma}}U$$

$$\frac{I}{A}=\sigma\frac{U}{l} \tag{6-5}$$

If we define I/A as the current density J (A/cm^2), then

$$J=\sigma F \tag{6-6}$$

On the other hand, the current density J, referring to the charges (Q) passing through a unit cross section within unit time (see Figure 6.2 for parameters of A, t and v), can also be expressed as,

$J=Q/(area \times time)=$ (carrier - density × elementary - charge × volume) $/(area \times time)=(n \cdot q \cdot A \cdot v \cdot t)/(A \,.\, t)=n \cdot q \cdot v$.

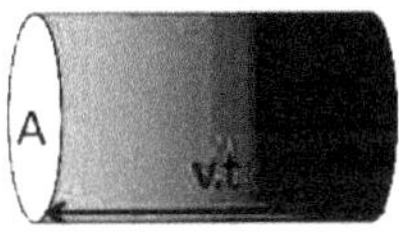

Figure 6.2 The charges with velocity of v, passing through a cross section area of A within time of t.

Therefore, the general equation for current density is:

$$J=nq\overline{v} \tag{6-7}$$

where n is the charge carrier density (carriers/cm^3), q is the elementary charge, and $\overline{v}$ is the average drift velocity (cm/s) of the charge carriers. Thus,

$$\sigma F=nq\overline{v} \tag{6-8}$$

Diffusion occurs as a result of temperature and concentration gradients, and drift occurs as a result of an applied electric or magnetic field. Conduction may occur as a result of diffusion, drift, or both, but drift is the dominant mechanism in electrical conduction.[5]

Define mobility (μ) of carrier as the carrier drift velocity (shown as Figure 6.3a) under unit electric field:

4 电导率和电阻率是表征材料电学性质的宏观参数。微观上，电导率指的是在电场下载流子迁移的程度，可以用载流子密度和迁移率表示。

5 温度和浓度梯度可以引起扩散，而迁移是由于外加电场或磁场引起的。传导可以通过扩散、迁移、或者两者同时产生，但是导电的主要机制是迁移。

$$\mu = \frac{\overline{v}}{F}\,(\mathrm{cm^2 \cdot V^{-1} \cdot s^{-1}}) \tag{6-9}$$

The mobility μ depends on atomic bonding, imperfections, microstructure, and, in ionic compounds, diffusion rates. This parameter will be further discussed in the later sections of intrinsic and extrinsic semiconductors. Combining Equation (6-8) and (6-9), one gets,

$$\sigma = nq\mu \tag{6-10}$$

The elementary charge q is a constant. From Equation (6-10), we know that we can manipulate the electrical conductivity of materials by (1) controlling the number of charge carriers in the material or (2) controlling the mobility.[6] Different type of materials hold their own features of carrier density and mobility, and exhibit different conductivity. As shown in Figure 6.3b, there are many free electrons in metal, which can move easily; in semiconductors and insulators (Figure 6.3c), if no impurities, valence bonds must be broken for an electron to be able to move to produce charge conduction; in doped semiconductor, electron and hole can be produced with the activation of thermal energy, then move under electric field; in many ionic bonded materials such as ceramics, when conduction does occur, it can be the result of electrons that "hop" from one defect to another, or the movement of entire ions via vacancy mechanism[7] (Figure 6.3d).

6 基本电荷 q 是常量。从公式（6-10）中可知我们可以通过以下方式控制材料的电导率：（1）控制材料中载流子数量，或者（2）控制载流子迁移率。

7 在很多离子键结合的材料中，如陶瓷，如果存在导电，可能是电子从一个缺陷“跳到”另一个缺陷，或者整个离子通过空位的移动。

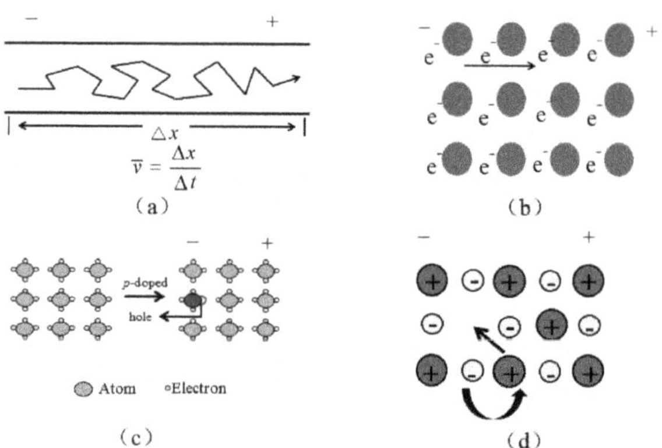

Figure 6.3 Charge carriers in different types of materials. (a) Schematic view for carrier drift velocity: charge carriers, such as electrons, are deflected by atoms or defects and take an irregular path through a conductor. The average rate at which the carriers move is the drift velocity. (b) Metals. (c) Intrinsic/extrinsic semiconductors and insulators. (d) Many ionic bonded materials.

The mobility is particularly important in metals, whereas the number of carriers is more important in semiconductors and insulators.[8] In semiconductors, electrons induce current as do positively charged carriers known as holes. Electrons and holes flow in opposite directions in response to an applied electric field and both contribute to the net current.[9] Thus, Equation (6-10) can be modified as follows to describe the conductivity of semiconductors:

$$\sigma = nq\mu_n + pq\mu_p \tag{6-11}$$

8 在金属中迁移率尤其重要，而载流子数量在半导体和绝缘体中比较重要。

9 在电场的作用下，电子和空穴相向流动，从而都对净电流产生贡献。

Here, μ_n and μ_p are the mobility of electrons and holes, respectively. The terms n and p represent the concentrations of free electrons and holes in the semiconductor.

6.1.3 Mean Free Path

Mean free path of electron refers to the average distance an electron travelled between two collisions. If the time span (τ) between two collisions and the average kinetic velocity (v) of the electrons are known, the mean freepath (l) can be calculated from the relationship:

$$l=\tau v \tag{6-12}$$

The magnitude of the mean free path is about 100 times distances between the atoms in the lattice. For Cu, the value is about 40nm. The larger the mean free path, the higher the electrical conductivity.

6.1.4 Materials with Different Electrical Properties

Electrical materials can be classified as (a) superconductors, (b) conductors, (c) semiconductors, and (d) dielectrics or insulators, depending upon the magnitude of their electrical conductivity. Materials with conductivity less than $10^{-12}\,\Omega^{-1}\cdot\text{cm}^{-1}$ are considered insulating or dielectric. Materials with conductivity less than $10^{3}\,\Omega^{-1}\cdot\text{cm}^{-1}$ but greater than $10^{-12}\,\Omega^{-1}\cdot\text{cm}^{-1}$ are considered semiconductors. Materials with conductivity greater than $10^{3}\,\Omega^{-1}\cdot\text{cm}^{-1}$are considered conductors. These are approximate ranges of values. Table 6.1 illustrated the electrical conductivity of different materials. These values are approximate and are for high-purity materials at 300 K (unless noted otherwise). One should also be noted that the values of conductivity for metals and semiconductors depend very strongly on temperature.[10]

We use the term "dielectric" for materials that are used in applications where the dielectric constant is important. The dielectric constant (ε) of a material is a microstructure sensitive property related to the material's ability to store an electrical charge.[11] We use the term "insulator" to describe the ability of a material to block the flow of DC or AC current, as opposed to its ability to store a charge. A measure of the effectiveness of an insulator is the maximum electric field it can sustain without an electrical breakdown.

10 需要注意的是，金属和半导体的电导率强烈依赖于温度。

11 材料的介电常数（ε）是一个微观结构敏感的性质，与材料储存电荷能力有关。

Table 6.1 Electrical conductivity of selected materials at T = 300 K[a]

Material Category	Name	Conductivity ($\text{ohm}^{-1}\cdot\text{cm}^{-1}$)
Super-con ductors	Hg	Infinite (under certain conditions such as low temperatures)
	Nb_3Sn	
	$YBa_2Cu_3O_{7-x}$	
	MgB_2	

continued table

Material Category	Name	Conductivity ($ohm^{-1}\cdot cm^{-1}$)
Metals	Na (Alkali metal)	2.13×10^{5}
	K (Alkali metal)	1.64×10^{5}
	Mg (Alkali earth metal)	2.25×10^{5}
	Ca (Alkali earth metal)	3.16×10^{5}
	Al (Group 3B metal)	3.77×10^{5}
	Ga (Group 3B metal)	0.66×10^{5}
	Fe (Transition metal)	1.00×10^{5}
	Ni (Transition metal)	1.46×10^{5}
	Cu (Group 1B metal)	5.98×10^{5}
	Ag (Group 1B metal)	6.80×10^{5}
	Au (Group 1B metal)	4.26×10^{5}
Semi-Conductors	Si (Group 4B element)	4×10^{-6}
	Ge (Group 4B element)	0.02
	GaAs (Compound)	2.5×10^{-9}
	AlAs (Compound)	0.1
	SiC (Compound)	10^{-10}
Insulators, Linear,and Nonlinear Dielectrics	Polyethylene (Polymer)	10^{-15}
	Polytetrafluoroethylene (Polymer)	10^{-18}
	Polystyrene (Polymer)	10^{-17} to 10^{-19}
	Epoxy (Polymer)	10^{-12} to 10^{-17}
	Alumina (Al_2O_3, Ceramic)	10^{-14}
	Silicate glasses (Ceramic)	10^{-17}
	Boron nitride (BN, Ceramic)	10^{-13}
	Barium titanate ($BaTiO_3$, Ceramic)	10^{-14}
	C (diamond, Ceramic)	$<10^{-18}$

[a]Unless specified otherwise, assumes high-purity material.

12 质子通常可以被认为起次要的作用，如保持电中性。中子有时候需要被考虑，例如在一些超导体材料中，临界温度取决于整个原子的质量。但是总的来说，电子能级是影响固体电学性质的关键。

As discussed previously, there are different types of materials considering electricity. The question is that what is account for these differences? In a common sense, the electrical properties of solids are mainly determined by the properties of electrons in them. Protons can usually be relegated to subordinate roles, like ensuring charge neutrality. Neutrons may sometimes need to be considered, as for example in some superconducting materials, in which the critical temperature depends on the total mass of the nucleus, but on the whole, the energy levels of electrons hold the key to the electrical properties of solids.[12] Modern theory of electron is the fundamental basis of material electricity and a key to understand materials with different electricity.

In this chapter, the different type electrical materials will be studied and explained by modern free electron theory and band theory.

6.2 Conductor

6.2.1 Metal Conductor

Ⅰ. Conducting Mechanism

Metals are excellent electrical conductors and their dominant carrier is free electron. Under the electric field, the free electrons, which are moving randomly due to the thermal activation, are accelerated and move to the direction opposite to the electric field.The number of charge carriers in metals is practically independent of temperature, but the conductivity depends on the material and is strongly influenced by temperature, purity, lattice defects, and constitution.[13] The specific electrical resistivity ρ is the reciprocal value of the electrical conductivity: $\rho = 1/\sigma$. A typical order of magnitude for metals is $\rho \sim 10^{-7}\ \Omega\text{m}$. Using classical mechanics, the expression for conductivity of an electron can be deduced as following:

13 金属中载流子的数量实际上与温度无关，但是电导率依赖于材料，并且强烈依赖于温度、纯度、晶格缺陷和组成。

Since in metals only electrons contribute to the electrical current, according to Equation (6-7), the current density is,

$$j=-nqv \qquad (6\text{-}13),$$

where n is the number of conduction electrons per unit volume, q is the elementary charge, and v is the average velocity of the electrons. Given the electric field is F, the average time between two collisions (relaxation time) is τ, electron mass is m, the force (f) a free electron feels is:

$$f=ma=m\frac{v}{\tau}=-eF \qquad (6\text{-}14)$$

From Equation (6-14) one get,

$$v=-\frac{eF\tau}{m} \qquad (6\text{-}15)$$

Combining with the definition of current density (Equation (6-13)), results in

$$j=-nev=n\frac{e^2F\tau}{m} \qquad (6\text{-}16)$$

Compare with Ohm's law, $j = \sigma F$, yields the electrical conductivity:

$$\sigma=\frac{e^2n\tau}{m}=\frac{e^2nl}{mv} \qquad (6\text{-}17)$$

where $l = \tau v$, is the mean free length of electrons.

As expressed in Equation (6-17), the conductivity of metal is proportional to free electron density n and mean free path of electron l.

Under classical free electron theory, the free electrons are treated as ideal gas and have no interaction between each other; their behaviors conform with Maxwell-Boltzmann statistics. Based on the above hypothesis, when electrical bias is applied, the free electrons in metal, moving randomly due to the thermal activation, are accelerated and have a direction movement opposite to the electric field.[14] When electron collides with lattice atom, movement is hampered, resulting in the decrease of electricity. This is the reason for resistance of metal. Though the conductivity for metals can be obtained based on classical free electron theory using Equation(6-17), its value is rather large than that of experimental result. This is because all the free electrons are assumed responsible for electrical conduction in classical model, leading to a larger values of carrier density.

14 基于上述假设，当施加电压时，金属中原本由于热激发而无序运动的自由电子，将被加速，并且沿着与电场相反的方向定向运动。

The quantum free electron theory concludes that electrons generally occupy the energy levels below Fermi level, and there are two paired electrons in one orbital.[15] Only electrons near Fermi level can be excited to higher energy levels by thermal activation and contribute to conductivity.Under this situation, conductivity is

15 量子自由电子理论得出的结论是，电子一般占据费米能级以下的能级，且一个轨道内含有两个成对的电子。

$$\sigma = \frac{e^2 n \tau_F}{m} = \frac{e^2 n l_F}{m v_F} \tag{6-18}$$

This has the same form as the classical one. But meanings of parameter are different: the n, τ_F, l_F, v_F are the density, relaxation time, the mean free path, and the drift velocity of electrons in the vicinity of Fermi level. Quantum free electron model can successfully explain the conductivity of alkaline metals. Other metals, such as transition metals, due to the complicated electronic structure, the electron distribution cannot be simply described by Fermi ball. Thus band theory need to be taken for electricity explanation.[16]

16 量子自由电子模型可以成功地解释碱金属的电导率。其他金属，例如过渡金属，由于复杂的电子结构，电子分布不能简单地描述为费米球。因此，需要引入能带理论来解释电学性质。

Under band theory, the conductivity of metal can be expressed as:

$$\sigma = \frac{e^2 n^* \tau_F}{m^*} = \frac{e^2 n^* l_F}{m^* v_F} \tag{6-19}$$

where n^* is the number of effective electron density, representing the number of real electrons per volume contributing to conductivity. m^* is the effective electron mass resulting from the interaction between lattice points and the electric field of other free electrons. This equation can be applied not only to metals, but also to nonmetal materials. For alkaline metals, $n^* = n$, $m^* = m$, the equation changes to Equation (6-18). Since different materials have different n^*, the variation of conductivity is very large:

(1) **Monovalent metals.** The alkaline metals of Li, Na, K, Rb, Cs in IA group, and transition metals of Cu, Ag, Au in IB group are monovalent

elements (Figure 6.4):

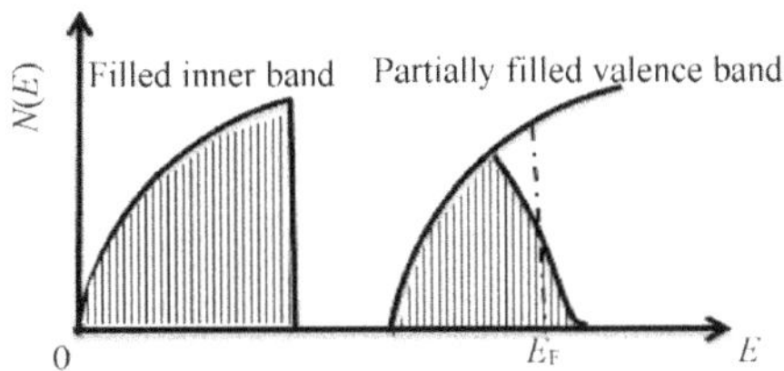

Figure 6.4 Density of state in monovalent metals.

In these materials, the outershell s valence band is half filled.The electrons in this band can move freely. Therefore, the quantity of the effective electrons for conduction is large, leading to good conductivity.[17] Their resistivity is in the range of $\rho = 10^{-6}–10^{-2}$ Ω·m.

17 这些材料中，最外层的 s 价电子能带是半充满的。这个能带中的电子可以自由移动。因此用于导电的有效电子数量很大，导致很好的导电性。

(2) **Divalent metals.** IIA group elements of Be, Mg, Ca, Sr, Ba and IIB group elements of Zn, Cd, Hg are divalent metals (Figure 6.5):

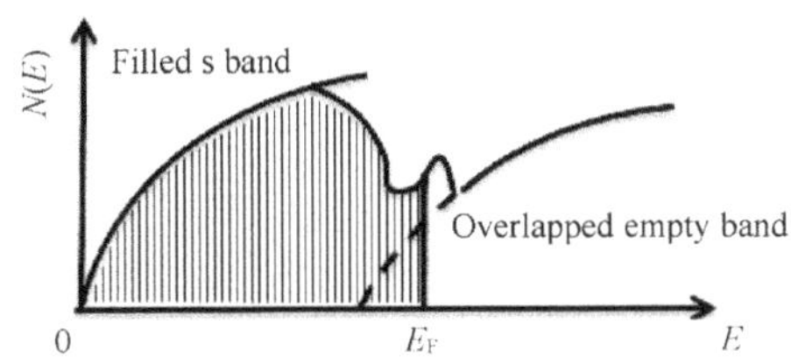

Figure 6.5 Density of state in divalent metals.

The s valence band is filled. These materials should be expected as insulators. However, due to the 3-dimensional crystal structure, bands are overlapped, electrons can fill in the empty conducting band. Thus bandgap does not exist and the number of effective electron for conduction is large.These materials are conductors.[18]

18 然而，由于三维的晶体结构，能带是相互重叠的，电子可以填入到空置的导带上。所以带隙不存在，可以导电的有效电子数量是巨大的。这些材料是导体。

(3) **Trivalent element materials.** IIIA group elements of Al, Ga, In and Tl are trivalent elements. Since they have filled s valence band, and partially filled p valence band. Electrons in the partially filled p valence band are conducting electrons. These materials are conductors.

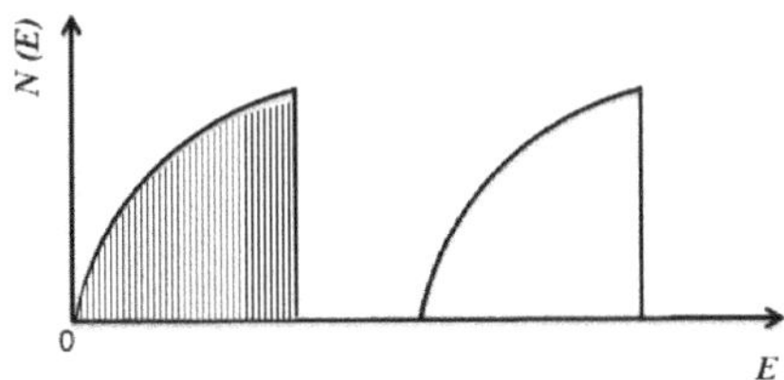

Figure 6.6 Density of state of quadrivalent element materials.

(4) **Quadrivalent elements.** The IVA group elements of Si and Ge are quadrivalent elements. They have filled valence band and empty conducting

band, and exhibit bandgap of E_g (Figure 6.6). The E_g of Si and Ge are 1.14 eV and 0.67 eV, respectively. These values are relatively small. At room temperature, valence electrons can be thermally excited to conducting band, resulting conductivity. Thus at low temperature, the number of effective electron for conduction is low, and these materials are insulators. At room temperature, they are semiconductors because some of the valence electrons can be thermally excited to conducting band.

(5) **Pentavalent metals**. As, Sb and Bi are pentads. Since each primitive cell has two atoms, there are 5 bonding valence-orbitals with 10 electrons, which are paired; and 5 antibonding valence-orbitals, which are empty. Therefore, effective electrons are rare. These materials are not good conductors, with conductivity of 4 magnitudes lower than a common metal. [19] Due to the low conductivity, and the similar temperature dependence to metals, these materials are called half metals.

19 由于每个原胞有两个原子，形成被 10 个成对价电子占据的 5 个成键价电子轨道，以及 5 个空置的反键价电子轨道。因此有效电子非常少，所以这些材料不是良导体，它们的电导率比一般的金属小 4 个数量级。

(6) **Ionic crystals.** The band structure of an ionic crystal is the same as quadrivalent elements, but the bandgap E_g is larger. The number of effective electrons is zero. These materials are insulators. For examples, in NaCl crystal, a 3s electron from Na shifts to 3p orbital of Cl,3s of Na^+ is empty, and 3p of Cl^- is full. The bandgap is 10 eV, which is too large to be thermally excited. Thus NaCl is an insulator.

Ⅱ. Crystalline Influences to Metal Conductivity

All transporting properties, including the electrical conductivity of crystalline solids, are not scalars but rather a symmetrical tensor of rank 2 because the transportation depends on the direction of crystallographic. [20] However, the conductivity of cubic crystals is the same in all directions. Only for this isotropic case is the conductivity represented by a single number, a scalar. The conductivity/resistivity of hexagonal (or lower symmetry crystal structure) metals are different along the *c* ($\rho_\perp$) and *a* ($\rho_\perp$) axis (Figure6.7). For an arbitrary direction inclined by the angle *a* to the basal plane, the resistivity can be expressed as:

20 所有的传输性能，包括晶态固体的电导率都不是标量而是 2 阶对称张量，这是因为传输依赖于晶向。

$$\rho_\alpha=\rho_\perp+(\rho_\parallel-\rho_\perp)\cdot\cos^2\alpha \tag{6-20}$$

For even lower crystal symmetry, the electrical conductivity will be different in all three spatial directions, for instance gallium (50.5; 16.1; 7.5)$\times10^{-8}$ Ωm. Materials usually are used as polycrystalline. In this case, the conductivity is an average value over all orientations and, therefore, would be isotropic except if the material has a pronounced crystallographic texture. [21] Here, we only consider the average condition without considering the crystal structure.

21 在这种情况下，电导率是所有方向上的平均值，因此，是各向同性的，除非材料有明显的晶体取向。

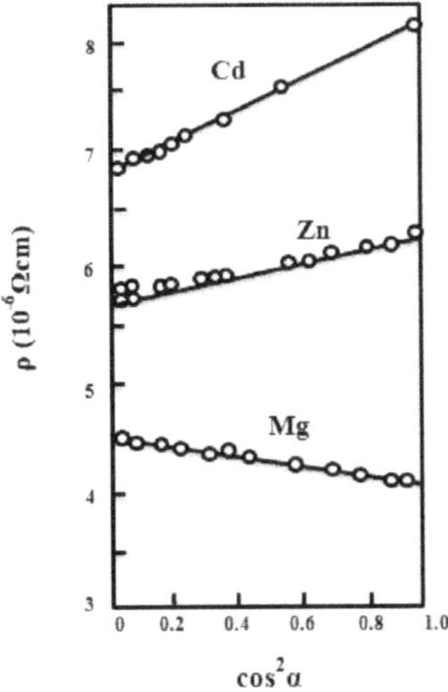

Figure 6.7 Electrical resistivity of some hexagonal metals as a function of the angle α with the basal plane. Resistivity change is linear with $\cos^2\alpha$.

Ⅲ. Origination of Resistance in Metal

From band theory, valence electrons of metal can move freely inside lattice. In ideal crystal lattice and at low temperature (e.g. 0 K), electron wave does not suffer from scattering, thus, resistance is zero. Destroying lattice periodicity and/or introducing lattice vibration (i.e. high temperature) result in electron scattering, which are resistance source. The influence factors of resistance include temperature, impurity and defect (vacancy, interstitial atoms, dislocation, and grain boundary etc.).

At high temperature, lattice vibration is strong, and is the dominant factor for resistance; at low temperature, impurity and defect are the dominant factors.[22] Figure 6.8 shows the relationship between resistivity, temperature and impurity/defect. For ideal conductor without impurity and defect, at 0 K, the resistance will be zero. Introducing defect, resistance appears at absolute temperature zero. And if both defect and impurity are introduced, the resistance will be larger.

22 高温下，晶格振动强烈，是产生电阻的主要因素；低温下，杂质和缺陷是主要因素。

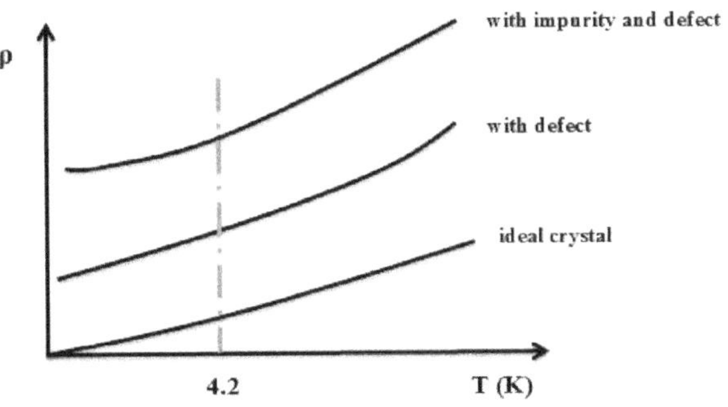

Figure 6.8 The relationship between resistivity, temperature and impurity/defect.

Matthiessen Rule

From Equation (6-19), define $\mu=1/l_F$ as scattering coefficient, then:

$$\rho=\frac{m^* v_F}{e^2 n^*}\mu \tag{6-21}$$

In real material, $\mu=\mu_T+\Delta\mu$, where μ_T is the electron scattering caused by lattice phonon, which is related to temperature. $\Delta\mu$ is the electron scattering caused by impurity and defects, which is related to impurity and defect concentration, and independent of temperature.

So,

$$\rho=\rho_0+\rho(T) \tag{6-22}$$

This means that resistance can be divided into two parts, temperature dependent part and temperature independent part. The temperature independent part is related to impurity and defect concentration, etc. This is Matthiessen Rule.[23]

23 这意味电阻可以分为两个部分：温度相关部分和温度不相关部分。温度无关部分与杂质和缺陷浓度等相关。这是马西森定律。

RRR (Residual Resistivity Ratio)

The resistivity at 4.2 K defines as the residual resistivity. The influence of impurities can be probed as residual resistivity because it dominates at low temperatures where the lattice vibrations are frozen. The residual resistivity increases with increasing concentration of impurity atoms. Under certain preparation procedure, the defect concentration is fixed, thus the residual resistance can be an indication of metal purity.[24]

24 在一定的制备工序下，缺陷浓度是确定的，所以残余电阻可以用来表示金属纯度。

RRR (Residual resistivity ratio) refers to the ratio of resistivity under 4.2 K to resistivity under 300 K:

$$RRR=\frac{\rho_{300\text{K}}}{\rho_{4.2\text{K}}} \tag{6-23}$$

The higher the RRR, the purer is the metal. RRR can be up to $10^4\sim10^5$.

Ⅳ. Influence of Temperature

Definition of Phonon: As shown in Figure 6.9, atoms in crystal keep on vibrating around its equilibrium position in the lattice. The atoms separate and aggregate alternatively within elastic range.These movements are termed as lattice vibration and possess wave property which can be termed as lattice wave with quantized energy. We called phonon.[25]

25 如图 6.9 所示，晶体中的原子在其晶格平衡位置附近保持振动。在弹性范围内，原子的分离和聚集运动交替进行。这些运动称为晶格振动，具有波性质，可以被称为具有量子化能量的晶格波。我们把它称作声子。

From Equation (6-10), we know that two factors account for conductivity, i.e., carrier density and mobility. In metals, among the two components of conductivity, only the mobility depends on temperature.

The mobility defines as,

$$\mu=\frac{v}{F}=\frac{v}{ma}=\frac{v}{m(v/\tau)}=\frac{\tau}{m} \tag{6-24}$$

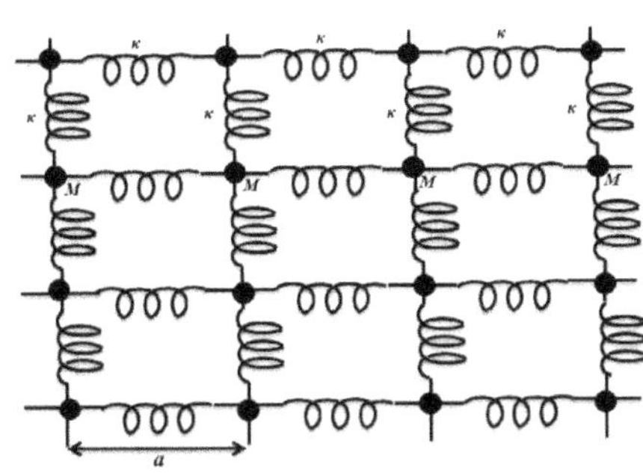

Figure 6.9 Lattice vibration(k: lattice vibration by energy of kT; M: metal atoms; a: lattice constant).

where v is velocity, F strength of electric field, m electron mass, τ time span between two collisions. According to Equation (6-24), the mobility is only affected via the time τ between collisions, which decreases with increasing numbers of scattering events. Such scattering centers are mainly lattice defects (impurity atoms) and lattice vibrations (phonons). Because,

$$\rho = \frac{1}{\sigma} = \frac{m \times v_F}{e^2 n \times l_F} \propto \frac{1}{l_F} \tag{6-25}$$

resistivity is proportional to the reciprocal of mean free path of the Fermi electron (l_F). For ideal crystal, since only phonon scattering exists, l_F is determined by phonon number, which increases with temperature.[26] Therefore, resistivity has different temperature-dependence in different temperature range:

26 对于理想晶体，由于只有声子散射存在，l_F 取决于声子数量，并随着温度的上升而增加。

At high temperature range of $T > 2T_D / 3$,

$$\rho \propto T \tag{6-26}$$

At low temperature of $T << T_D$:

$$\rho \propto T^5 \tag{6-27}$$

where T_D is the Debye temperature, i.e., the temperature that can result in phonons with wavelength equal to the distance between lattice atoms.

At extremely low temperature of below 2 K, electron to electron scattering is the main mechanism for resistance:[27]

27 在小于 2K 的极低温度下，电子散射是主要的电阻来源。

$$\rho \propto T^2 \tag{6-28}$$

At high temperatures the contribution of phonons dominates. Since the amplitude of lattice vibrations increases with temperature, the probability of collisions grows proportionally to temperature and, therefore, $\tau \sim 1/T$. Correspondingly, the resistivity in metals rises in proportion to temperature.[28] The lattice vibrations in dilute alloys are little influenced by solute atoms, this also explains Matthiessen's rule. With falling temperature, the lattice vibrations progressively decay. According to Debye the lattice vibrations decrease in proportion to T^3. The observed dependency,

28 在高温下，声子对电阻的贡献是主要的。由于晶格振动的振幅随温度增加，碰撞几率正比于温度，因此 $\tau \sim 1/T$。相应地，金属中电阻的增加与温度成正比。

proportional to T^5 is explained by the fact that the spectrum of exited phonons consists mainly of long-wave phonons, the momentum of which is so small that the electrons are scattered only through small angles, which is described by another T^2dependency.[29]

29 随着温度降低，晶格振动逐渐减小。根据 Debye 所述，晶格振动的降低正比于 T^3。观测到的 T^5 依赖关系，可用如下事实来解释：激发声子主要是长波声子，其动量很小，电子只是在很小角度时才受到晶格散射，这就可以用另外的 T^2 依赖关系描述。

TCR (Temperature Coefficient of Resistance)

Experimentally, an empirical formula can be used to express the relationship between resistance and temperature, neglecting the mechanism difference under different temperature range.[30] Using ρ_0 and ρ_T to express the resistivity under 0℃ and T℃, we have:

30 实验地，忽略不同温度区域时的不同机制，可用一个经验公式来表示电阻和温度的关系。

$$\rho_T = \rho_0(1+\alpha T) \tag{6-29}$$

where α is the temperature coefficient of resistance. This formula is applicable to most of metals under room temperature.Under temperature of 0-T℃, the average temperature coefficient of resistance is:

$$\bar{\alpha} = \frac{\rho_T - \rho_0}{\rho_0 T}(℃^{-1}) \tag{6-30}$$

α is not a constant, its value for T℃ can be expressed as:

$$\alpha_T = \frac{1}{\rho_T}\frac{d\rho}{dT}(℃^{-1}) \tag{6-31}$$

and can be determined by experiment.

Ⅴ. Functional Conducting Metals

- **Conducting Materials**

Table 6.2 shows the conductivity of many metals. As can be seen that both the electrical and thermal conductivity of Cu, Al, Ag are good, which are commonly used as conducting materials. Cu wire is generally made of electrolytic copper in order to improve its purity.

Table 6.2 The conductivity of metals

Thermal and Electric Conductivity Based on Copper = 100		
	Thermal	Electric
Aluminium	56	62
Brass	28	25
Bronze (Phosphor)	28	38
Copper	100	100
Gold	76	71
Iron	17	17
Lead	9	8
Mercury	2	2
Nickel	20	25
Platinum	18	16
Silver	108	105

continued table

Thermal and Electric Conductivity Based on Copper = 100		
Steel	13-17	15
Tin	17	15
Zinc	29	30

Al is a good conductive material, with relative conductivity of 62 and it is light weight and cheap. The density of Al is only 1/3 of Cu. However, the hardness of Al is low and Al cannot endure high temperature. Pure Al is generally not used as conductive materials. Silver has good electric conductivity with relative conductivity of 105. However, it is expensive and also soft. Gold is a good conductor but expensive and soft.[31]

- **Resistance Materials**

In some application, accurate resistances are required. These materials are generally special alloys with small temperature coefficient of resistance. e.g., Cu-Mn-Ni, Mn-Cu, Cu-Ni-Mn, Cu-Mn-Fe, Ag-Mn-Sn

- **High Temperature Heating Component or Electrode**

In some cases, circuit is work at high temperature, hence the components must be high temperature durable. Alloys of Ni-Cr and Fe-Cr-Al are two examples.

- **Electric Contact Materials**

Electric switch and relay need conductive materials that require two contacts. When current flows to the contact area, due to the uneven contact interface, as well as the foreign thin film formed in between the contact, resistance will be produced.[32] For good performance, small contact resistance, stable contact, wear proof, and diffusion-resistant are generally required. Common contact materials include Cu, Cu-Ag, Cu-Be, Ag, Cu-Ag-Pt, W-Ag, Pt-Ir, Ir-Os, Ir-Os-Pt, etc.

6.2.2 Transparent Conducting Oxide (TCO)

Transparent conductors are neither 100% optically transparent nor metallically conductive. From the band structure point of view, the combination of the two properties in the same material is contradictory: a transparent material (requiring band gaps larger than 3.3 eV) is an insulator which possesses completely filled valence and empty conduction bands; whereas metallic conductivity appears when the Fermi level lies within a band with a large density of states to provide high carrier concentration.[33] In this sense, transparent conductive oxides (TCOs) are exceptional materials.

Current transparent conducting oxides used in industry are primarily

31 Al 是良导体材料，其相对电导率为 62，它质轻又廉价。然而，Al 的密度是 Cu 的三分之一，Al 的硬度低，且不能够耐受高温，纯 Al 通常不被用作导电材料。银是电的良导体，相对电导率为 105，但是银昂贵且软。金是良导体，但是它也昂贵且软。

32 当电流通过接触区域时，由于接触界面的不平整型、以及在接触之间形成的外来薄膜，将产生电阻。

33 从能带结构的观点来看，一种材料同时具备这两个性质（导电和透明）是矛盾的：透明材料（要求能隙大于 3.3 eV）是绝缘体，它的价带是充满、导带是空的；而金属导电性的出现，缘于费米能级在能带内，且态密度很大，能够提供很高的载流子浓度。

n-type conductors, meaning their primary conduction is as donors of electrons. This is because electron mobility is typically higher than hole mobility, and the difficulty of finding shallow acceptors in wide band gap oxides to create a large hole population. Suitable p-type transparent conducting oxides are still being researched, though the best of them are still orders of magnitude behind n-type TCOs.

Ⅰ. Conducting Mechanism

Efficient transparent conductors find their niche in a compromise between a sufficient transmission within the visible spectral range and a moderate but useful in practice electrical conductivity.[34] This combination is achieved in several commonly used oxides – In_2O_3, SnO_2, ZnO and CdO. In the undoped stoichiometric state, these materials are insulators with optical band gap of about 3 eV. To become a transparent conducting oxide, these TCO hosts must be degenerately doped to displace the Fermi level up into the conduction band.

34 高效透明导体既可以在可见光范围内有足够的透光性，又拥有适中且满足实际需求的电导率。

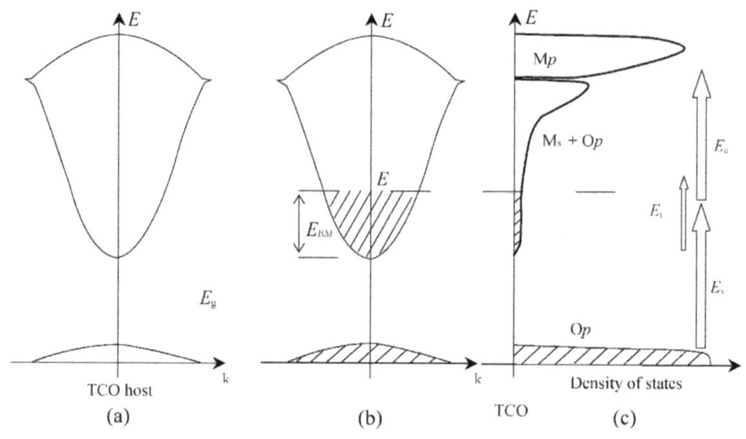

Figure 6.10 (a) Schematic electronic band structure of a TCO host – an insulator with a band gap E_g and a dispersed parabolic conduction band which originates from interactions between metal s and oxygen p states. (b) and (c) Schematic band structure and density of states of a TCO, where a degenerate doping displaces the Fermi level (E_F) via a Burstein-Moss shift, E_{BM}, making the system conducting. The shift gives rise to inter-band optical transitions from the valence band, E_V, and from the partially filled conduction band up into the next empty band, E_C, as well as to intraband transitions within the conduction band, E_i

The key attribution of any conventional *n*-type TCO host is a highly dispersed single free electron-like conduction band (Figure 6.10). Degenerate doping then provides both (i) the high mobility of extra carriers (electrons) due to their small effective mass and (ii) low optical absorption due to the low density of states in the conduction band.[35] The high energy dispersion of the conduction band also ensures a pronounced Fermi energy displacement up above the conduction band minimum, the Burstein–Moss (BM) shift. The shift helps to broaden the optical transparency window and

35 简并掺杂可以形成：（i）额外载流子（电子）的较高迁移率，缘于较小的有效质量；（ii）较低的光吸收，缘于导带中较低的态密度。

to keep the intense optical transitions from the valence band out of the visible range. This is critical in oxides which are not transparent throughout the entire visible spectrum, for example, in CdO where the optical (direct) band gap is 2.3 eV.

Achieving the optimal performance in a TCO is a challenging because of the complex interplay between the electronic and optical properties. The large carrier concentrations desired for a good conductivity may result in an increase of the optical absorption (i) at short wavelengths, due to inter-band transitions from the partially filled conduction band and (ii) at long wavelengths, due to intra-band transitions within this band. In addition, plasma oscillations may affect the optical properties by reflecting the electromagnetic waves of frequency below that of the plasmon. Furthermore, ionized impurity scattering on the electron donors (native point defects or substitutional dopants) have a detrimental effect on the charge transport, while the structural relaxation around the impurities may alter the electronic and optical properties of the host, leading to a nonrigid-band shift of the Fermi level.[36]

36 另外，电子给体(原生点缺陷或取代掺杂)上的电离杂质散射对电荷输运产生不利影响，而杂质周围的结构弛豫可能改变基质的电学的和光学的性质，导致费米能级的非刚性移动。

Ⅱ. Material Progress

The first TCO, In_2O_3:Sn (ITO), was reported by Rupperecht in 1954, followed by other TCOs (SnO_2 and ZnO). Although TCOs have been commercialized intensively as transparent window electrodes and inter connections, there is almost no application as transparent semiconductors because of the absence of *p*-type TCOs and the uncontrollability of conductance by applied voltage.

There have been some distinct advances in this area in recent years. One was the discovery of a *p*-type TCO, $CuAlO_2$, in 1997, which triggered the development of a series of *p*-type TCOs and transparent *pn* junction devices such as UV light emitting diodes (LEDs).[37] So far, the conduction type of TCO materials had been limited to n-type. This achievement has significantly changed our understanding of TCOs and has opened a new frontier, that of transparent oxide semiconductors (TOSs).

37 其中之一，是1997年*p*型透明导电氧化物，$CuAlO_2$的发现，引发了一系列*p*型透明导电氧化物和透明*pn*结器件的发展，如紫外发光二极管（LEDs）。

The second advance was the extension of TCO candidate materials. TCO materials are restricted to oxides of *p*-block metal cations such as In_2O_3, SnO_2, ZnO and CdO. However, two TCO materials which do not belong to *p*-block metal oxides were reported, i.e. TiO_2:Nb belonging to d-block metal oxides and electron-doped 12CaO $7Al_2O_3$ ($C_{12}A_7$) belonging to s-block metal oxides. One feature of both materials is that they are composed of elements that are abundant. In particular, $C_{12}A_7$ is made from the abundant and typical insulator CaO and Al_2O_3. This representative transparent insulating material, known as a constituent of commercial alumina cement, has been transformed to a transparent semiconductor, metal and eventually a superconductor by doping electrons to the sub-nanometer-sized cages constituting the unit cell of the crystal structure. Such a series of discoveries has opened a route to realize transparent semiconductors using naturally abundant oxides by successfully utilizing built-in nanostructures embedded in crystal structures.[38]

38 基于这一系列发现，通过成功地在晶体结构中嵌入内建纳米结构。开辟了一条利用自然界富含的氧化物得到透明半导体的途径。

To date, the industry standard in TCOs is ITO, or tin-doped indium-oxide. This material boasts a low resistivity of ~10^{-4} Ω·cm and a transmittance of greater than 80%. ITO has the drawback of being expensive. Indium, the film's primary metal, is rare, and its price fluctuates due to market demand. For this reason, doped binary compounds such as aluminium-doped zinc-oxide (AZO) and indium-doped cadmium-oxide have been proposed as alternative materials. AZO is composed of aluminium and zinc, two common and inexpensive materials, while indium-doped cadmium oxide only uses indium in low concentrations. Other novel transparent conducting oxides include barium stannate and the correlated metal oxides strontium vanadate and calcium vanadate.

6.3 Semiconductor

Semiconductor materials are a very important type of materials. They have an easily controlled electrical conductivity and provide the building blocks for many electronic devices. When properly combined, they can act as switches, amplifiers, or information storage devices, etc., which are the fundamental components for many electronic products and open the door of information ages.

Based on the different classification, there are elemental semiconductors and compound semiconductors; there are inorganic semiconductors and organic semiconductors; and there are also intrinsic semiconductors and extrinsic semiconductors.[39] Elemental semiconductors are found in Group 4B of the periodic table and include germanium and silicon. Compound semiconductors are formed from elements in Groups 2B and 6B of the periodic table (e.g., CdS, CdSe, CdTe, HgCdTe, etc.) and are known as II–VI (two–six) semiconductors. They also can be formed by combining elements from Groups 3B and 5B of the periodic table (e.g., GaN, GaAs,

39 根据不同的分类方法，可以分为单质半导体和化合物半导体；无机半导体和有机半导体；以及本征半导体和非本征半导体。

AlAs, AlP, InP, etc.). These are known as III–V (three–five) semiconductors. The above semiconductors are inorganic in nature and can be classified as inorganic semiconductors. Organic semiconductors refer to semiconductors that are made of organic materials. An intrinsic semiconductor is one with properties that are not controlled by impurities. An extrinsic semiconductor (*n*- or *p*-type) is preferred for devices, since its properties are stable with temperature and can be controlled using ion implantation or diffusion of impurities known as dopants.

6. 3. 1 Inorganic Semiconductor

Ⅰ. Intrinsic Semiconductor

When the atoms of a semiconductor come together to form a solid, two energy bands are formed. At 0 K, the energy levels of the valence band are completely filled, as these are the lowest energy states for the electrons. The valence band is separated from the conduction band by a bandgap. At 0 K, the conduction band is empty.

The energy gap E_g between the valence and conduction bands in semiconductors is relatively small. As a result, when temperature increases, some electrons possess enough thermal energy to be promoted from the valence band to the conduction band. The excited electrons leave behind unoccupied energy levels, or holes, in the valence band.[40] When an electron moves to fill a hole, another hole is created; consequently, the holes appear to act as positively charged electrons and carry an elementary charge. When a voltage is applied to the material, the electrons in the conduction band accelerate toward the positive terminal, while holes in the valence band move toward the negative terminal[41] (Figure 6.11). Current is, therefore, conducted by the movement of both electrons and holes.

40 结果是，当温度升高，一些电子拥有足够的热能，可以从价带上升到导带。电子激发后在价带留下一个未被占据的能级或者说空穴。

41 当在材料上施加电压时，导带电子加速向正极移动，同时价带空穴朝负极移动。

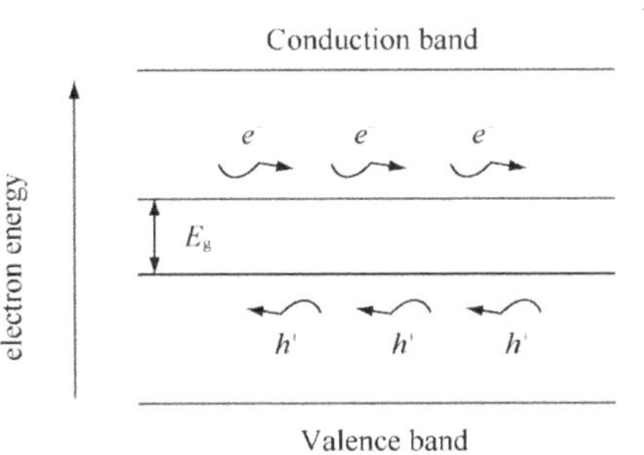

Figure 6.11 The movement of electrons in conduction band, and the opposite movement of holes in valence band, when a semiconductor is exerted under an electric field.

The conductivity is determined by the number of electrons and holes according to previously Equation (6-11). In intrinsic semiconductors, both inorganic and organic, for every electron promoted to the conduction band (LUMO for organic), there is a hole left in the valence band (HOMO for organic) (Figure 6.12), such that

$$n = p = n_i \tag{6-32}$$

where n_i is the intrinsic carrier concentration.

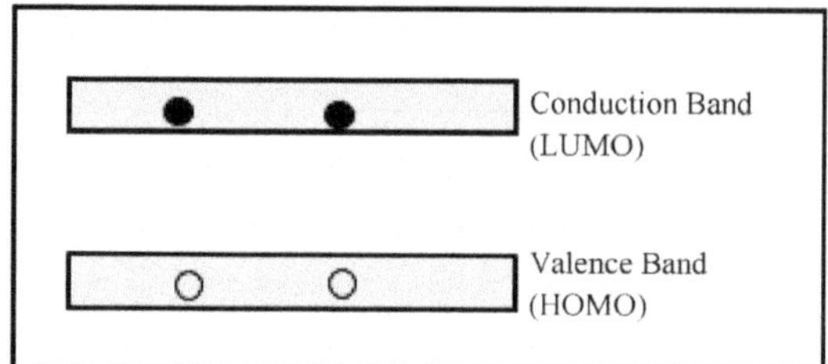

Figure 6.12 Electron and hole pair in intrinsic semiconductor.

Therefore, the conductivity of an intrinsic semiconductor is

$$\sigma = q n_i (\mu_n + \mu_p) \tag{6-33}$$

For both intrinsic and extrinsic (i.e., doped with impurities) semiconductors under a thermal equilibrium condition, the n and p product remains constant at any given temperature for a given semiconductor.[42] This is described by mass action law:

42 对于热平衡下的本征和非本征（掺杂了杂质）半导体，一定温度下 n 和 p 的乘积是常数。

$$n \times p = n_i^2 \tag{6-34}$$

The above equation is obvious for intrinsic semiconductor since $n = p = n_i$; in an extrinsic semiconductor, the increase of one type of carrier tends to reduce the number of the other type through recombination; thus the product of the two types of carriers will remain constant at a given temperature. This equation allows us to calculate n or p values at different temperatures.

- **Relationship Between Carrier Concentration and Temperature**

In intrinsic semiconductors, we control the number of charge carriers and, hence, the electrical conductivity by controlling the temperature. At absolute zero temperature, all of the electrons are in the valence band, whereas all of the levels in the conduction band are unoccupied (Figure 6.13 a). As the temperature increases, there is a greater probability that an energy level in the conduction band is occupied (and an equal probability that a level in the valence band is unoccupied, or that a hole is present)[43] (Figure 6.13b).

43 温度升高，导带中能级被占据的几率增加（价带中能级空置的几率，或者说空穴存在的几率相等地增加）。

The number of electrons in the conduction band, which is equal to the number of holes in the valence band, is given by

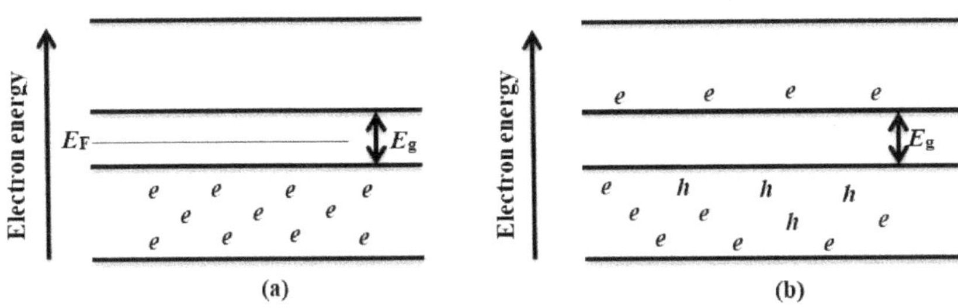

Figure 6.13 The distribution of electrons and holes in the valence and conduction bands (a)at absolute zero and (b) at an elevated temperature in an intrinsic semiconductor.

$$n = p = n_i = n_0 exp(\frac{-E_g}{2k_B T}) \tag{6-35}$$

where n_0 is given by

$$n_0 = 2\left(\frac{2\pi k_B T}{h^2}\right)^{3/2} (m_n^* m_p^*)^{3/4} \tag{6-36}$$

In these equations, k_B and h are the Boltzmann and Planck's constants and m_n^* and m_p^* are the effective masses of electrons and holes in the semiconductor, respectively. The effective masses account for the effects of the internal forces that alter the acceleration of electrons in a solid relative to electrons in a vacuum.[44] For Ge, Si, and GaAs, the room temperature values of n_i are 2.5×10^{13}, 1.5×10^{10}, 2×10^{6} cm^{-3}, respectively. Figure 6.14 shows examples of the relationship between carrier-density and temperature.

44 相对于真空电子，改变固体中电子加速度的内力效应，取决于有效质量。

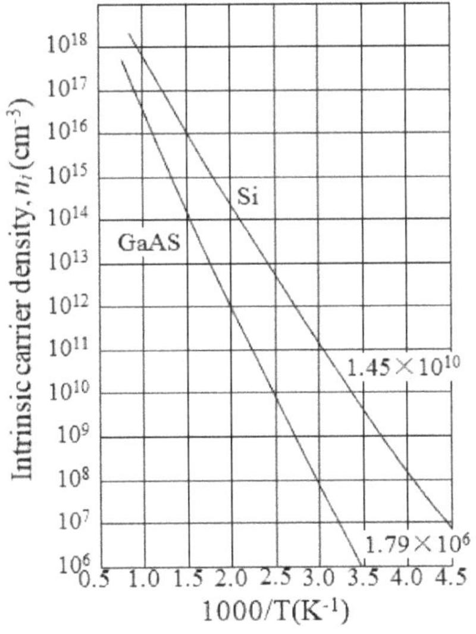

Figure 6.14 Intrinsic carrier densities in Si and GaAs as a function of the reciprocal of temperature.

Note that as for metals, the mobility of the carriers decreases at high

45 要注意到，与金属一样，高温下（半导体）载流子迁移率降低，但是这是一个比载流子数量的指数增加弱得多的依赖关系。

46 然而，即使在高温下，金属的电导率要比半导体的高几个数量级。

temperatures, but this is a much weaker dependence than the exponential increase in the number of charge carriers.[45] The increase in conductivity with temperature in semiconductors sharply contrasts with the decrease in conductivity of metals with increasing temperature (Figure 6.15). Even at high temperatures, however, the conductivity of a metal is orders of magnitudes higher than the conductivity of a semiconductor.[46]

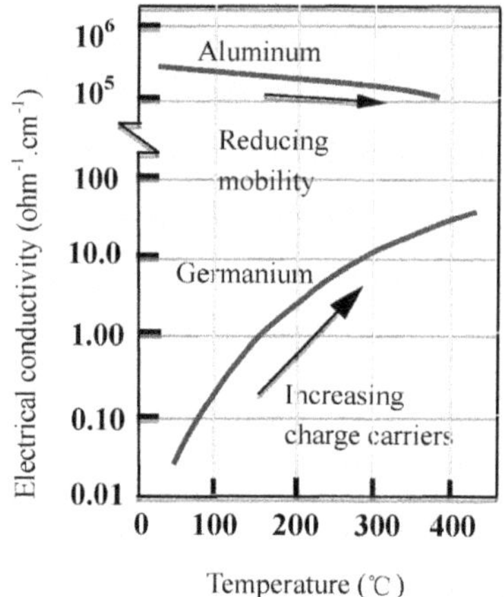

Figure 6.15 The electrical conductivity versus temperature for intrinsic semiconductors compared with metals. Note the break in the vertical axis scale.

- **Relationship Between Conductivity and Temperature**

Since the carrier density of an intrinsic semiconductor depends on temperature, its conductivity is also temperature dependent. Introducing Equation (6-35) to Equation (6-33), one has,

$$\sigma = n_0 q(\mu_n + \mu_p)\exp\left(\frac{-E_g}{2k_B T}\right) = \sigma_0 \exp\left(\frac{-E_g}{2k_B T}\right) \tag{6-37}$$

47 随着温度的升高，因为有更多的载流子可用来传导，所以半导体电导率也升高。

48 严格地讲，公式（6-37）中，能隙也是温度的函数，因为它依赖于随温度变化的晶格常数。然而，在通常使用的温度范围，这个效应是很小的，所以我们几乎总是不考虑这个因素。

Note that both n_i (Equation 6-35) and σ (Equation 6-37) are related to temperature by an Arrhenius equation, the $rate=\exp\left(\frac{-Q}{RT}\right)$. As the temperature increases, the conductivity of a semiconductor also increases because more charge carriers are available for conduction.[47] Figure 6.16 shows the relationship between conductivity and temperature. Strictly speaking, in Equation (6-37), the energy gap is also a function of temperature for the reason that it depends on the lattice constant, which does vary with temperature. That is, however, a small effect in the normally used temperature range, so we are nearly always entitled to disregard it.[48]

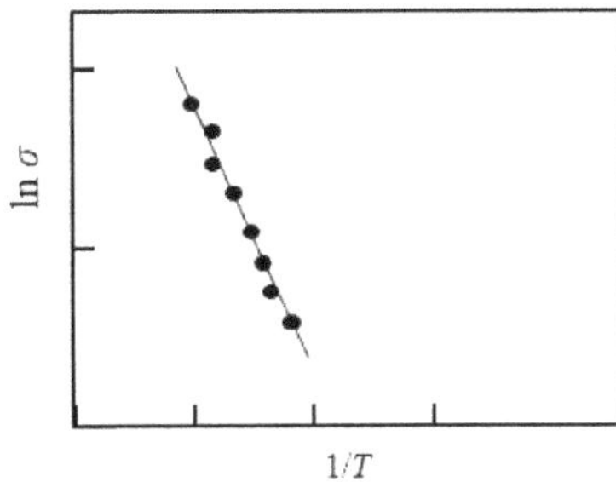

Figure 6.16 The relationship between conductivity and temperature in Si.

- **Relationship Between Mobility and Temperature**

As previously discussed, mobility is one of the two factors that determine conductivity ($\sigma = nq\mu$). And electron drift velocity is proportional to the applied electric field ($v = \mu F$). The proportionality factor (μ), which depends on the mean free time (τ) and the effective mass (m^*), is called mobility. Mobility is an important parameter for carrier transport because it describes how strongly the motion of an electron or a hole is influenced by an applied electric field.[49]

49 迁移率对于载流子输运来说是一个重要的参数，因为它描述了电场下，电子或空穴运动受到影响的程度。

$$\mu = \frac{v}{F} = \frac{q\tau}{m^*} \tag{6-38}$$

$$m^* = \frac{h^2}{\frac{\mathrm{d}^2 E}{\mathrm{d}k^2}} \tag{6-39}$$

Table 6.3 Properties of intrinsic semiconductor Si and Ge

material	E_g eV	μ_n $cm^2 \cdot V^{-1} \cdot s^{-1}$	μ_p $cm^2 \cdot V^{-1} \cdot s^{-1}$	n_i m^{-3}	resistivity Ωm
Si	1.12	1450	450	1.5×10^{16}	2300
Ge	0.66	3900	1900	2.4×10^{19}	0.46

Table 6.3 shows the mobility of intrinsic semiconductors of Si and Ge. In both Si and Ge the electron mobility is higher than hole mobility, due to the fact that in both materials the effective mass of electrons is smaller than that of holes.

The mobility is related directly to the mean free time between collisions, which in turn determined by the various scattering mechanisms. The two most important mechanisms are lattice scattering and impurity scattering.

Lattice scattering results from thermal vibrations of the lattice atoms at any temperature above absolute zero. These vibrations disturb lattice periodic potential and allow energy to be transferred between the carriers and the lattice. Since lattice vibration increases with temperature, lattice scattering becomes dominant at high temperatures; hence the mobility decreases with increasing temperature. Theoretical analysis shows that the mobility due to lattice scattering will decrease in proportion to $T^{-3/2}$ and at low temperature the mobility due to impurity scattering will increases with $T^{3/2}$(Figure 6.17).[50]

50 理论分析显示，随着温度降低缘于晶格散射的迁移率与 $T^{-3/2}$ 成比例；在低温下，随着温度升高，缘于杂质散射的迁移率与 $T^{3/2}$ 成比例。

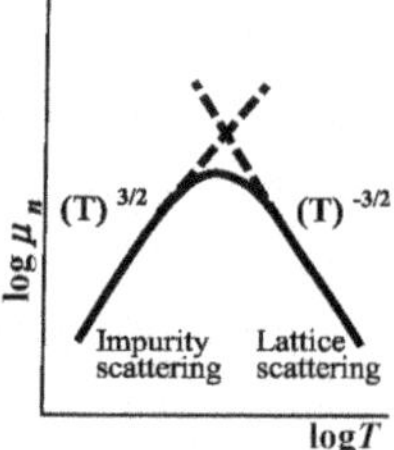

Figure 6.17 The theoretical temperature dependence of mobility due to both lattice and impurity scatterings.

The value of mobility is also subject to impurity (doping). This will be discussed in extrinsic semiconductor part.

Ⅱ. Doped/Extrinsic Semiconductors

The temperature dependence of conductivity in intrinsic semiconductors is nearly exponential, but this is not useful for practical applications. We cannot accurately control the behavior of an intrinsic semiconductor because slight variations in temperature can significantly change the conductivity. By intentionally adding a small number of impurity atoms to the material (called doping), we can produce an extrinsic semiconductor. The conductivity of the extrinsic semiconductor depends primarily on the number of impurity, or dopant, atoms and in a certain temperature range is independent of temperature.[51] This ability to have a tunable yet temperature independent conductivity is the reason why we almost always use extrinsic semiconductors to make devices.

51 非本征半导体的电导率主要取决于杂质、掺杂物、原子的数量，并且在一定温度范围内不随温度变化。

Doped semiconductor is a dilute substitution type solid solution, in which the host and the dopant have different valence state (Figure 6.18). In an *n*-type semiconductor, the dopant "donates" a free electron to the host matrix; while in a *p*-type semiconductor, an electron deficient dopant is employed.[52] Table 6.4 shows the dopant energy levels for Si and Ge with different dopants.

52 在 *n* 型半导体中，掺杂体向基体“给出”自由电子；在 *p* 型半导体中，使用的是缺电子的掺杂体。

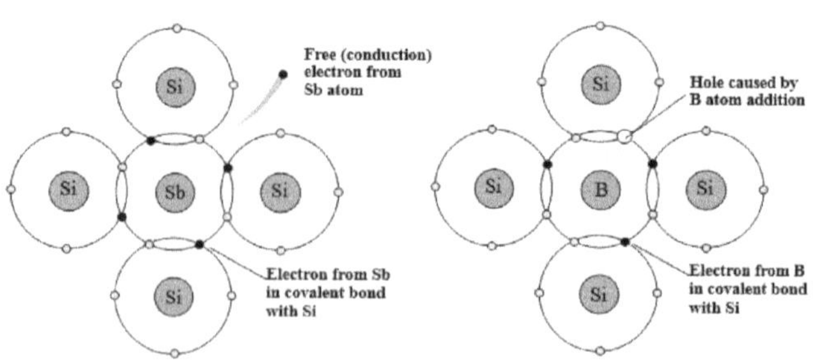

Figure 6.18 *n*-type and *p*-type doping in silicon.

Table 6.4 The donor and acceptor energy levels (in electron volts) when silicon and germanium semiconductors are doped

	Silicon		Germanium	
Dopant	E_d	E_a	E_d	E_a
P	0.045		0.0120	
As	0.049		0.0127	
Sb	0.039		0.0096	
B		0.045		0.0104
Al		0.057		0.0102
Ga		0.065		0.0108
In		0.016		0.0112

- ***n* –Type Semiconductor**

For *n*-type silicon, a controlled addition of a group V impurity, for example, antimony (Sb), arsenic (As), or phosphorus (P) is needed. Each group V atom will replace a silicon atom and use up four of its valence electrons for covalent bonding. There will, however, be a spare electron. It will no longer be so tightly bound to its nucleus as in a free group V atom, since the outer shell is now occupied (we might look at it this way) by eight electrons, so the dangling spare electron cannot be very tightly bound.[53] However, the impurity nucleus still has a netpositive charge to distinguish it from its neighbouring silicon atoms. Hence, we must suppose that the electron still has some affinity for its parent atom.[54]

Let us rephrase this somewhat anthropomorphic picture in terms of band theory. We have said the energy gap represents the minimum energy required to ionize a silicon atom by taking one of its valence electrons in intrinsic semiconductor. Now in an *n*-type one, the energy gap controlling conductivity is now E_d rather than E_g. The electron belonging to the impurity atom clearly needs far less energy than this to become available for conduction. We would expect its energy to be something like ΔE in Figure 6.19, which is typically of the order of 10^{-2} eV.When the donor electrons enter the conduction band, no corresponding holes are created. It is still the case that electron-hole pairs are created when thermal energy causes electrons to be promoted to the conduction band from the valence band; however, the number of electron-hole pairs is significant only at high temperatures.[55] The extra electron 'belonging' to the group V impurity is much more weakly bound to its parent atom than the electrons taking part in the covalent bond. This is equivalent to a donor level close to the conduction band in the band representation (Figure 6.19). At absolute zero temperature, an electron does occupy this donor level, and it is not available

53 它不再像一个自由的第V族原子里被紧紧束缚的电子一样，因为外壳层现在被 8 个电子占据（我们可以这样看待），所以悬浮的多余电子不再被紧紧地束缚。

54 因此，我们必须假设该电子对于它原本属于的原子仍然有一定的亲和。

55 虽然依然存在热能将电子从价带激发到导带产生电子空穴对的情况，但是只有在高温下电子空穴对的数目才变得重要。

for conduction. But at finite temperatures this electron needs no more than about 10^{-2} eV of energy to put it into the conduction band.[56] This phenomenon is usually referred to as an electron donated by the impurity atom. For this reason, E_d is called the donor level. If the impurity is less than, say, 1 in 10^6 silicon atoms, the lattice will be hardly different from that of a pure silicon crystal.

56 在绝对零度，电子占据这个给体能级，它不能够导电。但是在一定温度下，这个电子只需要不大于 10^{-2} eV 的能量就能够进入导带。

Figure 6.19 *n*-type semiconductor: Doping Si with Sb, P, and As, the ΔE is 0.04 eV, 0.044 eV, 0.049 eV, respectively.

- ***p*–Type Semiconductor**

If instead of a group V impurity by some group III atoms, for example, indium (In), aluminum (Al), or boron (B), there would be an electron missing from one of the covalent bonds (see Figure 6.20). A hole is created in the valence band that can be filled by electrons from other locations in the band. The holes act as "acceptors" of electrons. These hole sites have a somewhat higher than normal energy and create an acceptor level of possible electron energies just above the valence band.[57] An electron must gain an energy of E_a-E_{VB} in order to create a hole in the valence band. The hole then carries charge. This is known as a *p*-type semiconductor.

57 这些空穴能量位置略高于正常能量，并且在价带上方形成一个可接受电子的受体能级。

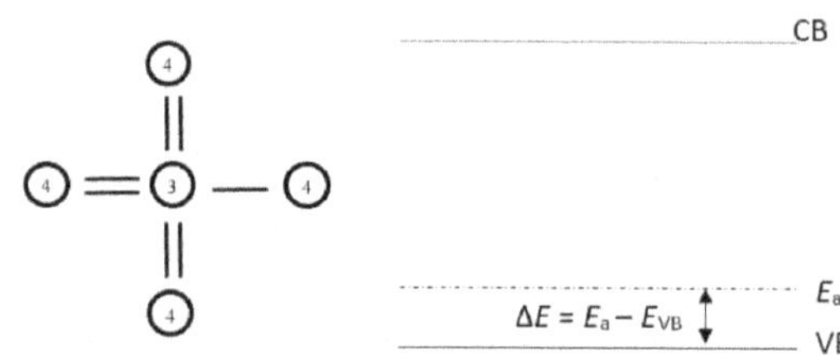

Figure 6.20 *p*-semiconductor of silicon. Doping Si with B, Al or Ga, the corresponding ΔE is 0.045 eV, 0.057 eV, 0.065 eV.

Before going further, let's say a few words about holes. There are three equivalent representations of holes, and you can always (or nearly always) look at them in the manner most convenient under the circumstances. You may think of a hole as a full-blooded positive particle moving around in the crystal, or as an electron missing from the top of the valence band, or as the actual physical absence of an electron from a place where it would be desirable to have one.[58] In the present case, the third interpretation is the most convenient one to start with. A group III atom has three valence electrons. So when it replaces a silicon atom at a certain atomic site, it will try to contribute to bonding as much as it can. However, it possesses only three electrons for four bonds. Any electron wandering around would thus be welcomed to help out. More aggressive impurity nuclei might even consider stealing an electron from the next site. The essential point is that a low energy state is available for electrons — not as low an energy as at the host atom but low enough to come into consideration when an electron has acquired some extra energy and feels an urge to jump somewhere. Therefore (changing now to band theory parlance), the energy levels due to group III impurities must be just above the valence band. Since these atoms accept electrons so willingly, they are called acceptors and the corresponding energy levels are referred to as acceptor levels.

- **Charge Neutrality**

In an extrinsic semiconductor, there has to be overall electrical neutrality. Thus, the sum of the number of donor atoms (N_d) and holes (p) per unit volume (both are positively charged) is equal to the number of acceptor atoms (N_a) and electrons (n) per unit volume (both are negatively charged):[59]

$$N_a+n=N_d+P \tag{6-40}$$

where n and p is concentration of electron and hole, respectively. N_a and N_d is the ionic concentration of acceptor and donor, respectively. If the extrinsic semiconductor is heavily n -type doped:

$$n\approx N_d \tag{6-41}$$

$$P=\frac{n_i^2}{n}\approx\frac{n_i^2}{N_d} \tag{6-42}$$

Similarly, if there is a heavily acceptor-doped (p-type) semiconductor, then,

$$p\approx N_a \tag{6-43}$$

$$n=\frac{n_i^2}{p}\approx\frac{n_i^2}{N_a} \tag{6-44}$$

Doping is important in semiconductor application, by adding a considerable amount of dopant, we can dominate the conductivity of a semiconductor by controlling the dopant concentration, avoiding the exponential change of conductivity in intrinsic ones.[60] This is why extrinsic semiconductors are most useful for making controllable devices such as transistors.

58 空穴有三种等效表示,在一定的环境下你可以总是（或者几乎总是）将它们看作最方便的那种。你可以认为空穴是在晶体周围移动的活跃的正粒子,或者是在价带顶丢失一个电子状态,或者在实际的物理空间缺少但却希望获得一个电子的位置。

59 因此，单位体积内给体原子数量与空穴数量之和（都是正电荷）同单位体积内受体原子数量与电子数量之和（都是负电荷）相等。

60 在半导体应用方面，掺杂是非常重要的，通过加入一定数量的掺杂物，我们可以通过控制掺杂浓度来控制半导体电导率，避免本征半导体电导率（随温度）指数性的变化。

- **Influence of Doping Concentration to Mobility**

Figure 6.21 shows the measured electron and hole mobility as a function of impurity concentration at room temperature for Ge and Si, respectively, which show the same trend. Note also that the mobility of electrons is greater than that of holes. Greater mobility is mainly due to the smaller effective mass of electrons.

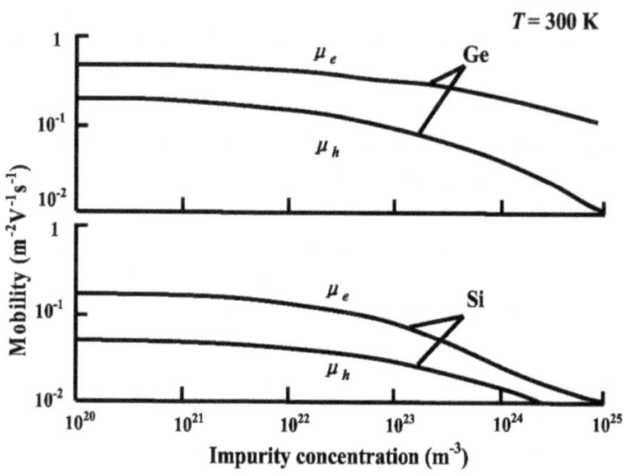

Figure 6.21 Influence of doping concentration on the hole and electron mobility of Ge and Si at 300K.

Mobility reaches a maximum value at low impurity concentrations; this corresponds to the lattice-scattering limitation. Both electron and hole mobility decrease with impurity concentration raising and eventually approach a minimum value at high concentrations.[61]

61 迁移率在低杂质浓度时达到最大值，这对应于晶格散射的极限。杂质浓度提高，电子和空穴的迁移率都下降，最终在高浓度时达到最小值。

Impurity scattering occurs when a charge carrier travels past an ionized dopant impurity (donor or acceptor). The charge carrier path will be deflected due to Coulombic interaction.The probability of impurity scattering depends on the total concentration of ionized impurities, that is, the sum of the concentration of negatively and positively charged ions. The mobility due to the impurity scattering can theoretically be shown to vary as $1/N_T$, where N_T is the total impurity concentration. The larger the impurity concentration, the lower the mobility:

$$\mu \propto \frac{1}{N_T} \tag{6-45}$$

Unlike lattice scattering, impurity scattering becomes less significant at higher temperature. At higher temperature, the carriers move faster, they remain near the impurity atom for a shorter time and are therefore less effectively scattered.[62]

62 不同于晶格散射，高温下杂质散射的影响变得不重要。高温时，载流子运动快，它们在杂质附近停留的时间更短，所以造成的有效散射较小。

Figure 6.22 shows the measured electron mobility as a function of temperature for silicon with five different donor concentrations. The insert shows the theoretical temperature dependence of mobility due to both lattice and impurity scatterings. For lightly doped samples (e.g. the sample with doping of 10^{14} cm^{-3}), the lattice scattering dominates, and the mobility decreases as the temperature increases.[63] For heavily doped samples, the effect of impurity scattering is most pronounced at low temperatures. The mobility increases as the temperature increases, as can be seen for the sample with doping of 10^{19} cm^{-3}. For a given temperature, the mobility decreases with increasing impurity concentration, because of enhanced impurity scatterings.[64]

63 对于轻微掺杂的样品（如掺杂为 10^{14} cm^{-3} 的样品），晶格散射是主要的，迁移率随温度升高而降低。

64 在一定温度下，由于杂质的散射，迁移率随着杂质浓度的升高而降低。

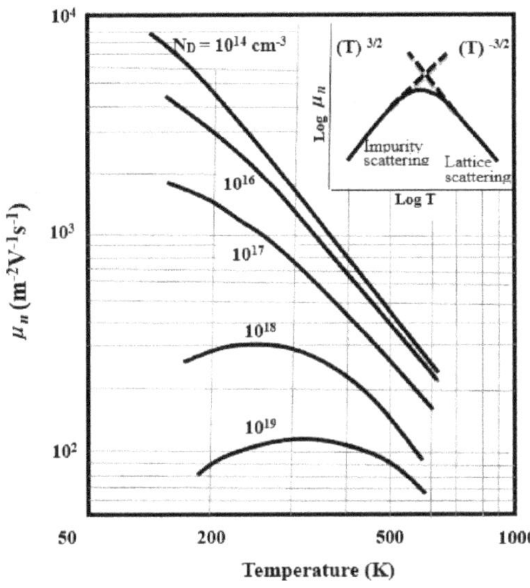

Figure 6.22 Influence of temperature to mobility for Si with five different donor concentrations.

- **Influence of Temperature to Fermi Level For *n* and *p*–Type Semiconductor**

The variation of the Fermi level as a function of temperature for an *n*-type and a *p*-type semiconductor is shown in Figure 6.23. The curves 1, 2, and 3 correspond to increasing impurity concentrations. $E_{Fi} = E_g/2$ is the intrinsic Fermi level for $m_h > m_e$ to which all curves tend at higher temperatures. For doped semiconductor, at low temperature, the Fermi level is in vicinity to the dopant level, i.e., E_D for *n*-type, E_A for *p*-type. With the increase of temperature, the Fermi level of the system is gradually approaching the intrinsic Fermi level. This is because, increasing temperature, the influence of the intrinsic charge, i.e., electron and hole, are gradually take effect, which finally dominate in electrical conductivity.

There is just one further point to note about the variation of energy gap with temperature. We have seen in the band structure that the interband gap is caused by the interaction energy when the electrons' de Broglie half-wavelength is equal to the lattice spacing.[65] It is reasonable to suppose that this energy would be greater at low temperatures for the following reason. At higher temperatures the thermal motion of the lattice atoms is more vigorous; the lattice spacing is thus less well defined and the interaction is weaker. This, qualitatively, is the case; for example, in germanium the energy gap decreases from about 0.75 eV at 4 K to 0.67 eV at 300 K.

65 在能带结构中我们可以了解，当电子的德布罗意半波长等于晶格间距时，电子与晶格之间的相互作能造成了能带之间的间隙。

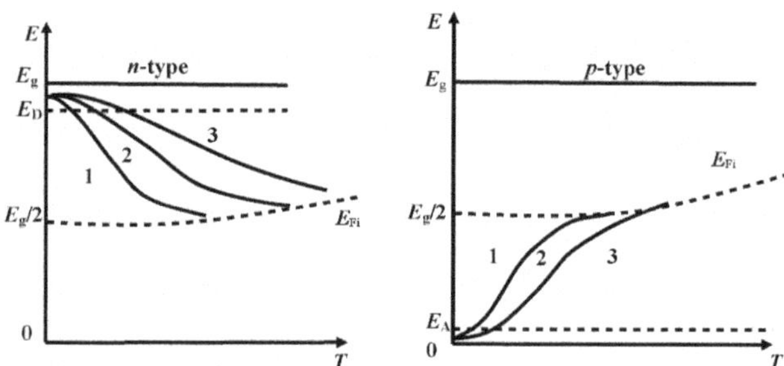

Figure 6.23 The variation of the Fermi level as a function of temperature for *n*-type and *p*-type semiconductor. The curves 1, 2, and 3 correspond to increasing impurity concentrations.

- **Effect of Temperature on an Extrinsic Semiconductor**

Taking an *n*-type semiconductor as an example, the changes in carrier concentration with temperature are shown in Figure 6.24. When the temperature is too low, the donor atoms are not ionized and hence the carrier concentration is very low. As temperature begins to increase, electrons from donor become available for conduction, carrier concentration increases.[66] At sufficiently high temperatures, all donors are ionized, the carrier concentration is nearly independent of temperature (region labeled as extrinsic). When temperatures become too high, the behavior approaches that of an intrinsic semiconductor since the valence electrons can be promoted to conduction band and carrier concentration increases dramatically accompanying the loss of the effect of dopants essentially.

66 当温度很低时，给体原子不能够被离子化，因此载流子浓度很低。当温度开始升高，给体的电子开始对传导起作用，载流子浓度增加。

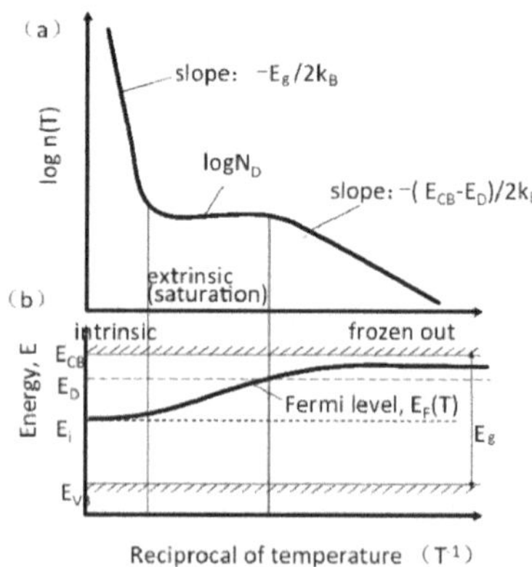

Figure 6.24 The changes in carrier concentration and Fermi level with temperature in *n*-type semiconductor.

Simultaneously taking account the effects of temperature on carrier density and the effects of dopant concentration on the mobility, as well as the temperature dependence of the bandgap, the changes of conductivity with temperature can be shown as Figure 6.25.[67] As can be seen, after the increase of conductivity with temperature (relates to the frozen out region in Figure 6.24), there is a bump curve which corresponds to the extrinsic region of Figure 6.24, followed by a steeper line. At even higher temperatures (not shown in Figure 6.25), the conductivity decreases again as scattering of carriers dominates.

67 同时考虑温度对载流子密度的影响、掺杂浓度对载流子迁移率的影响、以及带隙的温度依赖性，电导率随温度的变化可以表示为图6.25。

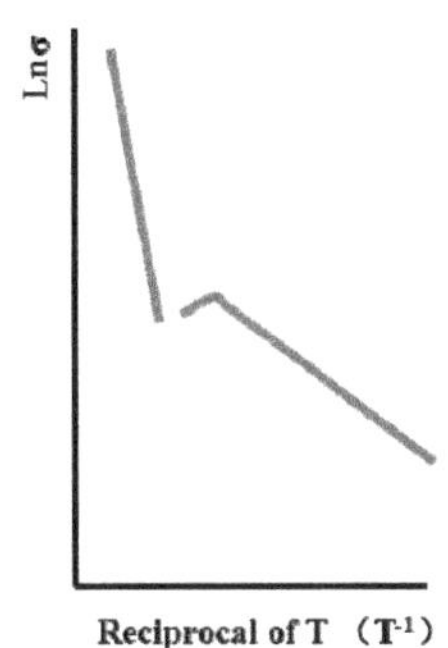

Figure 6.25 Changes of conductivity with temperature in doped semiconductor.

Finally, it should be pointed out that many other materials that are normally insulating (because the bandgap is too large) can be made semiconducting by doping. Examples of this include $BaTiO_3$, ZnO, TiO_2, and many other oxides. Thus, the concept of n - and p -type dopants is not limited to Si, Ge, GaAs, etc. We can dope $BaTiO_3$, for example, and make n- or p-type $BaTiO_3$. Such materials are useful for many sensor applications such as thermostats.

Ⅲ. Hall Effect

Applying a magnetic field to a conductor or semiconductor with current flow, and letting the magnetic field perpendicular to the direction of current flow, then in the third direction which is both perpendicular to the current and the magnetic field, there will build an electric field[68] (Figure 6.26). This phenomenon calls Hall effect.

68 如果在一个有电流通过的导体或者半导体上施加一个磁场，并且使磁场方向垂直于电流方向，则在垂直于电流和磁场的第三个方向上，就会产生一个电场。

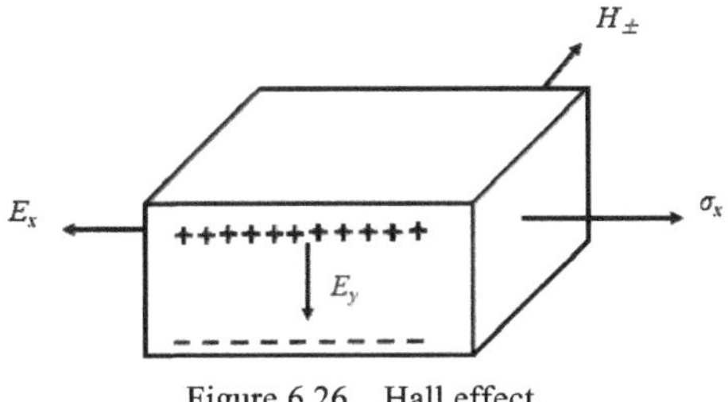

Figure 6.26 Hall effect.

Moving electrons (v) in magnetic field (B) will gain a Lorenz force (F), resulting in the movement of electrons toward the force direction. Hence one side of the material will possess negative charge and the opposite side possesses positive charge, leading to electric field.[69]

69 在磁场（B）中的移动电子（v）将会受到洛伦兹力(F)，结果使得电子沿力的方向运动。因此材料的一侧将会有负电荷，而相对的一侧有正电荷，这就形成了电场。

The force on an electron due to this magnetic field is:

$$\vec{F} = e(\vec{V} \times \vec{H}) \tag{6-46}$$

Under equilibrium condition, the value of the Hall voltage is:

$$E_y = V \cdot H$$

Hall coefficient (R_H) is termed as the Hall voltage under unit magnetic field and unit current density:

$$R_H = \frac{E_y}{J_x H_z} \tag{6-47}$$

where $\boldsymbol{E}_y$ is the electric potential difference in y direction by Hall effect, $\boldsymbol{J}_x$ is the current density in x direction, and $\boldsymbol{H}_z$ is the intensity of the magnetic field in z direction.

Since $E_y = v_x H_z$, $J_x = -env_x$(where n is electron density, e is the electron charge), we have:

$$R_H = \frac{v_x H_z}{-nev_x H_z} = -\frac{1}{ne} \tag{6-48}$$

Clearly, Hall coefficient is only related to electron density.

- **Application of Hall Effect**

Based on Hall effect experiment, it has been proved that: (1) Free electrons do exist in metal, which are the carriers of charges; (2) For conductive materials whose electric conduction is dominated by free electron, free electron density can be deduced by measuring R_H; (3) R_H can be used to judge the carrier type in semiconductor; (4) $R_H < 0$ free electron takes the role of electric conduction; (5) $R_H > 0$ hole is the component for electric conduction.[70] Experiment disclosed that in metals, there existed situation of $R_H > 0$. For instance, in Zinc and Iron the electric conduction component cannot be simply attributed to free electron. Due to the complexity of the electronic structure, hole may be the dominant factor for electric conduction.

70 基于霍尔效应实验，已经证明：(1)金属中的确存在自由电子，它们是电荷的载体；（2）对于自由电子起导电作用的导电材料，通过测量 R_H 可以推断自由电子密度；（3）R_H 可以用来判断半导体中的载流子类型；（4）当 R_H<0 时是自由电子导电；（5）当 R_H>0 时是空穴导电。

Ⅳ. *p*–n Junction

Creating an *n*-type region in a *p*-type semiconductor (or vice versa) will form a *p–n* junction (Figure 6.27). The *n*-type region contains a relatively large number of free electrons, whereas the *p*-type region contains a relatively large number of free holes. This concentration gradient causes diffusion of electrons from the *n*-type material to the *p*-type material and diffusion of holes from the p-type material to the n -type material.[71] At the junction where the p-and n-regions meet, free electrons in the n-type material recombine with holes in the p-type material. This creates a depleted region (W in Figure 6.27) at the junction where the number of available charge carriers is low, and thus, the resistivity is high. Consequently, an electric field develops due to the distribution of exposed positive ions on the n-side of the junction and the exposed negative ions on the p-side of the junction.The electric field counteracts further diffusion.[72]

71 n-型区域含有大量的自由电子，同时 p-型区域含有大量的自由空穴。它们之间的浓度梯度造成了电子从 n-型材料到 p-型材料以及空穴从 p-型材料到 n-型材料的扩散。

72 结果，p-n 结 n-型一侧暴露的正离子和 p-型一侧暴露的负离子形成一个电场。这个电场阻碍了(电荷的) 进一步扩散。

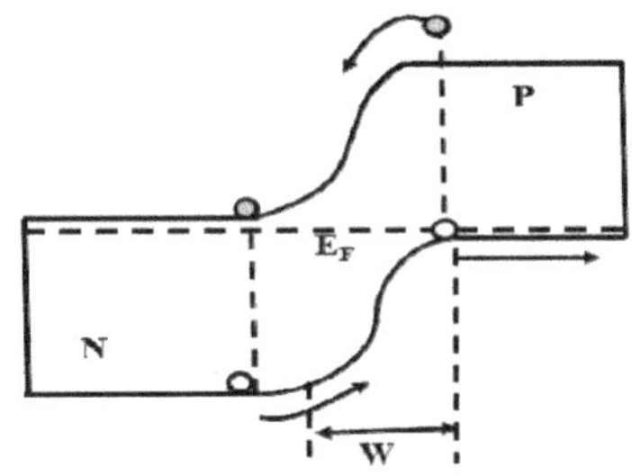

Figure 6.27 The p-n junction and depletion region W.

6.3.2 Organic Semiconductor

Ⅰ. Introduction

The range of materials suitable for producing various electronic devices has been steadily increasing. It is no great surprise that organic materials have also put in a claim to be represented. They were never completely disregarded, but having reached prominence around the year of 2000, they can no longer be ignored.

What are organics? Compounds with carbon atoms as their essential structural elements, forming solids as single crystals, polycrystals, or glasses are the organic solids. Normally organic solids are perfect insulators. There is, however, also a large number of organic semiconductors. Furthermore, there are organic solids with a high dark conductivity or with a quasi-metallic conductivity, and there are also organic superconductors.

In terms of semiconductor, organic solids of molecular crystals, such as radical-ion crystals, charge-transfer crystals, and thin films or layered structures of small-molecule/polymers, which include conjugated π-electron systems in their skeletal structures, can be semiconductors.[73] These are in turn primarily constructed of carbon atoms but often contain also N, O, S, or Se atoms. To this class belong in particular the aromatic hydrocarbons and alkenes (olefins) (Figure 6.28), but also N-, O- or S-containing heterocyclic compounds such as pyrrole, furane, thiophene, quinoxaline and others

73 就半导体而言，属于分子晶体多有机固体，如自由基离子晶体、电荷转移晶体，以及小分子/聚合物的薄膜或者层状结构，在它们的骨架中含有共轭π电子体系，可以是半导体。

(Figure 6.29). All of the molecules in figure 6.28 and 6.29 have a conjugated π-electron system. The regions of absorption shift towards longer wavelengths with increasing length of the conjugated electron chains. Many of these molecules are building blocks of still larger molecules, e.g. of dimers, oligomers, or polymers. Or else they are components for the side chains in polymers of ligands to central metal ions.[74]

74 这些分子中许多是更大分子的结构单元，例如二聚体，寡聚物或聚合物。或者，它们是聚合物的侧链组分作为中心金属离子的配位体。

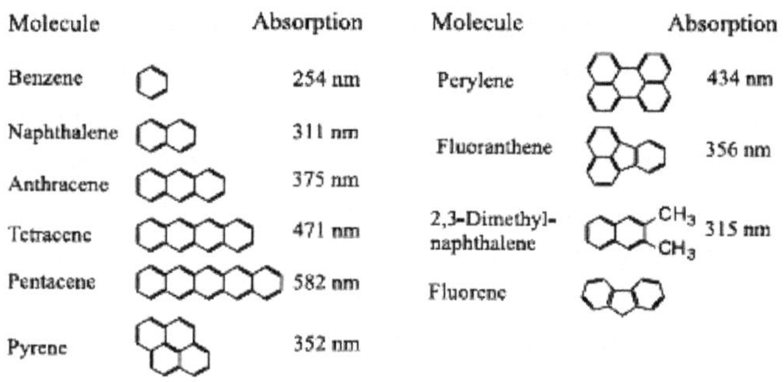

Molecule	Absorption	Molecule	Absorption
Benzene	254 nm	Perylene	434 nm
Naphthalene	311 nm	Fluoranthene	356 nm
Anthracene	375 nm	2,3-Dimethyl-naphthalene	315 nm
Tetracene	471 nm	Fluorene	
Pentacene	582 nm		
Pyrene	352 nm		

Figure 6.28 Molecular structures of some polyacene molecules, indicating the wavelengths of their lowest-energy optical absorption regions in solution at room temperature.

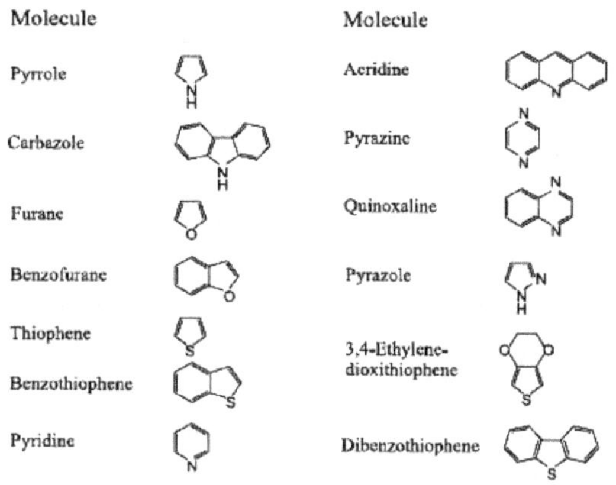

Molecule	Molecule
Pyrrole	Acridine
Carbazole	Pyrazine
Furane	Quinoxaline
Benzofurane	Pyrazole
Thiophene	3,4-Ethylene-dioxithiophene
Benzothiophene	
Pyridine	Dibenzothiophene

Figure 6.29 Some typical heterocyclic molecules.

To understand the electrical conductivity of an organic solid, let's start with the fundamental element in organic materials, carbon atom. The electron configuration of the free carbon atom in its ground state is $1s^2 2s^2 2p^2$. Carbon has four valence electrons due to the fact that the electron configurations in chemically-bonded carbon are derived from the configuration $1s^2 2s^1 2p^3$. From organic chemistry, we know that a so called double bond between two carbon atoms can form due to an sp^2 hybridization: three degenerate orbitals are constructed out of one s and two p orbitals.[75] They are coplanar and oriented at 120° relative to one another. Chemical bonds formed by these orbitals are called σ bonds; they are localized between the bonding C atoms. The fourth orbital, p_z, remains unchanged and is directed perpendicular to the plane of the 3 sp^2 orbitals,

75 根据有机化学知识，我们知道 sp^2 杂化，即一个 s 轨道和两个 p 轨道形成三个简并轨道，可以在两个 C 原子之间形成所谓的双键。

and thus to the plane of the C atoms. The p_z orbitals of neighboring atoms overlap. This leads to an additional bond, called π bond, and to a delocalized density of electrons above and below the plane of the molecule. This is the nodal plane for the π-electron density.[76]

76 相邻原子的 p_z 轨道互相重叠，导致一个额外的、称为 π 键的化学键，并且导致分子平面的上方及下方的电子离域。这个（平面）是 π 电子的节点平面。

Figure 6.30 shows the overall electron distribution in an aromatic molecule, anthracene. In addition to the total electron density, Figure 6.30 also shows two π orbitals, the energetically highest which is occupied in the ground state (HOMO) and the energetically lowest which is unoccupied in the ground state (LUMO). In comparison with the σ electrons, the contribution of the π electrons to bonding of the molecule is thus weak. Organic molecules and molecular crystals with conjugated π-electron systems therefore possess electronic excitation energies in the range of only a few eV and absorb or luminesce in the visible, the near infrared and/or the near ultraviolet spectral regions.[77] The electronic excitation energies of this absorption shift towards lower energies with increasing length of the conjugated system. The lowest electronic excitation states are excitations of the π electrons.

77 具有共轭 π 电子体系的有机分子或分子晶体具有仅仅几个 eV 的电子激发能，并且可以在可见光、近红外或近紫外光谱区产生光吸收或者光发射。

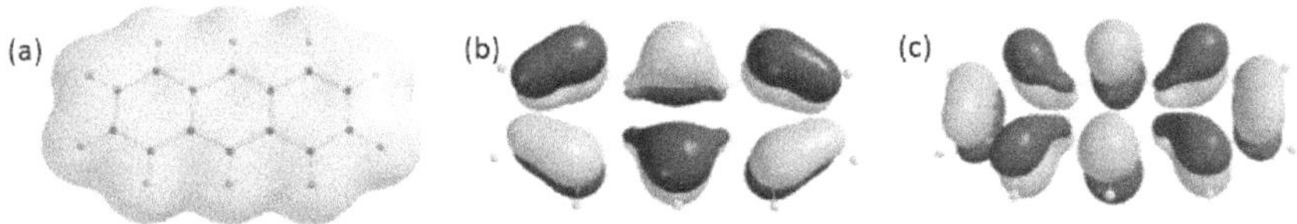

Figure 6.30(a) the overall distribution of the π electrons in the electronic ground state of the anthracene molecule, $C_{14}H_{10}$. The boundary was chosen so that ca. 90% of the total electron density was included. (b) The distribution of one π electron in the highest occupied molecular orbital (HOMO). (c) The distribution of one π electron in the lowest unoccupied molecular orbital (LUMO).

Ⅱ. Band like Versus Hopping Transport

All ultrapure organic molecular crystals which consist of only one type of molecules with conjugated π-electron systems are insulators at room temperature and below. This is however only true when the applied electric voltage, i.e. the "external" electric field strength, is low and the organic solid is not irradiated by light, electrons, or other high energy particles. It is also true for the non-crystalline organic solids or polymers with conjugated π-electron systems, e.g. also for non-doped polyacetylene. All these organic crystals or solids only become conduction of electrical current and are thus semiconductors when excess charge carriers are either produced by the internal photoelectric effect or are injected by applied electric field.[78] If, however, such organic crystals or solids contain defects, impurities, dopant atoms, or guest molecules at low concentrations, they can become conductors (semiconductors) at moderate temperatures. The cause of this is

78 只有通过内光电效应或者通过施加电场注入,产生过剩电荷载流子时,这些有机晶体或固体才会传导电流变成半导体。

that the defects can be thermally ionized and thus release mobile charge carriers. On the other hand, defects can act as traps for excess charge carriers and thereby also reduce the electrical conductivity.

The charge carrier mobility μ differ fundamentally in ultrapure aromatic molecular crystals from those in less-perfect organic crystals or disordered organic solids or polymers. In an ultrapure aromatic crystal, the mobility increases with decreasing temperature. This holds generally for high-purity single crystals. Precisely the positive temperature dependence of the mobility is exhibited by all disordered organic solids. This is also true for both disordered low-molecular films as well as polymer films. The values for the mobility in all non-crystalline solids are generally several orders of magnitude smaller than in crystals, and they decrease with decreasing temperature.[79] Furthermore, the mobility depends on the electric field strength F. The reason for the two fundamentally different temperature dependencies and the orders of magnitudes different of the mobility lies in processes during the motion of the charge carriers. In ultrapure crystals, the charge carriers have a quasi-momentum and are scattered by phonons. With decreasing temperature, the phonon density and its scattering probability decrease, therefore the mobility increase. One then refers to the process of electrical conductivity as band conduction. In contrast to ultrapure crystals, the charge carriers in disordered molecular solids are localized on the molecules and for transport, they must be thermally activated in order to hop from molecule to molecule.[80] Therefore, in the disordered molecular solids, the mobility becomes greater with increasing temperature. The process of electrical conduction is then termed hopping transporting. The boundary between band and hopping conduction is naturally not well defined. However, if the mean free path of the charge carriers is the order of only one lattice length, then the conduction will take place via a hopping process.

79 所有非晶体的迁移率通常都比晶体小几个数量级，且随着温度的下降而下降。

80 与超纯晶体相比，无序分子固体中的电荷载流子是在分子中定域的；为了传输，载流子必须被热激发，才能从一个分子跳到另一个分子。

Ⅲ. Doping

For inorganic semiconductors (e.g. Si), we know that doping with impurity concentration of up to 1 PPM can produce enormous ranges of both *p* or *n*-type of conductivities. Conjugate small-molecule/polymers, e.g., polyacetylene, can also be 'doped'. By *n* or *p*-type groups or more obviously by elements, for example Na, Ca as electron sources or sulphur for holes, *p* and *n*-types doped organic semiconductors can be obtained. This type of doping involves relatively massive impurities—from a solid fraction of one percent, up to forty percent.

6.3.3 Applications of Semiconductor

Using semiconductors, electronic devices, such as diodes, transistors, lasers, light-emitting diodes (LEDs) and photovoltaic devices, etc., can be built. We will discuss the fundamental of these applications in this section.

Ⅰ. Rectifier Diodes

Electrically, the *p-n* junction is conducting when the *p*-side is connected to a positive voltage. This forward bias condition is shown in Figure 6.31a. The applied voltage directly counteracts the electric field at the depleted region, making it possible for electrons from the *n*-side to diffuse across the depleted region to the *p*-side and holes from the *p*-side to diffuse across the depleted region to the *n*-side.[81] When a negative bias is applied to the *p*-side of a *p-n* junction (reverse bias), the *p-n* junction does not permit much current to flow. The depleted region simply becomes wider because it further depleted carriers. When no bias is applied, there is no current flowing through the *p-n* junction. The forward current can be as large as a few milli-amperes, while the reverse-bias current is a few nano-amperes.

81 外加电压直接抵消耗尽区的电场，使电子可以从 *n*-型一端扩散到 *p*-型一端以及空穴从 *p*-型一端扩散到 *n*-型一端。

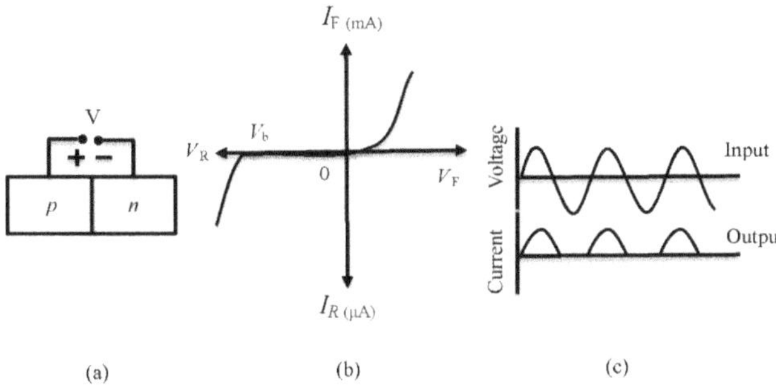

Figure 6.31 (a) A *p-n* junction under forward bias. (b) The current-voltage characteristic for a *p-n* junction. Note the different scales in the first and third quadrants. At sufficiently high reverse bias voltage, "breakdown" rectification occurs, and large currents can flow. Typically, this destroys the devices. (c) If an alternating signal is applied, rectification occurs, only half of the input signal can pass the rectifier.

The current–voltage (I–V) characteristics of a *p-n* junction are shown in Figure 6.31b. Because the *p-n* junction permits current to flow only in one direction, it will pass only half part of an alternating current, therefore converting the alternating current to direction current (Figure 6.31c). These devices are called rectifier diodes.[82]

82 由于*p-n*结只允许电流沿一个方向通过，它只可以通过一半的交流电，结果使交流转化为直流（图 6.31c）。这种器件叫整流二极管。

Ⅱ. Bipolar Junction Transistors

There are two types of electric devices based on *p-n* junctions, one is rectifier diode, another is transistor. The term transistor is derived from two words, "transfer" and "resistor." A transistor can be used as a switch or an amplifier. One type of transistor is the bipolar junction transistor (BJT). In the era of mainframe computers, bipolar junction transistors often were used in central processing units. A bipolar junction transistor is a sandwich of either *n-p-n* or *p-n-p* semiconductor materials, as shown in Figure 6.32a. There are three zones in the transistor: the emitter, the base, and the collector. As in the *p-n* junction, electrons are initially concentrated in the *n*-type material, and holes are concentrated in the *p*-type material.[83] Figure 6.32b shows a schematic diagram of an *n-p-n* transistor and its electrical circuit. The electrical signal to be amplified is connected between the base and the emitter, with a small voltage between these two zones. The output from the transistor, or the amplified signal, is connected between the emitter and the collector and operate at a higher voltage. The circuit is connected so that a forward bias is produced between the emitter and the base (the positive voltage is at the *p*-type base), while a reverse bias is produced between the base and the collector (with the positive voltage at the *n*-type collector).[84] The forward bias causes electrons to leave the emitter and enter the base.

83 晶体管中有 3 个区域：发射极、基极和集电极。如在 *p-n* 结中一样，电子最初在 *n* 型材料中聚集，空穴在 *p* 型材料中聚集。

84 电路形成后，发射极和基极之间会产生一个正向偏压(正电压在 *p*-型基极一端)，同时基极和集电极之间会产生一个相反的偏压（正电压在 *n*-型集电极一端）。

Electrons and holes attempt to recombine in the base; however, if the base is exceptionally thin and lightly doped, or if the recombination time is long, almost all of the electrons pass through the base and enter the collector.[85] The reverse bias between the base and collector accelerates the electrons through the collector, the circuit is completed, and an output signal is produced. The current through the collector (I_c) is given by

$$I_c = I_0 \exp\left(\frac{V_E}{B}\right) \tag{6-49}$$

where I_0 and B are constants and V_E is the voltage between the emitter and the base. If the input voltage V_E is increased, a very large current I_c is produced.

85 然而，如果基极非常薄并且轻微掺杂，或者复合时间很长，那么几乎所有电子通过基极进入集电极。

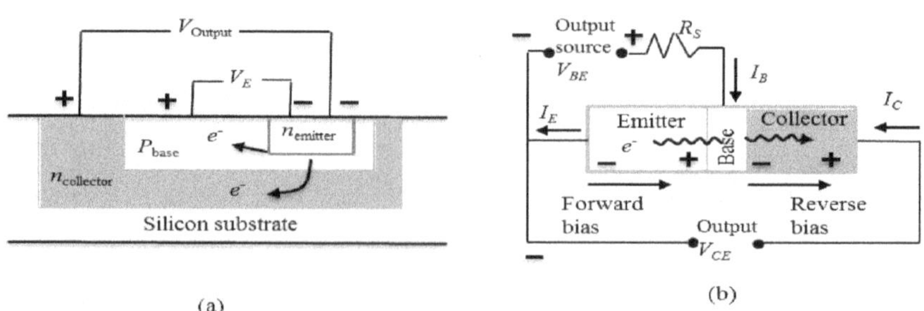

Figure 6.32 (a) Sketch of the cross-section of the transistor. (b) A circuit for a *n-p-n* bipolar junction transistor. The input creates a forward and reverse bias that causes electrons to move from the emitter, through the base, and into the collector, creating an amplified output.

Ⅲ. Field Effect Transistors

A second type of transistor, which is almost universally used today for information devices, is the field effect transistor (FET). As shown in Figure 6.33a, FET is a three-terminal system, including components of source, drain and gate electrodes, as well as dielectric and semiconductor. An FET device can comprise top gate or bottom gate. An example of top gate FET is MOS-FET device (Figure 6.33b, metal oxide semiconductor field effect transistor), which consists of two highly doped *n*-type regions (n^+) in a *p*-type substrate or two highly doped *p*-type regions in an *n*-type substrate. One of the *n*-type regions is called the source, another is called the drain. A potential is applied between the source and the drain with the drain region being positive, but in the absence of a third component of the transistor (an electrode called the gate), electrons cannot flow from the source to the drain under the action of the electric field through the low electron conduction *p*-type region.[86] The gate is separated from the semiconductor by a thin insulating layer of oxide and spans the distance between the two *n*-type regions. In advanced device structures, the insulator is only several atomic layers thick and comprises materials other than pure silica.

86 在源极和漏极施加一个电压并且漏极为正，如果晶体管缺少第三个组件（栅极电极），在传输能力较低的 *p* 型区域，电子不能在电场的作用下从源极流到漏极。

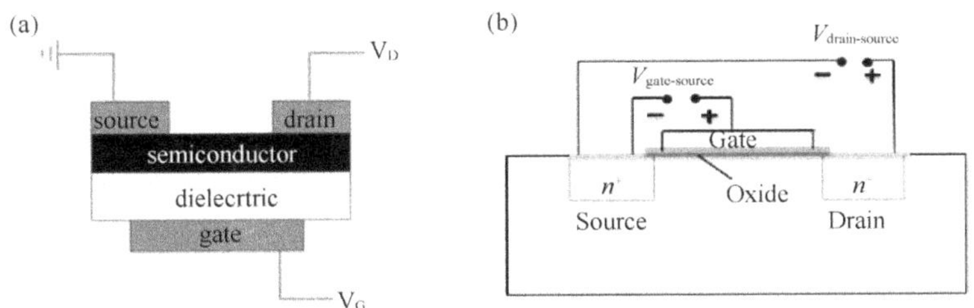

Figure 6.33(a) Cross-section of a field effect transistor with bottom gate. (b) Sketch of the cross-section of a metal of oxide semiconductor field effect transistor with top gate.

When a potential is applied between the gate and the source with the gate being positive, the potential draws electrons to the vicinity of the gate (and repels holes), but the electrons cannot enter the gate because of insulation layer.[87] The concentration of electrons beneath the gate makes this region (known as the channel) more conductive, so that a large potential between the source and drain permits electrons to flow from the source to the drain, producing a signal ("on" state). By changing the input voltage between the gate and the source, the number of electrons in the conductive path changes, thus also changing the output signal. When no voltage is applied to the gate, no electrons are attracted to the region between the source and the drain, and there is no current flow from the source to the drain ("off" state).

87 当在栅极和源极施加一个电压并且栅极为正时，电压使电子流向栅极附近（同时排斥空穴），但是由于绝缘层的存在，电子不能进入栅极。

An FET device can also be produced by organic materials (OFET), which usually adopts bottom gate structure. To begin with, it was only the active material (the channel between source and drain) that was made of organic semiconductors, but later it turned out to be possible to use heavily doped polymers also for the electrodes. The advantage of an all-polymer transistor is its low cost. These transistors are inherently slow on account of the low mobility of the organic materials. The main application envisaged for these transistors is as small plastic memory chips attached to various consumer products which is used for storing all kinds of information about the product, a lot more than is contained in present bar codes.[88] A further major advantage would be that these memory chips could be remotely interrogated by a radio frequency inidentification system.

88 可以预测，这些晶体管的主要应用是作为小型塑料存储芯片附加在各种消费品上，用来存储产品的各种信息，这比目前条形码包含的信息多得多。

Ⅳ. Photovoltaic Devices

Another important kind of *p-n* junction diode is a photovoltaic device. Figure 6.34a shows the cross section of a *p-n* junction with a thin *p* region so as to permit light to penetrate to the depletion region. The photons with energy greater than the band gap can create electron-hole pairs by promoting valence band electrons to the conduction band leaving an equal number of holes behind.[89] The electron-hole pairs created away from the depletion width are of little consequence and eventually merely re-combine. However, as is depicted in Figure 6.34b, the electron-hole pairs created in the depletion width are of great interest because the electrons can fall "downhill" into the CB of the *n*-type semiconductor while the holes flow "uphill" to the *p*-type side of the junction.[90] In this process, which is known as photovoltaic, current is created and photon energy is effectively transduced into electrical energy. To optimize a photovoltaic device, it is best to operate in the reverse bias region. In this case the downhill shape of the energy bands for both electrons and holes maximizes the current. Also the only current flowing in reverse bias is the saturation current. As was mentioned above, the thickness of the top layer needs to be minimized to allow maximum photon flux to the depletion region, and the area needs to be maximized to harvest as much of the light as possible. The depletion region is where the useful electron-hole pairs are created by the photons. This region can be maximized by using lightly doped semiconductors and even by the use of an intrinsic semiconductor that is insulating in between *p* and *n* regions. Such a diode is called a *p-i-n* diode.[91]

89 大于带隙能量的光子，通过激发价带电子到导带上，且在价带留下相等数量的空穴，从而产生电子空穴对。

90 在耗尽层中产生的电子空穴对，备受关注，因为这些电子可以落到 *n*-型半导体的 CB 一端，同时空穴上升到 *p*-型一端。

91 在耗尽区，光子能够产生有用的电子-空穴对。这一区域可以通过轻微掺杂的半导体，甚至在 *p* 和 *n* 之间使用绝缘的本征半导体来最大化。这样的二极管称为 *p-i-n* 型二极管。

In practice, besides *p-n* and *p-i-n* type photovoltaic devices to harvest solar energy (solar cell), other types of solar cells include organic solar cell (OPV), dye sensitized solar cell (DSSC), perovskite solar cell (PSC), etc. (Figure6.35).

These are beyond the content of this book and will not be discussed here.

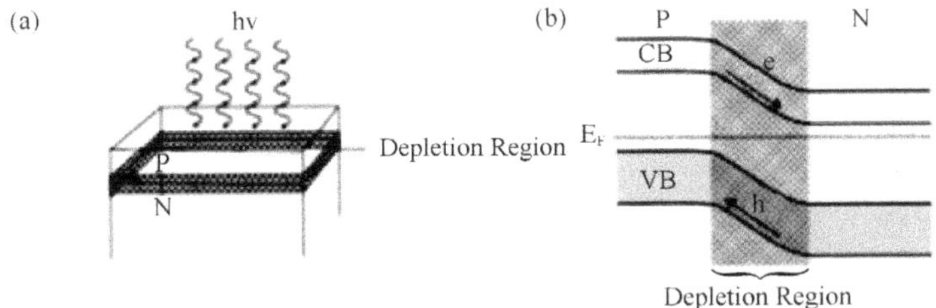

Figure 6.34 (a) Photocell *p-n* junction with a thin *p*-type region to allow light penetration. (b) The junction depletion region where photo-created electron-hole pairs are separated.

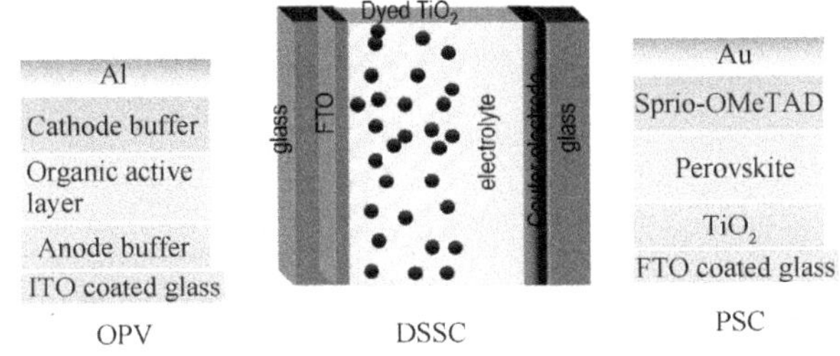

Figure 6.35 Photovoltaic device structures for OPV, DSSC and PSC.

Ⅴ. Light–Emitting Diodes (LED)

Inorganic semiconductor based LED is another important kind of *p-n* junction diode. In some situations, when a forward-biased potential of relatively high electrical field strength is applied across a *p-n* junction diode, visible light (or infrared radiation) is emitted.[92] This kind of devices, which conversing of electrical energy into light energy, is called a light-emitting diode (LED). The forward-biased potential attracts electrons on the *n*-side toward the junction, where some of them pass into (or are "injected" into) the *p*-side[93] (Figure 6.36a). Here, the electrons are minority charge carriers and therefore "recombine" with, or are annihilated by, the holes in the region near the junction. An analogous process occurs on the *p*-side—holes travel to the junction and recombine with the majority electrons on the *n*-side.

The elemental semiconductors silicon and germanium are not suitable for LEDs due to the detailed nature of their band gap structures. Rather, some of the III-V semiconducting compounds such as gallium arsenide (GaAs), and indium phosphide (InP), and alloys composed of these materials (e.g., $GaAs_xP_{1-x}$, where x is a small number less than unity) are frequently used.[94] The wavelength (i.e., color) of the emitted radiation is related to the band gap of the semiconductor (which is normally the same for both *n*- and *p*-sides of the diode). For example, red, orange, and yellow colors are possible for the GaAs–InP system. Blue and green LEDs have also been developed using (Ga,In) *N* semiconducting alloys. Thus, with this complement of colors, full-color displays are possible using LEDs.

92 在某些情况下，当 *p-n* 结两端施加一个相对较高的正偏压的电场时，可以产生可见光（或红外辐射）。

93 正向偏压吸引 *n*-型一端的电子，到 *p-n* 结处，其中一些电子进入（被注入）*p*-型一端。

94 由于元素半导体硅和锗的能隙结构，它们不适用于 LEDs。而一些 III-V 半导体化合物，例如砷化镓（GaAs）和磷化铟（InP），以及含有这些材料的合金（如 $GaAs_xP_{1-x}$，x 小于 1）经常被使用。

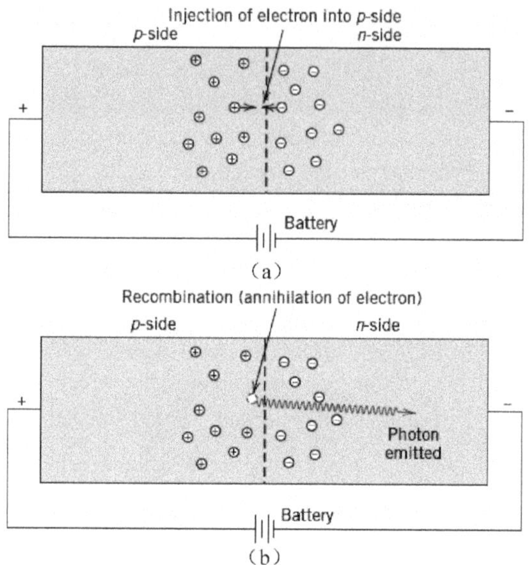

Figure 6.36 Schematic diagram of a forward-biased semiconductor *pn* junction showing (a) the injection of an electron from the *n*-side into the *p*-side, and (b) the emission of a photon of light as this electron recombines with a hole.

Excepting lighting application of LED, other important applications include digital clocks and illuminated watch displays, optical mice (computer input devices), and film scanners. In addition, they are more energy efficient than incandescent lights, generate very little heat, and have much longer lifetimes (because there is no filament that can burn out).[95]

95 另外，它们比白炽灯的能量利用率更高，产生很少的热量，同时有很长的寿命（由于没有能够烧断的灯丝）。

Similar to inorganic semiconductor based LEDs, LEDs using organic materials as active components are called as organic light-emitting diodes (OLEDs). The main asset of organics is their flexibility. It is unlikely that in the near future they would be able to compete with inorganic LEDs in efficiency, but they can on price and in applications where flexibility and large area are important requirements.[96] There will surely be demand for light sources which are large, cool and cheap, and can be fixed to curved surfaces. The ultimate would be the large television screen which can hang on a wall in the sitting room and when needed in another room it can be rolled up and swiftly moved.

96 不久的将来，在效率方面它还不能与无机LEDs相竞争，但是在价格和柔性与大面积为主的应用中，它们具有竞争力。

6.4 Superconductor

6.4.1 Phenomenon and Definition

In 1908, Kamerlingh Onnes of Netherland succeeded in liquefying helium and obtained low temperature of 1 K. He looked around for something worth measuring at that temperature range. His choice fell upon the resistivity of metals. He tried platinum first and found that its resistivity continued to decline at lower temperatures, tending to some small but finite value as the temperature approached the absolute zero. In 1911, with his second tried metal Hg, he found that the resistance of Hg showed quite unorthodox behavior with immeasurable level under temperature of about 4.2K (Figure 6.37). This is the beginning of super conduction.

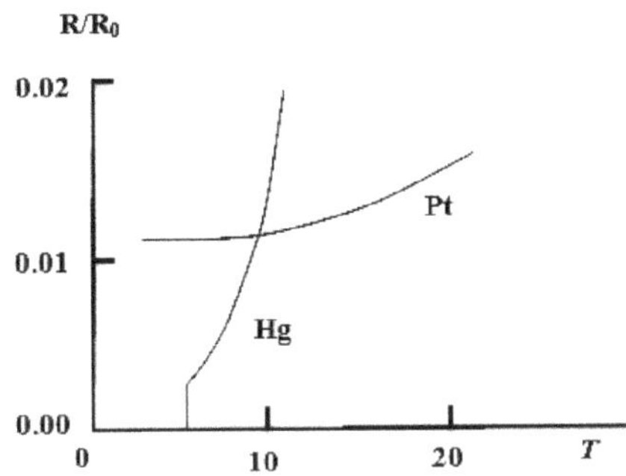

Figure 6.37 The resistance of platinum and mercury as a function of temperature (R_0 is the resistance at 0℃)

What is superconducting by definition? Substances, when cooled to very low temperature, its resistance decreases to zero (R = 0); and if a magnetic field is exerted to it, its own current will setup in the way of expelling the external magnetic field ($\chi_m = -1$).[97] This phenomenon is called **superconduction**. Compared to normal conductor, although their resistivity all decrease with temperature, the obvious difference is that below a critical temperature, the resistivity of a normal metal almost keeps constant, while that of a superconductor drops to zero (Figure 6.38).

97 当物质被冷却到非常低的温度时，其电阻值降低并趋于零(R = 0)，并且如果将其置于一个外加磁场中，物质自身将会产生电流来抵抗这一个外加磁场 ($\chi_m = -1$)。

The highest temperature that a material exhibits superconducting behavior is termed as superconducting transition temperature or critical temperature (T_C). The value of this critical temperature varies from material to material.[98] Conventional superconductors usually have critical temperatures ranging from around 20 K to less than 1 K. Cuprate superconductors can have much higher critical temperatures, and mercury-based cuprates have been found with critical temperatures in excess of 130 K.

98 一种材料表现其超导特性的最高温度被称为超导转变温度或者临界温度。不同的材料拥有不同的临界温度值。

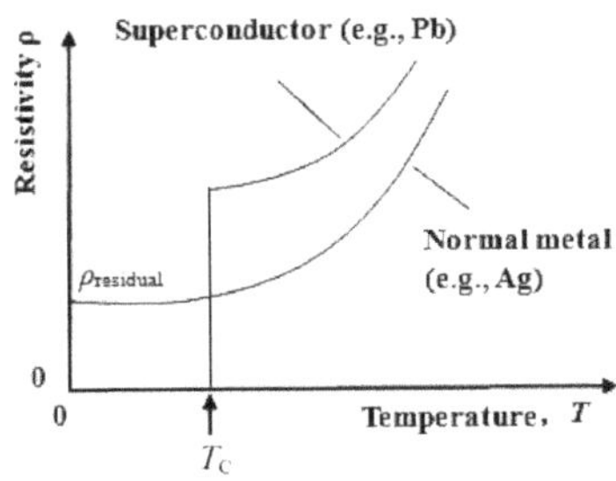

Figure 6.38 Differences between normal metals and superconductors.

The usual technique for superconducting test is to induce a current in a ring made of superconducting material under a magnetic field flux. In a normal metal the current would decay in about 10^{-12} s. In a superconductor the current can go round for a considerably longer time—measured not in picoseconds but in years.[99] One of the longest experiments was made somewhere in the United States, the current was going round and round for three whole years without any detectable decay. Unfortunately, the experiment came to an abrupt end when a research student forgot to fill up the Dewar flask with liquid nitrogen—so the story goes anyway. Experimental evidence points to a current lifetime of at least 100,000 years. Theoretical estimates for the lifetime of a persistent current can exceed the estimated lifetime of the universe, depending on the wire geometry and the temperature.

99 超导测试的一般技术手段是将由超导材料制成的一个圆环置于磁场下，使圆环中产生一个诱导电流。对于普通金属来说，这个电流将持续大约 10^{-12} 秒。对于超导体而言，这一电流能够持续一个相当长的时间——测量到的时间不是皮秒量级的，而是以年为单位。

Figure 6.39 shows history of superconductors progress. Here are some milestones:

1911, Netherlands, Onnes, Hg, T_C=4.2K

1941, Germany，Justi and Kramer, NbN, T_C = 15K

1953, USA, Hardy and Hulm, V_3Si, T_C = 17.1K

1957, BCS theory (1972 Nobel price): can not exceed 40 K

1973, Bell Lab, Nb_3Ge thin film, T_C = 23.2K

1986, IBM Zurich Laboratories, Muller and Bednorz, metal oxide $(La\text{-}xBa_x)CuO_4$，T_C > 77K (liquid N_2) Nobel price

1987, Chinese Academy of Science, Institute of Physics, Zhongxian Zhao et al, Y-Ba-Cu-O, T_C = 93 K

2006, Japan, Iron based superconductor, T_C ~ 25–40 K

The above mentioned superconducting materials are metals or inorganic materials. Actually, organic superconductors also have been discovered. For instance, the first organic superconductor, $(TMTSF)_2PF_6$, with T_C of 0.9 K, was found in 1980; In 1990, $K\text{-}(ET)_2Cu[N(CN)_2Cl$ was found having superconductivity below 12.8 K; In 1995, $(BEDT\text{-}TTF)_4[H_2O.Fe(C_2O_4)_3]PhCN$ was found to be the first paramagnetic organic super conductor; In 2003, $K\text{-}(BETS)_2FeBr_4$ was found to be the first antimagnetic organic super conductor. The highest temperature record for organic superconductor is from fullerene. It was found that 1991, K doped fullerene showed T_C = 18 K; in 2004, when fullerene doped with Chloroform and Bromoform, the T_C = 117 K.

It was found that some dielectric ceramics can be super conductors at much higher critical temperatures (e.g., $T_C \geqslant 80$ K). Unfortunately, however, the critical current of ceramic superconductors is much smaller than in metals, which is caused by grain boundaries in the oxide superconductors.[100]

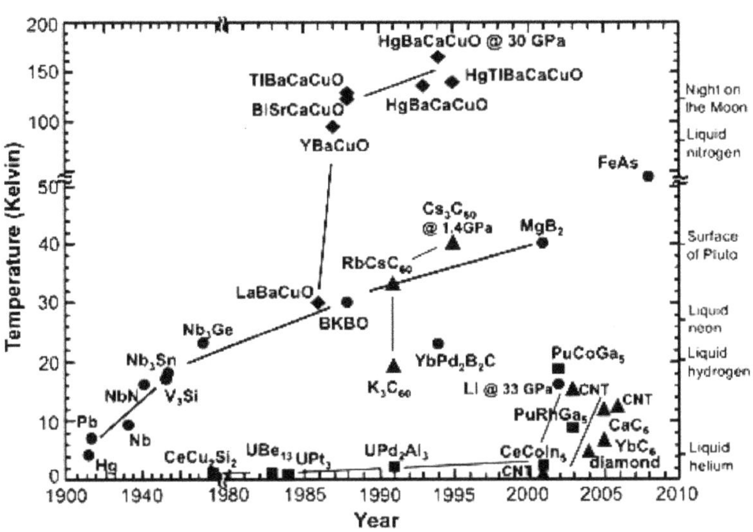

Figure 6.39 Milestones for discovery of superconducting materials

6.4.2 Meissner Effect

Besides the complete loss of electrical resistivity, the superconducting state also shows another defining property, namely the complete expulsion of a magnetic field from a material's interior (Figure 6.40). The superconductor behaves like an ideal diamagnet, the magnetic induction in its interior is zero. This effect is referred to as Meissner Ochsenfeld effect, or Meissner effect. It is not a consequence of the loss of resistivity and it is different from the behaviour of perfect conductor.[101]

Let's demonstrated these ideas through processes of moving a perfect conductor or a superconductor from point A to C by the paths of ABC and ADC (Figure 6.40a). For a perfect conductor (Figure 6.40b), at point A, there is no magnetic field and the temperature is higher than T_C. From A to B the temperature is reduced below the critical temperature; its electrical resistivity would completely disappear, but nothing else happens. Going from B to C means switching on the magnetic field. The changing flux creates an electric field that sets up a current opposing the applied magnetic field. This is just Lenz's law, and in the past we had referred to such currents as eddy currents. The essential difference now is the absence of resistivity. The eddy currents do not decay; they produce a magnetic field that completely cancels the applied magnetic field inside the material. The perfect conductor may regard as a perfect diamagnet.[102]

100 研究发现，一些介电陶瓷材料能够在较高的临界温度下（例如临界温度高于 80 K）表现为超导体。然而，不幸的是，由于氧化物超导体中晶界的缘故，陶瓷超导体的临界电流比金属的临界电流小得多。

101 超导态材料除了完全没有电阻外，也展现了另一种典型的性能，也就是其材料内部完全地排斥磁场（图 6.40）。超导体表现得像一个理想的抗磁体，其内部的磁诱导为零。这一效应被称为迈斯纳效应。这不是由电阻消失而导致的结果，并且也不同于完美导体的性质。

102 这一重要的区别是电阻的消失。这种涡流没有衰退，它们在材料内部产生了一个能够完全抵消附加磁场的一个磁场。完美的导体或许被认作为是一个完美的抗磁体。

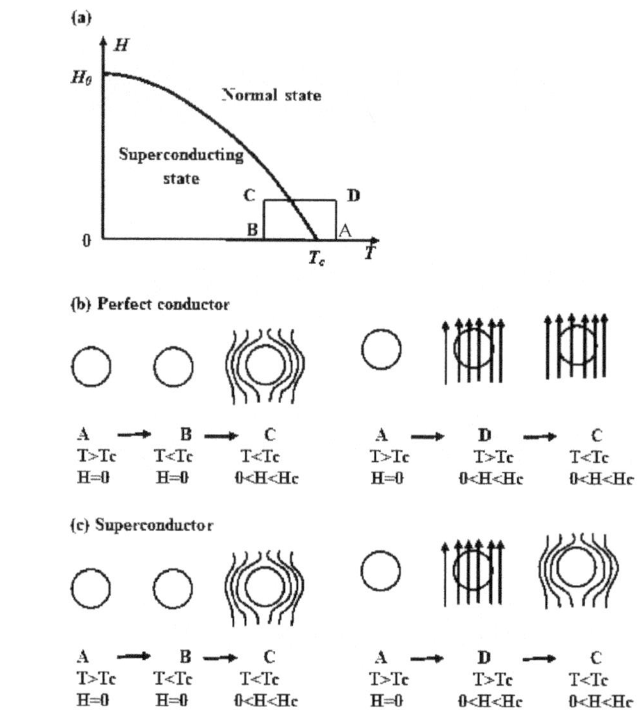

Figure 6.40(a) experimental paths both for perfect conductor and superconductor, from A-B-C, and A-D-C. (b) perfect conductor behaviors. (c) superconductor behaviors.

However, starting again at A with no magnetic field (Figure 6.40b) and proceeding to D puts the material into a magnetic field at a constant temperature. Assuming that our material is non-magnetic (superconductors are in fact slightly paramagnetic above their critical temperature, as follows from their metallic nature), the magnetic field will penetrate. Going from D to C means reducing the temperature at constant magnetic field. At point C the magnetic field should penetrate just as well as at D. Thus, for perfect conductor, the distribution of the magnetic field at C depends on the path we have chosen. If we go via B, the magnetic field is expelled; if we go via D, the magnetic field is the same inside as outside.

In contrast, for superconductor, things are different (Figure 6.40c). While the phenomena are the same as perfect conductor from A to B to C, the superconductor cooled in a constant magnetic field (from A to D to C) will set up its own current and expel the magnetic field when the critical temperature is reached.[103] The discovery of this effect by Meissner in 1933 showed superconductivity in a new light. It became clear that superconductivity is a new kind of phenomenon that does not obey the rules of classical electrodynamics.

It is worth to note that actually the external magnetic field does penetrate some distance into a superconductor, exponentially decaying with

103 相反的是，对于超导体，这一现象是不同的（图6.40c）。尽管从A到B再到C的现象同完美导体相似，但是被冷却的超导体当达到临界温度时，在一个恒定磁场下（从A到D再到C），自己将产生一个电流并且排斥外加的磁场。

characteristic length, λ, which is termed as the "penetration depth" or London penetration depth:

$$H(x)=H_0\exp(\frac{-x}{\lambda}) \tag{6-50}$$

Near 0 K, $\lambda = 10 \sim 100$ nm, and increases with temperature and magnetic field. When application of a magnetic field greater than H_c at T < T_C, the superconducting effect is destroyed due to the penetration depth, λ, becomes infinite, $H(x) = H_0$, magnetic field is penetrated through the sample.

6.4.3 Critical Magnetic Field

One of the applications of superconductor coming immediately to mind is the production of a powerful electromagnet. How nice it would be to have high magnetic fields without any power dissipation. The hopes of the first experimenters were soon dashed. They found out that above a certain magnetic field the superconductor became normal. Thus, in order to have zero resistance, not only the temperature but also the magnetic field must be kept below a certain threshold value. Experiments with various superconductors have shown that the dependence of the critical magnetic field on temperature is well described by the formula,[104]

$$H_c = H_0\left\{1-\left(\frac{T}{T_C}\right)^2\right\} \tag{6-51}$$

This relationship is plotted in Figure 6.41. It can be seen that the material is normal above the curve and superconductor below the curve. H_0 is defined as the magnetic field that destroys superconducting at absolute zero temperature. T_C is the beginning point of superconducting without a magnetic field (critical temperature).[105] The values of H_0 and T_C for a number of elemental superconductors are given in Table 6.5.

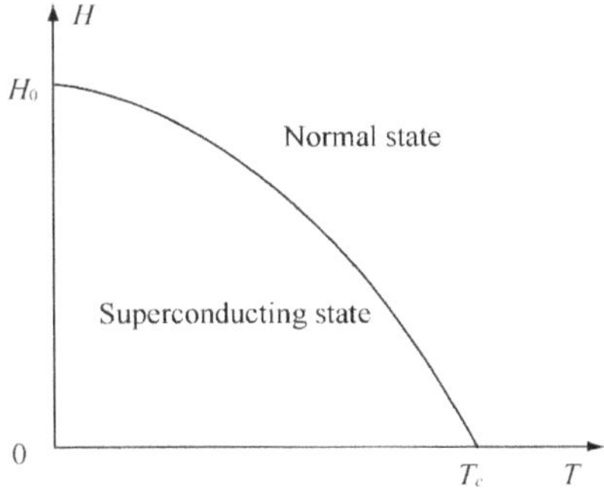

Figure 6.41 The critical magnetic field as a function of temperature.

104 因此，为了获得零阻值，不仅仅是温度，磁场也必须保持低于确定阈值。多种超导体的实验表明，临界磁场与温度的关系可以由一个公式很好地描述。

105 图 6.41 描绘了这种关系。可以看出，曲线以上的部分是材料的正常态，曲线以下的部分是超导态。H_0 被定义为是在绝对零度下破坏材料超导态的磁场。T_C 是没有磁场作用下超导态的起始点（临界温度）。

Table 6.5 The critical temperature and critical magnetic field of a number of elemental superconductors.

Element	T_C(K)	$H_0 \times 10^{-4}$ (Am^{-1})	Element	T_C(K)	$H_0 \times 10^{-4}$ (Am^{-1})
Al	1.19	0.8	Pb	7.18	6.5
Ga	1.09	0.4	Sn	3.72	2.5
Hg^{α}	4.15	3.3	Ta	4.48	6.7
Hg^{β}	3.95	2.7	Th	1.37	1.3
In	3.41	2.3	V	5.30	10.5
Nb	9.46	15.6	Zn	0.92	0.4

6.4.4 Critical Current Density (J_C)

Due to the zero resistance in superconductor, according to Ohm's law, a current in a superconductor should be infinite, which is not the real case. Superconductivity also disappears when current density exceeds a critical value, J_C, because the current itself creates a magnetic field which can exceed H_C. When the sum of external magnetic field and the electric current induced magnetic field exceeds H_C, the superconductive state will be destroyed. The current density at this moment is called critical current density (Figure 6.42).[106] At fixed temperature, without external magnetic field, J_C reaches maximum. The state of normal conduction is regained if either the temperature, or the current density, or an external magnetic field exceeds certain critical values, which are different for each material. Particularly, the T_C of a superconductor is very sensitive to synthesis technique and microstructure. This variability as well as the generally low value of J_C has greatly limited practical applications of these materials. For type I superconductors expected relationship between J_C and H_C has been found, while in type II superconductors, the correlation is more complex because of complex interactions between current, vortices and material microstructure.[107] Ceramic Hi-T_C superconductors show particular sensitivity of J_C to synthesis technique and microstructure. This variability (and generally low value of J_C) has greatly limited practical applications of these materials.

106 当电流密度超过一个临界值 J_C 时，超导态也会随之消失，因为电流本身会产生一个超过临界磁场 H_C 的磁场。当外加磁场和电流诱导磁场的总值超过 H_C 时，超导态也将会被破坏。在这一情况的电流密度被称为临界电流密度(J_C ,图 6.42)。

107 对于第一类超导体，J_C 和 H_C 之间的关系已经被发现了，但是对于第二类超导体材料，它们之间的关系由于电流，涡流和材料微观结构之间复杂的相互作用，变得更加复杂。

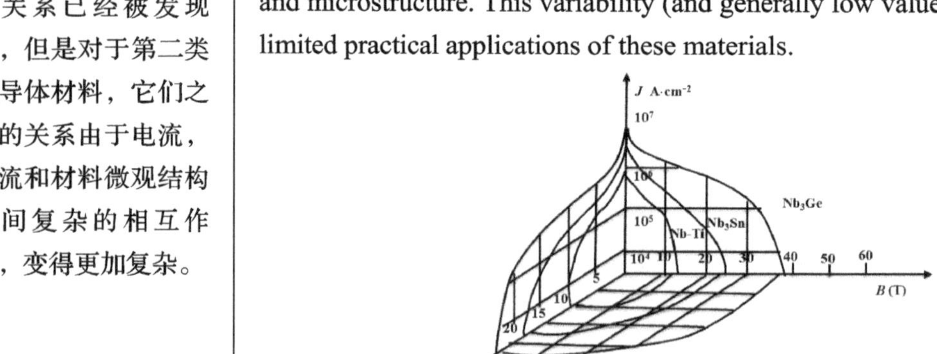

Figure 6.42 Graph of the critical values of current density J, magnetic field B and temperature T for some superconductors.

6.4.5 Thermal Dynamic Features

Let us look back at Figure 6.42 again. Above the curve the substance behaves in the normally way. However, as soon as we cross the curve, the properties of the substance become qualitatively different. Above the curve the substance is non-magnetic and has a finite electrical resistance, below the curve it becomes diamagnetic and possesses zero electric resistance.[108]

108 在曲线以上，物质表现为正常态。然而，一旦穿过这条曲线，物质的性能将会发生质变。曲线以上，物质是非磁性的，并有一个有限的电阻；曲线以下，物质将变为抗磁的，并且电阻为零。

If you think about it a little you will see that the situation is very similar to that you have studied under the name of 'phase change' or 'phase transition' in thermodynamics. The properties of the substance differ appreciably above and below the curve. We may talk about normal and superconducting phases when interpreting Figure 6.42.

Thus, the road is open to investigate the properties of superconductors by the well-established techniques of thermodynamics. Well, we must be careful; thermodynamics can be applied only if the change is reversible. Is the normal to superconducting phase change reversible? Fortunately, it is. Had we a perfect conductor instead of a superconductor the phase change would not be reversible, and we should not be justified in using thermodynamics.[109] Thanks to the Meissner effect, thermodynamics is applicable.

109 如果是完美的导体而不是超导体的话，这一相变是不可逆的，并且不能用热力学来解释。

The first law of thermodynamics:

$$dE=dQ-dW \tag{6-52}$$

Then there is the Gibbs function (to which we shall also refer as the Gibbs free energy) defined by:

$$G=E+PV-TS \tag{6-53}$$

E is the internal energy, W is the work, S is the entropy, P is the pressure, V is the volume, and Q is the heat.

An infinitesimal change in the Gibbs function gives

$$dG=dE-pdV+VdP-TdS-SdT \tag{6-54}$$

using Equation (6-52) and taking $dQ=YdS$, $dW=PdV$, Equation (6-54) reduces to:

$$dG=dQ-dW+pdV-Vdp-TdS-SdT \tag{6-55}$$

$$dG=VdP-SdT \tag{6-56}$$

Thus, for an isothermal isobaric process, $dG = 0$. The Gibbs function does not change while the phase transition takes place. In the case of the normal-to-superconducting phase-transition the variations of pressure and volume are small and play negligible roles, and so we can just as well forget them but, of course, we shall have to include the work due to magnetization.[110]

110 当发生相转变时，Gibbs 函数不会变化。在正常态到超导态的相变情况下，压力和体积的变化较小，起到了很小的作用，所以我们可以忽略这些变化，但是我们必须得将磁化做功考虑在内。

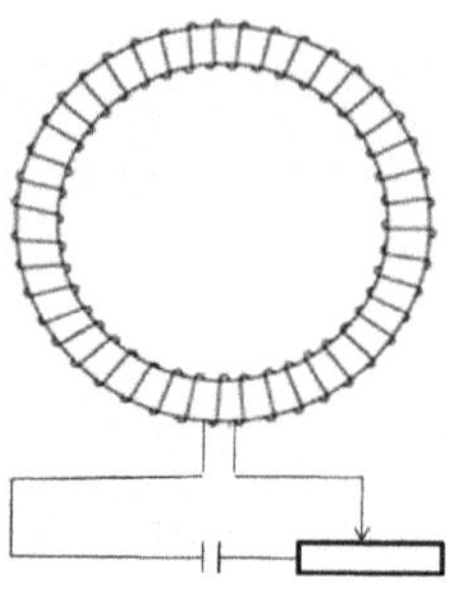

Figure 6.43 The magnetization of magnetic material in a toroid (for working out the magnetic energy).

In order to derive a relationship between work and magnetization let us investigate the simple physical arrangement shown in Figure 6.43. From the class of electricity that current doing work on a system in a time dt is

$$\mathrm{d}W=-UI\mathrm{d}t \tag{6-57}$$

where U is the voltage and I is the current, and the negative sign comes from the accepted convention of thermodynamics that the work done on a system is negative. Further, using Faraday's law, we have,

$$U = NA\frac{\mathrm{dB}}{\mathrm{d}t} \tag{6-58}$$

From Ampere's law,

$$HL=NI \tag{6-59}$$

here, A is the cross-section of the toroid, N is the number of turns, and L is the mean circumference of the toroid. Then we get,

$$\mathrm{d}W = -\left(NA\frac{\mathrm{d}B}{\mathrm{d}t}\right)\left(\frac{HL}{N}\right)\mathrm{d}t = -HLA\mathrm{d}B = -VH\mathrm{d}B \tag{6-60}$$

according to

$$B=\mu_0 (H+M) \tag{6-61}$$

we have

$$\mathrm{d}W= -VH\mu_0 (\mathrm{d}H+ \mathrm{d}M) = -\mu_0 VH\mathrm{d}H-\mu_0 VH\mathrm{d}M \tag{6-62}$$

The first term on the right-hand side of Equation (6-62) gives the increase of energy in the vacuum, and the second term is due to the presence of the magnetic material. Thus, the workdone on the material is:

$$\mathrm{d}W= -\mu_0 VH\mathrm{d}M \tag{6-63}$$

Hence, for a paramagnetic material the work is negative, but for a diamagnetic material (where M is opposing H) the work is positive, which means that the system needs to do some work in order to reduce the magnetic field inside the material.[111]

111 因此，对于顺磁性材料，做的是负功，但是对于抗磁性材料(磁感应与外磁场方向相反)，做的是正功，这意味着系统需要通过做功来降低材料内部的磁场。

In order to describe the phase transition in a superconductor, we have to define a 'magnetic Gibbs function'. Remembering that $P\mathrm{d}V$ gives positive work (an expanding gas does work) and $H\mathrm{d}M$ gives negative work, we have to replace PV in Equation (6-53) by $-\mu_0 VHM$. Our new Gibbs function takes the form,

$$G=E-\mu_0 VHM-TS \quad (6\text{-}64)$$

$$dG=dE-\mu_0 VHdM-\mu_0 VMdH-TdS-SdT \quad (6\text{-}65)$$

taking account of the first law for magnetic materials (again replacing pressure and volume by the appropriate magnetic quantities)

$$dE=Tds+\mu_0 VHdM \quad (6\text{-}66)$$

$$dG=-SdT+\mu_0 VMdH \quad (6\text{-}67)$$

This is exactly what we wanted. It follows immediately from the above equation that for a constant temperature and constant magnetic field process

$$dG=0 \quad (6\text{-}68)$$

G remains constant while the superconducting phase transition takes place. For a perfect diamagnet,

$$M=-H \quad (6\text{-}69)$$

which substituted into Equation (6-67) gives

$$dG=-SdT-\mu_0 VHdM \quad (6\text{-}70)$$

Integrating at constant temperature, we get,

$$G_S(H)=G_S(0)=+\frac{1}{2}\mu_0 H^2 V \quad (6\text{-}71)$$

G_S (0) is the Gibbs free energy at zero magnetic field, and the subscript s refers to the superconducting phase. Since superconductors are practically non-magnetic above their critical temperatures, we can write for the normal phase

$$G_N(H)=G_N(0)=G_N \quad (6\text{-}72)$$

In view of Equation (6-72) the Gibbs free energy of the two phases must be equal at the critical magnetic field, H_C, that is

$$G_S(H_C)=G_N(H_C) \quad (6\text{-}73)$$

Substituting Equation (6-71) and (6-72) into Equation (6-73) we get

$$G_N=G_S(0)+\frac{1}{2}\mu_0 H_C^2 V \quad (6\text{-}74)$$

Comparison of Equation (6-71) and (6-74) clearly shows that at a given temperature (below the critical one) the conditions are more favorable for the superconducting phase than for the normal phase, provided that the magnetic field is below the critical field.[112] There are three cases:

(i) If $H < H_C$, then $G_N > G_S(H)$

(ii) If $H > H_C$, then $G_N < G_S(H)$

(iii)If $H = H_C$, then $G_N = G_S(H)$

The substance will prefer the phase for which the Gibbs free energy is smaller, that is at temperature below T_C, in case (i) it will be in the superconducting phase, in case (ii) in the normal phase, and in case (iii) just in the process of transition[113].These results are schematically shown in

112 从（6-71）式与（6-74）式的比较可以清楚地看出，在一个给定的温度下（低于临界温度），假设磁场低于临界磁场，超导相的状态比正常相的状态更稳定。

113 物质将偏向于 Gibbs 自由能较低的相，也就是在低于临界温度下，在状态（i）是超导相，状态（ii）是正常相，而（iii）是相转变过程。

Figure 6.44.

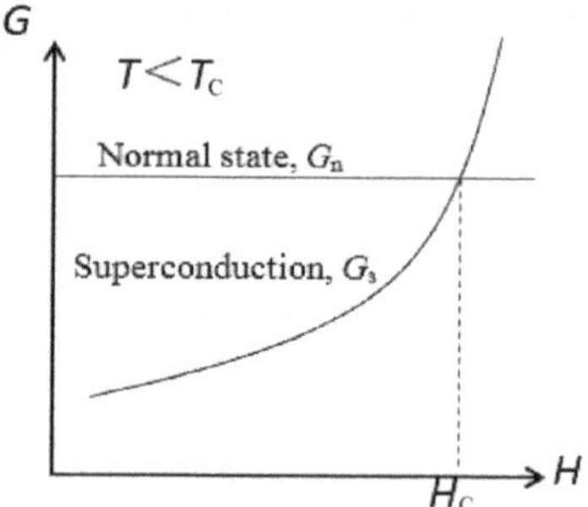

Figure 6.44 Free energy versus magnetic field for normal state and superconducting state under temperature below T_C.

If the transition takes place at temperature, T+dT, and magnetic field, Hc+dHc, then it must still be valid that[114]

114 如果转变发生在温度 T + dT，磁场 H_c + dH_c，则下面的（式子）成立。

$$G_S+dG_S= G_N+dG_N \tag{6-75}$$

where

$$dG_S=dG_N \tag{6-76}$$

using Equation (6-65), leads to

$$-S_S dT-\mu_0 VM_S dH_C =-S_n dT-\mu_0 VM_N dH_C \tag{6-77}$$

because

$$M_S=-H_C \text{ and } M_N= 0 \tag{6-78}$$

this reduces Equation (6-77) to

$$S_N - S_S = -\mu_0 VH_C \frac{dH_C}{dT} \tag{6-79}$$

according to Equation (6-79), the difference between the entropy of normal state and superconducting state can be shown as Figure 6.45.

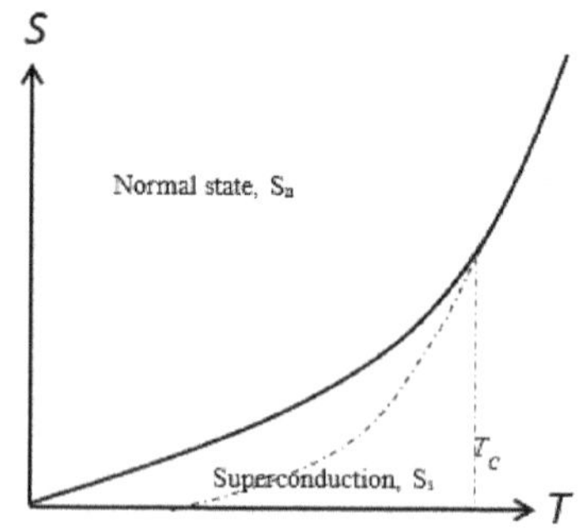

Figure 6.45 Entropy versus temperature for normal state and superconducting state in metal Tin.

The latent heat transition may be written in the form,

$$L=T(S_N-S_S) \tag{6-80}$$

and with the aid of Equation (6-79) this may be expressed as

$$L = -\mu_0 TVH_C \frac{dH_C}{dT} \tag{6-81}$$

At last, we arrived at a useful relationship. Now, we have a theory that connects the independently measurable variable L, V, H_C, and T. After getting them, dHc/dT can be determined from the H_C–T plot. Putting these values into Equation (6-81), the equation should be satisfied; and it is satisfied to a good approximation thus giving us experimental proof that we are on the right track. It is interesting to note that L vanishes at two extremes of temperature, namely, at $T = 0$ and at $T = T_C$ where the critical magnetic field is zero. A transition which takes place with no latent heat is called a second-order phase transition. In this transition entropy remains constant, and the specific heat is discontinuous.[115] Neglecting the difference between the specific heats at constant volume and constant pressure, we can write, in general for the specific heat as,

$$C = T\frac{dS}{dT} \tag{6-82}$$

Thus, using Equation (6-79), we have:

$$\begin{aligned} C_N - C_S &= T\left(\frac{dS_N}{dT} - \frac{dS_S}{dT}\right) \\ &= -VT\mu_0\left\{-H_C\frac{d^2H_C}{dT^2} + \left(\frac{dH_C}{dT}\right)^2\right\} \end{aligned} \tag{6-83}$$

where H_C=0 at $T = T_C$,

$$C_N - C_S = -\left\{VT\mu_0\left(\frac{dH_C}{dT}\right)^2\right\}_{T=T_C} \tag{6-84}$$

This is negative because the experimentally established Hc -T curves have finite slopes at $T = T_C$. It follows that in the absence of a magnetic field the specific heat has a discontinuity. This is borne out by experiments as well, as shown in Figure 6.46, where the specific heat of Tin is plotted against temperature. The discontinuity occurs at the critical temperature T_C = 3.72 K.[116]

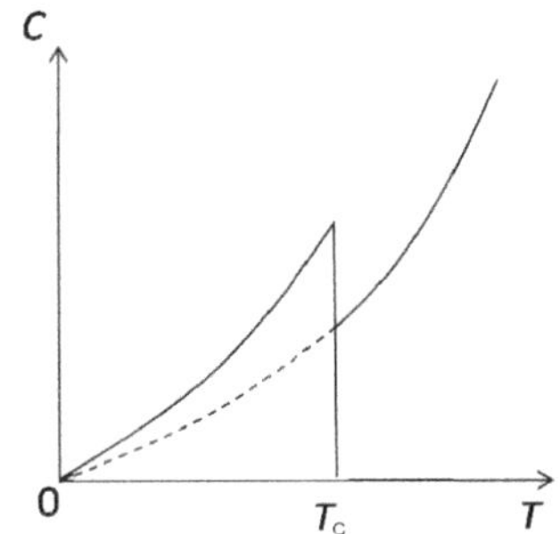

Figure 6.46 Specific heat capacity of Tin at normal conductor (below Tc is dashline), and superconductor (below Tc is solid line).

115 指出一个有趣的现象，在两个极端温度下，也就是 $T = 0$ 时和 $T = T_C$ 而临界磁场为零时，相变潜热 L 消失。没有潜热发生的相变被称为二阶相转变过程。在这一转变时，熵是一个常量，比热是不连续的。

116 这一个负值是因为在 $T = T_C$ 时，实验建立的 H_C-T 曲线具有有限的斜率。它遵循在没有外加磁场下比热是不连续的。这也由实验证实了，如图 6.46 描绘了锡的比热与温度的关系。这一不连续点发生在临界温度 T_C=3.72 K 时。

117 没有磁场的情况下，类似于正常态，在超导体中，其比热容正比于温度，并且在T_C时有一个突变。然而，相比于正常导体态，低于T_C时，超导体的比热容比正常态的要大。

Without magnetic field, similar to the normal state, in a superconductor, the specific heat capacity is proportional to temperature. There is an abrupt change at T_C. However, compared to normal conduction state, below T_C, the heat capacity of super conductor is larger than that for normal state.[117] Figure 6.47 shows the correlation between the heat capacity and resistivity of a superconductor. As can be seen that at the temperature above T_C, both resistivity and heat capacity increase with temperature. At T_C, resistivity drops to zero and there is an abrupt increase in heat capacity. The temperature dependence of heat capacity before and after T_C is different.

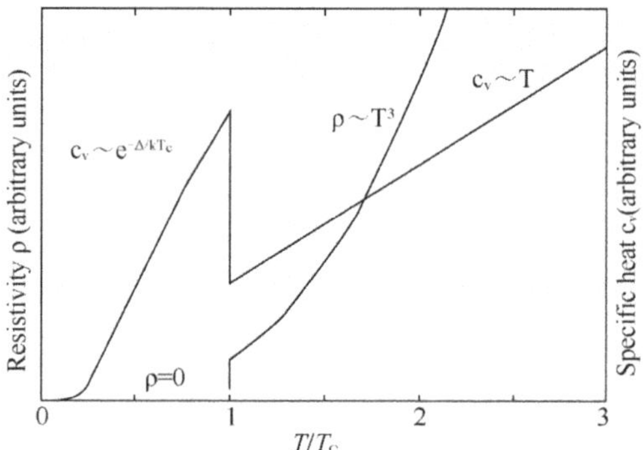

Figure 6.47 Behavior of heat capacity (c_v) and resistivity (ρ) at the superconducting phase transition.

6.4.6 Classification of Superconductor

We distinguish superconductors of type I (soft superconductors) and superconductors of type II (hard superconductors) according to their physical properties (Figure 6.48).

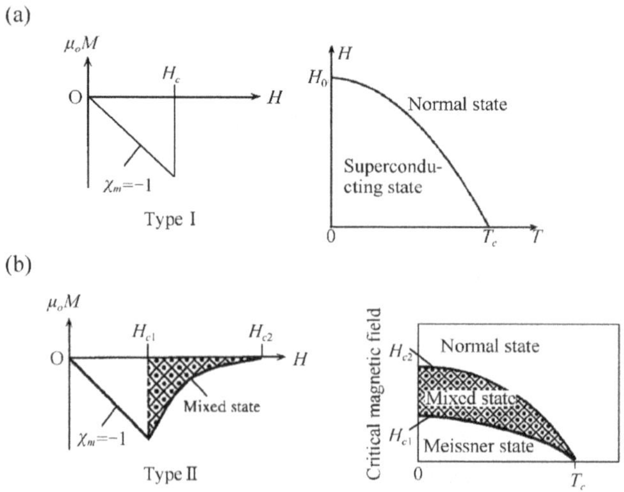

Figure 6.48 (a) Magnetization curve of a type I superconductor (perfect diamagnet) and the phase diagram. (b) Magnetization curve of a type II superconductor and the phase diagram. At field strength H_{C1} flux begins to penetrate into the sample.

Type I Superconductor: In superconductors with diamagnetic properties, for example, metals excluding V, Nb, and Ta, below T_C, when external magnetic field increases to a critical value of H_C, the induced magnetic field drops to zero abruptly, i.e., Meissner effect disappears, superconductivity is destroyed, and the material behaves as a normal conductor[118] (Figure 6.48a).The corresponding phase transformation diagram of type I superconductors can be described as the right of Figure 6.48a.

Type II Superconductor: In metal superconductors of V， Nb and Ta, as well as alloy and compound superconductors, below T_C, when the external magnetic field increases to H_{C1}, the induced magnetic field M starts to decease. After reaching H_{C2}, material becomes normal conductor (Figure 6.48b). H_{C1} and H_{C2} is called lower- and upper- critical magnetic field. H_{C2} can be 100 times greater than H_{C1}. Between H_{C1} and H_{C2}, material contains mixtures of super-conduction and normal conduction. This state is a mixed state, in which magnetic field can get through. In type II superconductor, up to H_{C1} the superconductor is diamagnetic, as shown in Figure 6.48b, where -M is plotted against the applied magnetic field. Above H_{C1} the magnetic field begins to penetrate (beyond the 'diamagnetic' penetration depth) and there is complete penetration at H_{C2}, where the material becomes normal.[119] The right hand of Figure 6.48b illustrates the corresponding phase transformation.

In fact, in type II superconductors, including all known high-temperature superconductors, an extremely small amount of resistivity appears at temperatures not too far below the nominal superconducting transition when an electric current is applied in conjunction with a strong magnetic field, which may be caused by the electric current. The resistance due to this effect is tiny compared with that of non-superconducting materials, but must be taken into account in sensitive experiments. However, as the temperature decreases far enough below the nominal superconducting transition, the resistance of the material becomes truly zero.

Hard superconductors (type II) allow the magnetic field to penetrate above a critical field strength H_{C1} into a thin surface layer of the material, but the material remains superconducting. Only for $H > H_{C2}$ the material becomes a normal conductor. The field H_{C2} can be two orders of magnitude larger than the critical field of a soft superconductor (type I). Therefore, hard superconductors are interesting for commercial applications, for instance, high-field superconducting magnets.[120]

118 在具有抗磁性的超导体中，例如除了 V，Nb，Ta 外的金属,在低于 T_C 温度下，当外磁场增加到临界磁场的值时，诱导磁场将迅速突然地降低为零，也就是说麦斯纳效应消失，超导体被破坏，材料表现为普通导体。

119 在第二类超导体中，磁场强度达到 H_{C1} 之前，超导体是抗磁的，如图 6.48b 描绘的-M 与施加磁场的关系曲线。高于 H_{C1} 时，磁场开始穿透（超过抗磁的穿透深度），并在 H_{C2} 时完全穿透，此时材料是正常态的。

120 硬超导体（第二类）在高于临界磁场强度时，允许磁场穿透材料薄的表层，但是材料仍为超导态。仅当 $H > H_{C2}$ 时材料才变为正常导体。H_{C2} 比软超导体（第一类）的临界磁场大两个数量级。因此，硬超导体对于商业应用是令人感兴趣的，比如高场超导磁体。

Superconductors also can be classified according to different categories:

By materials:

- Pure elements: such as lead or mercury (but not all pure elements, as some never reach the superconducting phase). Most of them are type I (except niobium, technetium, vanadium, silicon, and the above mentioned carbon allotropes).
- Allotropes of carbon: such as fullerenes, nanotubes, or diamond.
- Alloys: such as Niobium-titanium (NbTi), whose superconducting properties were discovered in 1962.
- Ceramics: which include the YBCO family, which are several yttrium-barium-copper oxides. They are the most famous high-temperature superconductors.

By operation theory:

- Conventional superconductors: those that can be fully explained with the BCS theory or related theories.
- Unconventional superconductors: those that failed to be explained using such theories.

This criterion is important, as the BCS theory is explaining the properties of conventional superconductors since 1957, but on the other hand there have been no satisfactory theory to explain fully unconventional superconductors. In most of cases type I superconductors are conventional, but there are several exceptions as niobium, which is both conventional and type II.[121]

121 这一准则是重要的，因为自从1957 年的 BCS 理论只解释了传统超导体的性能，但是另一方面没有满意地解释非传统超导体的理论。大多数情况下第一类超导体是传统的，但是也有几个例外，比如既是传统又是第二类超导体的铌。

By critical temperature:

- Low-temperature superconductors, or LTS: those whose critical temperature is below 77 K.
- High-temperature superconductors, or HTS: those whose critical temperature is above 77 K.

This criterion is used when we want to emphasize whether or not we can cool the sample with liquid nitrogen (whose boiling point is 77K), which is much more feasible than liquid helium (the alternative to achieve the temperatures needed to get low-temperature superconductors).

6.4.7 Theory about Superconducting

Ⅰ. Introduction

Superconductivity is, after all, such an extraordinary phenomenon, so much in contrast with everything we are used to. It is literarily out of this world. Our world is classical, but superconductivity is a quantum phenomenon — a quantum phenomenon on macroscopic scale. The wave functions, for example, that lead an artificial existence in quantum mechanics proper appear in superconductivity as measurable quantities.[122]

Thus, for all practical purposes we are faced with a real lossless phenomenon. It is so much out of the ordinary that no one quite knew how to approach the problem. Several phenomenological theories were born, but its real cause remained unknown for half a century. Up to 1957 it defied all attempts; so much so, that it gave birth to a new theorem, namely that 'all theories of superconductivity are refutable'. In 1957, Bardeen, Schrieffer, and Cooper produced a theory (called the BCS theory) that managed to explain all the major properties of superconductivity for all the superconductors known at that time. Unfortunately, the arrival of a host of new superconducting materials has cast fresh doubts on our ability ever to produce a complete theory.[123]

Ⅱ. BCS Theory

In a strongly simplified model superconductivity can be explained by the formation of pairs of two electrons with antiparallel momentum and antiparallel spin at low temperatures. The attractive interaction between the electrons occurs by the exchange of so-called virtual phonons through the crystal lattice. Such a Cooper pair forms a new particle (spin up, spin down) with zero total momentum and zero total spin.[124]

For explanation, we use a system containing seven electrons as example. Figure 6.49a shows the energy-momentum curve of an ordinary conductor with seven electrons sitting discreetly in their discrete energy levels. In the absence of an electric field the current from electrons moving to the right is exactly balanced by that from electrons moving to the left. Thus, the net current is zero.[125]

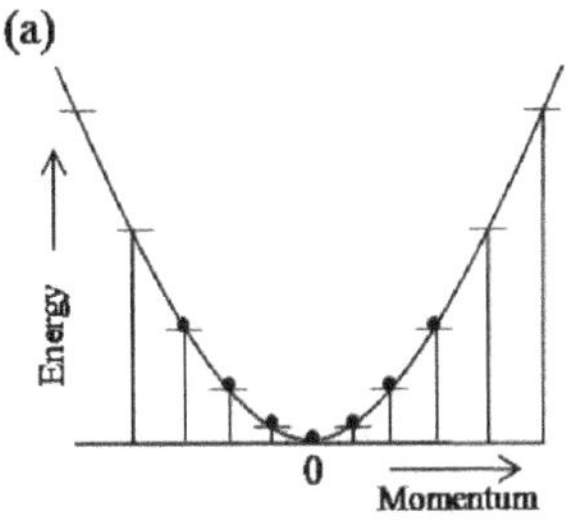

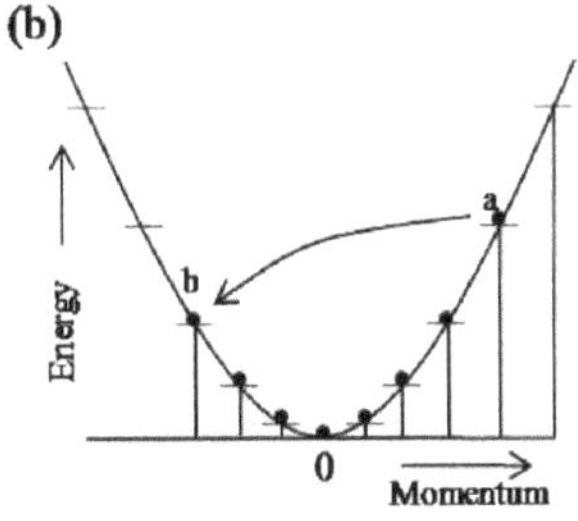

Figure 6.49 A one-dimensional representation of the energy-momentum curve for seven electrons in a conductor. (a) All electrons in their lowest energy states, the net momentum is zero. (b) There is a net momentum to the right as a consequence of an applied electric field.

When an electric field is applied, all the electrons acquire some extra

122 我们的世界是经典的，但是超导是一个量子现象——在宏观尺度上的量子现象。例如，波函数，即一个在量子力学中虚拟的存在，并且作为一个可测量的量适当地出现在了超导体中。

123 不幸的是，大量新超导材料的出现对于我们创造出一个完整理论的能力提出了新的疑问。

124 在一个极度简化的模型中，超导可以由低温下两个具有反平行动量和反平行自旋的电子形成的电子对来解释。这种电子间吸引的相互作用通过晶格间虚拟声子的交换作用产生。这一 Cooper 对形成了一个总动量总自旋为零的新粒子（一个自旋向上，一个自旋向下）。

125 没有电场作用下，电子向右移动产生的电流由向左移动产生的电流来平衡，因此，净电流为零。

126 由于与振动晶格、杂质原子或者其他不规则物的碰撞，较快速的电子被散射为较低的能态，直到原始的分布重新建立起来。对于我们这简化的模型，它说明了电子通过散射由能级 a 变为能级 b。

momentum, and this is equivalent to shifting the whole distribution in the direction of the electric field, as shown in Figure 6.49b. Now what happens when the electric field is removed? Owing to collisions with the vibrating lattice, with impurity atoms, or with any other irregularity, the faster electrons will be scattered into lower energy states until the original distribution is re-established. For our simple model, it means that the electron is scattered from the energy level of *a*, into energy level of *b*.[126]

In the case of a superconductor, it becomes energetically more favorable for the electrons to seek some companionship. Those of opposite momenta (the spins incidentally must also be opposite), pair up to form a new particle called a superconducting electron or, after its discoverer, a Cooper pair. This link between two electrons is shown in Figure 6.50a by an imaginary mechanical spring.

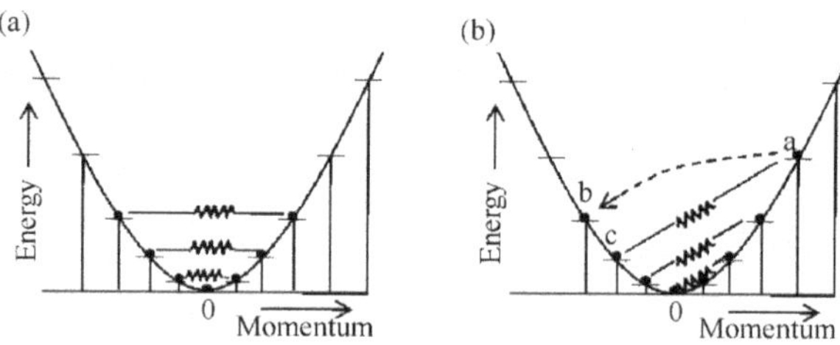

Figure 6.50 The energy-momentum curve for seven electrons in a superconductor. Those of opposite momenta pair up—that is represented here by a mechanical spring. (a) All pairs in their lowest energy states, the net momentum is zero. (b) There is a net momentum to the right as a consequence of an applied electric field.

An applied electric field will displace all the particles again, as shown in Figure6.50b. But when the electric field disappears, there is no change. Scattering from energy level of *a* to energy level of *b* is no longer possible because the electrons both at *b* and *c* would become pairless, which is energetically unfavorable.[127] One may imagine a large number of simultaneous scatterings that would just re-establish the symmetrical distribution of Figure 6.50a, but that is extremely unlikely. So the asymmetrical distribution will remain; there will be more electrons going to the right than to the left, and this current will persist forever—or, at least, for three years.

127 从能级 a 到能级 b 的散射不再是可能的，否则 a 处和 b 处的都将不成对，这在能量上是不利的。

In a Cooper pair, the two constituents of the particle move with v and $-v$, respectively; thus the velocity of the center of mass is zero. According to wave particle duality ($\lambda = h/p$), the wavelength associated with the new particle is infinitely long. And this is valid for all superconducting electrons.[128] It does not quite follow from the above simple argument (but it comes out from the theory) that all superconducting electrons behave in the same way. This is, for our electrons, a complete break with the past.

128 根据波粒二象性（$\lambda = h/p$），这一新粒子的波长是无限长的，并且这对所有超导电子都是有效的。

A particle with a total spin zero, is a Boson rather than a Fermion and, therefore, is not constrained by Fermi-statistics. Instead, all Cooper pairs can occupy the same quantum state. Since all Cooper pairs occupy the same quantum state they move as a whole in phase with zero-point lattice vibrations. In the presence of an electric field the entity of all Cooper pairs assumes a slightly higher energy state, in which all pairs share the same momentum, which corresponds to a superconducting current.[129] The superconducting state requires the same quantum mechanical state for all pairs. Therefore, a single pair cannot interact with the crystal lattice, i.e. to exchange momentum, without leaving the superconducting state, for which a finite energy has to be expended, namely the interaction energy between Cooper pairs. Therefore, below *Tc* only the absorption of large energies in electrical or magnetic fields will eventually destroy the superconducting property.

129 在有电场存在的情况下，假设所有Cooper对存在于一个稍高的能态，所有的电子对拥有相同的动量，形成超导电流。

The microscopic theory is part of quantum field theory and has something to do with Green's functions and has more than its fair share of various operators. We shall not say much about this theory, but we should just like to indicate what is involved. The fundamental tenet of the theory is that superconductivity is caused by a second-order interaction between electrons and the vibrating lattice.[130] This is rather strange. After all, we do know that thermal vibrations are responsible for the presence of resistance and not for its absence. This is true in general; the higher the temperature the larger the electrical resistance. Below a certain temperature, however, and for a select group of materials, the lattice interaction plays a different role. It is a sort of intermediary between two appropriately placed electrons. It results in an apparent attractive force between the two electrons, an attractive force larger than the repulsive force, owing to the Coulomb interaction. Hence, the electron changes its character. It stops obeying Fermi-Dirac statistics, and any number of electrons (or more correctly any number of electron pairs) can be in the same state.[131]

130 这一理论的基本原则是超导是由电子和振动晶格间的二级相互作用引起的。

131 它导致了两电子间明显的吸引力，这种吸引力大于由库仑相互作用的排斥力。因此，电子改变了它的特性。不遵从费米狄拉克统计，并且任何电子（或者更准确地说是任何电子对）都变成同一态。

Do we have any direct experimental evidence that superconductivity is caused by electron-lattice interaction? Yes, the so-called isotope effect. The critical temperature of a superconductor depends on the total mass of the nucleus. If we add a neutron (that is, use an isotope of the material) the critical temperature decreases. A simple explanation of the interaction between two electrons and the lattice is shown in Figure 6.51a. The first electron moving to the right causes the positive lattice ions to move inwards, which then attract the second electron. Hence, there is an indirect attraction between the two electrons.[132]

132 第一个向右移动的电子引起了晶格正离子的内部移动，从而吸引第二个电子。因此，两电子间有一个间接的吸引力。

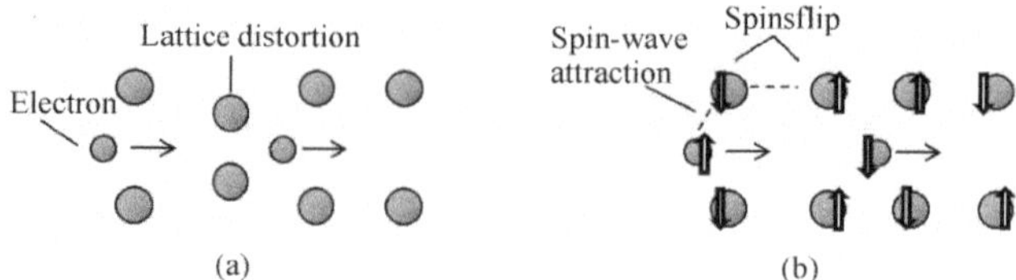

Figure 6.51 Interactions leading to Bose-Einstein condensation (a) between the lattice and electrons, (b) between spin waves and electrons.

Are there any other kind of interactions resulting in superconductivity? Nobody knows for certain, but it may be worthwhile to describe briefly one of the mechanisms proposed for explaining the behavior of the recently discovered oxide superconductors. It is electron attraction mediated by spin waves. As may be seen in Figure 6.51b, an electron with a certain spin disrupts the spin of an ion, which causes the spin of its neighboring ion to flip, which then attracts a second electron of opposite spin. It has been suggested recently that induced magnetic fluctuations may also be responsible for the pairing mechanism.[133]

133 如图 6.51b 所示，一个具有某种自旋的电子扰乱了一个离子的自旋，导致了近邻离子自旋的翻转，并从而吸引自旋相反的第二个电子。最近有建议认为诱导磁场的波动可能也是电子配对的成因。

Ⅲ. Band Gap of Superconductors

The electron pairs have a slightly lower energy and leave an energy gap above them on the order of 0.001 eV which inhibits the kind of collision interactions which lead to ordinary resistivity. For temperatures such that the thermal energy is less than the band gap, the material exhibits zero resistivity. This explanation happens to be correct and is in agreement with the predictions of the BCS theory. The width of the gap can be deduced from measurements on specific heat, electromagnetic absorption, or tunneling. The gap does not appear abruptly; it is zero at the critical temperature and rises to the value of $3.5kT_C$ at absolute zero temperature. The temperature variation is very well predicted by the BCS theory, as shown in Figure 6.52 for these superconductors. Although the electron pairing due to phonon exchanges (BCS theory) explains superconductivity in conventional superconductors, but it does not explain superconductivity in the newer superconductors that have a very high critical temperature.[134]

134 尽管基于声子交换产生的电子配对（BCS 理论）解释了传统超导体的超导性，但是不能够解释有很高临界温度的新型超导体的超导性。

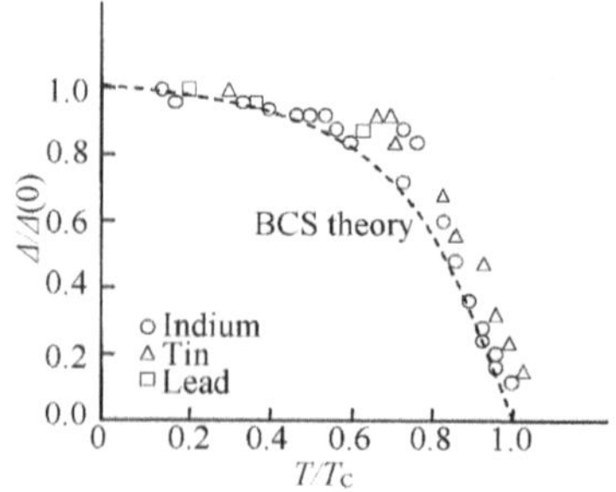

Figure 6.52 The temperature variation of the energy gap (related to the energy gap at $T = 0$) as a function of T/T_C.

6.4.8 Tunnelling/Josephson Effect

When we put a thin insulator between two identical superconductors, the energy diagram (Figure 6.53a) looks very similar to those we encountered when studying semiconductors. The essential difference is that in the present case the density of states is high just above and below the gap, as shown in Figure 6.53b. An applied voltage produces practically noncurrent until the voltage difference is as large as the gap itself; the situation is shown in Figure 6.54. If we increase the voltage further, electrons from the left-hand side may tunnel into empty states on the right-hand side, and the current rises abruptly as shown in Figure 6.55.[135]

135 重要的区别是在这种情况下态密度仅在间隙上下部分较高，如图 6.53b 所示。直到外加的电势差与间隙一样大时才产生电流，这一情况表示在图 6.54 中。如果进一步增加电压，电子将会从左端隧穿到右端的空态上，并且电流会急剧升高，如图 6.55 所示。

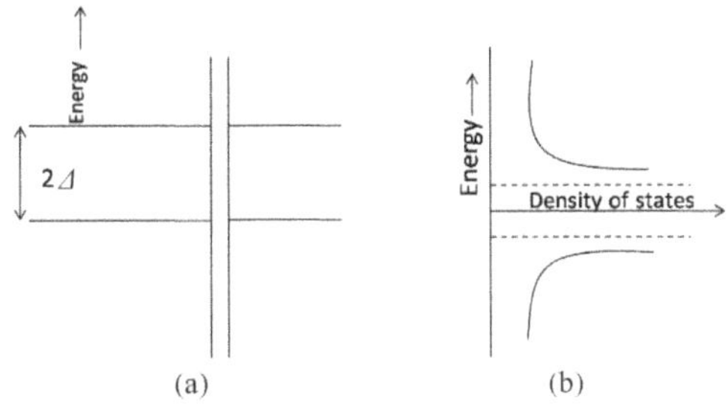

Figure 6.53 (a) Energy diagram for two identical superconductors separated by a thin insulator. (b) The density of states as a function of energy.

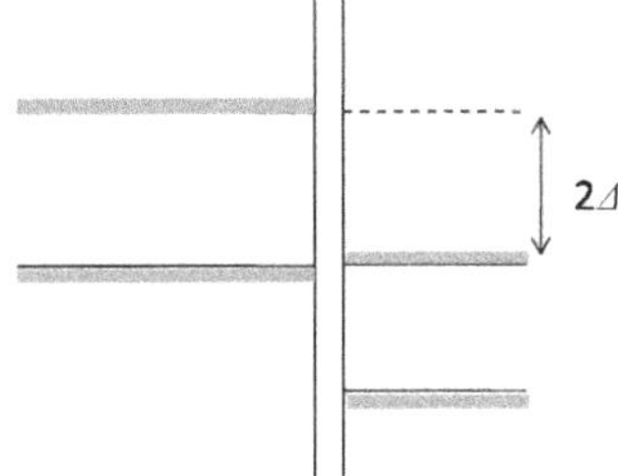

Figure 6.54 The energy diagram of sample in Fig. 6.45a when a voltage of 2Δ/e is applied.

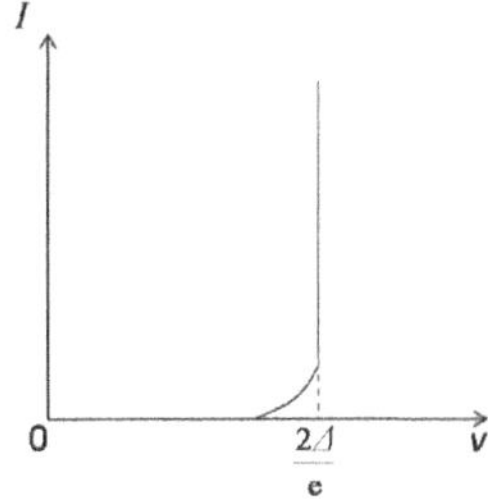

Figure 6.55 The current as a function of voltage for a junction between two identical superconductors by a thin insulator.

An even more interesting case arises when the two superconductors have different gaps. Since the Fermi level is in the middle of the gap (as for intrinsic semiconductors) the energy diagram at thermal equilibrium is as shown in Figure 6.56a. There are some electrons above the gap (and holes below the gap) in superconductor A but hardly any (because of the larger gap) in superconductor B. When a voltage is applied, a current will flow and will increase with voltage (Figure 6.57) because more and more of the thermally excited electrons in superconductor A can tunnel across the insulator into the available states of superconductor B.[136] When the applied voltage reaches $(\Delta_2-\Delta_1)/e$ (Figure 6.57), it has become energetically possible for all thermally excited electrons to tunnel across. If the voltage is increased further, the current decreases because the number of electrons capable of tunneling is unchanged, but they now face a lower density of states. When the voltage becomes greater than $(\Delta_2+\Delta_1)/e$ the current increases rapidly because electrons below the gap can begin to flow.

136 因为（像本征半导体一样）超导体费米能级位于间隙的中间，热平衡下的能级图如图 6.56a 所示。在超导体 A 中，间隙上端存在电子，下端是空穴。但是在超导体 B 中几乎没有电子（由于较大的间隙）。当施加一个电压时，会有电流产生并随着电压而增加（图 6.57）因为在超导体 A 中越来越多的热激发电子隧穿过绝缘体进入超导体 B 中的空态。

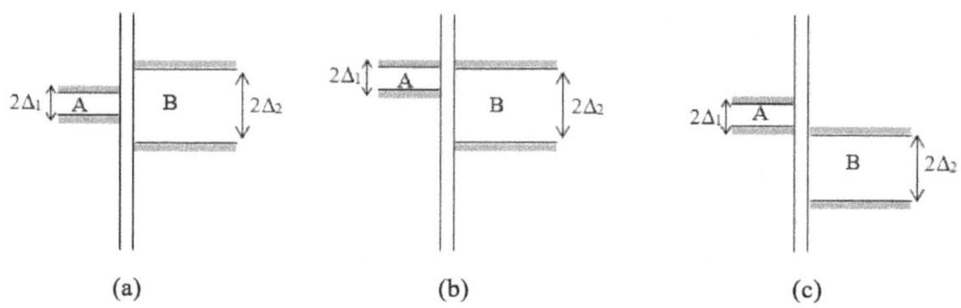

Figure 6.56 Energy diagrams for two different superconductors separated by a thin insulator. (a) $U = 0$, (b) $U = (\Delta_2 - \Delta_1)/e$, (c) $U = (\Delta_1 + \Delta_2)/e$.

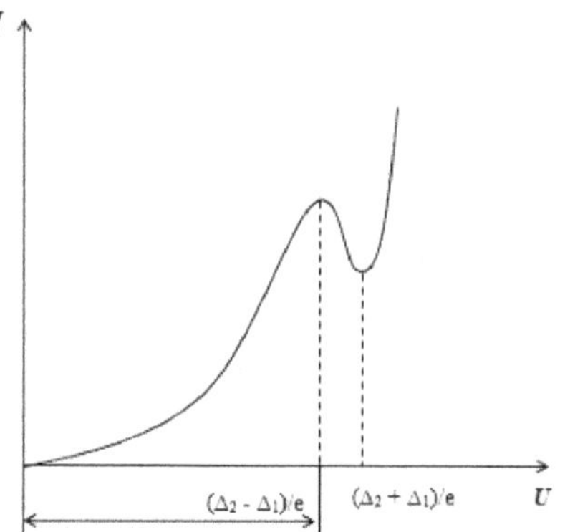

Figure 6.57 The current as a function of voltage for a junction between two different superconductors separated by a thin insulator. There is a negative resistance region of $(\Delta_2 - \Delta_1/e < U < (\Delta_2 + \Delta_1)/e$.

Thus, a tunnel junction comprised of two superconductors with different band gaps may exhibit negative resistance, similarly to the semiconductor tunnel diode. Unfortunately, the superconducting tunnel junction is not as useful because it works only at low temperatures. The tunnelling we have just described follows the same principles we encountered when discussing semiconductors.[137] There is, however, a tunneling phenomenon characteristic of superconductors, and of superconductors alone; it is the so-called superconducting or Josephson

137 因此，类似于半导体隧穿二极管，由两个具有不同带隙的超导体构成的隧穿结展现出负电阻。不幸的是，由于超导体只能在低温下工作，超导体的隧穿结不能很好的应用。我们刚刚描述的隧穿效应，和我们在讨论半导体时遇到的，遵循相同的原理。

tunneling (discovered theoretically by Josephson, a Cambridge graduate student at the time) which takes place when the insulator is very thin. It displays a number of interesting phenomena, as briefly described here:

(a) For low enough currents there can be a current across the insulator without any accompanying voltage; the insulator turns into a superconductor. The reason is that Cooper pairs (not single electrons) tunnel across.

(b) For larger currents there can be finite voltages across the insulator. The Cooper pairs descending from the higher potential to the lower one may radiate their energy according to the relationship,

$$h\omega=(2e)U_{AB} \quad (6\text{-}85)$$

U_{AB} is the DC voltage between the two superconductors, and ω is the angular frequency of the electromagnetic radiation. Thus, we have a very simple form of a DC tunable oscillator that could work up to infrared frequencies. It is not, however, very practical because, besides the low temperatures needed, it can produce only a minute amount of power. A direct transition may be caused between the Josephson characteristics and the 'normal' tunneling characteristics by the application of a small magnetic field (Figure 6.58).

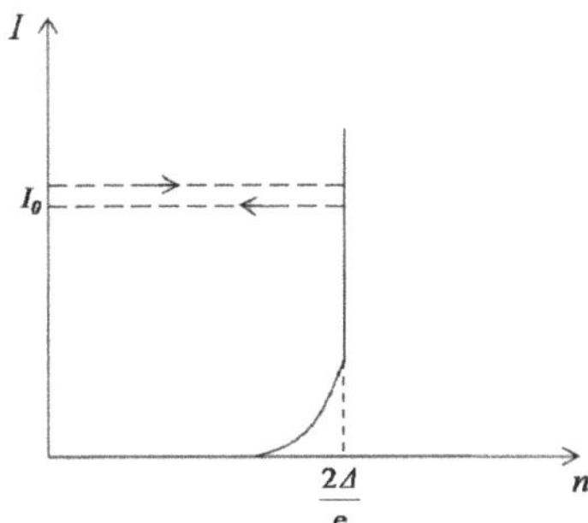

Figure 6.58 The current as a function of voltage for a junction which may display both 'normal' and Josephson tunneling. I_0 is the current flowing without any accompanying voltage.

(c) When two Josephson junctions are connected in parallel (Figure 6.59) the maximum supercurrent that can flow across them is a periodic function of the magnetic flux,

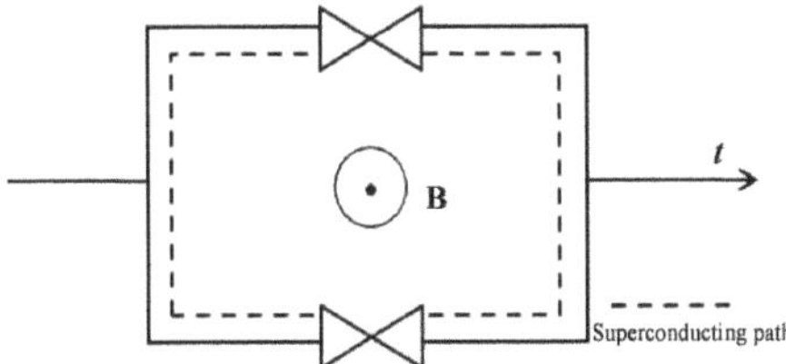

Figure 6.59 Two Josephson junctions in parallel connected by a superconducting path.

$$I_{max} = 2I_J \left| \cos \frac{\Phi}{\Phi_0} \right| \quad (6\text{-}86)$$

I_J is a constant depending on the junction parameters, Φ is the enclosed magnetic flux, and Φ_0 is the so-called flux quantum equal to $h/2e = 2 \times 10^{-15}$ Wb.

The application of a small magnetic field causes a transition between the Josephson and 'normal' tunneling characteristics. Once this extra magnetic field is removed, the voltage returns to zero. Josephson device can be used in many applications, such as super high speed switch. Combining with magnetic field, applications spread in a variety.[138]

138 小磁场的添加造成了约瑟夫森和正常隧穿特性间的转变。一旦移除这一外加磁场，电压将变为零。约瑟夫森器件能够用在许多应用中，比如超高速开关。结合了磁场，其应用扩展到多个领域。

6.4.9 Applications of Superconductor

Superconductivity was a scientific curiosity for a long time, but by now there are some actual applications like producing high magnetic fields for Magnetic Resonance Imaging, and there are lots of potential applications. The big prize would of course be superconductive lines for power transmission which would eliminate resistance losses. But quite apart from applications.

Ⅰ. High–Field Magnets

For the moment the most important practical application of superconductors is in producing a high magnetic field. There is no doubt that for this purpose a superconducting solenoid is superior to conventional magnets.[139] A magnetic flux density of 20 T can be produced by a solenoid not larger than about 12 cm × 20 cm. A conventional magnet capable of producing one-third of that flux density would look like a monster in comparison and would need a few megawatts of electric power and at least a few hundred gallons of cooling water per minute.What sort of materials do we need for obtaining high magnetic fields? Obviously, type II superconductors remain superconducting up to quite high magnetic fields. However, high magnetic fields are allowed only at certain points in the superconductor that are surrounded by current vortices. When a DC current flows (so as to produce the high magnetic field in the solenoid) the vortices experience a J×B force that removes the vortices from the material. To exclude the high magnetic field costs energy, and the superconductor consequently becomes normal, which is highly undesirable, the problem is to keep the high magnetic fields inside.[140] This is really a problem similar to the one we encountered in producing 'hard' magnetic materials, where the aim was to prevent the motion of domain walls. The remedy is similar; we

139 目前，超导体最重要的实际应用是制造一个高强的磁场。毫无疑问，在这方面，超导螺线管比传统的磁体要更有优势。

140 为了排除高磁场消耗能量，以及由此带来的不希望的结果超导体变为正常态，问题的关键是要保持内部的高磁场。

must have lots of structural defects; that we must make our superconductor as 'dirty' and as 'non-ideal' as possible. The resulting materials are, by analogy, called hard superconductors. Some of their properties are shown in Table 6.6.[141]

141 解决方法是相似的；必须要有许多的结构缺陷；必须使得超导体尽可能地“肮脏”和“不理想”。类似地，最终的材料叫作硬超导体。它们的性能如表 6.6 所示。

Table 6.6 The critical temperature and critical magnetic field (at T= 4.2 K) of the more important hard superconductors

Material	T_C(K)	$H_C \times 10^{-7}$ (Am^{-1})
Nb-Ti	9	0.9
$Pb_{0.9}Mo_{5.1}S_6$	14.4	4.8
V_3Ga	14.8	1.9
NbN	15.7	0.8
V_3Si	16.9	1.8
Nb_3Sn	18.0	2.1
Nb_3Ga	20.2	2.6
$Nb_3(Al_{0.7}Ge_{0.3})$	20.7	3.3
Nb_3Ge	22.5	2.9

There is, however, a further difficulty with vortices. Even if they do not move out of the material, any motion represents resistance loss, causing heating, and making the material become normal at certain places. To avoid this, a good thermal conductor and poor electrical conductor, copper—yes, copper—is used for insulation, so that the heat generated can be quickly pass away. It must be noted that it is not particularly easy to produce any of these compounds, different techniques may easily lead to somewhat different values of T_C and H_C. The two superconductors used in practical devices are the ductile Nb-Ti alloy and the brittle intermetallic compound Nb3Sn, the latter one being used at the highest magnetic fields.

Ⅱ. Switches and Memory Elements

The use of superconductors as switches follows from their property of becoming normal in the presence of a magnetic field. We can make a superconducting wire resistive by using the magnetic field produced by a current flowing in another superconducting wire.[142] Memory elements based on such switches have indeed been built, but they were never a commercial success.

142 我们能够用一根超导体金属丝中的电流产生的磁场来使得另一根超导体金属丝有电阻。

Superconducting memory elements based on the properties of Josephson junctions have a much better chance. As we have mentioned before, the junction has two stable states, one with zero voltage and the other one with a finite voltage. It may be switched from one state into the other one by increasing or decreasing the magnetic field threading the junction.[143] The advantage of this Josephson junction memory is that there

143 正如我们之前提到的，约瑟夫森结有两个稳定的态，一个是零电压，另一个是有限的电压。这两个态可以通过增强或减弱穿过结的磁场从一个态到另一个态进行转换。

is no normal to superconducting phase transition necessary, only the type of tunneling is changed, which is a much faster process. Switching times as short as 10 ps have been achieved.

Will it ever be worthwhile to go to the trouble and expense of cooling memory stores to liquid helium temperatures? So far computer manufacturers have been rather reluctant (understandably, it is a high risk business) to introduce superconducting memories. It is difficult to predict, but the latest members of the family, Rapid Single Flux Quantum (RSFQ) devices, might have a chance to be introduced in practice sometime in the future when high speed will be the principal requirement. The basic architecture is a ring containing a Josephson junction. A large number of such rings coupled magnetically make up the device that can serve both as a memory element (it stores a single flux quantum) and a logic device. The latter property is due to the fact that voltage pulses can travel from element to element extremely rapidly along such line.[144] The highest speed observed so far at which these devices can operate is 770 GHz. Apart from speed a further advantage is the quantized nature of the storage mechanism providing protection both against noise and cross talk.

144 大量的这种磁性偶合环可做成能够记忆的单元（储存单磁通量）和逻辑器件。后者的性能是由于电压脉冲能够沿着这样的线快速地在单元间传输。

Ⅲ. Magnetometers

A further important application of Josephson junctions is in a magnetometer called SQUID (Superconducting Quantum Interference Device). Its operation is based on the previously mentioned property that the maximum supercurrent through the two junctions in parallel is dependent on the magnetic flux enclosed by the loop. It follows from Equation (6-86) that there is a complete period in I_{max} while Φ varies from 0 to Φ_0. Thus, if we can tell to an accuracy of 1% the magnitude of the supercurrent, and we take a loop area of $1cm^2$, the smallest magnetic field that can be measured is 10^{-12} T. Commercially available devices (working on roughly the same principle) can offer comparable sensitivities.

Although the Josephson junction does many things superlatively well, like other topics in superconductivity, its applications (so far) are few. However, it is worth mentioning two sensitive magnetometer applications which would be quite impossible with classical devices.[145]

145 尽管约瑟夫森结表现出了相当好的性能，像超导的其它课题，它的应用（目前为止）很少。然而，值得一提的是两个灵敏磁力计的应用，这一应用在传统的器件中是不可能实现的。

A major preoccupation of the military is to keep watch on nuclear submarines. The difficulty is that water is such a good absorber of microwaves, light, and sound, which are traditionally used to locate targets. However, underwater caches of superconducting magnetometers can detect small perturbations of the Earth's magnetic field as a submarine arrives in the locality. They have to be connected to a surface buoy containing a transmitter which informs boffins in bunkers what is going past.

A more definite and much safer application is one which Oxford's Laboratory of Archaeology works on and publishes freely. The silicaceous and clay-like materials in pottery are mildly paramagnetic. When they are fired in kilns, the high temperature destroys the magnetism, and as they cool the permanent dipoles re-set themselves in the local magnetic field of the Earth.[146] When an archaeologist uncovers an old kiln, he can measure this magnetism in the bricks and thus find the direction of the Earth's field when the kiln was last fired. The variation of the Earth's field and angle of dip has been determined for several thousand years at some places. Thus, it is possible to date kilns by accurate measurements. As large ceramic articles have to be kilned standing on their bases, accurate measurements of the dip angle can also date cups and statues, if their place of origin is known. With very sensitive magnetometers, this measurement can be done on a small, unobtrusive piece of ceramic removed from the base of the statue. It is a method considerably used by major museums.[147]

146 当它们在烧窑里烧制时，高温破坏了磁性，当它们冷却后，永久偶极在地球的磁场作用下重新排列。

147 使用非常灵敏的磁力计，可以测量雕塑底部的很小的不起眼的陶瓷碎片。这一方法被相当多地应用于大博物馆中。

Ⅳ. Metrology

We mentioned earlier that one can determine one of our fundamental constants (velocity of light) with the aid of lasers. It turns out that Josephson junctions may be used for determining another fundamental constant. The relevant formula is Equation (6-86). By measuring the voltage across the junction and the frequency of radiation, h may be determined. As a result, the accepted value of Planck's constant changed recently from 6.62559 to 6.626196×10^{-34} J·s.

Ⅴ. Suspension Systems and Motors

Frictionless suspension systems may be realized by the interaction between a magnetic flux produced externally and the currents flowing in a superconductor. If the superconductor is pressed downwards, it tries to exclude the magnetic field, hence the magnetic flux it rests on is compressed, and the repelling force is amplified. Noting further that it is possible to impart high speed rotation to a suspended superconducting body, and that all the conductors in the motor are free of resistance, it is quite obvious that the ideal of a hundred-percent-efficient motor can be closely approximated.[148]

148 值得进一步注意的是将高速旋转体安置于一个悬挂的超导体是可能的，马达中所有的导体都是没有阻力的，显而易见，实际中马达的效率可以达到近似于100%的理想效率。

Ⅵ. Radiation Detectors

The operation of these devices is based on the heat provided by the incident radiation. The superconductor is kept just above its critical temperature, where the resistance is a rapidly varying function of temperature. The change in resistance is then calibrated as a function of the incident radiation.

Ⅶ. Heat Valves

The thermal conductivity of some superconductors may decrease by as much as two orders of magnitude, when made normal by a magnetic field. This phenomenon may be used in heat valves in laboratory refrigeration systems designed to obtain temperatures below 0.3 K.

6.5 Dielectric Material

6.5.1 What is Dielectric Material?

Materials used to insulate an electric field from its surroundings are required in a large number of electrical and electronic applications. Electrical insulators obviously must have a very low conductivity, or high resistivity, to prevent the flow of current. Insulators must also be able to withstand intense electric fields.[149] Insulators are produced from ceramic and polymeric materials in which there is a large energy gap between the valence and conduction bands; however, the high-electrical resistivity of these materials is not always sufficient. At high voltages, a catastrophic breakdown of the insulator may occur, and current may flow. For example, the electrons may have kinetic energies sufficient to ionize the atoms of the insulator, thereby creating free electrons and generating a current at high voltages. In order to select an insulating material properly, we must understand how the material stores, as well as conducts, electrical charge.[150] Porcelain, alumina, cordierite, mica, and some glasses and plastics are used as insulators. The resistivity of most of them is > 10^{14} Ωcm, and the breakdown electric fields are ~5 to 15 kV·mm^{-1}.

Dielectric materials refer to materials that exhibit polarization rather than long-range charge migration under electric field. Here we need to know the differences between dielectric materials and insulators.[151] We use the term "dielectric" for materials that are used in applications where the dielectric constant is important. The dielectric constant (ε) of a material is a microstructure sensitive property related to the material's ability to store an electrical charge. We use the term "insulator" to describe the ability of a

149 被用作隔绝周围电场的材料在电学和电子应用领域有很大的需求。显然电绝缘体必须要有一个很低的电导率，或者是高的电阻率来阻止电流的流动。绝缘体也必须能够耐受强电场。

150 在高电压下，绝缘体灾难性的雪崩可能会发生，并且会有电流流动。例如，在高电压下，电子可能会有足够的动能来电离绝缘体中的原子，因此产生自由电子从而产生电流。为了选择合适的绝缘材料，我们必须了解材料如何存储和传导电荷。

151 介电材料指的是在电场下表现出极化而不是长程电荷迁移的材料。这里，我们需要了解介电材料和绝缘材料间的区别。

material to stop the flow of DC or AC current, as opposed to its ability to store a charge. A measure of the effectiveness of an insulator is the maximum electric field it can support without an electrical breakdown.[152]

152 我们用术语“绝缘体”来描述材料阻碍直流或交变电流通过的能力，而不是其存储电荷的能力。绝缘体效果的量度方法是其不发生电致击穿的最大电场。

The behaviors of a dielectric can be explained adequately in terms of electromagnetic theory with the energy gap defined in an insulator. From the viewpoint of optical properties, an insulator can be defined as a material in which electron-hole pairs are not created by visible light, which follows in the electromagnetic radiation range from 400 to 700 nm. Since a photon of 400 nm wavelength has an energy of about 3 eV, we may say that a dielectric has an energy gap in excess of that value.

There are two types of dielectric materials: polar and nonpolar dielectric materials. The former refers to materials in which the statistic centers of the positive and negative charges do not overlap; the latter refers to material in which the statistic centers of the positive and negative charges are overlap.[153]

153 有两种类型的介电材料：极性和非极性介电材料。前者指的是材料中正负电荷的中心没有重叠；后者指的是正负电荷中心重叠的材料。

6.5.2 Characterization

Ⅰ. Capacitance and Dielectric Constant

Capacitance is the ability of a media to store an electrical charge under unit electric field. Any object that can be electrically charged exhibits capacitance. A common form of energy storage device is a parallel-plate capacitor. Capacitor plates in vacuum (Figure 6.60a) possesses the electrical quantity of Q:

$$Q = \sigma A = \varepsilon_0 F A = \varepsilon_0 \frac{U}{\mathrm{d}} A \tag{6-87}$$

where σ is the charge density (the number of charges per unit area), A is area of plate; d is the distance between the two plates; F is the intensity of electric field; U is the voltage between two plates; ε_0 is the vacuum dielectric constant.

Then the capacitance C_0 is given by:

$$C_0 = \frac{Q}{U} = \frac{\varepsilon_0 \left(\dfrac{U}{d}\right) A}{U} = \frac{\varepsilon_0}{d} \tag{6-88}$$

In a parallel plate capacitor in vacuum, the capacitance is directly proportional to the surface area of the conductor plates and inversely proportional to the separation distance between the plates.[154]

154 对于真空中的平行平板电容器，其电容直接与导体板的表面积成正比，与两板间距成反比。

Inserting a dielectric material between a plate capacitor (Figure 6.60b), polarization occurs in the dielectric material: negative charges are induced at position close to the positive plate, and positive charges are induced in vicinity to the negative plate. Thus, the surface charge of the plate increases to:

$$Q' = \sigma' A = \varepsilon_0 \varepsilon_r F A = \varepsilon_0 \varepsilon_r \frac{U}{d} A \qquad (6\text{-}89)$$

σ' is the charge density (the number of charges per unit area).

The capacitance changes to:

$$C = \frac{Q'}{U} = \varepsilon_0 \varepsilon_r \frac{U}{d} A \frac{1}{U} = \varepsilon \frac{A}{d} \qquad (6\text{-}90)$$

where ε_r is relative dielectric constant, $\varepsilon = \varepsilon_r \varepsilon_0$ is material dielectric constant.

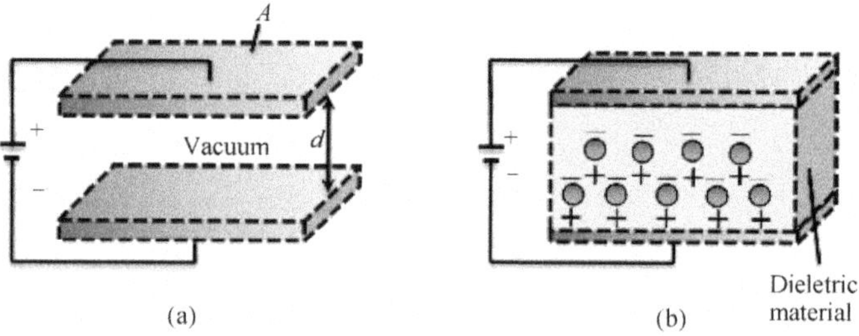

Figure 6.60 (a) A charge can be stored on the conductor plates in a vacuum. (b) When adielectric is placed between the plates, the dielectric polarizes, and additional charge is stored.

Ⅱ. Dipole Moment

A dipole refers to a pair of opposite charges separated by a certain distance, which is generally produced by the process called polarization.[155]

155 偶极指的是一对分离一定距离的相反电荷，通常是在极化过程中产生的。

Under an external electric field, polar or non-polar dielectric molecules will possess dipole moment μ:

$$\mu = q \times d \qquad (6\text{-}91)$$

where q is electric charge, d is the distance between the positive and negative centers (see Figure 6.61)

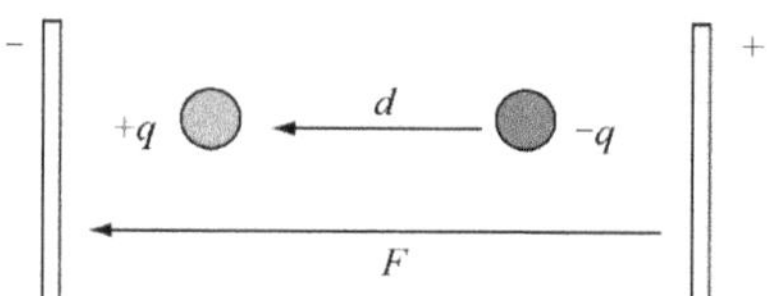

Figure 6.61 The definition of dipole.

Ⅲ. Refractive Index

The refractive index (n) is defined as the ratio of the velocity of light in a vacuum (c) to that in the material (v),

$$n = \frac{c}{v} = \sqrt{\varepsilon_r \mu_r} \qquad (6\text{-}92)$$

Since for most materials that transmit light, $\mu_r = 1$, therefore,

$$n = \sqrt{\varepsilon_r \mu_r} \approx \sqrt{\varepsilon_r} \qquad (6\text{-}93)$$

Conventionally, we talk of dielectric constant (or permittivity) for the lower frequencies in the electromagnetic spectrum and of refractive index for light. The above equation shows that they are both a measure of the polarizability of a material in an alternating electric field.[156]

156 通常，我们在电磁谱中低频区讨论的是介电常数（或者电容率），而对于光（高频场）讨论的是折射率。上面的公式表示这些都是描述材料在交变电场下极化情况的。

A characteristic of non-polar materials is that, the refractive index at optical wavelengths is approximately equal to the square root of the relative dielectric constant at low frequencies. This is illustrated in Table 6.7.

Table 6.7 Refractive index of materials with feature of polar, nonpolar and weakly polar

Material	Refractive index	(Refractive index)^2	Dielectric constant measured at 10^3Hz
	Non-polar		
C(diamond)	2.38	5.66	5.68
H_2(liquid)	1.11	1.232	1.23
	Weakly polar		
polythene	1.51	2.28	2.30
Ptfe(ploy-tetra Fljuoro-ethylene)	1.37	1.89	2.10
CCl_4(carbon-tetrachloride)	1.46	2.13	2.24
Paraffin	1.48	2.19	2.20
	Polar		
NaCl(rocksalt)	1.52	2.25	5.90
TiO_2(rutile)	2.61	6.8	94.0
SiO_2(quartz)	1.46	2.13	3.80
Al_2O_3(cerramic)	1.66	2.77	6.5
Al_2O_3(ruby)	1.77	3.13	4.31
Sodium carbonate	1.53	2.36	8.4
Ethanol	1.36	1.85	24.30
Methanol	1.33	1.76	32.63
Acetone	1.357	1.84	20.7
Soda glass	1.52	2.30	7.60
Water	1.33	1.77	80.4

As can be seen that most transparent dielectrics, polar or not, have a refractive index of around 1.4 to 1.6; only extreme materials like liquid hydrogen, diamond, and rutile show appreciable deviation.

6.5.3 Polarization Mechanism

Ⅰ. Definition of Polarization Intensity and Dielectric Susceptibility

When we apply stress to a material, some level of strain develops. Similarly, when we subject materials to an electric field, the atoms, molecules, or ions respond to the applied electric field (F). Thus, the

157 当我们对材料施加力时，一定程度的应力产生。相似的，当我们给材料施加一个电场时，其原子分子或者离子会对外加的电场（F）做出反应。因此，材料就被极化了，也就是说产生了偶极。

material is said to be polarized and concomitantly, a dipole is produced.[157]

Polarization means any separation of charges (e.g., between the nucleus and electron cloud) or any mechanism that leads to a change in the separation of charges that are already present (e.g., movement or vibration of ions in an ionic material). Polarization intensity ($\vec{P}$) is defined as the vector sum of the dipole moments within unit volume:

$$\vec{P}=\frac{\sum\vec{\mu}}{\Delta V} \tag{6-94}$$

Experiment found that the polarization intensity was proportional to the effective electric field (F, external electric field plus the electric field produced by charge polarization) and to the vacuum dielectric constant:

$$P=\chi_e\varepsilon_0F=\sigma' \tag{6-95}$$

where the proportionality factor, χ_e, is termed as dielectric susceptibility. σ' is the surface charge density of the dielectric material. It can be deduced as follow. For plate capacitor:

$$F=F_0-F' \tag{6-96}$$

F_0 is the external electric field, F' is the electric field caused by polarization from dielectric material, whose direction is opposite to F_0.

Suppose the density of free electron is σ,

$$\sigma=\varepsilon_0F_0 \tag{6-97}$$

$$F_0=\frac{\sigma}{\varepsilon_0} \tag{6-98}$$

$$F=F_0-F'=\frac{\sigma}{\varepsilon_0}-\frac{\sigma'}{\varepsilon_0}=\frac{\sigma-P}{\varepsilon_0} \tag{6-99}$$

For isotropic infinite dielectric material, the inside effective electric field (F) decreases to $1/\varepsilon_r$ of the external field

$$F=\frac{F_0}{\varepsilon_r}=\frac{\sigma}{\varepsilon_r\varepsilon_0} \tag{6-100}$$

$$\sigma=F\varepsilon_r\varepsilon_0 \tag{6-101}$$

Thus, plus previous expression of

$$F=F_0-F'=\frac{\sigma}{\varepsilon_0}-\frac{\sigma'}{\varepsilon_0}=\frac{\sigma-P}{\varepsilon_0} \tag{6-102}$$

we get: $$P=\sigma-\varepsilon_0F=\varepsilon_0\varepsilon_rF-\varepsilon_0F=(\varepsilon_0\varepsilon_r-\varepsilon_0)F=\chi_e\varepsilon_0F \tag{6-103}$$

where F is the strength of the electric field (V/m), and x_e is the dielectric susceptibility:

$$\chi_e=\varepsilon_r-1 \tag{6-104}$$

For materials that polarize easily, both the dielectric constant and the capacitance are large and, in turn, a large quantity of charge can be stored. In addition, Equation (6-103) suggests that polarization increases, at least until all of the dipoles are aligned, as the voltage (expressed by the strength of the electric field) increases. The dielectric constant of vacuum is one, or its dielectric susceptibility is zero.[158] This makes sense since a vacuum does not contain any atoms or molecules.

158 对于易极化的材料，介电常数和电容都很大，也就是能够储存大量的电荷。此外，公式（6-103）表明极化增加，电压（即电场强度）也随之增加，直到所有的偶极都向同一方向排列。因为真空中不包含任何原子或分子，真空介电常数为1，或者真空电极化率为零。

Ⅱ. Linear and Nonlinear Dielectrics

In linear dielectrics, P is linearly related to F and χ_e is constant. This is similar to how stress and strain are linearly related by Hooke's law. In linear dielectrics, χ_e remains constant with changing F. In materials such as $BaTiO_3$, the dielectric constant changes with F, and hence, Equation (6-103) cannot be used. These materials in which P and F are not related by a straight line are known as nonlinear dielectrics or ferroelectrics. These materials are similar to elastomers for which stress and strain are not linearly related and a unique value of the Young's modulus cannot be assigned.[159]

159 这些 P 和 F 不呈线性关系的材料是非线性介电材料或者是铁电材料。这些材料与弹性体相似，弹性体的压力和应力不是线性相关的，并且其杨氏模量的值也不能被确定。

In addition to being nonlinear relation of F and χ_e, ferroelectric materials demonstrate a spontaneous nonzero polarization even when the applied field E is zero. The distinguishing feature of ferroelectrics is that the spontaneous polarization can be reversed by a suitably strong applied electric field in the opposite direction, the polarization is therefore dependent not only on the current electric field but also on its history, yielding a hysteresis loop.[160]

160 铁电材料显著的特性是在适当强的反向外加电场下，其自发极化会被改变，因此这种极化不仅与当前电场有关而且与其经历的电场情况有关，产生出一种迟滞回线。

Typically, materials demonstrate ferroelectricity only below a certain phase transition temperature, called the Curie temperature, T_C, and are paraelectric above this temperature: the spontaneous polarization vanishes. Many ferroelectrics lose their piezoelectric properties above T_C completely, because their paraelectric phase has centrosymmetric crystallographic structure.

Ⅲ. Polarization Mechanism

There are four primary polarization mechanisms: (1) electronic polarization, (2) ionic polarization, (3) molecular polarization, and (4) space charge (Figure 6.62). Their occurrence depends upon the electrical frequency of the applied field, just like the mechanical behavior of materials depends on the strain rate. If we apply a very rapid rate of strain, certain mechanisms of plastic deformation are not activated. Similarly, if we apply a rapidly alternating electric field, some polarization mechanisms may be unable to induce polarization in the material. According to the frequency of the applied field, the produced dipole can be classified as high frequency dipolar referring to vibration, and low frequency dipolar referring to rotation.[161]

161 相似的，如果施加一个快速变化的电场，一些极化机制或许不能够在材料中诱发极化。根据施加电场的频率，产生的偶极能够分为意味着振动的高频偶极和意味着旋转的低频偶极。

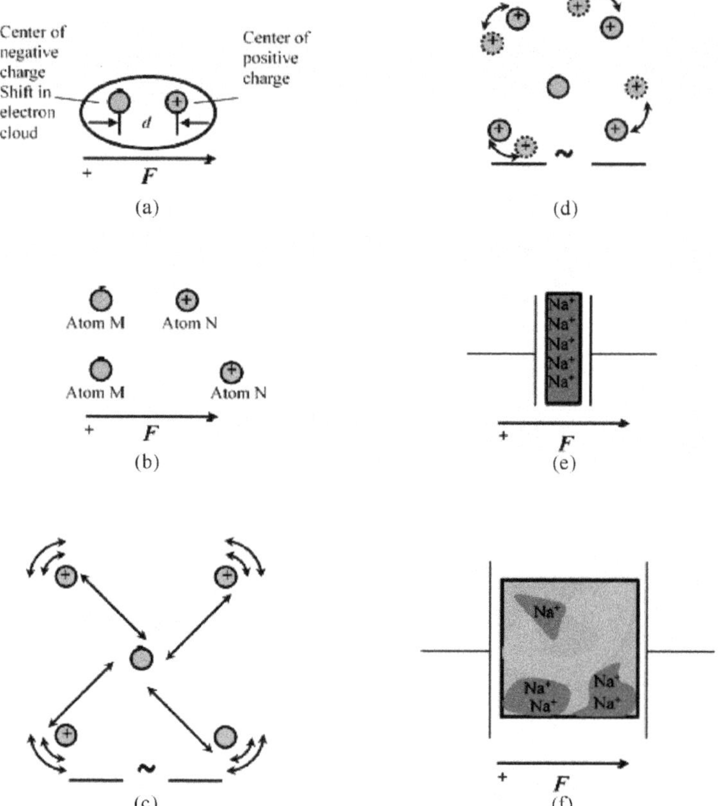

Figure 6.62 Different types of polarization: (a) electronic, (b) atomic or ionic, (c) high-frequency dipolar or orientation (present in ferroelectrics), (d) low-frequency dipolar (present in linear dielectrics and glasses), (e) interfacial-space charge at electrodes, and (f) interfacial-space charge at heterogeneities such as grain boundaries.

- **Electronic Polarization**

Electronic polarization is omnipresent since all materials contain atoms or ions surrounded by electron clouds. As electrons are very light, they have a rapid response to the change of field. The electron cloud gets displaced from the nucleus in response to the field seen by the atoms. The separation of charges creates a dipole moment (Figure 6.62a and Figure 6.63). This mechanism can survive at the highest electrical frequencies (~10^{15} Hz) since an electron cloud can be displaced rapidly, back and forth, as the electrical field switches.[162]

162 这种（极化）机制在最高电场频率下可以存在，因为当电场转换时电子云可以被迅速地前后移动。

According to Born atom model, the average dielectric susceptibility for electronic type polarization (χ_{electron}) can be expressed as

$$\chi_{\text{electron}} = \frac{4}{3}\pi\varepsilon_0 R^3 \tag{6-105}$$

where R is the radius of atom or ion.

Electrons have a rapid response to field changes, the polarization can be set up or erased around 10^{-15}~10^{-16} s.

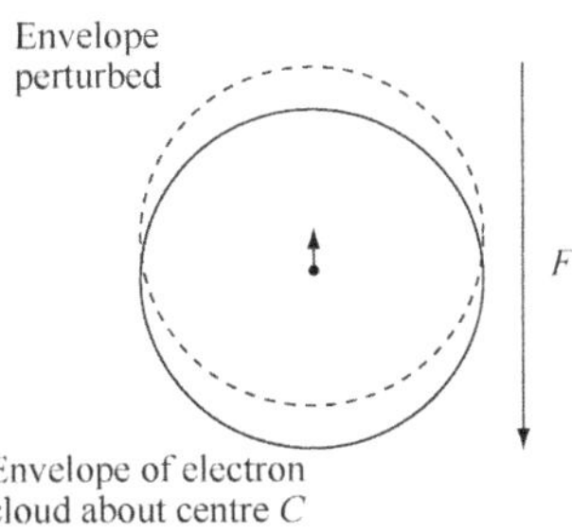

Figure 6.63 Displacement of electron cloud about an atom centred on C by a distance d induced by an applied electric field.

Larger atoms and ions have higher electronic polarizability (tendency to undergo polarization), since the electron cloud is farther away from the nucleus and held less tightly. This polarization mechanism is also linked closely to the refractive index of materials, since light is an electromagnetic wave for which the electric field oscillates at a very high frequency (10^{14}~10^{16} Hz). The higher the electronic polarizability, the higher the refractive index.[163] We use this mechanism in making "lead crystal" which is really an amorphous glass that contains up to 30% PbO. The large lead ions (Pb^{2+}) are highly polarizable due to the electronic polarization mechanisms and provide a high-refractive index when high enough concentrations of lead oxide are present in the glass.

163 因为电子云距离核更远，束缚力变小，更大的原子和离子具有更高的电子极化（更强的极化趋势）。这种极化机制也与材料的折射率密切相关，因为光是一种电场在非常高频率(10^{14}～10^{16}Hz)下振动的电磁波。电子极化程度越高，则折射率越大。

- **Ion (Molecular) Polarization**

Bonds between atoms are stretched by applied electric fields when the lattice ions are charged, resulting in polarization (Figure 6.62b).

According to classical elastic vibration theory, the dielectric susceptibility by ion polarization can be estimated as:

$$\chi_{\mathrm{i}} = \frac{a^3}{n-1} 4\pi\varepsilon_0 \qquad (6\text{-}106)$$

where a is the average distance between the positive and negative ion without electric field, n, the repulsive parameter, and with values of 7 ~ 11 in ionic crystal.

Since ion quality is larger than electron quality, the build-up or erase of the polarization is slow (10^{-12} ~ 10^{-13} s).

- **Orientation Polarization**

Physically, we may consider the dipole moments as trying to line up but, jostled by their thermal motion, not all of them succeed (Figure 6.64).

This occurs in liquids or gases when whole molecules, having a permanent or induced dipole moment, move into line with the applied field (Figure 6.62c-d). Known from Boltzmann statistics that in thermal equilibrium the number of molecules with an energy E is proportional to $\exp(-E/kT)$, so at any finite temperature besides the lowest energy state, other energies and orientations will also be present.[164]

164 在液体或气体中，当具有永久或者诱导偶极矩的所有分子移动成与外加电场同一方向时（图6.62c-d）产生取向极化。根据波尔兹曼统计可知，在热平衡下，拥有能量 E 的分子数正比于 exp(-E/kT)，因此在任何有限温度下除了最低能量态，其他的能量和取向也存在。

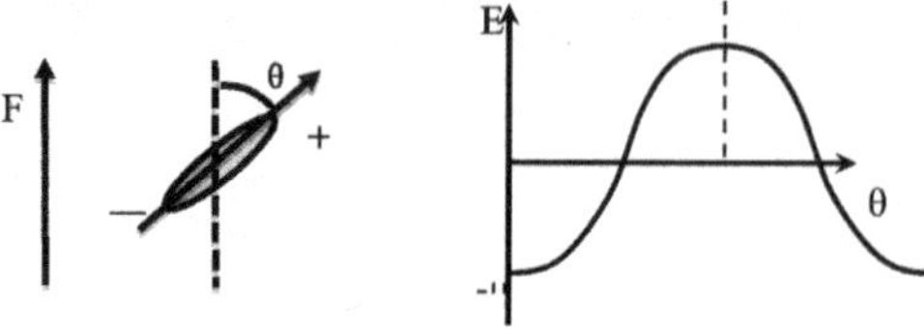

Figure 6.64 (a) A dipole moment trying to line up under electric field. (b) Energy distribution of a dipole under electric field.

The energy (E) of a dipole in an electric field (F) is:

$$E=-\mu F\cos\theta \tag{6-107}$$

where μ is the value of the dipole moment; F is the electric field strength; θ is the angle between the dipole and the electric field.

By mathematic calculation, the polarizability (termed as average dipole moment of a molecular dipole under electric field, F), can be deduced as:

$$\overline{P}=\frac{net\ dipole\ moment\ of\ the\ assembly}{total\ number\ of\ dipoles}=\frac{\Sigma\mu}{total\ number\ of\ dipoles}$$

$$=\frac{\mu_0^2 F}{3kT} \tag{6-108}$$

This means that the orientation polarizability is proportional to its original dipole moment and the external electric field, and inversely proportional to the absolute temperature.[165]

165 这意味着取向极化与其原始偶极和外加电场成正比，与绝对温度成反比。

The dielectric susceptibility of orientation type is:

$$\chi_{\mathrm{d}}=\frac{\mu_0^2}{3kT} \tag{6-109}$$

μ_0 is the molecular dipole moment.

The build-up and erase of an orientation polarization is a slow process (10^{-2}–10^{-10} s).

Orientation susceptibility is generally 2 orders of magnitude higher than the electronic susceptibility. For polar materials, besides orientation polarization, under electric field, molecules are also subjected to electronic or ionic polarization. The total susceptibility should be the sum of these three effects.[166]

166 取向极化率普遍比电子极化高两个数量级。对于极性材料，除了取向极化，在电场下，分子也受制于电子或离子极化。总的极化率应是这三种效应的总和。

6.5.4 Complex Dielectric Constant and Dielectric Loss

In engineering practice, the dielectric constant is often divided into real and imaginary parts, which is called complex dielectric constant.[167]

167 在实际应用中，介电常数经常被分成实部和虚部两个部分，称为复介电常数。

Ⅰ. Complex Conductivity

On an ideal plate capacitor, exerting an alternating electric field with angular frequency of $\omega = 2\pi f$, that is:

$$U = U_0 e^{i\omega t} \tag{6-110}$$

The charge (Q) on the plate electrode is:

$$Q = C_0 U = C_0 U_0 e^{i\omega t} \tag{6-111}$$

The loop current is:

$$I_C = \frac{dQ}{dt} = iwC_0 U_0 e^{i\omega t} = i\omega C_0 U \tag{6-112}$$

The conclusion is that the phase of the current (I_C) in a capacitor is 90° ahead of that of voltage (U)[168] (Figure 6.65).

168 结论是电容器的电流相位比电压相位超前90°。

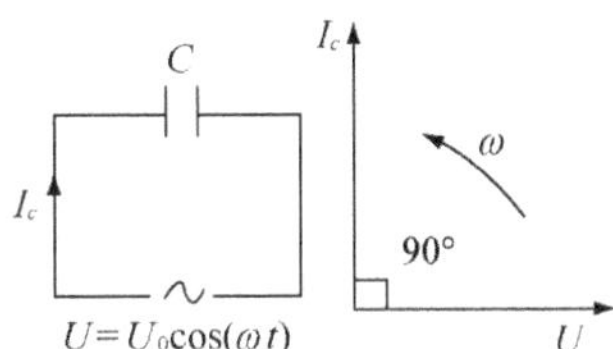

Figure 6.65 The phase relationship between the current and voltage of an ideal plate capacitor under sine electric field.

Suppose in between the plates an ideal dielectric with dielectric constant of ε_r was inserted,

$$C = \varepsilon_r C_0 \tag{6-113}$$

$$I'_C = i\omega CU = i\omega\varepsilon_r C_0 U = \varepsilon_r I_C \tag{6-114}$$

Real dielectric is not ideal, having some leakage (Figure 6.66). The leakage current has the same phase as the voltage.[169] Thus,

$$I_T = I'_C + I_L = i\omega CU + GU = (i\varepsilon C + G)U \tag{6-115}$$

169 真正的介电材料不是理想的，而是有一定的漏电（图 6.66）。该漏电电流与电压的相位相同。

where I_T is the total current, I_C' is the ideal part, I_L is the conductance part, which has the same phase as the voltage, with value of $G = \sigma\frac{A}{d}$, where σ is conductivity and A is area.

Because: $$C = \varepsilon_r C_0 = \varepsilon_r C_0 = \varepsilon_r \varepsilon_0 \frac{A}{d} = \varepsilon\frac{A}{d} \tag{6-116}$$

So: $$I_T = (i\omega\varepsilon + \sigma)\frac{A}{d}U \tag{6-117}$$

Define $\sigma^* = i\omega\varepsilon + \sigma$ as complex conductivity, the current density is expressed as $J = \sigma^* F$.

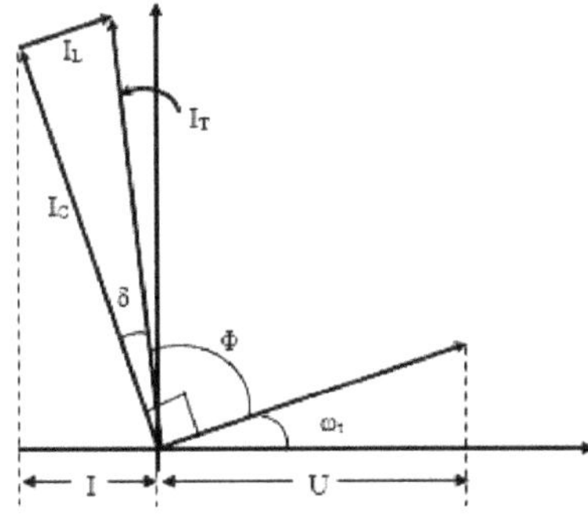

Figure 6.66 The I-U relationship of a real dielectric under AC electric field.

Ⅱ. Complex Dielectric Constant

Similar to complex conductivity,

$$I_T = (i\omega\varepsilon + \sigma)\frac{A}{d}U = \omega(i\varepsilon + \sigma / \omega)\frac{A}{d}U = i\omega(\varepsilon - i\frac{\sigma}{\omega})\frac{A}{d}U \quad (6\text{-}118)$$

The dielectric constant can also be divided into:

ideal part: $$\varepsilon = \varepsilon'\varepsilon_0 \quad (6\text{-}119)$$

and loss part: $$-(\sigma / \omega) = \varepsilon''\varepsilon_0 \quad (6\text{-}120)$$

So the complex dielectric constant can be expressed as:

$$\varepsilon^* = \varepsilon' - i\varepsilon'' \quad (6\text{-}121)$$

Most materials are polarizable in several different ways. As each type has a different frequency response, the dielectric constant will vary with frequency in a complicated manner. For example, at the highest frequencies (light waves) only the electronic polarization will 'keep up' with the applied field. Thus, we may measure the electronic contribution to the dielectric constant by measuring the refractive index at optical frequencies.[170] An important dielectric, water has a dielectric constant of about 80 at radio frequencies, but its refractive index is 1.3, not $(80)^{1/2}$. Hence we may conclude that the electronic contribution is about 1.7, and the rest is probably due to the orientation polarizability of the H_2O molecule.

170 大多数材料可以有几种不同的极化方式。因为每一方式都有不同的频率响应，介电常数将以一种复杂的方式随着频率而变化。比如，在高频下（光波），只有电子极化和外加电场保持同步。因此，我们可通过测量光频率下的折射率来测量电子极化对介电常数的贡献。

The general behavior of complex dielectric constant is shown in Figure 6.67. At every frequency where ε' varies rapidly, there tends to be a peak of the ε'' curve. In some cases, this is analogous to the maximum losses that occur at resonance in a tuned circuit: the molecules have a natural resonant frequency because of their binding in the crystal, and they will absorb maximum energy from an electromagnetic wave at this frequency. Another case is the 'viscous lag' occurring between the field and the polarized charge which is described by the Debye equations. A consequence of all this is that materials that transmit light often absorb strongly in the ultraviolet and infrared regions, for example most forms of glass.[171] Radio reception indoors is comparatively easy because (dry) bricks transmit wireless waves but absorb light, we can listen in privacy. The Earth's atmosphere is a most interesting dielectric. Of the fairly complete spectrum radiated by the Sun, not many spectral bands reach the Earth. Below 10^8 Hz the ionosphere absorbs or reflects, between 10^{10} and 10^{14}Hz there is molecular resonance absorption in H_2O, CO_2, O_2, and N_2, and above 10^{15} Hz there is a very high scattering rate by molecules and dust particles. The visible light region (about 10^{14} - 10^{15} Hz) has, of course, been of greatest importance to the evolution of life on Earth. One wonders what we would all be like if there

171 另一种情况是发生在电场和极化电荷之间的“粘性滞后”，可用德拜方程来描述。所有这些造成的结果是透光材料，比如大多数玻璃，在紫外和红外区域，通常有很强的吸收。

had been just a little more dust around, and we had had to rely on the 10^8 - 10^{10} Hz atmosphere window for our vision.

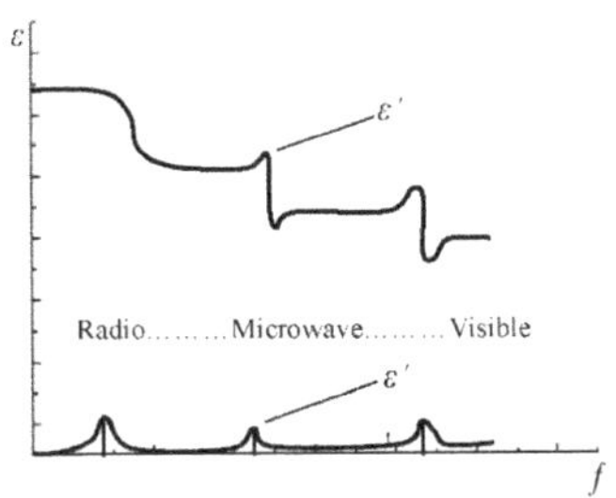

Figure 6.67 Typical variation of ε' and ε'' with frequency.

Ⅲ. Dielectric Loss

When polarization sets in, charges move (ions or electron clouds are displaced). If the electric field oscillates, the charges move back and forth. These displacements are extremely small (typically <1Å), however, they cause dielectric losses, which are often measured by a parameter known as tanδ.[172]

As previously shown in Figure 6.66, the total current of a dielectric consists of two parts: no energy loss part (I_C, corresponding to relative dielectric constant, ε_r') and energy loss part (I_L, corresponding to relative loss factor, ε_r''). If no loss, the current would be 90° ahead of the electric field. However, due to the loss of current, the current phase changes to Φ, having loss of δ.[173] The loss tangent is defined as,

$$\tan\delta=\frac{\varepsilon''}{\varepsilon'}=\frac{I_L}{I_C} \tag{6-122}$$

When we are interested in extremely low loss materials, such as those used in microwave communications, we refer to a parameter known as the dielectric quality factor (Q_d ~ 1/tanδ).

The dielectric loss is similar to the viscous deformation of a material and depends strongly on electrical frequency and temperature. For different application, tanδ is differently expected. When serving as insulator, in order to reduce energy loss, prevent material damage due to heat, e.g., deterioration of polymer, tanδ is desired to be small. For high frequency heating or drying, tanδ is desired to be large. For example, if we want to store a charge, as in a capacitor, dielectric loss is not good, however, if we want to use microwaves to heat up our food, dielectric losses that occur in water contained in the food are great![174]

Ⅳ. Dielectric Loss Due to Frequency Response

As shown in figure 6.67, the frequency variation of dielectric constant

172 当极化发生时，电荷移动（离子和电子云被转移）。如果电场振荡，电荷将来回移动。然而，这些位移非常地小（大约小于 1Å），它们造成了介电损失，这一损失通常用 tanδ 这个参量来测量。

173 如之前的图 6.66 所示，介电材料的总电流由两部分组成：无能量损失部分（I_C，与相对介电常数相对应，ε_r'）和能量损失部分（I_L，与相对损失因子对应，ε_r''）。如果没有损失，电流将与电场方向呈 90°夹角。但是，由于电流损失，电流相改变 Φ，损失 δ。

174 对于不同的应用，tanδ 预期的值也不同。当作为绝缘体时，为了减少能量损失，阻止材料由于热产生的损坏，比如聚合物的劣化，tanδ 的值需要尽可能小。对于高频的加热或干燥，tanδ 的值需要尽量大。比如，如果我们想要在电容器中储存电荷，介电损失是无益的；但是如果我们想要用作微波来加热食物，发生在水中的介电损失是有益的。

is a complicated affair. Debye equation describes how materials with orientation polarizability behave in the region where the dielectric polarization is 'relaxing':[175]

175 如图 6.67 所示，介电常数的频率变化是一个复杂的事情。德拜方程描绘了在介电极化弛豫的区间内具有取向极化材料的行为。

$$\varepsilon' = \varepsilon_\infty + \frac{\varepsilon_S - \varepsilon_\infty}{1+\omega^2\tau^2} \tag{6-123}$$

$$\varepsilon'' = \frac{\omega\tau}{1+\omega^2\tau^2}(\varepsilon_S - \varepsilon_\infty) \tag{6-124}$$

$$\tan\delta = \frac{\varepsilon''}{\varepsilon'} = \frac{(\varepsilon_S - \varepsilon_\infty)\omega\tau}{\varepsilon_S + \varepsilon_\infty\omega^2\tau^2} \tag{6-125}$$

where ε_∞ is the dielectric constant at optical wavelength, ε_S is the dielectric constant under static or low frequency state, ω is the angular frequency of the alternative electric field, and τ is the relaxation time. These equations agree well with experimental results. Their general shape is shown in Figure 6.68. Notice particularly that ε'' has a peak at $\omega\tau = 1$, where the slope of the ε' curve is a maximum.

Taking the derivative of the tanδ equation (Equation6-125), when d(tanδ)/dω = 0, we got maximum loss condition:

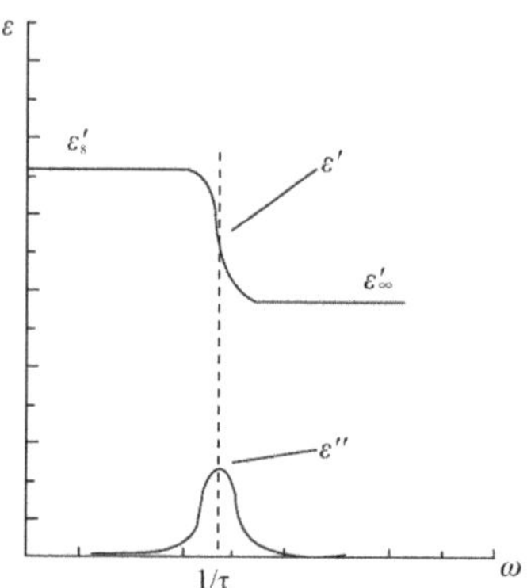

Figure 6.68 Frequency variation predicted by the Debye equations.

$$\omega = \frac{\sqrt{\dfrac{\varepsilon_S}{\varepsilon_\infty}}}{\tau} \tag{6-126}$$

$$\tan\delta = \frac{\varepsilon_S - \varepsilon_\infty}{2\varepsilon_S}\sqrt{\frac{\varepsilon_S}{\varepsilon_\infty}} \tag{6-127}$$

Other two ultimate states are shown in Figure 6.69:

(1) When $\omega \to 0$, all polarization mechanism can catch up with the change of the electric field. Thus, no dielectric loss exists, all losses should be ascribed to leak current.[176]

176 当 $\omega \to 0$ 时，所有的极化机制都能够跟得上电场的变化。因此没有介电损耗，所有损耗都缘于漏电流。

$$\omega\tau \ll 1$$

$$\varepsilon' = \varepsilon_\infty + \frac{\varepsilon_S - \varepsilon_\infty}{1+\omega^2\tau^2} \sim \varepsilon_\infty + (\varepsilon_S - \varepsilon_\infty) \to \varepsilon_S \tag{6-128}$$

$$\varepsilon'' = \frac{\omega\tau}{1+\omega^2\tau^2}(\varepsilon_S - \varepsilon_\infty) \to 0 \tag{6-129}$$

(2) When $\omega\to\infty$, $\tan\delta\to0$, no dielectric loss again. Because at very high frequency, all the polarization processes can't follow the electric field change, thus no such a dielectric loss exists.[177]

177 当 $\omega\to\infty$，$\tan\delta\to0$，又没有介电损耗。因为在非常高的频率下，所有的极化过程都跟不上电场的变化，所以没有介电损耗存在。

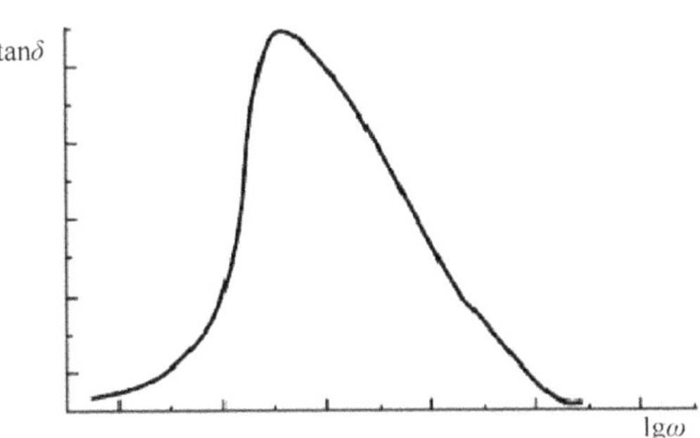

Figure 6.69 The dielectric loss versus frequency

6.5.5 Dielectric Breakdown

Electric breakdown is a subject to which it is difficult to apply our usual scientific rigor. A well-designed insulator (in the laboratory) breaks down in service if the wind changes direction or if a fog descends. An oil-filled high-voltage condenser will have bad performance, irrespective of the oil used, if there is 0.01 % of water present. The presence of grease, dirt, and moisture is the dominant factor in most insulator design. The flowing shapes of high-voltage transmission line insulators are not entirely due to the fact that the ceramic insulator firms previously made chamber pots, the shapes also reduce the probability of surface tracking. The onset of dielectric breakdown is an important economic as well as technical limit in capacitor design. Generally, one wishes to make capacitors with a maximum amount of stored energy. Since the energy stored per unit volume is $\frac{1}{2}\varepsilon F^2$, the capacitor designers value high breakdown strength (F_b) even more highly than high dielectric constant (ε).[178]

178 介电击穿的起始电压在电容器设计上是一个重要的经济和技术上的限制。通常来说，就是希望制成的电容器具有最大储能量。因为单位体积的储能为$\frac{1}{2}\varepsilon F^2$，所以电容器的设计者们相对于高的介电常数更看中材料高的击穿强度。

In general, breakdown is manifested by a sudden increase in current when the voltage exceeds a critical value U_b as shown in Figure 6.70. Below U_b there is a small current due to the few free electrons that must be in the conduction band at finite temperature. When breakdown occurs it does so very quickly, typically in 10^{-8} s in a solid.

There are three main mechanisms that are usually blamed for dielectric breakdown: (1) intrinsic, (2) thermal, and (3) discharge (avalanche) breakdown.

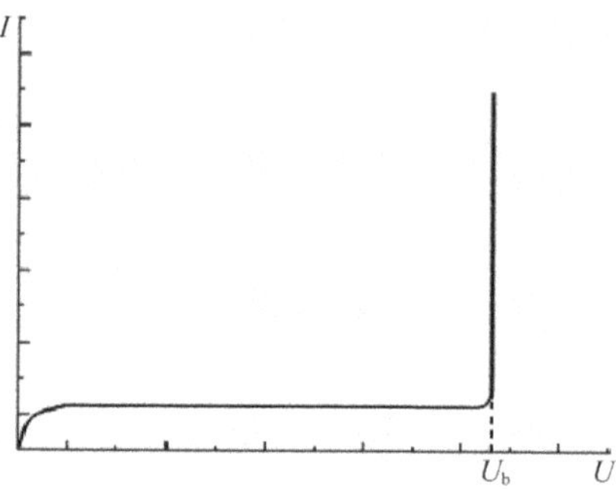

Figure 6.70 Current voltage characteristics for an insulator. The current increases very rapidly at the breakdown voltage, U_b.

Ⅰ. Intrinsic Breakdown

When the few electrons present are sufficiently accelerated (and lattice collisions are unable to absorb the energy) by the electric field, they can ionize lattice atoms. The minimum requirement for this is that they give to the bound (valence) electron enough energy to excite it across the energy gap of the material.[179] The critical requirement for intrinsic breakdown:

$$A(F,E_0) = B(T,E_0) \tag{6-130}$$

where $A(F, E_0)$ is the rate for electron obtaining energy from electric field, $B(T, E_0)$ is the rate for electron losing energy to lattice, F is the electric field strength, and E_0 is the electron energy. Intrinsic breakdown is independent of sample shape, only related to the intrinsic property of material.

179 当这些存在的极少电子经电场加速到足够的速度时（且晶格振动不能吸收这个能量），它们可以使晶格原子电离。电离需要的最小能量是使得价电子能够被激发超过材料带隙的能量。

Ⅱ. Thermal Breakdown

This occurs when the operating or test conditions heat the lattice. For example, an AC test on a material in the region of its relaxation frequency, where ε'' is large would cause heating by the dielectric loss rather than by accelerating free electrons. The heated lattice ions could then be more easily ionized by free electrons, and hence the breakdown field could be less than the intrinsic breakdown field measured with DC voltages.[180] The typical polymer, polyethylene, has a breakdown field of 3 to 5×10^8 Vm^{-1} for very low frequencies, but this falls to about 5×10^6 Vm^{-1}at around 10^6 Hz, where a molecular relaxation frequency occurs. Ceramics such as steatite and alumina exhibit similar effects.

180 被加热的晶格离子能够更容易被自由电子电离，因此热击穿所需的电场可能比通过施加直流电测得的本征击穿电场强度要小。

If it was not for dielectric heating effects, breakdown fields would be higher at high frequencies simply because the free electrons have only half a period to be accelerated in one direction.[181] A typical breakdown time was 10^{-8} s, so we might suppose that at frequencies above 10^8 Hz breakdown would be somewhat inhibited. This is true, but a fast electron striking a lattice ion still has a greater speed after collision than a slow one, and some of these fast electrons will be further accelerated by the field. Thus, quite

181 如果不是介电热效应、高频时的击穿电场会更高，仅仅因为在一个方向上自由电子只有半个周期的时间来加速。

spectacular breakdown may sometimes occur at microwave frequencies (10^{10} Hz) when high power densities are passed through the ceramic windows of klystrons or magnetrons.[182] Recent work with high-power lasers has shown that dielectric breakdown still occurs at optical frequencies. In fact, the maximum power available from a solid-state laser is about 10^{12}W from a series of cascaded neodymium glass amplifier lasers. The reason why further amplification is not possible is that the optical field strength disrupts the glass laser material.

Ⅲ. Discharge (Avalanche) Breakdown

By field emission (electron accelerated by electric field) or ion collision, electrons are accumulated gradually, leading to electric injection barrier changes, and causing tunneling effect. The tunneling electrons are accelerated by electric field, causing further collision, then electrical breakdown. Since collision is involved, thinner dielectrics are more difficult to breakdown compared to thicker dielectrics.[183] Discharge breakdown usually occurs in materials such as mica or porous ceramics, where there is occluded gas. The gas often ionizes before the solid breaks down. The gas ions can cause surface damage, which accelerates breakdown. This shows up as intermittent sparking and then breakdown as the test field is increased.

6.6 Thermoelectricity of Material

6.6.1 Thermoelectric Potential and Absolute Thermoelectric Coefficient

It was found experimentally that a conductor or semiconductor creates voltage when there is a different temperature on each side. Conversely, when a voltage is applied to it, it creates a temperature difference. This can be explained in atomic scale (see Figure 6.71). Between the two ends of a conductor or semiconductor, an applied temperature gradient causes charge carriers to diffuse from the hot side to the cold side, resulting in electric potential.[184]

By definition, the thermoelectric effect is the direct conversion of temperature differences to electric voltage and vice-versa.

Obviously, the bigger temperature gradient, the more charge diffusion, and resulting in larger electric potential. On the other hand, when

182 击穿时间普遍为 10^{-8}秒，所以我们假设在电场频率高于 10^{8} 赫兹时，击穿将在某种程度上被禁阻。这是正确的，但是高速电子在撞击晶格离子后仍比低速电子的速度快，其中一些快电子将被电场进一步地加速。因此，当高密度功率通过的电子速调管或者磁控管的陶瓷窗口时，有时很可能观测到在微波频率(10^{10}Hz)下的击穿。

183 由于场发射（电子由电场加速）或者离子碰撞，电子逐渐聚集，导致了电注入势垒改变，造成隧穿效应。隧穿电子被电场加速，造成进一步的碰撞，然后形成电致击穿。因为碰撞的参与，薄的介电材料比厚的介电材料更难击穿。

184 实验发现，导体或半导体的两端存在温差时，会产生电压。相反的，当施加一个电压时，材料两端会产生温度差。这可以从原子尺度上来解释（图 6.71）。在导体或半导体的两个端点之间，施加的温度梯度造成了载流子从热端到冷端的扩散，导致了电势差。

185 显然，温度梯度越大，电荷扩散越多，导致了更大的电势差。另一方面，当热电势差建立时，将会有一个电场来阻止电荷载流子的扩散。最终达到平衡条件，在给定的温度梯度下有一个确定的电势差。

thermoelectric potential is built, there will be an electric field that prevents the charge carrier diffusion. Eventually, equilibrium condition is achieved, with definite electric potential under fixed temperature gradient.[185] This means different material possesses different thermoelectric potential at the same temperature gradient. Thereby, the absolute thermoelectric coefficient is defined:

$$S = \frac{dU}{dT} \tag{6-131}$$

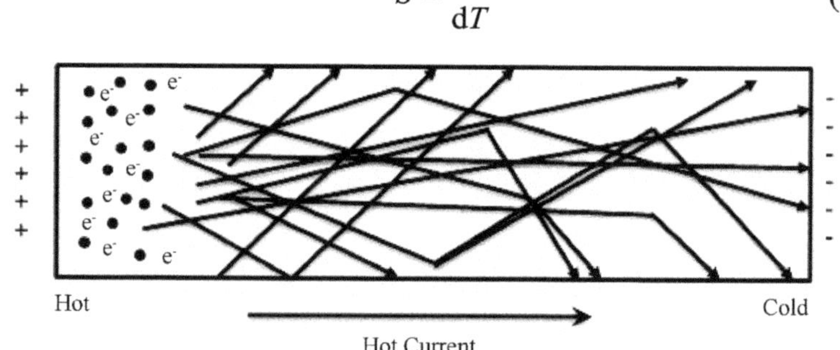

Figure 6.71 Thermoelectric potential

where U is the thermoelectric field, T is temperature. By quantum dynamic theory, the high temperature thermoelectric coefficient was deduced by Mott and Jones as,

$$S = \frac{\pi^2 k^2 T}{3e}\left\{\frac{\partial[ln\sigma(E)]}{\partial E}\right\} | E = E_F \tag{6-132}$$

where k is Boltzmann constant, e is electron charge, σ is electric conductivity, E is energy, E_F is Fermi energy.

186 热电效应不同于焦耳热，即当电压施加于有电阻的材料而产生热效应。前者是通过热能和电能交换产生的热力学可逆的过程，但是焦耳热是一个不可逆的过程，热量都散布到了周围的环境中。

Thermoelectric effect is different to Joule heating that is generated whenever a voltage is applied across a resistive material. The former is a thermodynamically reversible process characterizing with thermal and electrical energy exchange, whereas Joule heating is not a reversible process which is a heat dissipation to environment.[186]

6.6.2 Classification

Thermoelectric effect encompasses the Seebeck effect and Peltier effect.

Ⅰ. Seebeck Effect (T. J. seebeck, 1821)

In 1821, Thomas Seebeck found that an electric current would flow continuously in a closed circuit made up of two dissimilar metals, if the junctions of the metals were maintained at two different temperatures (see Figure 6.72).

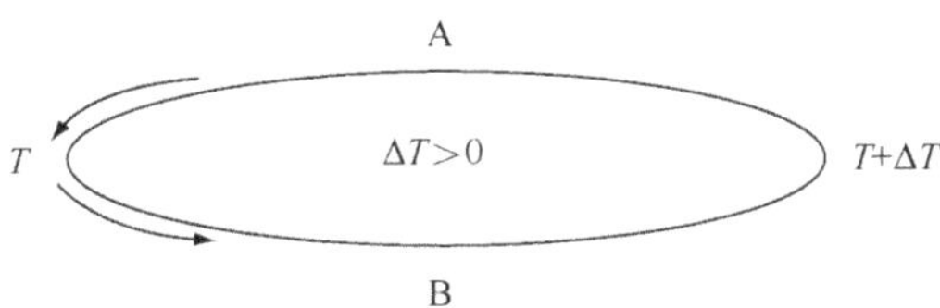

Figure 6.72 Seebeck effect: temperature different creates electric potential.

The thermoelectric potential (U_{AB}) is:

$$U_{AB} = S_{AB}\Delta T \tag{6-133}$$

$$S_{AB} = S_A - S_B \tag{6-134}$$

S_{AB} is the relative Seebeck coefficient between material A and B, S_A and S_B is the absolute Seebeck coefficient, and ΔT is temperature difference.

Ⅱ. Peltier Effect(J. C. A. Peltier, 1834)

In 1834, a French watchmaker and part time physicist, Peltier found that an electrical current would produce a temperature gradient at the junction of two dissimilar metals (Figure 6.73).

The Peltier effect is the reverse process of the Seebeck effect.

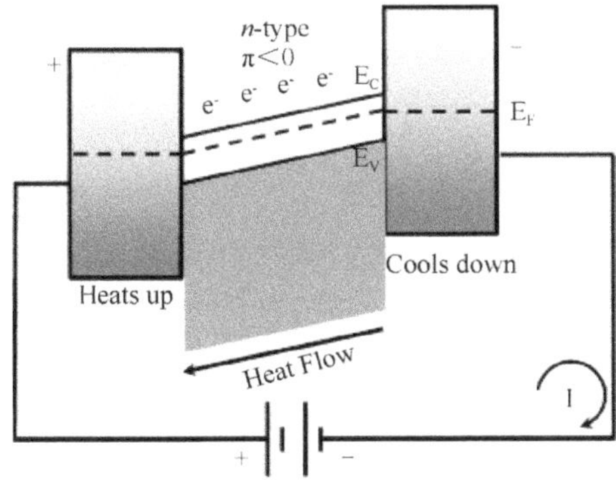

Figure 6.73 High energy electrons move from right to left under electric potential (note: Thermal current and electric current flow in opposite directions).

6.6.3 Applications

Thermoelectric effect can be used to generate electricity, measure temperature or change the temperature of objects. Because the direction of heating and cooling is determined by the polarity of the applied voltage, thermoelectric devices can be used as temperature controllers.[187]

Figure 6.74 shows the schematic structure for a thermoelectric cooling machine.The cooling capacity of the thermoelectric cooler is generally very small, so should not be large cooling capacity. However, due to its flexibility, simple and convenient hot and cold with a cold place to switch easily, very suitable for miniature refrigeration field or have special requirements.[188]

187 热电效应可以用作发电，测量温度或者改变物体的温度。因为加热和冷却的方向是由外加电场的两极所决定，所以热电装置可以被用作温控器件。

188 热电制冷机的冷却能力普遍非常的小,所以不能具有较大的冷却容量。但是由于其灵活性,可以简单方便的加热和冷却,冷缺位置能容易地转变成加热位置,很适合于微型制冷领域或者特殊的需求。

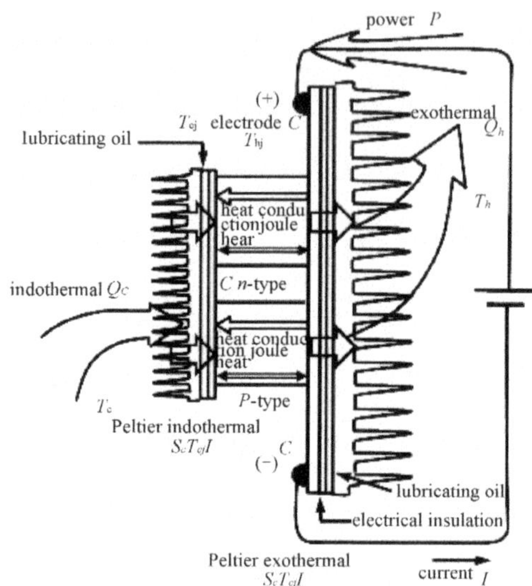

Figure 6.74 Cooling mechanism by thermoelectric effect.

189 热电效应也能被用于热电偶。因为热电势能的值依赖于材料的本质和连接点温度，并与其形状尺寸无关，因此在热电回路中，在其两端插入一个温度相同的导体，对于原本的热电势能没有任何影响。

Thermoelectricity can also be used in at thermal coupler. Since the value of the thermoelectric potential depends on material nature and the contact temperatures, and is independent of the shape and size of the materials, therefore, within the thermoelectric circuit, inserting a conductor with identical temperature at its two ends, there will be no influence to the original thermoelectric potential.[189] A thermometer for practical measurement of temperature can be a junction of specific alloys which have a predictable and repeatable relationship between temperature and voltage. Different alloys are used for different temperature ranges. Properties such as resistance to corrosion may also be important when choosing a type of thermocouple. One type of vacuum gauge is based on thermal coupler, and Figure 6.75 shows its corresponding device structure.

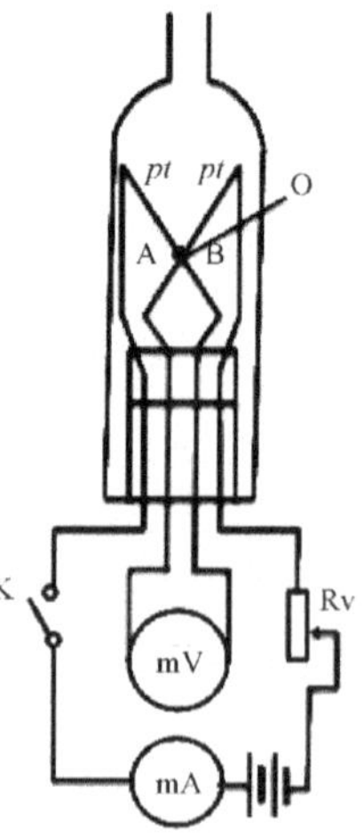

Figure 6.75 The schematic structure of a vacuum gauge. p_t is thermal heater, A and B are thermal couplers, O is the contact point.

Using thermoelectricity, one kind of power generator can also be produced (Figure 6.76).

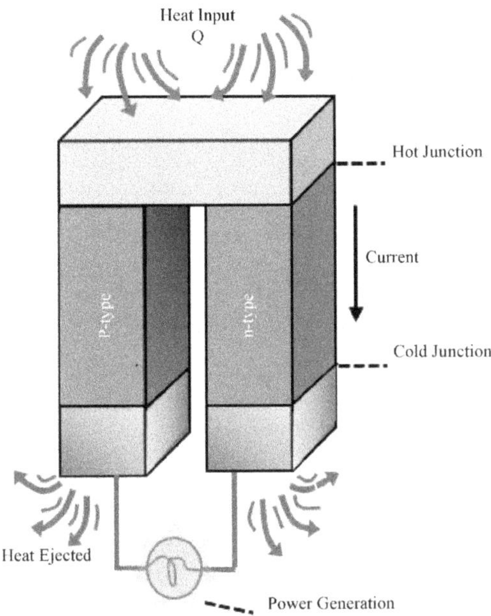

Figure 6.76 Power generator by thermoelectric effect.

本章小结

1. 内容概要

本章介绍了材料的电学性质。本章除了导体、半导体以及超导体外，还讲述了介电和热电材料的相关知识；同时，对基于材料不同电学性质的各种应用也进行了介绍；其中，对有机材料的相应性能及应用进行了讨论。

2. 基本概念

欧姆定律、电导率、电阻率、平均自由程、马西森定律、声子、电阻温度系数、功能性导电金属、透明导电氧化物、半导体、本征半导体、非本征半导体、p-型半导体、n-型半导体、霍尔效应、p-n 结、有机半导体、二极管、晶体管、超导体、临界温度、临界磁场强度、麦斯纳效应、临界电流密度、转变潜热、隧穿效应、BCS 理论、Cooper 电子对、介电材料、电容、介电常数、偶极矩、折射率、极化强度、电极化率、复介电常数、介电损耗、介电击穿、热电势、热电系数、塞贝克效应、帕尔帖效应。

3. 主要公式

（1）电场强度：$F=\dfrac{U}{l}$

（2）欧姆定律：$I=\dfrac{U}{R}$

（3）电阻：$R=\rho\dfrac{l}{A}=\dfrac{l}{\sigma A}$

（4）电功率：$P=UI=I^2R$

（5）电流密度：$J=\sigma F$，$J=nq\bar{v}$

（6）载流子迁移率：$\mu=\dfrac{v}{F}=\dfrac{v}{ma}=\dfrac{v}{m(v/\tau)}=\dfrac{\tau}{m}$

（7）电导率：$\sigma=nq\mu$

（8）半导体的电导率：$\sigma=nq\mu_{\mathrm{n}}+pq\mu_{\mathrm{p}}$

（9）金属电导率：$\sigma=\dfrac{e^2n\tau}{m}=\dfrac{e^2nl}{mv}$

（10）能带理论下的电导率：$\sigma=\dfrac{e^2n^*\tau_{\mathrm{F}}}{m^*}=\dfrac{e^2n^*l_{\mathrm{F}}}{m^*v_{\mathrm{F}}}$

（11）平均自由程：$l=\tau v$

（12）各向异性电阻率：$\rho_\alpha=\rho_\perp+(\rho_\parallel-\rho_\perp)\cdot\cos^2\alpha$

（13）马西森定律：$\rho=\rho_0+\rho(T)$

（14）剩余电阻率：$RRR=\dfrac{\rho_{300\mathrm{K}}}{\rho_{4.2\mathrm{K}}}$

（15）电阻温度系数：$\bar{\alpha}=\dfrac{\rho_T-\rho_0}{\rho_0 T}$ $\quad\alpha_{\mathrm{T}}=\dfrac{1}{\rho_{\mathrm{T}}}\dfrac{\mathrm{d}\rho}{\mathrm{d}T}$

（16）质量作用定理：$n \times p = n_i^2$

（17）本征半导体导带/价带中的电子/空穴：$n = p = n_i = n_0 \exp(\frac{-E_g}{2k_B T})$

（18）霍尔电压：$E_y = V \cdot H$

（19）霍尔系数：$R_H = \frac{E_y}{j_x H_z}$

（20）三极管通过集电极的电流：$I_C = I_0 \exp\left(\frac{V_E}{B}\right)$

（21）磁场穿透深度：$H(x) = H_0 \exp\left(\frac{-x}{\lambda}\right)$

（22）临界磁场强度：$H_C = H_0 \left\{\left\{1 - \frac{T}{T_C}\right\}^2\right\}$

（23）正常态与超导态下熵的关系：$S_N - S_S = -\mu_0 V H_C \frac{dH_C}{dT}$

（24）转变潜热：$L = T(S_n - S_S)x$，$S_N - S_S = -\mu_0 V H_C \frac{dH_C}{dT}$

（25）正常态与超导体热容的关系：$C_N - C_S = T\left(\frac{dS_S}{dT} - \frac{dS_S}{dT}\right) = -VT\mu\left\{-H_C \frac{d^2 H_C}{dT^2} + \left(\frac{dH_C}{dT}\right)^2\right\}$

（26）Cooper 电子对辐射能与电势差的关系：$h\omega = (2e)U_{AB}$

（27）电量公式：$Q = \sigma A = \varepsilon_0 F A = \varepsilon_0 \frac{U}{d} A$

（28）真空电容：$C_0 = \frac{Q}{U} = \frac{\varepsilon_0 \left(\frac{U}{d}\right) A}{U} = \frac{\varepsilon_0 A}{d}$

（29）介电材料电量：$Q' = \sigma' A = \varepsilon_0 \varepsilon_r F A = \varepsilon_0 \varepsilon_r \frac{U}{d} A$

（30）介电材料电容：$C = \frac{Q'}{U} = \varepsilon_0 \varepsilon_r \frac{U}{d} A \frac{1}{U} = \varepsilon \frac{A}{d}$

（31）偶极矩：$\mu = q \times d$

（32）折射率：$n = \frac{c}{v} = \sqrt{\varepsilon_r \mu_r}$

（33）极化强度：$\vec{P} = \frac{\sum \vec{\mu}}{\Delta V}$，$P = \chi_e \varepsilon_0 F = \sigma'$

（34）电极化率：$\chi_e = \varepsilon_r - 1$

（35）波尔原子模型下的电子平均电极化率：$\chi_{electron} = \frac{4}{3} \pi \varepsilon_0 R^3$

（36）离子的电极化率：$\chi_i = \frac{a^3}{n-1} 4\pi\varepsilon_0$

（37）取向电极化率：$\chi_d = \frac{\mu_0^2}{3kT}$

（38）复电导率：$\sigma^{*}=i\omega\varepsilon+\sigma$

（39）复介电常数：$\varepsilon^{*}=\varepsilon'-i\varepsilon''$

（40）介电损耗：$\tan\delta=\dfrac{\varepsilon''}{\varepsilon'}=\dfrac{I_{\mathrm{I}}}{I_{\mathrm{C}}}$

（41）最大介电损耗条件：$\omega=\dfrac{\sqrt{\dfrac{\varepsilon_{\mathrm{S}}}{\varepsilon_{\infty}}}}{\tau}$　　$\tan\delta=\dfrac{\varepsilon_{\mathrm{S}}-\varepsilon_{\infty}}{2\varepsilon_{\mathrm{S}}}\sqrt{\dfrac{\varepsilon_{\mathrm{S}}}{\varepsilon_{\infty}}}$

（42）塞贝克热电势差：$U_{\mathrm{AB}}=S_{\mathrm{AB}}\Delta T$　　$S_{\mathrm{AB}}=S_{\mathrm{A}}-S_{\mathrm{B}}$

Vocabulary

absolute zero temperature	绝对零度
active energy	活化能
alternative electric field	交变电场
anisotropic	各向异性的
avalanche Breakdown	雪崩击穿
battery	电池
behavior	行为，特性
capacitance	电容
capacitor plate	平板电容器
carrier	载流子
catastrophic	灾难性的
circumference	圆周
complex conductivity	复电导率
complex dielectric constant	复介电常数
complexity	复数
concentration	浓度
conductive materials	导电材料（导体、半导体）
conductivity	电导率
conductor	导体
continuous transition	连续转变
Cooper pair	库帕对
cosine	余弦
critical current density	临界电流密度
critical magnetic field	临界磁场
critical temperature	临界温度
defect	缺陷

density	密度
depletion region	耗尽区
diamagnet	抗磁体
dielectric breakdown	介电击穿
dielectric constant	介电常数
dielectric loss	介电损耗
dielectric material	介电材料
dielectric susceptibility	电极化率
diffraction	衍射
diffusion	扩散
dipole	偶极
dipole moment	偶极矩
direction	方向
divalent element	二价元素
dominate	主导
dopant	掺杂
drift velocity	迁移速率
electrical contact material	电接触材料
electrical field strength	电场强度
electrical generator/power generator	发电机
electrical quantity	电量
electrode	电极
electrolyte	电解质
electron	电子
electronic polarization	电子极化
electronic structure	电子结构
elliptical	椭圆的
energy band	能带
entropy	熵
equation	方程，等式
exert to	施加
expel force	施加于
extrinsic	外在的，非本征的
factor	因子
fast ion	快离子
flow through	流过
free electron model	自由电子模型
free energy	自由能
gas constant	气体常数
gate	栅极

germanium	锗
get rid of	除掉
ground state	基态
half metal	半金属
Hall coefficient	霍尔系数
Hall effect	霍尔效应
heterogenous	异质的
hole	空穴
homogenous	同质的，同种类的
identical	相同
impurity	杂质
incidence	入射
influence	影响
initial	最初的
insulator	绝缘体
integer	整数
intensity	强度
interaction	相互作用
interstitial space	间隙
intrinsic	本征的
ionic crystal	离子晶体
ionic polarization	离子极化
irreversible	不可逆的
isobaric process	等压过程
isotropic	各向同性的
Joule heat	焦耳热
lattice binding	晶格束缚
lattice point wave	点阵波
lattice wave	晶格波
loss angle	损耗角
loss factor	损耗因子
maglev train	磁悬浮列车
magnetic field	磁场
Matthiessen rule	马西森定律
mean free path	平均自由程
Meissner effect	迈斯纳效应
melting point	熔点
mercury, Hg	汞
mobility	迁移率
molecules	分子

monoenergetic	单能的
monovalent element	一价元素
motion	运动
negative charge	负电荷
neutralize	中和
neutron	中子
normal	法线
nucleus	原子核
obtain	获得
occlude	使闭塞、堵塞
optical frequency	光频
orientational polarization	取向极化
overlap	重叠
parabola	抛物线
Pauli exclusion principle	泡利不相容原理
Peltier effect	珀尔帖效应
penetration depth	穿透深度
pentad	五价元素
perfect conductor	完美导体
perimeter	周长
period	周期
periodicity	周期性
permittivity	介电常数
polar	极性的
polarization	极化
polarization intensity	极化强度
positive charge	正电荷
power generator	发电机
privacy	不受干扰
proportionality factor	比例系数
proton	质子
prove	证明
quadrivalent element	四价元素
real and imaginary parts	实部和虚部
rectifier diodes	整流二极管
reflect	反射
refractive index	折射率
relative dielectric constant	相对介电常数
relaxation time	弛豫时间
represent	代表

residual resistivity	剩余电阻
residual resistivity ratio	剩余电阻率，3R
resistance	电阻
resistance material	电阻材料
resistivity	电阻
scattering coefficient	散射系数
Seebeck coefficient	塞贝克系数
selection rule	选择定则
semiconductor	半导体
sensor	传感
set up (erase)	产生(消除)
several order of magnitudes higher	高出几个数量级
silicon	硅
specific heat capacity	比热容
stannate	锡酸盐
strengthen	加强
structure	结构
subject to	受……影响，受……管制
super conducting transition temperature	超导转变温度
super conduction	超导
superconducting cable	超导电缆
surface charge density	表面电荷密度
temperature	温度
temperature coefficient of resistance	电阻温度系数
temperature dependent	温度依赖
thermal coupler	热电偶
thermoelectric cooling	热电制冷
thermoelectric potential	热电势
thermoelectricity	热电性
threshold	阈值
toroid	环形线圈
track	轨道
transition	跃迁
trivalent element	三价元素
tunneling effects	隧道效应
vacuum capacitance	真空电容
vacuum dielectric constant	真空介电常数
valence electron	价电子
velocity	速度
vortex state	涡流态

Problems

Section one: Basic concepts for electricity

1. Determine the cross-sectional area of an electrical transmission line 1500 m long that will carry a current of 50 A with no more than 5×10^5 W loss in power, employing aluminium with conductivity of $3.77 \times 10^5 \Omega^{-1}cm^{-1}$.

2. Assuming that all of the valence electrons contribute to current flow, (a) calculatethe mobility of an electron in copper and (b) calculate the average drift velocity forelectrons in a 100 cm copper wire when 10 V are applied. (Note: copper is FCC with lattice parameter of copper is 3.6151×10^{-8} cm, it possesses one valence electron, its conductivity is $5.98 \times 10^5 \Omega^{-1}cm^{-1}$).

3. A 10 mm cube of germanium passes acurrent of 6.4 mA when 10 mV is appliedbetween two of its parallel faces. Assuming that the charge carriers areelectrons that have a mobility of 0.39 $m^2V^{-1}s^{-1}$, calculate the density ofcarriers.What is their collision time if theelectron's effective mass in germanium is 0.12 m_0, where m_0 is the free electronmass?

4. Using the concept of current density, electric field, mobility and resistivity, based on the Ohm's law, deduce the relationship between the conductivity and mobility.

5. True or false questions

   (1) If only the free carrier density is high, material exhibits high conductivity.

   (2) The conductivity of semiconductors is lower than metal.

   (3) Free electrons in metal possess wave particle duality.

Section two: Conductors

1. Using band theory, explain the electricity of element materials with valence electron of 1 to 5, as well as that of ionic crystal.

2. What is RRR? Explain the meaning of its value.

3. Please compare the advantages of iron, copper, aluminium and gold when served as conducting materials

4. True or false question:

   (1) According to Ohm law, the electric current is proportional to the electric field.

   (2) The movement of valence electrons in a metal is only influenced by lattice vibration, but is not affected by impurity and defect.

   (3) The conductivity of metals decreases with the increase of temperature.

   (4) The order of metal conductivity is Al < Au < Cu < Ag.

   (5) Phonon relates to the vibration of lattice, its energy is quantized and increases with temperature.

5. Please state the origination of resistance in metal.

6. Describe the origination of phonon in metal crystal, point out its relationship with temperature.

7. The current density in a pure copper wire is 10 $A \cdot mm^{-2}$. The density of Cu is 8.93×10^3

kg·m^{-3}. Its molar weight is 63.6.

(1) Calculate the density of conduction electrons in the metal if we assume that there is one conduction electron per Cu atom.

(2) Calculate the drift velocity at the given current density.

Section three: Semiconductors

1. True or false questions.

(1) The intrinsic semiconductors only include silicon and germanium.

(2) Mobility decreases with impurity concentration.

(3) In semiconductor, carrier concentration decreases with temperature.

(4) In semiconductor, conductivity decreases with temperature.

(5) With the increase of doping concentration, the carrier concentration increases.

2. Explain the following terms: Hall Effect, intrinsic semiconductor, extrinsic semiconductor.

3. At 300 K, the intrinsic carrier density of Si is 1.5×10^{16} m^{-3}, and the electron and hole mobility of intrinsic Si is 0.135 m^2·V^{-1}·s^{-1} and 0.048 m^2·V^{-1}·s^{-1}, respectively. The bandgap of Si is 1.1 eV. Find the resistivity of Si at 300 K and at 150 K.

4. In a *n*-type semiconductor, the energy levels are described as following figure, where CB, E_d, and VB are conducting band, donor level, and valence band, respectively. Please point out: (1) the main (major, majority) carrier; (2) which energy band (level) is responsible for carrier conduction, give your reason.

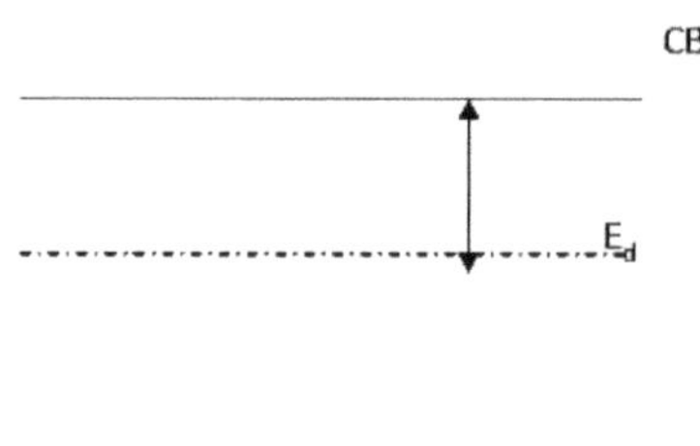

5. At 300 K, the intrinsic carrier density of Si is 1.5×10^{16} m^{-3}, and the electron and hole mobility of intrinsic Si is 0.135 m^2·V^{-1}·s^{-1} and 0.048 m^2·V^{-1}·s^{-1}, respectively. When Si is doped with P at concentration of 10^{21} m^{-3} at 300 K, assuming that all P are ionized at 300 K and the mobilities of electron and hole are independent of dopant, find the majority and minority carrier concentrations, and deduce the resistivity of the doped Si.

6. Find the electron and hole concentrations in silicon at 300 K for boron doping at 3×10^{16}/ cm^{3}and 2.9× 10^{16}cm^{-3} phosphorous, knowing that the n_i of Si at 300 K is 1.5×10^{10} cm^{-3}. *Note:* You cannot assume that $N_A >> N_D$ for this question, only that both N_A and $N_D >> n_i$.

7. Find the density of electrons in the conduction band in GaAs doped with $N_A=10^{16}$cm^{-3} at room temperature, the intrinsic carrier density of GaAs is known as 2 × 10^6 cm^{-3}. Suppose that all the acceptors are reduced.

8. Find the number of holes in the valence band of InP doped with $N_D = 10^{14}$ cm^{-3} at 0 K.

9. In inorganic semiconductor, why the electron mobility is generally larger than hole mobility?

10. The figure shows the temperature dependence of carrier concentration in a n-type semiconductor. Please describe the temperature dependent feature of carrier concentration and

explain the reasons.

11. As we know, the bandgap, the mobilities of electron/hole, and the resistivity of Si and Ge are 1.14 and 0.67 eV, 0.135/0.048 and 0.39/0.19 $cm^2 \cdot s^{-1} \cdot V^{-1}$, 2300 and 0.46 $\Omega \cdot m$, respectively. It seems that Ge possesses higher mobility and lower resistivity. But why Si is more popularly used than Ge in semiconductor industry?

12. Give the equation showing the relationship between electrical conductivity, mobility, and carrier density. And describe the temperature dependence of the conductivity of semiconductors.

13. Please define mobility, and state the dependence of mobility on temperature in semiconductors.

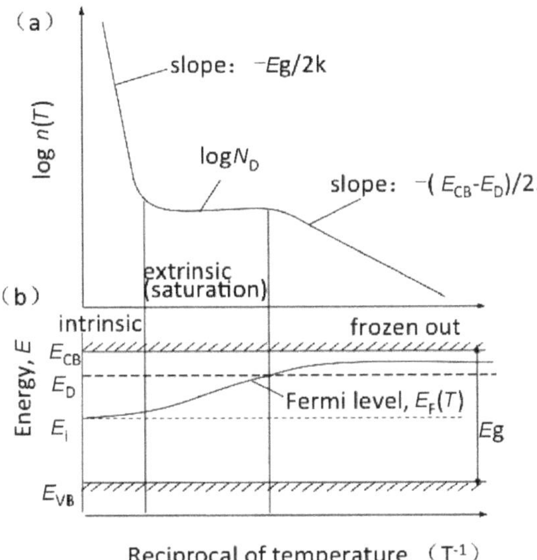

14. If both electrons and holes are presentthe conductivities. This is because under the effect of an applied electric field the holes and electrons flow in opposite directions, and a negative charge moving in the (say) $+z$ -direction is equivalent to apositive charge moving in the $-z$ -direction. Assume that in a certain semiconductorthe ratio of electron mobility, μ_e , to holemobility, μ_h , is equal to 10, the density of holes is $N_h = 10^{20}$ m^{-3}, and the density of electrons is $Ne = 10^{19} m^{-3}$. The measured conductivity is 0.455 $ohm^{-1}m^{-1}$. Calculate the mobility of electron and hole, respectively.

15. For germanium at 25°C, estimate (a) the number of charge carriers, (b) the fraction of the total electrons in the valence band that are excited into the conduction band.

16. Design a *p*-type semiconductor based on silicon, which provides a constant conductivity of 100 ohm^{-1} cm^{-1} over a range of temperatures. Compare the required concentration of acceptor atoms in Si with the concentration of Si atoms.

Section four: Superconductors

1. State the advantages, and the current and potential applications of superconductor
2. Please summarize the main points of BCS Theory for superconduction
3. Please tell true or false about the following statements:

   (1) In a super conductor, below T_C and H_C, the free energy is lower in the state of super conductor than in the normal state.

(2) In a super conductor, below T_C, the specific heat capacity is lower in the state of super conductor than in the normal state.

(3) Superconducting quantum interference device utilizes the superconductor feature of Josephson Effect.

(4) The zero resistance of a superconductor is due to the formation of Cooper pair by the assisting of lattice phonon.

(5) When a superconductor in a fixed magnetic field, is cooled to temperature below T_C, it will expel the magnetic field.

4. Please explain why for a superconductor below T_C, the entropy of the super conduction state is lower than that of the normal conducting state.

5. Under a fixed external magnetic field, when a substance is cooled to temperature below T_C, please explain the influence of the external magnetic field to (1) perfect conductor, (2) superconductor.

6. Explain the following terms: (1) Superconductivity, (2) Meissner effect.

Section five: Dielectric property and thermoelectricity

1. True or false questions.

(1) In nonpolar dielectric materials, the statistic centers of the positive and negative charges of molecules overlap.

(2) Compared to polar dielectric materials, the dielectric constant of a nonpolar dielectric material is lower.

(3) Thermoelectric effect is irreversible.

(4) For a parallel plate capacitor in vacuum, the capacitance is higher in the one employing aluminum as the electrodes than the one with silver as electrodes.

(5) In a dielectric material, applying an alternative electric field, the phase of charging current will be the same as that of the electric field.

(6) The refractive index relates to the polarization of dielectric material under optical frequency that only responses to the electric polarization.

2. Please give the definition about dielectric Materials. Tell three basic types of polarization, and give the order based on their response time.

3. What is thermoelectrical effect? Please give two examples of the application of the thermoelectrical effect.

4. What is a dipole? Please point out the direction a dipole moment

5. What is polarization intensity? What is the relationship between polarization intensity and the effective electric field (please give equation which includes dielectric susceptibility, vacuum dielectric constant and electric field)?

6. Please deduce the relationship between the electric susceptibility and the dielectric constant.

7. What is the origination of dielectric loss?

8. What is dielectric breakdown? Please schematically draw the I-V curve, and please enumerate the types of dielectric breakdown.

Chapter 7
Magnetic Properties

This chapter deals with material magnetic properties. After historical introduction about the phenomena and application of magnetism, the mechanism for producing magnetic properties is presented. Then the characterization of magnetism is discussed and the types of magnetic materials are classified. Finally, functional magnetic materials are presented.

7.1 Introduction

Magnetic properties have already been observed early from the ancient times. For examples, Chinese used magnetism of compass for direction determination around 2500 BC, the fact that lodestone can attract iron was known to Thales of Miletus in the sixth century BC, and written by Englishman in 1600 that "By good luck the smelters of iron or diggers of metal discovered magnetite as early as 800 BC".

Nowadays, magnetism is widely and intensively studied, and has a variety of applications in many fields. For instance, magnetic materials are used in electrical motors, generators, and transformers. Most of data storage media of electronic device (computer hard disks, video and audio cassettes, etc.) is based on magnetic materials. Magnets are also used in loudspeakers, telephones, televisions, and video recorders. Superconductors can also be viewed as magnetic materials. Magnetic materials, such as iron oxide (Fe_3O_4) particles, are used to make exotic compositions of "liquid magnets" or ferrofluids. The same iron oxide particles are also used to bind DNA molecules, cells, and proteins. Some of research and application fields of magnetism are depicted in Figure 7.1.[1]

1 磁性材料，如氧化铁（Fe_3O_4）颗粒，用来做"液体磁铁"或者铁磁流体的外来成分。同样的氧化铁颗粒也被用来与 DNA 分子、细胞和蛋白质结合。一些磁性研究领域如图 7.1 所示。

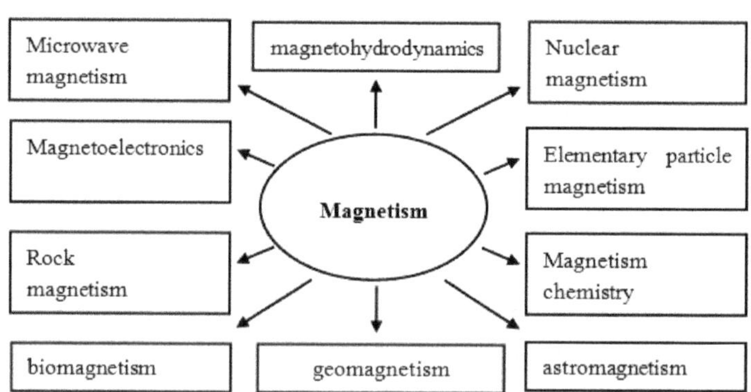

Figure 7.1 Research and application fields of magnetism

In fact, magnetism is the basic property of materials and it is similar to the material properties of quality and electricity. Simply, it can be said that every material possesses magnetism, but not all substances are magnets.[2]

2 事实上，磁性是材料的基本特性，它与材料的质量和电性类似。简单地说，每一种材料都具有磁性，但是不是所有的物质都是磁体。

7.2 Characterization and Principles for Magnetism

7.2.1 Magnetic Field Intensity and Magnetic Force—Macroscopic Approach

Ⅰ. Magnetic Field Intensity

A current I in a linear conductor with unlimited-length will produce magnetic field (Figure 7.2). The magnetic field intensity H at a distance of r, can be expressed as: [3]

3 无限长直线导体中的电流 I 会产生磁场（图 7.2）。距离为 r 时的磁场强度 H 可以表达为：

$$H = \frac{I}{2\pi r} \tag{7-1}$$

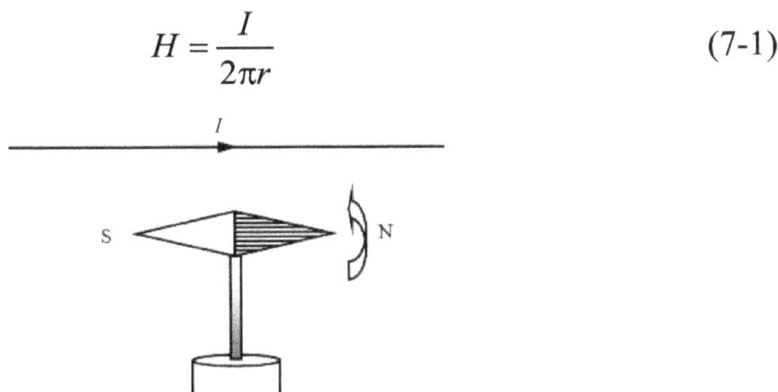

Figure 7.2 A linear conductor with unlimited-length attracts a magnetic compass.

Similarly, when an electric current is passed through a coil, a magnetic field H is produced (Figure 7.3), with the strength of the field given by[4]

4 相似地，当电流通过一个线圈时，会产生磁场 H（图 7.3），场强可以被表示为：

$$H = \frac{nI}{l} \tag{7-2}$$

where n is the number of turns, l is the length of the coil, and I is the current.

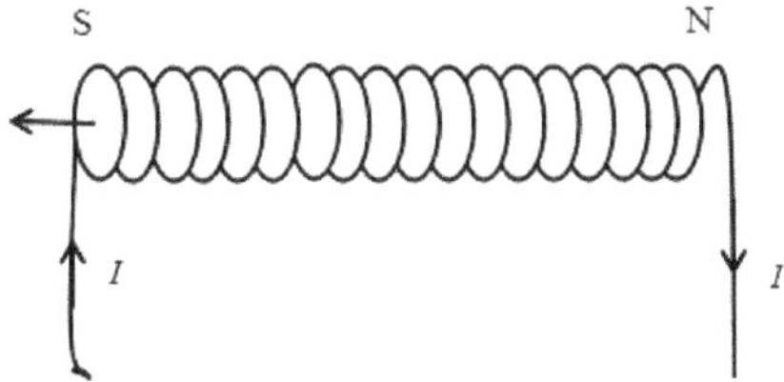

Figure 7.3 An electric current passing through a coil produces a magnetic field. Equation (7-1) and (7-2) connect current I to magnetic field intensity H.

Ⅱ. Force Exerted by External Magnetic Field

Similar to electric charge, there is magnetic charge that can be defined as the basic unit of the magnetic. If there are q_1 and q_2 two magnetic charges at the two opposite magnetic poles, then the force for each other will be: [5]

5 类似于电荷，磁荷可以定义为磁性的基本单位。如果在相反的磁极有 q_1 和 q_2 两个磁荷，它们之间的作用力是：

$$F = k\frac{q_1 q_2}{r^2} \tag{7-3}$$

where r is distance between q_1 and q_2, k is *a* constant.

Put a magnetic charge q in an external magnetic field H. It will subject to a magnetic force:

$$F = qH \tag{7-4}$$

The magnetic force obtainable using a permanent magnet exposed in a magnetic filed is given by

$$F = \frac{\mu_0 M^2 A}{2} \tag{7-5}$$

where A is the cross-sectional area of the magnet, M is the magnetization referring the strength of magnetic flux induced when this magnet is put in a magnetic field (discussed later), and μ_0 is a constant which is called permeability of vacuum.

7.2.2 Magnetic Dipole, Magnetic Moment and Magnetization—Microscopic Approach

Ⅰ. Magnetic Dipole

When a mass point, comprising of a positive and a negative charge, moves within limited space along the electric field, a dipole directing from the negative towards the positive charges formed.[6] The dipole moment is defined as: (Figure 7.4a)

6 当一个由正电荷和负电荷组成的质点，在电场作用下移动了有限的空间，从负电荷指向正电荷的偶极子就形成了。

$$\mu = qd \tag{7-6}$$

where q is electric charge; d is the separation between positive and negative charges.

Similarly, as shown in Figure 7.4b, magnetic dipole is referred to a pair of opposite magnetic poles that can move together in external magnetic field. The magnetic dipole moment is described as:

$$\overrightarrow{P}_{\mathrm{m}} = qd \tag{7-7}$$

where q is magnetic charge, d is the separation. Here $\overrightarrow{P_m}$ means the maximum magnetic dipole moment when the direction of dipole is the same as magnetic field H (Figure 7.4b). A normal case of a magnetic dipole in a magnetic field is shown in Figure 7.4c.

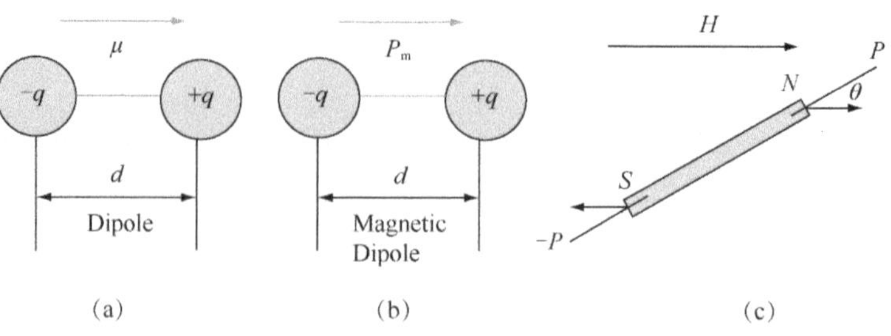

Figure 7.4 (a) An electrical dipole. (b) A magnetic dipole. (c) A magnetic dipole with angle of θ to the magnetic field H.

Ⅱ. Magnetic Moment

The magnetic moment of a magnet is a quantity that determines the torque it will experience in an external magnetic field. A loop of electric current, a bar magnet, an electron (revolving around a nucleus), a molecule, and a planet all have magnetic moments.[7]

7 磁体的磁矩是决定其在外磁场下受到扭矩的量。电流回路、条形磁体、电子（围绕原子核旋转）、分子以及行星都具有磁矩。

The magnetic moment may be considered to be a vector having a magnitude and direction. The direction of the magnetic moment points from the south to north pole of the magnet. The magnetic field produced by the magnet is proportional to its magnetic moment.

The magnetic behavior of materials can be traced to the structure of atoms. The orbital motion of the electron around the nucleus and the spin of the electron about its own axis (Figure 7.5) cause separate magnetic moments. These two motions (i.e., spin and orbiting) contribute to the magnetic behavior of materials.[8] Besides electron spin and the electron orbital momentum, the nucleus of the atom consists of protons and neutrons, which also have a spin. Due to the relatively small value of nucleus magnetic moment, it is not considered in atom magnetism.

8 材料的磁性行为可以追溯到原子结构。围绕原子核的电子轨道运动和电子的自旋（图 7.5）分别产生磁矩。这两种运动（即自旋和轨道运动）对材料的磁性行为有贡献。

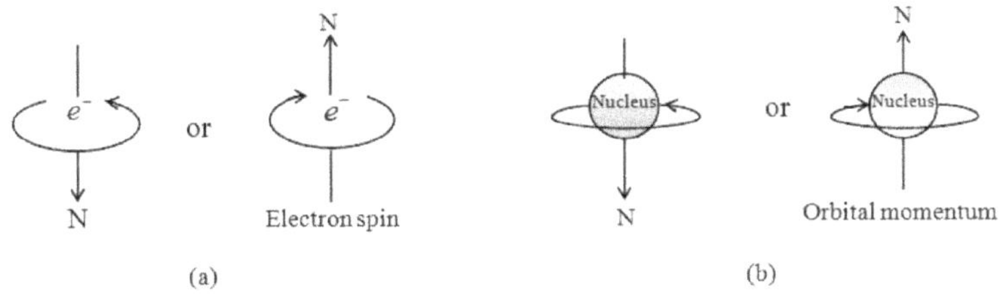

Figure 7.5 Origin of magnetic dipoles: (a) The spin of the electron produces a magnetic field with the direction dependent on the quantum number m_s. (b) Electrons orbiting around the nucleus create a magnetic field around the atom.

- **Magnetic Moment of Electron Orbital Motion**

If we imagine the atoms as systems of electrons orbiting round protons, they can certainly give rise to magnetism, which is referred as angular magnetism. As shown in Figure 7.6a, an electric current I going round in a plane will produce a magnetic moment,[9]

9 如果我们想象原子是一个围绕着质子做轨道运动的电子系统，它们肯定会引起磁性，这被称为角动量磁性。如图 7.6a 所示，平面上的环形电流 I 会产生磁矩。

$$\mu_L = IS \tag{7-8}$$

The direction of μ_L can be determined by right hand rule (note: the direction of current I is opposite to electron movement). S is the area of the current loop.

If the current is caused by a single electron rotating with an angular frequency ω (Figure 7.6b), the current is:

$$I = \frac{dq}{dt} = \frac{e}{\frac{2\pi}{\omega}} = e\frac{\omega}{2\pi} \tag{7-9}$$

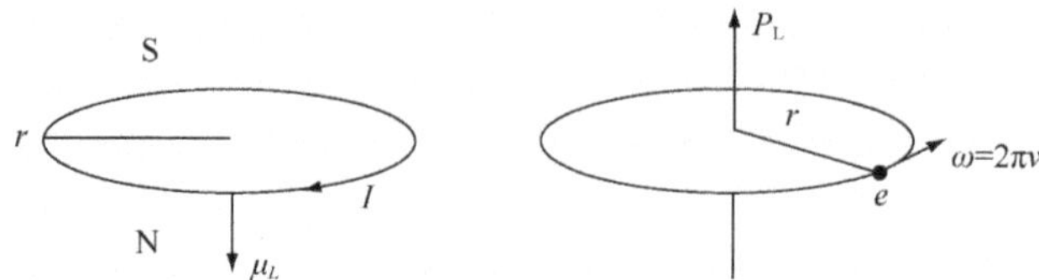

Figure 7.6 (a) An electric current going round in a plane and its corresponding magnetic moment μ_L. (b) The angular moment of the rotating electron (P_L).

Then the magnetic moment of electron orbital motion becomes:

$$\mu_L = IS = e\frac{\omega}{2\pi}\pi r^2 = \frac{e}{2m}m\omega r^2 \tag{7-10}$$

Here, r is the radius of the current circle.

Introducing the angular moment (Figure 7.6b),

$$\begin{aligned}\vec{P}_L &= \vec{r}\times m\vec{v}\\ P_L &= mvr\end{aligned} \tag{7-11}$$

The magnetic moment becomes:

$$\mu_L = \frac{e}{2m}m\omega r^2 = \frac{e}{2m}P_L \tag{7-12}$$

Since the angular momentum of orbital is quantized, μ_L should also be a quantized value:

$$\begin{aligned}P_L &= \frac{h}{2\pi}\sqrt{L(L+1)}\\ \mu_L &= \mu_B\sqrt{L(L+1)}\end{aligned} \tag{7-13}$$

where L is the angular quantum number of an atom; μ_B is the constant which representing the minimum magnetic moment, called Bohr magneton.

$$\mu_B = \frac{qh}{4\pi m_e} = 9.274\times10^{-24}\,\mathrm{A}\cdot m^2 \tag{7-14}$$

where q is electron charge, h is Planck's constant, and m_e is the mass of the electron. The Bohr magneton is directed along the axis of electron spin.

The relationship between the total orbital angular momentum and its magnetic moment of an atom is illustrated in Figure 7.7.

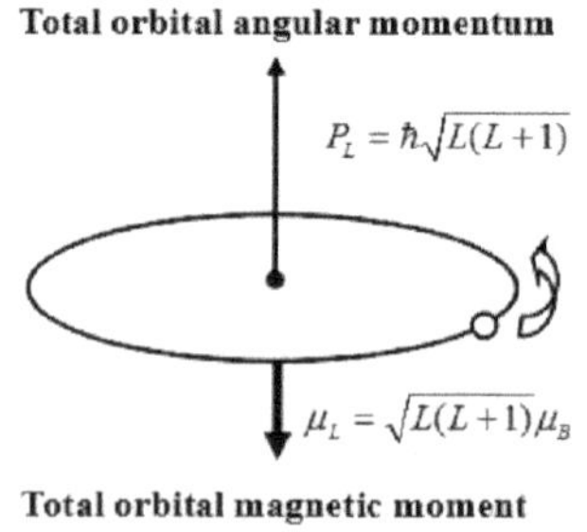

Total orbital angular momentum

$P_L = \hbar\sqrt{L(L+1)}$

$\mu_L = \sqrt{L(L+1)}\mu_B$

Total orbital magnetic moment

Figure 7.7 Total orbital angular momentum P_L of an atom and its corresponding magnetic momentμ_L. L is total orbital quantum number.

To angular magnetic moment of an electron, if a magnetic field is applied, what happens?

Consider a magnetic dipole that happens to be at an angle θ to the direction of the magnetic field (Figure 7.8).The magnetic field produces a torque:

$$\vec{\mu}_{m} \times \vec{B} \tag{7-15}$$

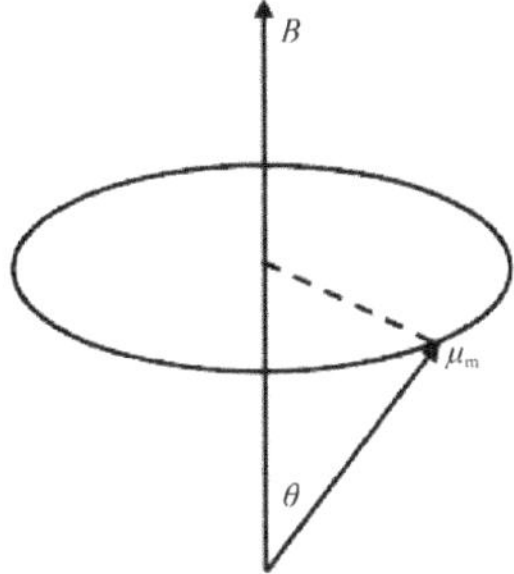

Figure 7.8 A magnetic dipole (μ_m) under magnetic field (B).

Since the torque is perpendicular both to $\vec{\mu}_m$ and $\vec{B}$, the change in the angular momentum will also be perpendicular, causing the magnetic dipole to precess around the magnetic field. From kinetics we can easily show that the frequency of precession is.[10]

$$\omega_L = \frac{eB}{2m} \tag{7-16}$$

10 由于扭矩与 μ_m 和 B 都是垂直的，角动量的变化也将是垂直的，引起磁偶极子沿磁场进动。从动力学角度，我们可以很容易地表示该进动的频率：

Looking back to Equation (7-12), if the magnetic moment is induced by the external magnetic field, ω_L can be used to replace ω in the equation. Therefore, the induced magnetic moment of orbiting electron is:

$$\mu_{ind} = \frac{e}{2m} m\omega r^2 = \frac{er^2}{2}\frac{eB}{2m} = \frac{e^2 r^2 B}{4m} \tag{7-17}$$

- **Magnetic Moment of Electron Spin**

When the electron spins, there is a magnetic moment associated with that motion. Measured by experiment, the spin magnetic moment of electron is approximately equal to Bohr magneton[11]:

$$\mu_s = \pm \mu_B$$

11 当电子自旋时，就存在一个与该运动相关联的磁矩。实验测得，电子的自旋磁距大约等于玻尔磁子。

Its direction depends on the direction of spin, the spin direction which is the same as the external magnetic field is generally taken as positive direction.

Actually, the above value only represents the z components of the electron-spin magnetic moment. For an electron of an atom, the spin vector $s = \pm \frac{1}{2}$, and the total spin magnetic moment can be expressed as:

$$|\mu_s| = 2\mu_B |s| = 2\mu_B \sqrt{s(s+1)} = \sqrt{3}\mu_B \tag{7-18}$$

Since there are many electrons in one atom, if the total spin of atom is S, the total magnetic moment caused by electron spin is:

$$|\mu_s| = 2\mu_B\sqrt{S(S+1)} \tag{7-19}$$

- **Magnetic Moment of Nucleus**

Since the protons in the nucleus possess positive charge, their spin can also produce magnetic moment. The mass of a proton is about one thousand times to that of an electron, hence its magnetic moment is 3 orders magnitude smaller than electron. It is negligible in magnetic moment consideration for atoms.[12] However, the magnetic energy level of nucleus can be utilized to analyze material structure (e.g., bonding, magnetic moment structure, etc.). The most frequently used instrument utilizing nucleus magnetism is the nuclear magnetic resonance, NMR.

12 由于原子核中的质子具有正电荷，它们的自旋也可以产生磁矩。质子的质量约为电子的一千倍，因此它的磁矩比电子的小三个数量级。在考虑原子磁矩时可以忽略。

- **Magnetic Atoms**

As previously stated, magnetic moment of atom includes orbital and spin magnetic moments of electron, as well as spin magnetic moment of nucleus. Generally speaking, the magnetism of a material depends on the characteristics of the atoms inside.

In atoms with even number of electrons, the electrons are paired with opposite spin directions. These paired electrons have opposite spin magnetic moments as well as opposite orbital magnetic moments, resulting in canceling effect. Hence, the net magnetic moment of an atom is zero and it is generally called non-magnetic atom, the material is non-magnetic material.[13]

13 含有偶数个电子的原子，电子是自旋反向配对的。这些成对电子有相反的自旋磁矩，以及相反的轨道磁矩，产生抵消效应。因此，原子的净磁矩为零，它通常被称为非磁性原子，这种材料是非磁性材料。

In atoms, if the number of electrons is odd, or even if the number of the electrons is even, due to some special reasons, the magnetic moment from electrons can't be completely canceled out, the total magnetic moment of the atom will not be zero. This kind of atoms are called magnetic atoms. Some elements, such as transition elements (3d, 4d, 5d partially filled), the lanthanides (4f partially filled), and actinides (5f partially filled), have a net magnetic moment due to an unpaired electron. For example, the electron distribution of an isolated iron atom is: $1s^2 2s^2 2p^6 3s^2 3p^6 3d^6 4s^2$. For 3d orbital, the electron occupation is shown in Figure 7.9.

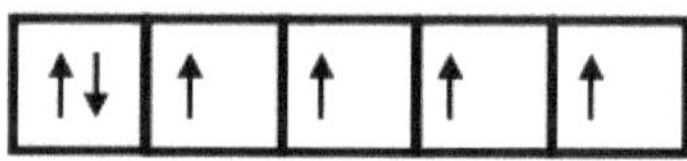

Figure 7.9 The electron occupation of the 3d orbital in iron

Hence, the total spin magnetic moment is $4\mu_B$. This is the main contribution to atom magnetic moment. Fe atom is magnetic.

In many crystal, the magnetic moments exist in free individual atoms, however, when the atoms form crystalline materials, the energy levels may split, Hund's rule may break, the spin magnetic moment may disappear, resulting in zero magnetic moment.[14] For example, the Fe^{+2} ion has a net magnetic moment (four times the magnetic moment of an electron), however, in $FeCl_2$ crystal with symmetric cubic crystalline, the d electrons will distribute according to equilibrium of the columbic force, with all the six electrons paired, and the spin magnetic moment disappears (Figure 7.10).

14 在许多晶体中，虽然磁矩存在于孤立的原子中，然而，当原子形成晶体材料时，能级可能会发生分裂，Hund 规则可能不适用，自旋磁矩可能会消失，导致零磁矩。

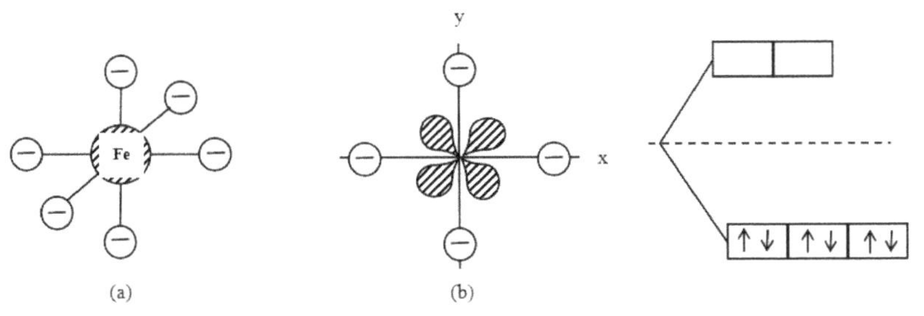

Figure 7.10 The electron occupation of d orbital in $FeCl_2$ crystal with symmetric cubic crystalline.

Certain elements, such as the transition metals, have an inner energy level that is not completely filled, the elements of scandium (Sc) through copper (Cu) are typical. Their electronic structures are shown in Table 7.1. Except for chromium and copper, the valence electrons in the 4s level are paired. The unpaired electrons in chromium and copper are canceled by interactions with other atoms. Copper also has a completely filled 3d shell and thus does not display a net magnetic moment.[15] The electrons in the 3d level of the other transition elements do not enter the shells in pairs. Instead, as in manganese, the first five electrons have the same spin. Only after half of the 3d level is filled do pairs with opposing spins form. Therefore, each atom in a transition metal has a permanent magnetic moment, which is related to the number of unpaired electrons. Each atom behaves as a magnetic dipole.[16]

15 除铬和铜，其他元素的 4s 价电子是成对的。铬和铜中未成对的价电子由于与其他原子的（价电子）相互作用，而被抵消。铜原子的 3d 层也是完全填充的，因而不表现出净磁矩。

16 只有在 3d 轨道半满后，自旋相反电子的配对才会形成。因此，过渡金属的每一个原子都具有一个永久磁矩，这个磁矩大小与未成对电子数量相关。每个原子都表现为磁偶极子。

Table 7.1 The electron spins in the 3d energy level in transition metals with arrows indicating the direction of spin

Metal	3*d*					4*s*
Sc	↑					↓↑
Ti	↑	↑				↓↑
V	↑	↑	↑			↓↑
Cr	↑	↑	↑	↑	↑	↑
Mn	↑	↑	↑	↑	↑	↓↑
Fe	↓↑	↑	↑	↑	↑	↓↑
Co	↓↑	↓↑	↑	↑	↑	↓↑
Ni	↓↑	↓↑	↓↑	↑	↑	↓↑
Cu	↓↑	↓↑	↓↑	↓↑	↓↑	↑

Ⅲ. Magnetization

The response of the atom to an applied magnetic field depends on how the magnetic dipoles of each atom react to the field. The degree of magnetization can be described by magnetization intensity (M), meaning the number of magnetic moment per volume:

$$\vec{M} = \frac{\sum \vec{m}_{\mathrm{m}}}{V} \tag{7-20}$$

where V is volume, $\sum \vec{m}_{\mathrm{m}}$ is the sum of the vectors of the magnetic moments.

Most of the transition elements (e.g. Cu, Ti) react in such a way that the sum of the individual atoms' magnetic moments is zero. The atoms in nickel (Ni), iron (Fe), and cobalt (Co), however, undergo an exchange interaction, whereby the orientation of the dipole in one atom influences the surrounding atoms to have the same dipole orientation, producing a desirable amplification of the effect of the magnetic field. In the case of Fe, Ni, and Co, the magnetic moments of the atoms line up in the same directions, and these materials are known as ferromagnetic.[17] In certain materials, such as BCC chromium (Cr), the magnetic moments of atoms at the center of the unit cell are opposite in direction to those of the atoms at the corners of the unit cell. Thus, the net magnetic moment is zero. In these materials, the so called super exchange effect exists. These exchange and super exchange effects will be discussed later.

17 然而，在镍（Ni）、铁（Fe）和钴（Co）材料中的原子将进行交换作用，即一个原子的偶极子取向影响周围原子的偶极子取向，使得它们取向相同，由此产生希望的磁场放大效应。对于铁、镍和钴，其原子磁矩在同一方向上排列，这些材料被认为是铁磁性的。

7.2.3 Magnetic Inductance, Permeability and Magnetic Susceptibility

Ⅰ. Magnetic Inductance and Permeability

When a magnetic field is applied in a vacuum, lines of magnetic flux are induced. The number of lines of flux, called the flux density, or inductance B, is related to the applied field by

$$B = \mu_0 H \tag{7-21}$$

where B is the inductance, H is the magnetic field, and μ_0 is a constant called the magnetic permeability of vacuum. If H is expressed in units of Oersted, then B is in gauss and μ_0 is 1 gauss/Oersted. In an alternate set of units, H is in A/m, B is in Tesla (also called Weber/m^2), and μ_0 is $4\pi\times10^{-7}$ Weber (Am) (also called H/m).

When we place a material within the magnetic field, the magnetic-flux density is determined by the manner in which the induced and permanent magnetic dipoles interact with the field.[18] The flux density now is

$$B = \mu H \tag{7-22}$$

18 当我们在磁场中放置一个材料时，磁通密度取决于诱导以及永久磁偶极子与磁场相互作用的方式。

where μ is the permeability of the material in the field. If the magnetic moments reinforce the applied field, then $\mu > \mu_0$, a greater number of lines of flux that can accomplish work are created, and the magnetic field is magnified. If the magnetic moments oppose the field, $\mu < \mu_0$.

We can describe the influence of the magnetic material by the relative permeability μ_r, where

$$\mu_r = \frac{\mu}{\mu_0} \tag{7-23}$$

A large relative permeability means that the material amplifies the effect of the magnetic field. Thus, the relative permeability has the same importance that conductivity has in dielectrics.[19] A material with higher magnetic permeability (e.g., iron) will carry magnetic flux more readily. We will learn later that the permeability of ferromagnetic or ferrimagnetic materials is not constant and depends on the value of the applied magnetic field (H).[20]

19 较大相对磁导率意味着材料对磁场的放大效应。因此，相对磁导率与介电材料中的电导率相同重要。

20 在后续的学习中我们将知道，铁磁性或亚铁磁性材料的磁导率不是常数，而是依赖于所施加磁场的大小（H）。

The magnetization M, which is defined in previous section (i.e., the sum of magnetic moment in unit volume), represents the increase in the inductance due to the core material, so we can rewrite the equation for inductance as

$$B = \mu H = \mu_0 H + \mu_0 M = \mu_0 \mu_r H \tag{7-24}$$

This means that the magnetic induction strength inside a material can be divided into two parts: (a) The vacuum magnetic induction strength ($\mu_0\vec{H}$) and (b) the resulted magnetic field from magnetization ($\mu_0\vec{M}$) (Figure 7.11).[21]

21 这意味着材料内部的磁感应强度可以被分为两个部分：（a）真空磁感应强度($\mu_0\vec{H}$)和（b）磁化导致的磁感应强度($\mu_0\vec{M}$)（图 7.11）

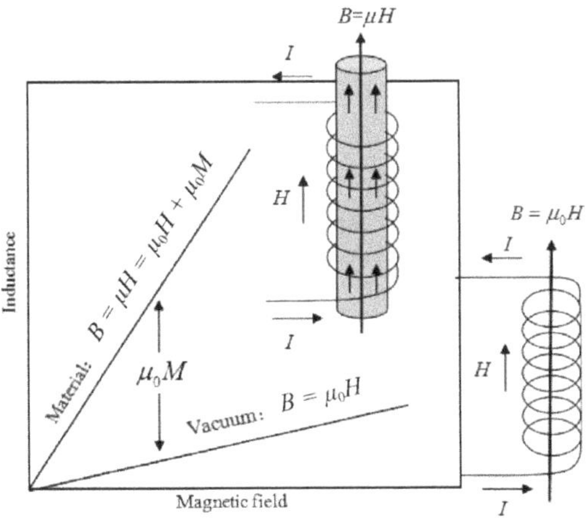

Figure 7.11 A current passing through a coil sets up a magnetic field H with a flux density B. The flux density is higher when a magnetic core is placed within the coil.

Ⅱ. Magnetic Susceptibility

The magnetic susceptibility χ_m, which is the ratio between magnetization and the applied field, gives the amplification produced by the material. We have:

$$B = \mu H = \mu_0 H + \mu_0 M = \mu_0 H + \mu_0 \chi_m H = \mu_0 (1 + \chi_m) H \tag{7-25}$$

Therefore,

$$\chi_m = \frac{\mu}{\mu_0} - 1 = \mu_r - 1 \tag{7-26}$$

Both μ_r and χ_m refer to the degree to which the material enhances the magnetic field. Normally, we are interested in producing a high inductance B or magnetization M. This is accomplished by selecting materials that have a high relative permeability or magnetic susceptibility.

7.2.4 Magnetism Classification

Magnetism of material is originated from atoms. The net magnetic moment of an atom depends on the orientation of its electron orbits and spin. If the total angular momentum L and the total spin momentum S are zero, for example for symmetry reason in filled shells or subshells, the atomic magnetic moment will be zero.[22] Even if the atom possesses a net magnetic moment, the magnetization of the solid is zero for most materials owing to random orientation of the atoms in the absence of an external magnetic field.

Under external magnetic field, according to the value of magnetic susceptibility, magnetic materials can be classified into five groups:

Ⅰ. Diamagnetism

The electrons in the shells and subshells of the atoms appear in pairs. Consider a pair of electrons with equal angular momentum l, the two electrons "rotate" in opposite directions, i.e., their angular magnetic quantum numbers are $\pm m_l$, and their electron spins have opposite directions (spin up and spin down), which can be expressed as $m_s = \pm 1/2$. Such electron pair has zero magnetic moment.[23] All filled shells and subshells consist of such electron pairs have no resulting magnetic moment. A magnetic field acting on this kind of atoms induces a magnetic dipole for the entire atom by influencing the magnetic moment caused by the orbiting electrons. These dipoles oppose the magnetic field, causing the magnetization to be less than zero. This behavior, called diamagnetism, gives a relative permeability of about 0.99995 (or a negative susceptibility approximately -10^{-6}, note the negative sign).

The diamagnetic effect appears in all materials but is generally completely screened by other magnetic effects unless the resulting magnetic moment of the atoms is zero. Hence diamagnetic solids have filled electron shells or subshells.[24]

22 材料的磁性源于原子。原子的净磁矩取决于它的电子轨道和自旋的取向。例如，在满壳层或者亚满壳层中，由于对称的缘故，如果总角动量和总自旋动量都为零，原子的磁矩将为零。

23 考虑到一对具有相同角动量 l 的电子，两个电子的旋转方向相反，即角动量磁量子数为$\pm m_l$，并且电子自旋方向相反（上旋和下旋），可表示为$m_s = \pm 1/2$。这样电子对的磁矩为零。

24 抗磁效应存在于所有材料中，但一般是被其它磁效应屏蔽的，除非原子磁矩为零。因此，抗磁性固体具有填满的外壳层或亚外壳层。

The negative direction of the magnetization can be understood as follows. Consider a single closed coil which suddenly is exposed to a perpendicular magnetic field. An electric field is induced which cause an induction current in the coil in such a direction that it gives a magnetic field opposite to the external magnetic field (Lentz' law). If the electric resistance in the coil is zero, the current continues to flow. The coil corresponds to a magnet with its magnetic moment in the opposite direction to the magnetizing field. Alternatively, we can say that the time-dependent magnetic field induces an electric field, which accelerates or retards the electrons in their orbits slightly. The result is that their magnetic moments no longer cancel but have a resultant in the direction opposite to the magnetizing field, i.e., the magnetization (*M*) direction is opposite to the direction of applied field (*H*).[25]

The diamagnetic susceptibility for atom with filled electrons can be expressed as:

$$\chi_m^d = -\frac{ne^2\mu_0}{6m}\sum_{i=1}^{z}\overline{r_i^2} \tag{7-27}$$

where μ_0 is the vacuum permeability, n is the number of atoms per unit volume, e and m are charge and mass of the electron respectively, Z is the number of electrons of the atom, and r_i^2 is the square radius of the i-th electron of the atom. Owing to its origin, it can be named orbital magnetic susceptibility. It should be noted that χ_m^d is independent of temperature for diamagnetic substances.

If there are conducting electrons in the solid, such as metals, the diamagnetic susceptibility by the conducting electrons is:

$$\chi_m^{de} = -\frac{1}{2}\times\frac{n\mu_B^2}{kT_F}\left(\frac{m}{m^*}\right)^2 \tag{7-28}$$

where n is the number of free electrons per unit volume, μ_B is Bohr magneton, k is Boltzmann constant, $E_F = kT_F$ (E_F is the Fermi level), m and m^* are the mass of the free electron and the effective mass of conducting electron respectively.

About half of the metals are diamagnets at room temperature, such as copper, silver, cadmium, gold, mercury and zinc, bearing an electron structure with no unpaired electrons. Superconductors are perfect diamagnets; they lose their superconductivity at higher temperatures or in the presence of a high magnetic field.[26]

Ⅱ. Paramagnetism

When materials with a net magnetic moment in each atom are put in a

25 结果是它们的磁矩不再取消，而是有一个与磁化场强方向相反的磁矩，即磁化（*M*）方向与施加场（*H*）方向相反。

26 室温下大约一半的金属是抗磁体，如铜、银、镉、金、汞、锌，它们的电子结构没有未成对电子。超导体是完美抗磁体；在较高温度下或在强磁场下它们会失去超导性。

magnetic field, the magnetic dipoles align with the field, causing a positive magnetization. Because the dipoles do not interact, extremely large magnetic fields are required to align all of the dipoles. In addition, the effect is lost as soon as the magnetic field is removed. This effect is called paramagnetism.[27] Only those materials in which the paramagnetic effects dominate over the diamagnetic effects are called paramagnetic and show paramagnetic properties.

27 因为磁偶极子之间没有相互作用，需要非常大的磁场才能够使所有的磁偶极排列成一个方向。此外，当磁场被移除时，（磁偶极的排列）效果就消失了。这种效应被称为顺磁性。

The paramagnetism in solids consists of three effects. One is the preferred orientation of permanent magnetic moment in atoms, the second emanates from the contribution from conducting electrons which are common for the crystal lattice in metals, and the third is Van Vleck paramagnetism.

- **Paramagnetism of the Permanent Resultant Atomic Orbital Magnetic Moment**

Atoms with a resulting magnetic moment behave like small magnets. Under the influence of the external magnetic field, they orient themselves in directions closer to the magnetic field than their original positions. The reorientation is counteracted by the scattering of the atoms and their magnetic moments, owing to interaction with the electron gas and electron–phonon collisions, until equilibrium is achieved. The stronger the external magnetic field is, the stronger will be the orientation of the small magnets in the direction of the external field. Hence the weak total magnetic moment of a piece of paramagnetic material is proportional to the external magnetizing field.[28]

28 外磁场越强，取向于磁场方向上的小磁铁就越强。因此，顺磁性材料较弱的总磁矩与外部磁场成正比。

The alignment of the permanent magnetic moments of the atoms is counteracted by the kinetic motion of the atoms. Hence the paramagnetic susceptibility depends on temperature. The higher the temperature is, the smaller will be χ_m. This rule was discovered by P. Curie in 1895 and called Curie Law:

$$\chi_m = \frac{C}{T} \tag{7-29}$$

C is Curie constant, and has the following expression:

$$C = \frac{n\mu_0\mu_m^2}{3k} \tag{7-30}$$

where n is the number of atoms per unit volume, μ_m is the permanent magnetic moment of atom. By determining the $\chi_m \sim T$ data experimentally (Figure 7.12), C can be obtained as slope. Then with Equation (7-30), the permanent magnetic moment of atom can be calculated.

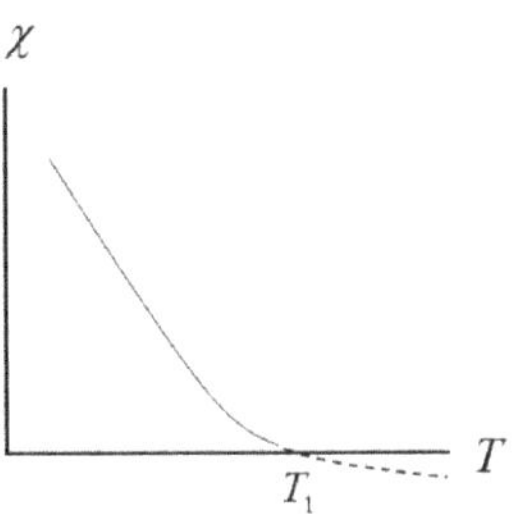

Figure 7.12 Susceptibility of paramagnetic solids versus temperature. A negative of χ means that the diamagnetic effect, which is always present, dominates over the paramagnetic effect.

The magnetic susceptibility (χ_m) of paramagnetic materials is positive and lies between 10^{-4} and 10^{-5}. Ferromagnetic and ferrimagnetic materials above the Curie temperature also exhibit paramagnetic behavior.[29]

29 顺磁性材料的磁化率(χ_m)是正值，且介于 10^{-4} 和 10^{-5} 之间。铁磁性和亚铁磁性材料在居里温度之上也表现出顺磁性。

- **Paramagnetism of Free Electrons in Metals**

In a magnetic field, the degeneracy between free electrons in the conducting band is broken and the electrons have slightly different energy levels. The electrons with positive spin relative to the direction of the magnetic field B lower their energy in the magnetizing field by the amount $\mu_B \cdot B$ and electrons with negative spin relative to B increase their energy by the amount $\mu_B \cdot B$ in the magnetic field.[30] A necessary condition for equilibrium is that the maximum energy level is the same for both spin-down and spin-up electrons. This is achieved by transfer of some spin-down electrons to the spin-up side. The shaded areas in Figure 7.13 represent the number of spin-down and spin-up electrons per unit volume. Obviously there are more spin up than spin down electrons. The excess spin gives rise to a resulting magnetic moment. This is the explanation of paramagnetism of the conduction electrons.[31]

30 在磁场下，导带中自由电子的简并被打破，电子有略微不同的能级。与磁场 B 方向相同的正自旋电子在磁场磁化下降低了$\mu_B \cdot B$ 能量；与磁场 B 相反的负自旋电子增加了$\mu_B \cdot B$ 能量。

31 图 7.13 的阴影面积，表示单位体积内上自旋和下自旋电子的数量。显然，上自旋电子比下自旋电子多。多出来的自旋产生了磁矩。这是传导电子顺磁性的解释。

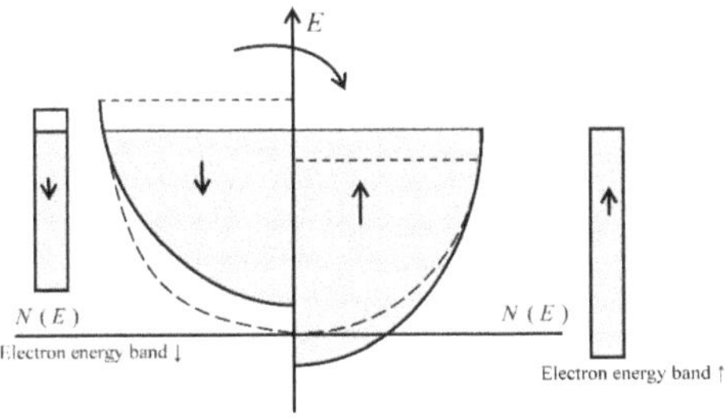

Figure 7.13 Density of Fermi electron energy states of spin-up and spin-down electrons of paramagnetic atoms in the magnetic field. The tops of the energy bands have the same energy. The energy difference between the bottoms of the electron bands is $2\mu_B B$.

The magnetic moment of a free electron is μ_B. Hence the net magnetic moment of the free electrons per unit volume will be:

$$M = \mu_B(n_\uparrow - n_\downarrow) \tag{7-31}$$

According to free electron approximate, the paramagnetic susceptibility for conducting electrons at 0 K is:

$$\chi_m^{pe} = -\frac{3}{2} \times \frac{n{\mu_B}^2}{kT_F} \tag{7-32}$$

Comparing with Equation (7-28), it can be seen that in terms of free electron the paramagnetic susceptibility is three times of the diamagnetic susceptibility. Thus, free electrons generally exhibit paramagnetism.[32]

32 与方程(7-28)相比，可以看出自由电子的顺磁磁化率是抗磁磁化率的三倍。因此，自由电子一般表现出顺磁性。

Considering both the diamagnetic and paramagnetic susceptibilities of conducting electrons, the total susceptibility can be approximately estimated as:

$$\chi_m^e = \chi_m^{de} + \chi_m^{pe} = \frac{3n{\mu_B}^2}{2kT_F}\left[1 - \frac{1}{3}\left(\frac{m}{m^*}\right)^2\right] \tag{7-33}$$

When $\frac{1}{3}(\frac{m}{m^*})^2$ is smaller than 1, the conducting electrons are paramagnetic, otherwise, they will be diamagnetic.

- **Paramagnetism of Van Vleck Effect**

Due to the deformation of electron cloud caused by external magnetic field, excited state can be mixed with ground state, resulting in the disturbing of electron state, as well as the electron spin state, thus producing paramagnetism. This effect can apply to either orbital electrons or the conducting electrons in energy band.[33]

33 因为外磁场的存在导致电子云的变形，使得激发态与基态混合，导致电子态以及电子自旋状态受到干扰，从而产生顺磁性。这种效应可以适用于轨道电子或者能带中的导电电子。

Ⅲ. Ferromagnetism, Antiferromagnetism and Ferrimagnetism

A characteristic and common feature of ferromagnetic, antiferromagnetic and ferrimagnetic materials is that there exists strong coupling between the magnetic moments of neighboring atoms, which is strong enough to persist in spite of thermal excitation.[34] In ferromagnetic materials, the atomic magnetic moments of equal size are oriented roughly in the same direction (Figure 7.14a). In antiferromagnetic materials, the atomic magnetic moments are coupled roughly in opposite directions (Figure 7.14b) and cancel, which makes the material un-magnetic. Ferrimagnetic materials are a type of antiferromagnetic material where the two types of magnetic moments have different sizes and can't cancel completely (Figure 7.14c). Consequently, such materials are magnetic.

34 铁磁、反铁磁和亚铁磁性材料的共同特征是相邻原子的磁矩之间存在强耦合，这种互作用非常强，足以抵抗热激发的影响。

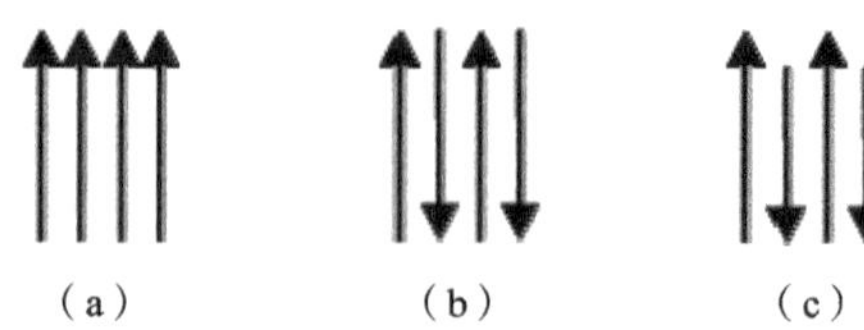

Figure 7.14 (a) Ferromagnetic materials. (b) Antiferromagnetic materials. (c) Ferrimagnetic materials.

- **Ferromagnetism**

Ferromagnetic behavior is caused by the unpaired electrons in the 3d level of iron, nickel, and cobalt. Similar behavior is found in a few other materials, including gadolinium (Gd), yttrium (Y), dysprosium (Dy).

In ferromagnetic materials, the permanent unpaired dipoles easily line up with the imposed magnetic field due to the exchange interaction, or mutual reinforcement of the dipoles. Large magnetizations are obtained even for small magnetic fields, giving large susceptibilities approaching 10^6. Similar to ferroelectrics, the susceptibility of ferromagnetic materials depends upon the intensity of the applied magnetic field. This is similar to the mechanical behavior of elastomers with the modulus of elasticity depending upon the level of strain. Above the Curie temperature, ferromagnetic materials behave as paramagnetic materials.

- **Antiferromagnetism**

In materials, such as MnO, the magnetic moments produced in neighboring dipoles line up in ferromagnetism opposition to one another in the magnetic field, even though the strength of each dipole is very high. These materials are antiferromagnetic and have almost zero magnetization.[35] The magnetic susceptibility is positive and very small. Besides MnO, other metal oxides, such as NiO, Cr_2O_3, $ZnFeO_4$, and CoO, are also antiferromagnetic materials.

- **Ferrimagnetism**

In ceramic materials, different ions have different magnetic moments. In a magnetic field, the dipoles of cation A may line up with the field, while dipoles of cation B oppose the field. Because the strength or number of dipoles is not equal, a net magnetization results.[36] The ferrimagnetic materials can provide good amplification of the imposed field. These materials show a large, magnetic-field dependent magnetic susceptibility similar to ferromagnetic materials. They also show Curie-Weiss behavior (similar to ferromagnetic materials) at temperatures above the Curie temperature. Most ferrimagnetic materials are ceramics and are good insulators of electricity. Thus, in these materials, electrical losses (known as eddy current losses) are much smaller compared to those in metallic ferromagnetic materials. Therefore, ferrites are used in many high-frequency applications. Magnetite and ferrite belong to this group.

Finally, we want to point out that when the grain size of ferromagnetic materials falls below a certain critical size, these materials behave as if they are paramagnetic. The magnetic dipole energy of each particle becomes comparable to the thermal energy. This small magnetic moment changes its direction randomly (as a result of the thermal energy). Thus, the material

35 在如氧化锰一样的材料中，虽然每个偶极子的磁矩都很大，但是相邻偶极子产生的磁矩在磁场中以互为反向的铁磁性排列。这些材料是反铁磁性的，几乎零磁化。

36 在陶瓷材料中，不同离子具有不同磁矩。在磁场中，正离子 A 的磁偶极可能沿着磁场方向排列，而正离子 B 的磁偶极排列与磁场反向。由于不同磁偶极的数量和强度都不相同，导致净磁化不为零。

behaves as if it has no net magnetic moment. This is known as superparamagnetism. Thus, if we produce iron oxide (Fe_3O_4) particles with size within 3 to 5 nm, they behave as superparamagnetic materials. Such iron-oxide superparamagnetic particles are used to form dispersions in aqueous or organic carrier phases or to form “liquid magnets” or ferrofluids.[37] The particles in the fluid move in response to a gradient in the magnetic field. Since the particles form a stable sol, the entire dispersion moves and, hence, the material behaves as a liquid magnet.[38] Such materials are used as seals in computer hard drives and in loudspeakers as heat transfer (cooling) media. The permanent magnet used in the loudspeaker holds the liquid magnets in place. Superparamagnetic particles of iron oxide (Fe_3O_4) also can be coated with different chemicals and used to separate DNA molecules, proteins, and cells from other molecules.

37 因此，如果我们制备的氧化铁(四氧化三铁)颗粒，尺寸在3～5 nm之间，它们将具有超顺磁材料行为。这些氧化铁超顺磁粒子可应用在液体或有机载体中形成分散相，或者用来形成磁流体或铁磁流体。

38 流体中这些颗粒在磁场梯度下产生运动。由于这些粒子构成了稳定的溶胶，整个分散相项一起移动，材料表现出磁流体行为。

- **Coupling between the Magnetic Moments of Adjacent Atoms**

As previously discussed, in ferro-, antiferro-, and ferri- magnetic materials, there exists strong magnetic coupling between adjacent atoms, which is generally believed to be caused by the coupling of the spin vectors of neighboring atoms. A schematic example was given in Figure 7.15. As shown in the figure, the hybridization of the 4s and 3d electrons in ferromagnetic atoms (Figure 7.15 a and b) results in an excess of spin-up electrons (Figure 7.15c), and gives a resulting magnetic moment.

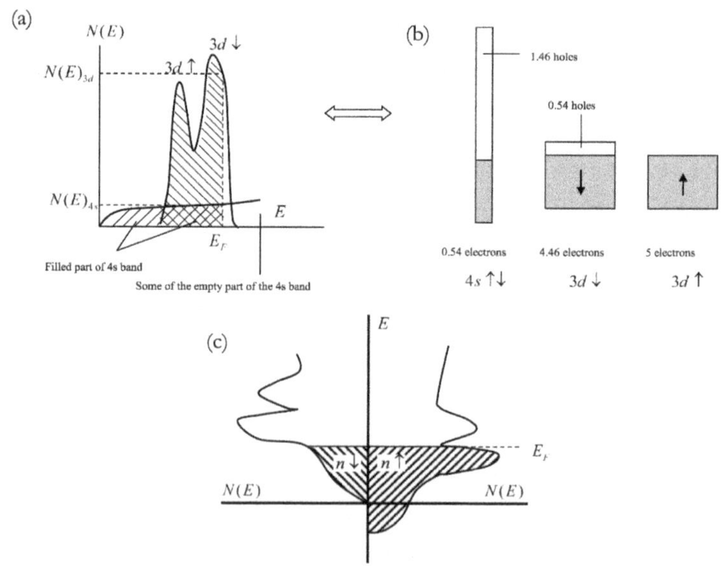

Figure 7.15 (a) Density of the electron energy states in the 4s and 3d bands as a function of the energy E in Ni at absolute zero temperature. (b) Schematic electron distribution in the 4s and 3d sub-bands in Ni at absolute zero temperature. The shaded areas of the 4s and 3d bands between $E = 0$ and $E = E_F$ represent the number of electrons per unit volume in the 4s and 3d bands (0.54 and 9.46 electrons, respectively). The figure is not drawn to scale. (c) Density of electron energy states in a ferromagnetic metal as a function of energy.

The coupling of electron spins is described by the so-called exchange integral, which is a function of the symmetric and asymmetric combinations of the wave of the two electrons.[39] The unpaired electrons in an atom form a resultant spin vector $\vec{S}$. The study of the exchange integral shows that the energy state with lowest possible energy is that with highest possible value of $\vec{S}$. This statement (Hund's rule) explains why atoms with incompletely filled d shells, for example the ferromagnetic transition metals, have high resulting spins and consequently large magnetic moments.

39 电子自旋的耦合，可以通过所谓的交换积分描述，它是 2 个电子波之间的对称性和非对称性的组合函数。

When two atoms are approaching to each other, or N atoms form crystal, by exchanging electrons in two adjacent atoms $\vec{\mathrm{i}}$ and $\vec{\mathrm{j}}$, due to the spin exchange integral, the energy of the system may reduce. The exchange integral represents the difference between the two energy states which correspond to parallel and antiparallel spins.[40]

40 当两个原子互相靠近时，或者 N 原子形成晶体时，通过交换相邻原子 i 和 j 的电子，由于自旋交换积分，系统能量可能降低。交换积分表示平行和反平行自旋能量态间的差值。

The exchange energy for two adjacent atoms with spin vectors of $\vec{S}_1$ and $\vec{S}_2$ is

$$E_{\mathrm{ex}} = 2JS_1 \cdot S_2 \tag{7-34}$$

where J is the exchange integral, which is determined by the initial and the final states, and represents the degree of electron exchange.

The quantum mechanical results:

When $J > 0$, for lowering system energy (i.e., $E_{\mathrm{ex}} < 0$), their spins should be in the same direction ($\vec{S}_1 \cdot \vec{S}_2 > 0$), hence the material is ferromagnetic. With $J < 0$, when spins are in opposite direction ($\vec{S}_1 \cdot \vec{S}_2 < 0$), $E_{\mathrm{ex}} < 0$, system energy is low, and the material is antiferromagnetic.

Therefore, ferromagnetism requires not only the permanent magnetic moment in atoms, but also the large exchange integral among the adjacent atomic magnetic moments. The exchange integral in a ferromagnet is positive ($J > 0$), the spin direction among adjacent atoms are the same, and the atom magnetic moments are lined up, leading to magnetic domain and very large magnetic susceptibility. Above Curie temperature (T_C), the exchange effect is destroyed by thermal energy, the material changes to paramagnetism. The susceptibility of a ferromagnetic material above T_C is given by the following equation, known as Curie-Weiss law (Figure 7.16):

$$\chi_{\mathrm{m}} = \frac{C}{T - T_{\mathrm{C}}} \tag{7-35}$$

Here, C is a constant that depends upon the material, T_C is the Curie temperature, and T is the temperature above T_C. Essentially, the same equation also describes the change in dielectric permittivity above the Curie temperature of ferroelectrics. Similar to ferroelectrics, ferromagnetic materials show the formation of hysteresis loop domains and magnetic domains.[41]

41 本质上，同样的公式也可以描述铁电材料高于居里温度时的介电常数变化。类似于铁电材料，铁磁材料表现出迟滞回线和磁畴。

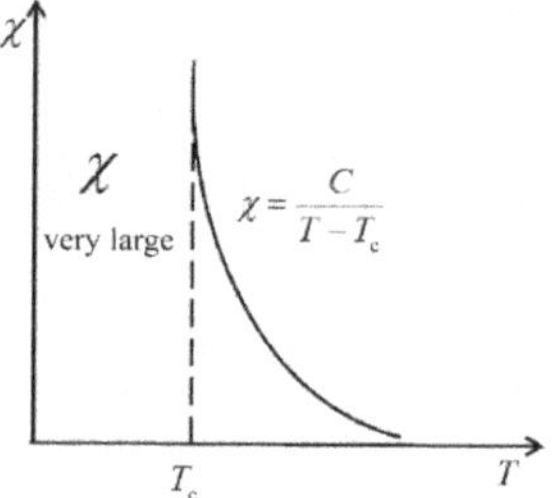

Figure 7.16 Susceptibility versus temperature of a ferromagnetic material.

In antiferromagnetic materials, exchange integral $J<0$, spin directions among adjacent magnetic atoms should be opposite for lowering system energy, and atom magnetic moments oppositely line up.[42] At very low temperature, the oppositely line-up magnetic moment almost completely cancels each other. The χ_m is positive and nearly approaches to zero. When T increases, χ_m also increases until the temperature arrives Néel point T_N. Above T_N, the exchange effect is destroyed, material changes to paramagnetism (Figure 7.17).

42 在反铁磁体材料中，交换积分 $J<0$，为了降低体系能量相邻磁原子的自旋方向应该相反，同时原子间的磁矩应反向排列。

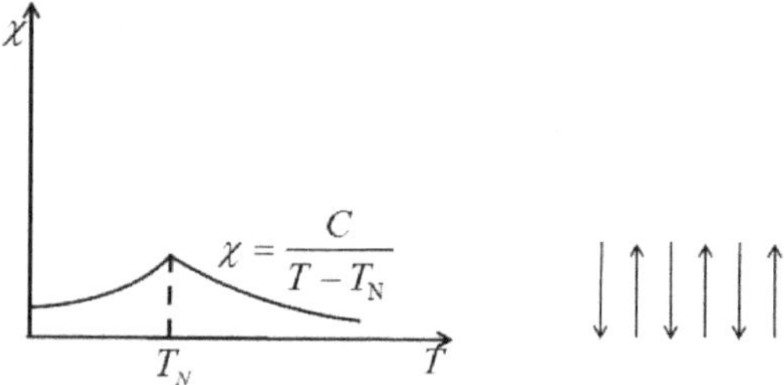

Figure 7.17 Susceptibility versus temperature of an antiferromagnetic material.

In some cases, the antiparallel magnetic moments of antiferromagnets can also be explained by super exchange effect, such as in MnO crystal (Figure 7.18). Mn is a magnetic atom and O atoms act as spacer. The magnetic moments of Mn is antiparallel between two Mn atomic planes in (111) direction. The antiparallel magnetic moments can't be explained by exchange effect directly because the distance between the adjacent two layers of magnetic moments in MnO crystal is longer than that in ferromagnetic materials with magnetic dipoles adjacent to each other. The exchange interaction between such long distance magnetic moments is termed as super exchange effect.[43] According to this theory, the exchange process can be accomplished via the excited state of the adjacent cation. Taking MnO crystal as an example, in the excited state, O^{2-} transfers a

43 有些情况，反铁磁体中反平行排列的磁矩也可以通过超交换作用解释，如 MnO 晶体（图 7.18）。Mn 原子是磁性原子，O 原子作为间隔。Mn 原子的磁矩在 2 个 Mn 原子平面的（111）方向反平行。由于这些相互反平行 Mn 原子的距离比铁磁性材料中相邻平行磁矩之间的距离长，不能直接用交换作用解释。这种长距离磁矩交换作用称为超交换作用。

2p-electron to the adjacent Mn^{2+}, resulting in O^-, and Mn^+.[44] Now the arrangement of the electron spin is shown in Figure 7.19.

44 根据这一理论，交换过程可以通过相邻阳离子的激发态完成。以 MnO 晶体为例，在激发态，O^{2-} 转移一个 2p 电子到相邻 Mn^{2+}，形成 O^- 和 Mn^+。

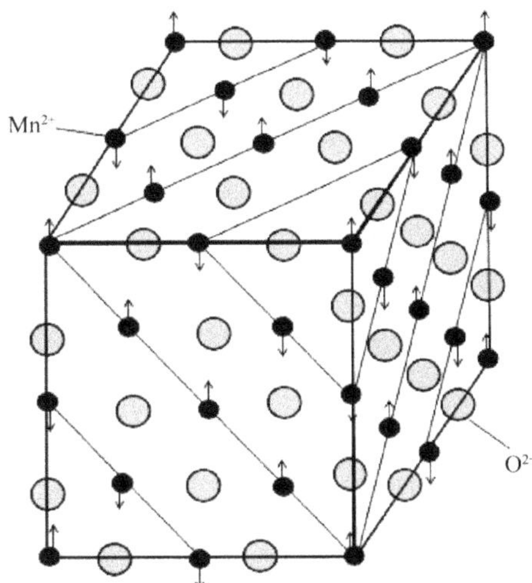

Figure 7.18 The crystal structure of MnO consists of alternating layers of {111} type planes of oxygen and manganese ions. The magnetic moments of the manganese ions in every other (111) plane are oppositely aligned. Consequently, MnO is antiferromagnetic.

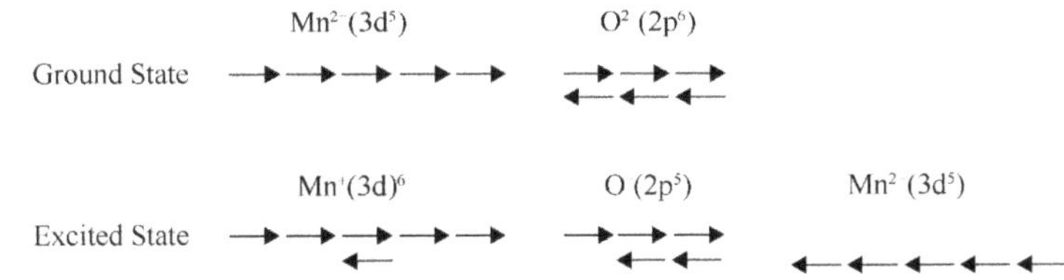

Figure 7.19 (a) Spin alignment of Mn^{2+} in MnO lattice points. (b) Super exchange in ions of MnO crystal.

The spin direction of O^- is the same as the left hand Mn^+. When the spin direction of the right hand Mn^{2+} is opposite to the left hand Mn^+O^-, the energy of the system is low. In this case the spin of the right hand Mn^{2+}couples with the left hand Mn^+ in long distance via O^-. This is due to the 2p electron from O^{2-}.[45] Since the shape of p electron in space is ∞ like, 180° angle between M – O – M results in the strongest exchange effect. When φ = 90°, the exchange effect is the weakest. Since the increase of temperature helps the excitation of O ion, below the Curie temperature, the susceptibility of an antiferromagnet will increase with temperature.

45 O^-的自旋方向和左手边 Mn^+的自旋方向一样。当右手边 Mn^{2+} 的自旋方向和左手边 Mn^+O^-的自旋方向相反时，系统能量降低。此时，右手边 Mn^{2+}和左手边 Mn^+的自旋，通过 O^-，发生了长距离耦合。这源于 O^{2-}的 2p 的电子。

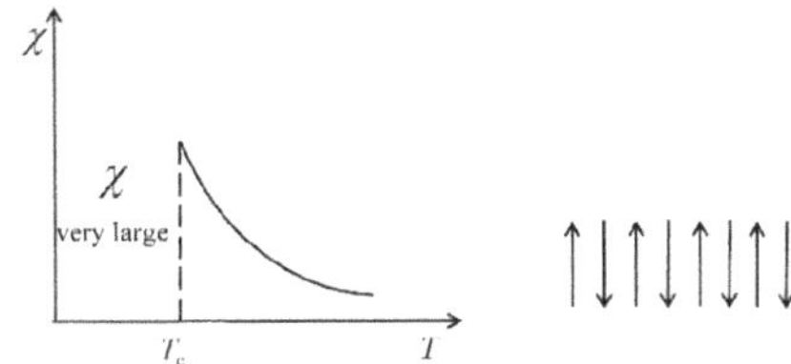

Figure 7.20 Schematic view of ferrimagnetism.

46 对于亚铁磁体，交换积分 $J<0$，磁矩是反向平行的。但是这些反向平行的磁矩具有不同的大小，导致非零的磁性。

For ferrimagnetism, the exchange integral $J<0$, and the magnetic moments are antiparallel. However, the oppositely paralleled magnetic moments are with different values, resulting in non-zero magnetism.[46] Above Tc, the exchange effect is destroyed, material changes to paramagnetism (Figure 7.20). In Figure 7.21 the summarization of various types of magnetism is given.

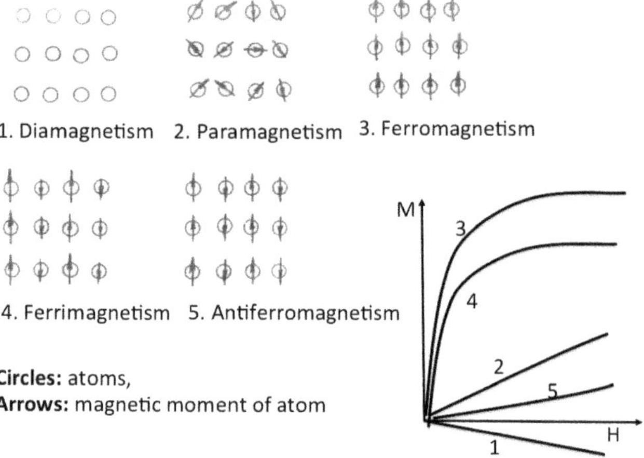

Figure 7.21 Summarization of the main magnetisms.

7.3 Features of Ferromagnetic Materials

7.3.1 Curie Point

47 对每一种铁磁性材料，在一个特征温度下，原子磁偶极之间的强耦合作用消失，铁磁性材料变为顺磁性材料（图 7.22）。

The properties of a ferromagnet depend on temperature. At higher temperatures, the thermal motion in a ferromagnet becomes more and more violent. At a certain characteristic temperature for each ferromagnetic material, the strong coupling between the magnetic moments of the atoms disappears and the ferromagnetic material becomes paramagnetic (Figure 7.22).[47]

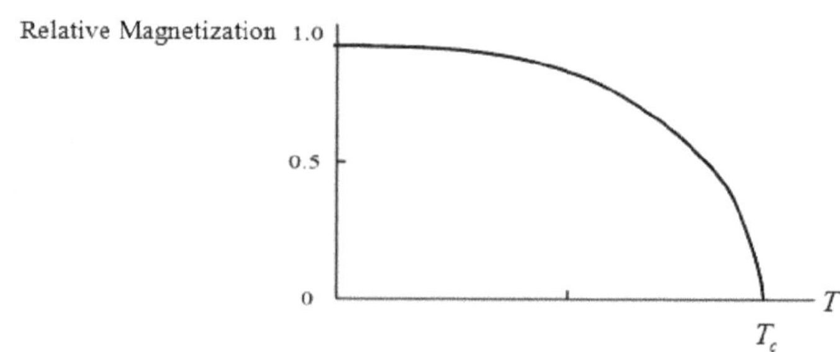

Figure 7.22 Saturation magnetization as a function of temperature for a ferromagnetic material.

The transition temperature Tc is called the Curie point. Some values for ferromagnetic metals are given in Table 7.2.

Table 7.2 Curie points of some common ferromagnetic metals.

Metal	Curie point (K)
Fe	1043
Co	1390
Ni	630

7.3.2 Order–Disorder Transformations

The transformations of ferromagnetic solids at the Curie point represent a special type of order–disorder transformation.[48] Figure 7.23 shows Cp of nickel as a function of T. The curve has a typical anomaly at the Curie point of Ni, 630 K. The discontinuity of Cp appears at the transition from an ordered structure with Weiss domains to a random, disordered structure without coupling between the spins of the atoms.

48 铁磁性固体在居里温度下的转变是一种特殊的有序-无序转变。

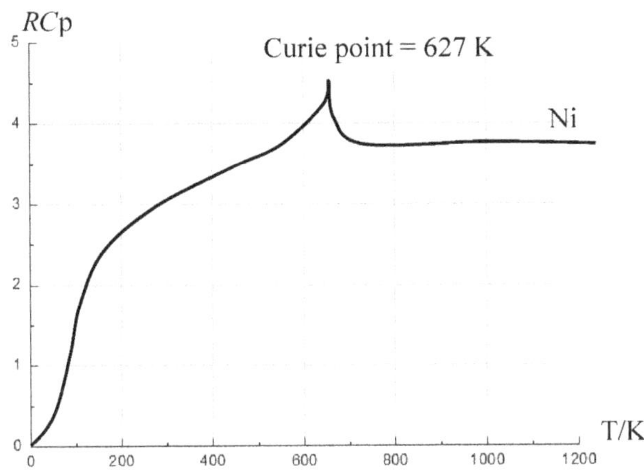

Figure 7.23 The molar heat capacity at constant pressure of Ni as a function of temperature.

Figure 7.24 shows experimental values of Cp of iron as a function of T. The left of the curve (including the peak) corresponds to α-Fe, which has a BCC structure. The middle lower part corresponds to γ-Fe, which has an FCC structure. At temperatures up to 1183 K the stable form of iron is the α-phase. In the temperature interval 1183–1673 K the stable form with lowest energy is γ-Fe. It can be seen that discontinuities appear in the transitions $\alpha \rightarrow \gamma$ (1183 K) and $\gamma \rightarrow \delta$ (1673 K). These transformations are true structure changes of the crystal lattice and cause an abrupt change in the heat capacity of Fe but no peak in the curve. The anomalous peak in the α-curve at 1043 K corresponds to the transformation from an ordered magnetic structure of α-Fe to a disordered nonmagnetic structure.[49] The

49 这些转变是晶格点阵中真实的结构改变，同时导致了Fe热容的突变，但是曲线中没有峰出现。在1043 K时在曲线α-相部分的异常峰表示α-Fe从有序磁性结构向无序非磁性结构的转变过程。

disorder depends on random orientation in space of the magnetic moments of the atoms when the Weiss domains disappear.

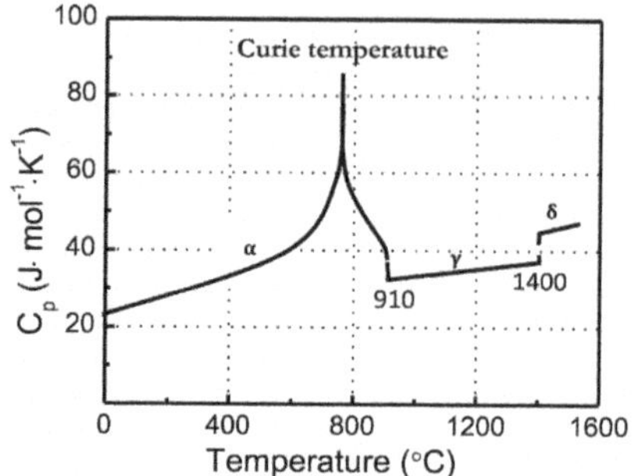

Figure 7.24 The molar heat capacity at constant pressure of Fe as a function of temperature.

7.3.3 Magnetic Domain

Ⅰ. Definition

50 因为在铁磁材料中交换能 $J>0$，所以存在许多排列规则的磁矩区域。在铁磁材料中排列规则的磁矩小区域被定义为为磁畴（Weiss 畴）。

Since the exchange integral in a ferromagnet greater than zero ($J>0$), there are a lot of magnetic moments lineup area in it. By definition, the small area of lined up magnetic moments in a ferromagnet is called magnetic domain (or Weiss domains),[50] as shown in Figure 7.25.

Figure 7.25 Schematic view of domains in a magnetic material. The small compasses indicate the direction of atomic magnetic moments.

In a ferromagnetic material that has never been exposed to a magnetic field, the individual domains have a random orientation. Because of this, the net magnetization in the virgin ferromagnetic or ferrimagnetic material as a whole is zero. This happens in freshly smelted iron.

Ⅱ. Domain Wall

Domain wall, also called Bloch wall, refers to the boundary between two domains, which separates the individual magnetic domains.

The domain walls are generally narrow zones in which the direction of the magnetic moment gradually and continuously changes from that of one domain to that of the next (Figure 7.26).[51] The domains are typically very small, about 0.005 cm or less, while the domain walls are about 100 nm thick.

On the basis of the orientation of two adjacent domains, there are 180° domain walls referring to boundaries that separate two antiparallel arranged magnetic domains, and 90° domain walls referring to boundaries that separate two perpendicular arranged magnetic domains,[52] see Figure 7.26b.

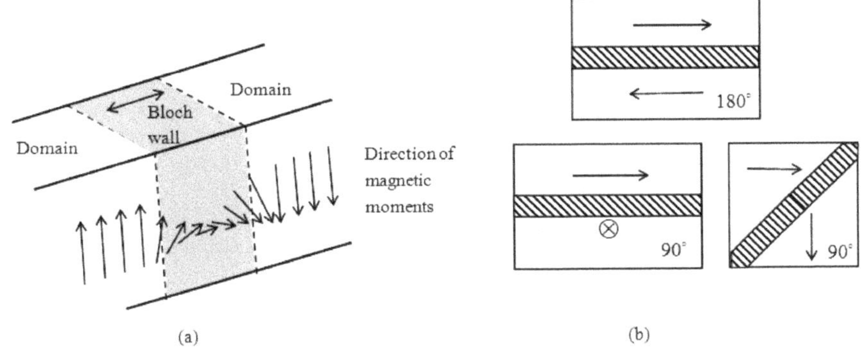

Figure 7.26 (a) The magnetic moments change direction continuously across the boundary between domains. (b) 180° domain wall and 90° domain wall.

Ⅲ. Structure of Magnetic Domain

Ferromagnet can be spontaneously magnetized, forming magnetic domains. There are several types of domain structure as discussed below.

For single crystal with only one domain (Figure 7.27a), the whole crystal can be spontaneously magnetized to saturation point. The magnetization is usually along the easy direction, both the exchange energy and the anisotropy energy are the lowest. But magnetic poles and magneto static energy will be produced, the energy for demagnetization is the largest.[53]

If the single crystal contains two magnetic domains (Figure 7.27b), domain wall energy appears. Compared to (a), the demagnetization energy is lower. In this case, the two domains tend to be divided into small domains by demagnetization process (Figure 7.27c). The more the domains, the smaller the demagnetization energy, but the domain wall energy increases. When domain wall energy and the demagnetization energy reach the equilibrium state, the separation of domain stops, reaching the equilibrium domain structure.[54] To further decrease demagnetization energy, closed domains with triangular shape may form. But this increases the anisotropy energy (Figure 7.27d). In real single crystals, this type of domain structure was found in some cases (Figure 7.28).

51 畴壁通常是指 2 个不同方向的磁矩之间的狭窄区域，在畴壁内磁矩逐渐且连续地由一个方向转变为另一个方向。

52 基于 2 个相邻磁畴的取向，存在 180°畴壁和 90°畴壁，前者指将 2 个反平行磁畴隔开的界限，后者是指将 2 个相互垂直磁畴隔开的界限。

53 对于只有一个磁畴的单晶（图 7.27a），整个晶体能够自发地磁化到饱和点。磁化一般沿着容易的方向，即交换能和各向异性能都是最低的。但是会产生磁极和静磁能，去磁化能量是最高的。

54 磁畴越多，去磁化能就越小，但畴壁能会增加。当畴壁能和去磁化能达到平衡时，磁畴的分裂停止并达到平衡的磁畴结构。

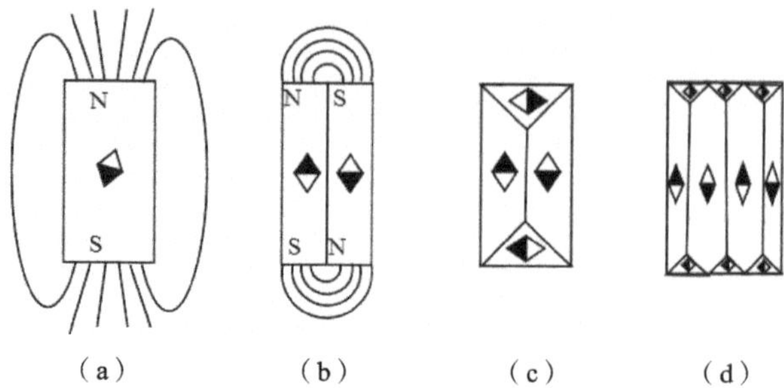

Figure 7.27 Magnetic domain in single crystal: (a) one domain, (b) two domains, (c) four domains, (d) multiple domains.

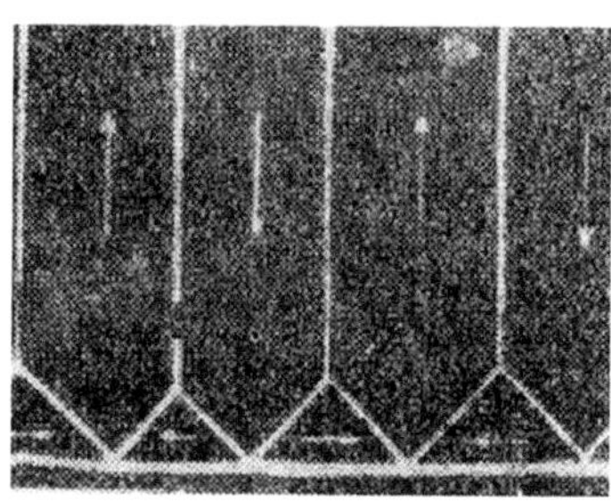

Figure 7.28 The powder pattern for single crystal of Fe-Si alloy.

The domain structures for polycrystalline ferromagnetic are complicate, with each grain containing several domains (Figure 7.29). The crystal boundary, second phase, crystal defects, impurity, stress, and segregation will influence the domain structure.[55] But the direction of the magnetization strength within a domain is all along the easy way.

55 多晶铁磁材料的磁畴结构非常复杂，每个晶粒都包含很多磁畴（图 7.29）。晶界、第二相、晶体缺陷、杂质、应力和偏析都会影响磁畴结构。

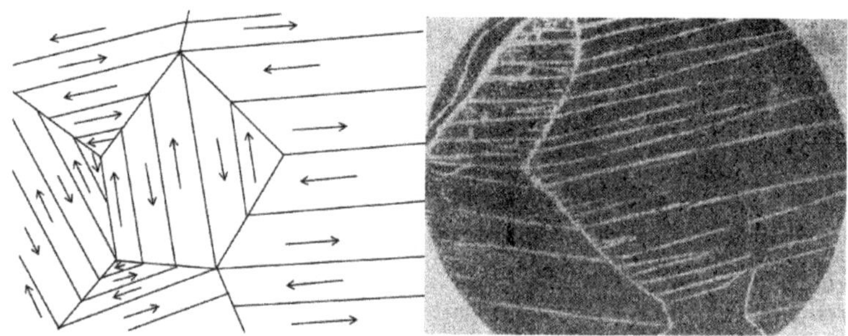

Figure 7.29 (a) Schematic drawing of domains in polycrystalline materials. (b) Powder pattern for polycrystalline of Fe-Si alloy.

7.3.4 Features of Ferromagnetic Materials in Magnetic Fields

Ⅰ. Magnetization and Susceptibility

If ferromagnetic materials are exposed to a magnetic field H, they become magnetized, i.e. the Weiss domains align themselves more or less in the direction of the H field and contribute to the magnetic field. As discussed previously, the magnetic inductance B is defined by the relationship:

$$B = (1+\chi)\mu_0 H = \mu\mu_0 H \tag{7-36}$$

B is not proportional to H because the relative permeability (or you can say the susceptibility) varies strongly with the magnetizing field.

Consider a single crystal of a ferromagnetic solid where the Weiss domains have random orientations (Figure 7.30a, for simplicity only a few of them are shown). If a weak magnetizing field is applied in a direction that coincides with one of the crystal axes, the Bloch walls of the Weiss domains become displaced.[56] The domains in the field direction grow at the expense of the others (Figure 7.30b). The process is reversible for small displacements. When the magnetic field decreases, the domain walls become displaced in the reverse directions. The magnetic order of the domains is counteracted by the thermal motion. The equilibrium depends on the strength of the magnetic field H. When the magnetic field is increased, the domains in the field direction continue to grow by domain wall displacements.[57] If the magnetic field is strong enough, the domains in the field direction dominate completely (Figure 7.30c). Figure 7.30d shows the B-H curve for a ferromagnet. At $H = 0$ the specimen is un-magnetized, i.e. the Weiss domains have random orientations. Region I in Figure 7.30d of the curve represents the process of reversible displacements of the domain boundaries. Further displacements of the domain boundaries are irreversible (region II). The third region represents rotation of Weiss domains from easy directions to the direction of the H field. The curve is called the virgin curve.

56 如果施加一个与晶轴方向一致的弱磁场，磁畴的布洛赫壁将会移动。

57 磁畴的磁有序性与热运动是抗衡的。平衡态依赖于磁场强度 H。当磁场强度增加,与磁场方向一致的磁畴将通过畴壁的位移而持续增加。

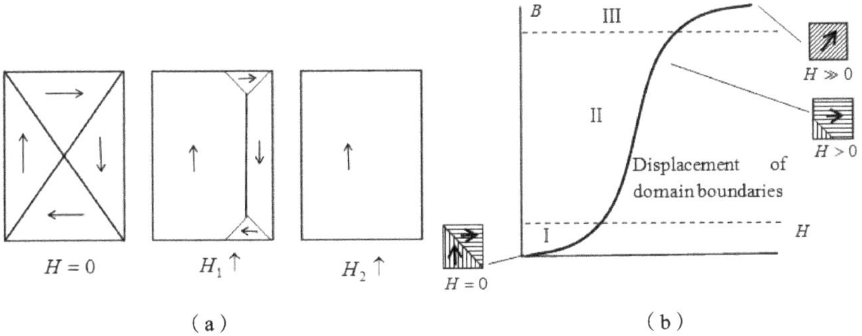

Figure 7.30 Magnetizing a ferromagnetic crystal in the direction of one of the crystal axes under different external magnetic field. (a) $H = 0$, (b) small H, (c) large H, (d) B-H curve for a ferromagnet. With increase of H, magnetization increases, leading to higher magnetic susceptibility. The arrows represent the directions of the magnetic moments of the Weiss domains.

Ⅱ. Saturation of Magnetization

When a magnetic field is imposed on the material, domains that are nearly lined up with the field grow at the expense of unaligned domains. In

58 为了使磁畴增长，布洛克壁必须移动，外磁场提供了该移动所需的力。

59 当磁场强度增强时，优势取向的磁畴比较容易增长，同时磁导率增加。

order for the domains to grow, the Bloch walls must move and the field provides the force required for this movement.[58] Initially, the domains grow with difficulty, and relatively large increases in the field are required to produce even a little magnetization. This condition is indicated in Figure 7.31 by a shallow slope, which is the initial permeability of the material. As the field increases in strength, favorably oriented domains grow more easily, with permeability increasing as well.[59] A maximum permeability can be defined as shown in the figure. Eventually, the unfavorably oriented domains disappear, and rotation completes the alignment of the domains with the field. The saturation magnetization, produced when all of the domains are oriented along with the magnetic field, is the greatest amount of magnetization that the material can obtain. Under these conditions, the permeability of these materials becomes quite small.

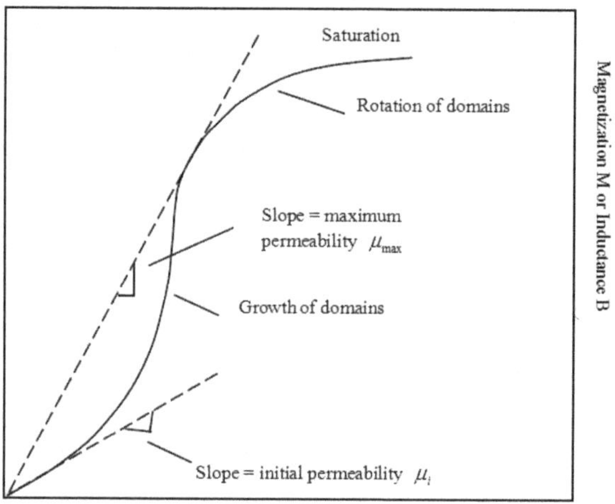

Figure 7.31 When a magnetic field is first applied to a magnetic material, magnetization initially increases slowly, then more rapidly as the domains begin to grow. Later, magnetization slows, as domains must eventually rotate to reach saturation. Notice the permeability values depend upon the magnitude of *H*.

Ⅲ. Hysteresis Loop, Remanance and Coercivity

60 随着磁场强度*H*的增加，样品将被磁化，从零一直增加到磁化饱和，即磁化强度变为常数。

The easiest way to magnetize a ferromagnetic specimen is to place it in a homogeneous magnetizing field *H* caused by a coil. The *H* field is easily changed by varying the current *I* in the coil ($H = nI$, where n = number of turns per unit length). The specimen is magnetized by increasing the *H* field from zero until saturation magnetization has been achieved, i.e. becomes constant.[60]

If the coil is fed with alternating current, the magnetizing field *H* also alternates with the same frequency as the current. The specimen has to alter its direction of saturation magnetization over and over again. The remagnetization is shown in Figure 7.32.

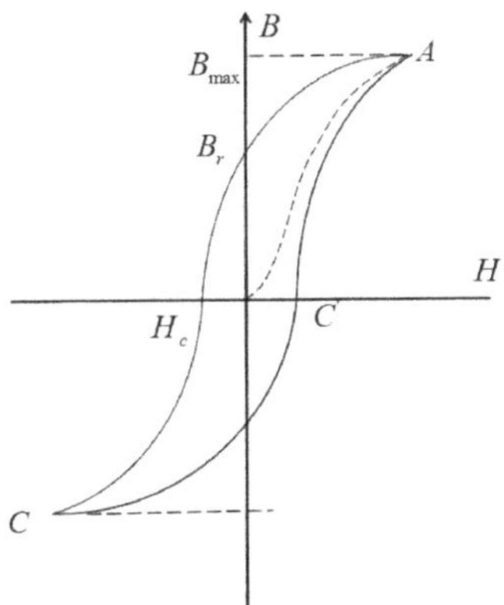

Figure 7.32 Hysteresis loop of a ferromagnetic material. B_r is residual magnetization and *Hc* is coercitive force.

The dashed curve represents the virgin curve discussed above. At point A, the specimen is fully magnetized in the direction of the *H* field. When the *H* field decreases, becomes zero and then negative, the *B* field does not follow the virgin curve but the AB_rH_c C curve. When the *H* field decreases again and changes direction once more, the *B* field follows the curve CC′A back to the starting point. The closed curve is known as the hysteresis loop.[61] B_{max} is the saturation magnetization. After saturation, reduce *H* gradually, *B* decrease too. This process is called demagnetization. The curve intersects the *B* axis at $B = B_r$ when $H = 0$, while *B* decreases from B_{max} to B_r, is the residual magnetization (remanence). For negative *H* value, *B* continues to decrease. The value of *H* that corresponds to $B = 0$ is H_c, which is called the coercitive force or coercivity. If the magnetic material is exposed to H_c the residual magnetization B_r disappears. The area inside the hysteresis loop represents the heat losses or hysteresis losses. So-called eddy currents are induced in the metal by the incessant changes of the magnetic field during the magnetization–remagnetization cycles.[62] The eddy currents generate heat in the metal. The heat is emitted to the surroundings.

Figure 7.33 shows the influence of temperature to hysteresis loop. When the temperature of a ferromagnetic or ferrimagnetic material is increased, the added thermal energy increases the mobility of the domains, making it easier for them to become aligned, but also preventing them from remaining aligned when the field is removed.[63] Consequently, saturation magnetization, remanance, and the coercive field are all reduced at high temperatures.

61 当磁场强度 *H* 再次降低并又一次改变方向，磁感应场强遵循曲线 CC′A 返回起始点。这一闭合曲线被称为磁滞回线。

62 所谓的涡流是金属在磁化-去磁循环的连续变化磁场中产生的感应电流。

63 当铁磁体和亚铁磁体材料的温度升高时，增加的热能使磁畴的迁移性增加，使之更容易对齐，同时增加的热能也使得磁场移除时，磁畴的对齐难以保持。

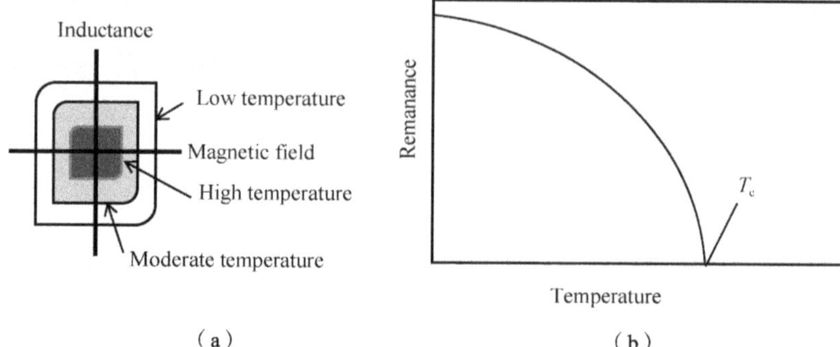

Figure 7.33 The effect of temperature on (a) the hysteresis loop and (b) the remanance. Ferromagnetic behavior disappears above the Curie temperature.

7.4 Functional Magnetic Materials

Ⅰ. Definitions of Hard and Soft Magnetic Materials

Ferromagnetic and ferrimagnetic materials are classified as magnetically soft or magnetically hard depending upon the shape of the hysteresis loop (Figure 7.34a). Generally, if the coercivity value is greater than ~10^4 Am^{-1}, we consider the material as magnetically hard. If the coercivity values are less than 10^3 Am^{-1}, we consider the materials as magnetically soft. Figure 7.34b shows classification of different commercially important magnetic materials. Note that while the coercivity is a strongly microstructure-sensitive property, the saturation magnetization is constant (i.e., it is not microstructure dependent) for a material of a given composition.[64] This is similar to the way the yield strength of metallic materials is strongly dependent on the microstructure, while the Young's modulus is not. Many factors, such as the structure of grain boundaries and the presence of pores or surface layers on particles, affect the coercivity values.

64 需要注意，矫顽力是一个对微观结构极敏感的性质，但是组成确定材料的饱和磁化强度是常数（即不依赖于微观结构）。

The coercivity of single crystals depends strongly on crystallographic directions. There are certain directions along which it is easy to align the magnetic domains. There are other directions along which the coercivity is much higher. Coercivity of magnetic particles also depends upon shape of the particles. This is why in magnetic recording media we use acicular and not spherical particles. This effect is also used in Fe-Si steels, which are textured or grain oriented so as to minimize energy losses during the operation of an electrical transformer.

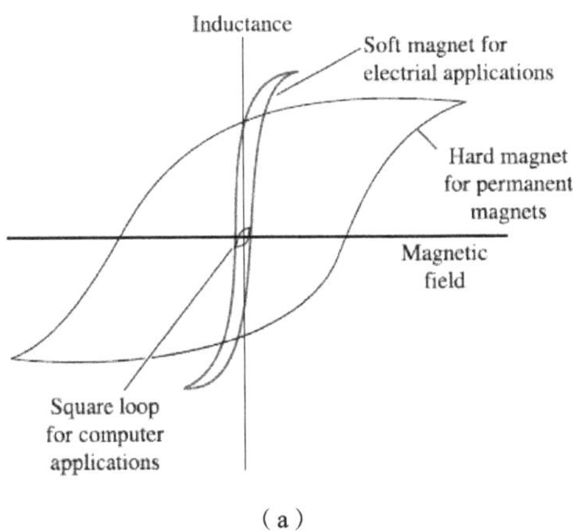

(a)

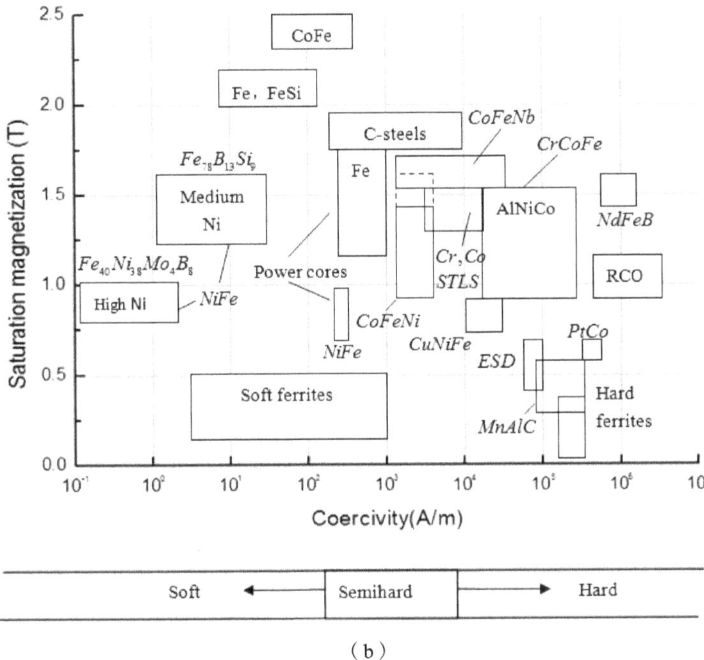

(b)

Figure 7.34 (a) Comparison of the hysteresis loops for three applications of ferromagnetic and ferrimagnetic materials. (b) Saturation magnetization and coercivity values for different magnetic materials.

Ⅱ. Permanent Magnets

Strong permanent magnets, often called hard magnets, refer to those that after the magnetization, removing the external magnetic field, the magnetism of the material can be remained quite strong. Requirements of hard magnets are as following:

- High remanance (stable domains).
- High permeability.
- High coercive field.
- Large hysteresis loop.

- High power (or *BH* product).

The record for any energy product is obtained for $Nd_2Fe_{14}B$ magnets with an energy product of ~445 kJ m^{-3} [~56 Mega-Gauss-Oersteds (MGOe)]. These magnets are made in the form of a powder by the rapid solidification of a molten alloy. Powders are either bonded in a polymer matrix or by hot pressing, producing bulk materials.[65] The energy product increases when the sintered magnet is "oriented" or poled. Corrosion resistance, brittleness, and a relatively low Curie temperature of ~312°C are some of the limiting factors of this extraordinary material.

65 这些磁体是粉末形式，通过快速固化熔融合金形成的。粉末与聚合物基质结合或通过热压制得块体材料。

The power of the magnet is related to the size of the hysteresis loop, or the maximum product of *B* and *H*. The area of the largest rectangle that can be drawn in the second or fourth quadrants of the *B–H* curve is related to the energy required to demagnetize the magnet.[66] For the product to be large, both the remanance and the coercive field should be large.

66 磁体的磁能与其磁滞回线的尺寸相关，或是与 *B* 和 *H* 乘积的最大值相关。*B-H* 曲线在第二或第四象限能够得到的最大矩形面积与磁体去磁所需能量相关。

In many applications, we need to calculate the lifting power of a permanent magnet. The magnetic force obtainable using a permanent magnet is given by

$$F = \frac{\mu_0 M^2 A}{2} \tag{7-37}$$

where, *A* is the cross-sectional area of the magnet, *M* is the magnetization, and μ_0 is the magnetic permeability of free space.

One of the most successful examples of the contributions by materials scientists and engineers in this area is the development of strong rare earth magnets. Permanent magnets are used in many applications including loudspeakers, motors, generators, holding magnets such as the encapsulation strip for refrigerator, mineral separation, and bearings. Typically, they offer a nonuniform magnetic field, however, it is possible to use geometric arrangements known as Halbach arrays to produce relatively uniform magnetic fields.

Ⅲ. Soft Magnetic Materials

Soft magnetic materials refer to the magnetic materials having high magnetic permeability, low coercivity and low remanance. Applications include cores for electromagnets, electric motors, transformers, generators, ferromagnetic head, record magnetic head, and other electrical equipment. Because these devices utilize an alternating field, the core material is continually cycled through the hysteresis loop. Note that in these materials the value of relative magnetic permeability depends strongly on the strength of the applied field.

Soft magnetic materials often have the following characteristics:

- High-saturation magnetization;
- High permeability;
- Small coercive field;
- Small remanance;
- Small hysteresis loop;
- Rapid response to high-frequency magnetic fields;
- High electrical resistivity.

High saturation magnetization permits a material to do work, while high permeability permits saturation magnetization to be obtained with small imposed magnetic fields.[67] A small coercive field also indicates that domains can be reoriented with small magnetic fields. A small remanence is desired so that almost no magnetization remains when the external field is removed. These characteristics also lead to a small hysteresis loop, therefore minimizing energy losses during operation.

If the frequency of the applied field is so high that the domains cannot be aligned in each cycle, the device may heat due to dipole friction. In addition, higher frequencies naturally produce more heating because the material cycles through the hysteresis loop more often, losing energy during each cycle.[68] For high frequency applications, materials must permit the dipoles to be aligned at exceptionally rapid rates.

Energy can also be lost by heating if eddy currents are produced. During operation, electrical currents can be induced into the magnetic material. These currents produce power losses and Joule, or I^2R, heating.[69] Eddy current losses are particularly severe when the material operates at high frequencies. If the electrical resistivity is high, eddy current losses can be held to a minimum. Soft magnets produced from ferromagnetic ceramic materials have a high resistivity and therefore are less likely to heat than metallic ferromagnetic materials.

67 高饱和磁化强度的材料可以用来做功，而高磁导率使得材料能够在很小的磁场下达到饱和磁化。

68 如果外加磁场的频率过高使得磁畴无法在每一个循环中排列齐整，器件有可能因偶极摩擦而发热。此外，高频磁场自然产生多的热量，因为材料通过磁滞回线的循环更加频繁，每次循环时都会损失能量。

69 如果涡流产生，能量也可能通过热量而损失。过程中，磁性材料中能够产生电流。这些电流产生能耗或者焦耳热 I^2R。

Ⅳ. Data Storage Materials

Magnetic materials are used for data storage. Data is stored by magnetizing the material in a certain direction. For example, if the "north" pole is up, the bit of information stored is 1. If the "north" pole is down, then a 0 is stored.

For this application, materials with a square hysteresis loop, a low remanence, a low saturation magnetization, and a low coercive field are preferable. Hard ferrites based on CrO_2, acicular iron particles, and γ-Fe_2O_3 satisfy these requirements. The stripe on credit cards and bank machine cards are made using γ-Fe_2O_3 or Fe_3O_4 particles. The square loop ensures that a bit of information placed in the material by a field remains stored; a steep and abrupt change in magnetization is required to remove the information from storage in the ferromagnet. Furthermore, the magnetization produced by small external fields keeps the coercive field (H_c), saturation magnetization, and remanance (B_r) low.

Many new alloys based on Co-Pt-Ta-Cr have been developed for the manufacture of hard disks. Computer hard disks are made using sputtered thin films of these materials. Many different alloys, such as those based on nanostructured Fe-Pt and Fe-Pd, are being developed for data storage applications. More recently, a technology known as spintronics (spin-based electronics) has evolved. In spintronics, the main idea is to make use of the spin of electrons as a way of affecting the flow of electrical current (known as spin-polarized current) to make devices such as field effect transistors (FET).[70] The spin of the electrons (up or down) is also being considered as a way of storing information. A very successful example of a real-world spintronic-based device is a giant magnetoresistance (GMR) sensor that is used for reading information from computer hard disks.

70 在自旋电子学中，其主旨是利用电子自旋作为一种影响电流流动的方式（称为自旋-极化电流），从而制造器件，例如场效应晶体管（FET）。

本章小结

1. 内容概要

本章讲述材料的磁性。首先，从历史角度介绍了磁性的现象和应用；然后，阐述了磁性产生的机理；接着，介绍了磁性的表征参数和磁性材料的分类；最后，讨论了功能磁性材料。

2. 基本概念

磁性、磁场强度、磁力、透磁率（磁导率）、磁偶极子、磁矩、磁性原子、磁化强度、磁感应系数、磁化率、抗磁性、顺磁性、居里定律、铁磁性、反铁磁性、亚铁磁性、居里温度、有序-无序转变、磁畴、畴壁、磁滞回线、剩磁、矫顽力、永久磁体、软磁材料、硬磁材料。

3. 主要公式

（1）直导线的磁场强度：$H = \dfrac{I}{2\pi r}$

（2）线圈的磁场强度：$H = \dfrac{nI}{l}$

（3）环形电流诱导的磁矩：$\mu_{\mathrm{L}} = IS$

（4）两磁荷间的磁力：$F = k\dfrac{q_1 q_2}{r^2}$

（5）磁荷在外磁场作用下的磁力：$F = qH$

（6）永久磁体在磁场下受到的磁力：$F = \dfrac{\mu_0 M^2 A}{2}$

（7）电偶极矩：$\mu = qd$（q 为电荷）

（8）磁偶极矩：$P_{\mathrm{m}} = qd$（q 为磁荷）

（9）环行电子的电流：$I = \dfrac{\mathrm{d}q}{\mathrm{d}t} = \dfrac{e}{\dfrac{2\pi}{\omega}} = e\dfrac{\omega}{2\pi}$

（10）环行电子的磁矩：$\mu_{\mathrm{L}} = IS = e\dfrac{\omega}{2\pi}\pi r^2 = \dfrac{e}{2m}m\omega r^2$

（11）电子轨道角动量：$P_{\mathrm{L}} = r \times mv = mr^2\omega$

（12）电子轨道角动量磁矩：$\mu_{\mathrm{L}} = IS = e\dfrac{\omega}{2\pi}\pi r^2 = \dfrac{e}{2m}m\omega r^2 = \dfrac{e}{2m}P_{\mathrm{L}}$

（13）量子化的电子轨道角动量和磁矩：$P_L = \dfrac{h}{2\pi}\sqrt{L(L+1)}$，$\mu_L = \mu_{\mathrm{B}}\sqrt{L(L+1)}$

（14）玻尔磁子：$\mu_{\mathrm{B}} = \dfrac{qh}{4\pi m_{\mathrm{e}}} = 9.274 \times 10^{-24}\,\mathrm{A \cdot m^2}$

（15）磁扭矩：$\boldsymbol{\mu}_{\mathrm{m}} \times \boldsymbol{B}$

（16）外磁场下电子轨道角动量磁矩的进动频率及诱导磁矩：

$$\omega_{\mathrm{L}}=\frac{eB}{2m}，\quad \mu_{\mathrm{ind}}=\frac{e}{2m}m\omega r^2=\frac{e^2r^2B}{4m}$$

（17）电子自旋磁矩：$\mu_{\mathrm{s}}=\pm\mu_{\mathrm{B}}$

（18）总电子自旋旋磁矩：$|\mu_{\mathrm{s}}|=2\mu_{\mathrm{B}}\sqrt{S(S+1)}$

（19）磁化强度：$M=\dfrac{\Sigma m_{\mathrm{m}}}{V}$

（20）真空下磁感应强度：$B=\mu_0H$

（21）材料的磁感应强度：$B=\mu H=\mu_0H+\mu_0M=\mu_0(1+\chi_{\mathrm{m}})H$

（22）磁性材料的相对磁导率：$\mu_{\mathrm{r}}=\dfrac{\mu}{\mu_0}$

（23）磁化率：$\chi_{\mathrm{m}}=\mu_{\mathrm{r}}-1$

（24）原子的抗磁磁化率：$\chi_m^d=-\dfrac{ne^2\mu_0}{6m}\sum_{i=1}^{z}\overline{r_l^2}$

（25）导带电子的抗磁磁化率：$\chi_m^{de}=-\dfrac{1}{2}\dfrac{n\mu_{\mathrm{B}}^2}{kT_{\mathrm{F}}}\left(\dfrac{m}{m^*}\right)^2$

（26）居里定律：$\chi_{\mathrm{m}}=\dfrac{C}{T}$

（27）居里常数：$C=\dfrac{n\mu_0\mu_m^2}{3k}$

（28）单位体积内自由电子的净磁矩：$M=\mu_{\mathrm{B}}(n_{\uparrow}-n_{\downarrow})$

（29）0K 时导带电子的顺磁磁化率：$\chi_m^{pe}=-\dfrac{3}{2}\times\dfrac{n\mu_B^2}{kT_{\mathrm{F}}}$

（30）导带电子的总磁化率：$\chi_m^e=\chi_m^{de}+\chi_m^{pe}=\dfrac{3n\mu_B^2}{2kT_{\mathrm{F}}}\left[1-\dfrac{1}{3}\left(\dfrac{m}{m^*}\right)^2\right]$

（31）原子间的自旋交换能：$E_{\mathrm{ex}}=2JS_1\cdot S_2$

（32）居里-外斯定律：$\chi_{\mathrm{m}}=\dfrac{C}{(T-T_C)}$

Vocabulary

antiferromagnet	反铁磁体
Bloch walls	布洛赫壁
Bohr magneton	玻尔磁子
coercivity	矫顽力
compass	指南针
Curie temperature	居里温度
diamagnet	抗磁体

diamagnetism	抗磁性
electromagnetic induction	电磁感应
electron spin	电子自旋
ferrimagnet	亚铁磁体
ferrimagnetism	亚铁磁性
ferromagnet	铁磁体
hard magnet	硬磁体
magnetic charge	磁荷
magnetic dipole	磁偶极矩
magnetic domain	磁畴
magnetic flux line	磁通量线；磁通线
magnetic hysteresis loop	磁滞回线
magnetic inductance	磁感应强度
magnetic moment	磁矩
magnetic permeability	磁导率,磁透率
magnetic property	磁学性质
magnetic susceptibility	磁化率
magnetism	磁性
magnetite	磁铁矿
magnetization	磁化，磁化强度
mass point	质点
orbital angular momentum	轨道角动量
orbital magnetic moment of electron	电子轨道磁矩
paramagnetism	顺磁性
remanance	剩磁
soft magnet	软磁
solenoid	螺线管
spinel	尖晶石
Weiss domain (magnetic domain)	磁畴

Problems

1. Normally we disregard the magnetic moment of the nucleus, why? In what application does the nuclear magnetic moment become important?

2. What are the two electron motions that are important in determining the magnetic properties of materials?

3. Calculate the maximum, or saturation, magnetization that we expect in iron. The lattice parameter of BCC iron is 2.866 Å. (note: in one BCC unit cell, there are two Fe atoms, and the electronic configuration of iron is $3d^6 4s^2$)

4. Define the following terms: (1) magnetization, (2) magnetic induction, (3) magnetic susceptibility, and (4) magnetic permeability.

5. Derive the equation $\chi_m = \mu_r - 1$.

6. True or false questions:

(1) The spin magnetic moment of an electron is approximately equal to Bohr magneton.

(2) Two magnetic moments with opposite direction will cancel each other.

(3) In ferromagnetic materials, there exists strong coupling among adjacent magnetic moments.

(4) A diamagnetic material exhibits no permanent magnetic moments in atoms.

(5) In a paramagnetic material, there is no permanent magnetic moments in atoms.

(6) In a magnetic atom, there are generally unpaired electrons.

(7) Among all materials, the magnetic susceptibility in ferromagnetic one is the highest.

(8) Below the critical temperature and critical magnetic field, the susceptibility of a superconductor is -1.

(9) In antiferromagnetic materials, there exist strong magnetic moment couplings.

(10) Generally, the susceptibility of an antiferromagnetic material is smaller than that of a paramagnetic one.

7. Is the magnetic permeability of ferromagnetic or ferrimagnetic materials constant? Explain.

8. What are the major differences between ferromagnetic and ferrimagnetic materials?

9. A magnetic material has a coercive field of 167 A/m, a saturation magnetizationof 0.616 tesla, and a residual inductance of 0.3 tesla. Sketch the hysteresis loop for the material.

10. Define the soft and hard magnetic materials. Draw a typical M-H loop for comparison.

11. Above the temperature of ___________, ferromagnetic materials loss their ferromagnetism.

12. In the heat capacity versus temperature curve for a ferromagnetic material, at Curie point, there is a ___________

13. True or false questions.

(1) The magnetic susceptibility of a paramagnetic material is decreases with temperature.

(2) Above the Curie temperature (T_C), the magnetic susceptibility of a ferromagnetic material is larger than those below T_C.

(3) A ferromagnet with only one domain present macro magnetism and there are magne to static energy.

(4) If there are a lot of magnetic domains in a ferromagnet and these domains are randomly distributed, magnetism will present.

(5) A diamagnet is put in external magnetic field (H), the value of magnetic induction is almost proportional to H.

Chapter 8 Optical Properties

In this chapter, the knowledge about optical properties, i.e., the responses of a material upon exposed to electromagnetic radiation, in particular, to visible light, will be discussed. Firstly, focusing on light, some basic concepts are introduced. Then, the interaction between light and materials is interpreted in the aspects of macro-phenomena and micro theories, followed by the introduction of the optical behaviors of metallic and non-metallic materials, such as absorption, scattering, reflection, and transmission, etc., Emphasis is given to absorption and scattering. Finally, two typical materials, optical fiber and plastic optics, are explained with their application.

8.1 Light

The perception of light is the principal means by which we know the world. From the single celled creatures seen under a microscope, to the most distant stars seen through a telescope, it is light that informs us.[1] The incredible variety of forms and the structure of our universe are revealed by light. Yet the nature of light eluded scientists and natural philosophers through most of recorded history. Only since the mid-nineteenth century have physicists begun to understand this most fundamental natural phenomenon. The understanding physicists have today required the development of electromagnetic theory, relativity, and quantum mechanics.[2] Even with the full array of techniques of modern physics there remain unsolved problems having to do with the interaction of light with matter.

1 从在显微镜下看到的单细胞生物，到透过望远镜看到的最遥远的星辰，都是我们通过光来感知的。

2 当今物理学家对光的理解依靠的是电磁理论、相对论和量子力学的发展。

8.1.1 History of Optical Science

What was known about light in the seventeenth century? First of all, it travelled in a straight line and Galileo (1564–1642) tried unsuccessfully to measure its speed. Second, it was reflected off smooth surfaces follow the laws of reflection. Third, it changed direction when it passed from one medium to another (refraction), the explanations for this phenomenon were not so obvious, but they were proposed by Snell (1591–1626) and were later confirmed by Descartes (1596–1650). Fourth, what we now call Fresnel diffraction had been discovered by Grimaldi (1618–63) and by Hooke (1635–1703). Finally, double refraction had been discovered by Bartholinus (1625–98). It was on the basis of these phenomena that a theory of light had to be constructed. The last two facts were particularly puzzling. Why did shadows reach a limiting sharpness as the size of the source became small, and why did fringes appear on the light side of the shadow of a sharp edge? And why did light passing through a crystal of calcite produce two images while light passing through most other transparent materials produced only one?[3] Two explanations were put forward: corpuscles and waves, and an acrimonious controversy resulted. Newton (1642–1727) threw his authority behind the theory that light is corpuscular, mainly because his first law of motion said that if no force acts on a particle it will travel in a straight line;[4] he assumed that the velocity of the corpuscles was large enough that gravitational bending would be negligible.[4] Double refraction he explained by some asymmetry in the corpuscles, so that their directions depended upon whether they passed through the crystal forwards or sideways. He

3 为什么光通过一个方解石晶体产生两个图像，而光通过大多数其他透明材料只产生一个？

4 牛顿（1642–1727）以其权威来压制理论，认为光是粒子的，主要是由于他的第一运动定律指出如果没有力作用在粒子上，它将沿一条直线运动；他假设粒子的速度足够大，引力引起的弯曲可以忽略。

envisaged the corpuscles as resembling magnets and the word 'polarization' is still used even though this explanation has long been discarded.

Diffraction, however, was difficult. Newton realized its importance and was aware of what are now known as Newton's rings, and he saw that the fringes formed in red light were separated more than those formed in blue light. He was also puzzled by the fact that light was partly transmitted and partly reflected by a glass surface; how could his corpuscles sometimes go through and sometimes be reflected?[5] He answered this question by propounding the idea that they had internal vibrations that caused 'fits of reflection' and 'fits of transmission'; in a train of corpuscles some would go one way and some the other. He even worked out the lengths of these 'fits' (which came close to what we now know as half the wavelength). But the idea was very cumbersome and was not really satisfying. His contemporary Huygens (1629–95) was a supporter of the wave theory. With it he could account for diffraction and for the behavior of two sets of waves in a crystal, without explaining how the two sets arose. Both he and Newton thought that light waves, if they existed, must be like sound waves, which are longitudinal. If they had thought of transverse waves, the difficulties of explaining double refraction would have disappeared.[6]

5 他也困惑于这样一个事实：在玻璃表面，光被部分地透射和部分地被反射；粒子是如何有时穿过有时被反射呢？

6 如果他们有横波这一想法，那么解释双折射的困难就消失了。

Newton's authority kept the corpuscular theory going until the end of the eighteenth century, however, by then ideas were coming forward that could not be suppressed. In 1801 Young (1773–1829) demonstrated interference fringes between waves from two sources – an experiment so simple to carry out and interpret that the results were incontrovertible. In 1815 Fresnel (1788–1827) worked out the theory of the Grimaldi–Hooke fringes and in 1821 Fraunhofer (1787–1826) invented the diffraction grating and produced diffraction patterns in parallel light for which the theory was much simpler. These three men laid the foundation of the wave theory that is still the basis of what is now called physical optics.

The defeat of the corpuscular theory, at least until the days of quantum ideas, came in 1818. In that year, Fresnel wrote a prize essay on the diffraction of light for the French Académie des Sciences on the basis of which Poisson (1781–1840), one of the judges, produced an argument that seemed to invalidate the wave theory by *reductio ad absurdum*. Suppose that a shadow of a perfectly round object is cast by a point source; at the periphery all the waves will be in phase, and therefore the waves should also be in phase at the centre of the shadow, and there should therefore be a bright spot at this point. Absurd! Then Fresnel and Arago (1786–1853) carried out the experiment and found that there really was a bright spot at the centre. The triumph of the wave theory seemed complete.

8.1.2 Light Wave-Electromagnetic Wave

The greatest step towards understanding the optical waves came from a completely different direction - the theoretical study of magnetism and electricity.[7]

7 理解光波最重要的一步，来自一个完全不同的方向——磁和电的理论研究。

In 1865, Maxwell (1831-79) was inspired to combine his predecessors' observations in mathematical form by describing the region of influence around electric charges and magnets as an 'electromagnetic field' and expressing the observations in terms of differential equations. In manipulating these equations, he found that they could assume the form of a transverse wave equation, a result that had already been guessed by Faraday in 1846. The velocity of the wave could be derived from the known magnetic and electric constants, and was found to be equal to the measured velocity of light; thus light was established as an electromagnetic disturbance. A key to Maxwell's success was his invention of the concept of 'field', which is a continuous function of space and time representing the mutual influence of one body on another, a prolific idea that has dominated the progress of physics ever since then.[8] This began one of the most brilliant episodes in physics, during which different fields and ideas were brought together and related to one another.

8 麦斯威尔成功的一个关键是他发明的“场”的概念，这是一个代表粒子之间在空间和时间维度上的连续函数：一个主导着后续物理学进展的多产创意。

In the classical sense, electromagnetic radiation is considered to be wavelike, consisting of electric and magnetic field components that are perpendicular to each other and also to the direction of propagation[9] (Figure 8.1). Light, heat (or radiant energy), radar, radio waves, and x-rays are all forms of electromagnetic radiation. Each is characterized primarily by a specific range of wavelengths and also according to the technique by which it is generated. The electromagnetic spectrum of radiation spans the wide range from γ-rays (emitted by radioactive materials) having wavelengths on the order of 10^{-12} m (10^{-3} nm) through X-ray, ultraviolet, visible, infrared, and finally radio waves with wavelengths as long as 10^5 m. This spectrum is shown on a logarithmic scale in Figure 8.2.

9 在经典理论中，电磁辐射以波的形式存在，有电场和磁场分量，它们互相垂直并且垂直于传播方向。

Visible light lies within a very narrow region of the spectrum, with wavelengths ranging between about 0.4 μm to 0.7 μm. The perceived color is determined by wavelength, for example, radiation having a wavelength of approximately 0.4 μm appears violet, whereas green and red occur at about 0.5 and 0.65 μm, respectively. The spectral ranges for several colors are included in Figure 8.2. White light is simply a mixture of all colors.

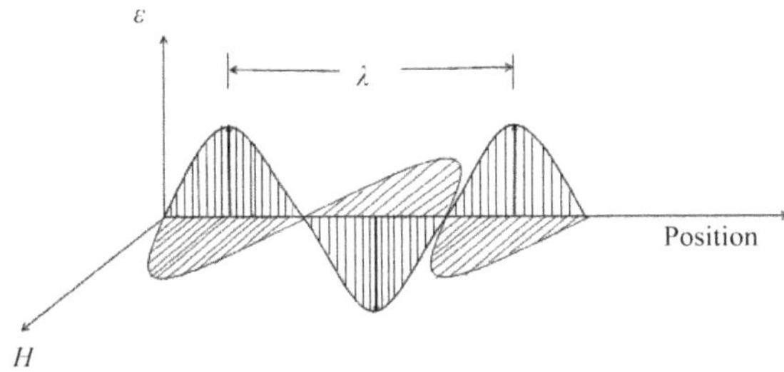

Figure 8.1 An electromagnetic wave showing electric field ε and magnetic field H components and the wavelength λ.

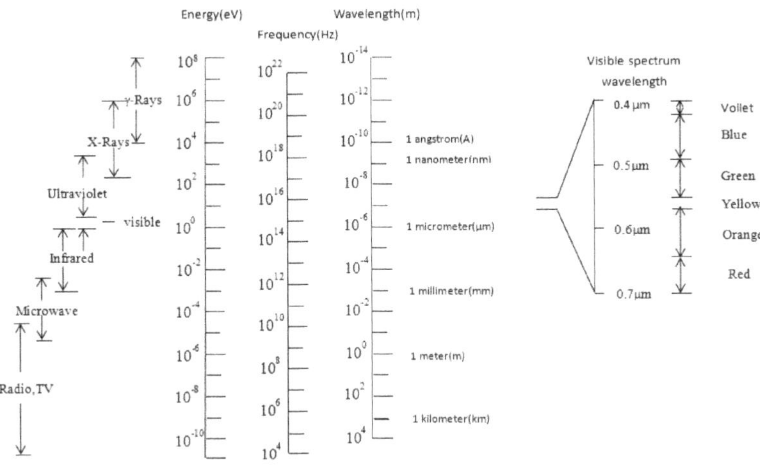

Figure 8.2 The spectrum of electromagnetic radiation, including wavelength ranges for the various colors in the visible spectrum.

In 1678 Römer (1644–1710) realized that an anomaly in the times of successive eclipses of the moons of Jupiter could be accounted for by a finite speed of light, and deduced that it must be about 3×10^8 ms^{-1}. In 1726, Bradley (1693–1762) made the same deduction from observations of the small ellipses that the stars describe in the heavens; since these ellipses have a period of one year they must be associated with the movement of the Earth. It was not, however, until 1850 that direct measurements were made, by Fizeau (1819–96) and Foucault (1819–68), confirming the estimates obtained by Römer and Bradley. Knowledge of the exact value was an important confirmation of Maxwell's theory of electromagnetic waves, which allowed the wave velocity to be calculated from the results of laboratory experiments on static and current electricity.[10] In the hands of Michelson (1852–1931) their methods achieved a high degree of accuracy—about 0.03%. Subsequently much more accurate determinations have been made, and the velocity of light in vacuum has now become one of the fundamental constants of physics, replacing the standard meter. This

10 准确数值的获得是对麦克斯韦电磁波理论的一个重要肯定，它允许通过静止和流动电荷的实验室结果来计算出电磁波的波速。

velocity, c, is related to the electric permittivity of a vacuum ε_0 and the magnetic permeability of a vacuum μ_0 through:

$$c = \frac{1}{\sqrt{\varepsilon_0 \mu_0}} \tag{8-1}$$

Light that is transmitted into the interior of transparent materials experiences a decrease in velocity, the velocity of light in non-vacuum media is:

$$v = \frac{1}{\sqrt{\varepsilon\mu}} = \frac{c}{\sqrt{\varepsilon_r \mu_r}} \tag{8-2}$$

where ε and μ are the relative electric permeability and relative magnetic permeability of media, respectively.

Sometimes it is more convenient to view electromagnetic radiation from a quantum-mechanical perspective, in which the radiation, rather than consisting of waves, is composed of groups or packets of energy called photons.[11] The energy E of a photon is said to be quantized, or can only have specific values, defined by the relationship

$$E = h\nu = \frac{hc}{\lambda} \tag{8-3}$$

where h is Planck's constant. Thus, photon energy is proportional to the frequency of the radiation, or inversely proportional to the wavelength. Photon energies are also included in the electromagnetic spectrum (Figure 8.2).

When describing optical phenomena involving the interactions between radiation and matter, an explanation is often facilitated if light is treated in terms of photons. On other occasions, a wave treatment is preferred; both approaches are used in this discussion, as appropriate.

11 有时，从量子力学的角度来看电磁辐射更为方便，即将电磁辐射看作是由一组或一包能量而不是由波组成。这个能量称为光子。

8.2 Interaction of Light with Materials

In this section the general physics of the interaction of light with matter is briefly presented. A detailed insight into theoretical electrodynamics cannot be given here. The interested reader might refer to standard textbooks on electrodynamics e.g. (J. Jackson: *Classical Electrodynamics* (Wiley, New York 1975), L. D. Landau, E. M. Lifshitz: *The Classical Theory of Fields* (Addison Wesley, New York 1971)).

8.2.1 Macro Phenomena

When light proceeds from one medium into another (e.g., from air into a solid substance), several things happen. Some of the light radiation may be transmitted through the medium, some will be absorbed, and some will be reflected or scattered at the interface between the two media.[12] The intensity I_0 of the beam incident to the surface of the solid medium must equal the sum of the intensities of the transmitted, absorbed, reflected and scattered beams, denoted as I_T, I_A, I_R and I_S, respectively, that is

$$I_0 = I_T + I_A + I_R + I_S \tag{8-4}$$

Radiation intensity, expressed in watts per square meter, corresponds to the energy being transmitted per unit of time across a unit area that is perpendicular to the direction of propagation.

An alternate form of Equation (8-4) is

$$T + A + R + S = 1 \tag{8-5}$$

where T, A, R, and S represent, respectively, the transmissivity (I_T/I_0), absorptivity (I_A/I_0), reflectivity (I_R/I_0), and scattering rate (I_S/I_0) or the fractions of incident light that are transmitted, absorbed, reflected and scattered by a material. Their sum must equal unity because all of the incident light is transmitted, absorbed, reflected or scattered. Materials that are capable of transmitting light with relatively little absorption and reflection are transparent—one can see through them. Translucent materials are those through which light is transmitted diffusely, that is, light is scattered within the interior to the degree that objects are not clearly distinguishable when viewed through a specimen of the material. Materials that are impervious to the transmission of visible light are termed opaque. Bulk metals are opaque throughout the entire visible spectrum—that is, all light radiation is either absorbed or reflected. However, electrically insulating materials can be made to be transparent. Furthermore, some semiconducting materials are transparent, whereas others are opaque.

12 一些光辐射可能透过介质，一些将被吸收，一些则两个介质之间的界面被反射或散射。

8.2.2 Mechanisms

8.2.2.1 Maxwell Explanation (Dielectric Function)

The starting point for an analysis of any interaction between electromagnetic waves with matter is Maxwell's equations. The static interaction for the dielectric displacement and the magnetic induction is described by

$$\begin{aligned} \nabla \cdot D &= \rho \\ \nabla \cdot B &= 0 \end{aligned} \tag{8-6}$$

whereas the dynamic interaction of the electric and magnetic fields is given by

$$\begin{aligned} \nabla \times E &= -\dot{B} \\ \nabla \times H &= j + \dot{D} \end{aligned} \tag{8-7}$$

E and B are the electric and magnetic fields; D and H are the electric displacement and the auxiliary magnetic fields; ρ and j are the charge and the current density. Material equations are needed to close Maxwell's equations

$$\begin{aligned} D &= \varepsilon_0 E + P \\ B &= \mu_0 H + M \end{aligned} \tag{8-8}$$

where P and M are the polarization and magnetization densities. ε_0 is the vacuum permittivity and μ_0 is the vacuum permeability.

The complete optical properties for any spatial combination of matter are included in the solution of Equation (8-6) and (8-7), which are closed by using the material Equations (8-8) and by using appropriate boundary conditions. For only a few special cases such a solution can be solved directly.

Ⅰ. In Vacuum

If we want to solve Equation (8-6) and (8-7) in infinite vacuum, we have the following boundary conditions and material equations

$$P(r)=0,\ \ M(r)=0,\ \ \rho(r)=0,\ \ j(r)=0 \tag{8-9}$$

where r = (x, y, z) are the three spatial coordinates. With these simplest possible boundary conditions, the Equation (8-8) are

$$\begin{aligned} D &= \varepsilon_0 E \\ B &= \mu_0 H \end{aligned} \tag{8-10}$$

After applying a few vector operations, one gets the wave equation for the electromagnetic field E in vacuum

$$\Delta E - \mu_0 \varepsilon_0 \ddot{E} = 0 \tag{8-11}$$

An identical wave equation can be derived for the magnetic field B. Equation (8-11) immediately defines the speed of light c (in vacuum) as shown in Equation (8-1).

Equation (8-11) can generally solve by all fields which fulfil $E(r,t) = E_0 \cdot f(kr \pm \omega t)$ involving any arbitrary scalar function f. The most common systems of function f are plane waves

$$E_s(r,t) = E_0 \,\mathrm{Re}\left(e^{-i(kr-\omega t)}\right) \tag{8-12}$$

These plane waves, with a time and spatial dependent phase θ = kr, are described by a wave vector $\vec{k}$, an angular frequency ω, and a corresponding wavelength λ =2π/k= 2πc/ω, where $k = |\vec{k}|$ is the absolute value of the wave vector. Describing an arbitrary field E interim of plane waves is identical to decomposing this electrical field into its Fourier components.

Ⅱ. In an Ideal Transparent Medium

We can describe an ideal transparent material by simply replacing the speed of light in vacuum by that of the medium.

$$c \to \frac{c}{n} \tag{8-13}$$

where n is the (in this case only real) refractive index of the material. In fact, most parts of an optical design can be done by treating optical glasses as such ideal transparent materials. Even though such an ideal material cannot exist in reality, optical glasses come very close to it (for electromagnetic radiation in the visible range). For such an ideal material the wave equation (8-11) reads

$$\Delta E - \frac{n^2}{c^2}\ddot{E} = 0 \tag{8-14}$$

It is solved again by plane transverse waves. Where the speed of light is now reduced to the speed of light in the transparent medium $c_{\text{med}} = c/n$ and the wavelength of the light wave is reduced to $\lambda_{\text{med}} = \lambda/n$.

Ⅲ. In an Isotropic, Homogeneous Medium

We now consider wave propagation in an ideal optical material. This is a nonmagnetic, homogeneous, isotropic, perfectly insulating medium, which is further a perfectly linear optical material.[13] Considering time dependence including retardation in the material Equation (8-7) leads to:

$$D(r,t) = \varepsilon_0 E(r,t) + P(r,t) \tag{8-15}$$

The polarizability is related to the electric field via the susceptibility χ. In the case of a homogeneous isotropic material χ is a scalar function.

$$P(r,t) = \int \mathrm{d}r' \int_{-\infty}^{t} \mathrm{d}t' \chi(r-r', t-t') E(r,t') \tag{8-16}$$

Fourier transformation in time and space de-convolutes the integral and leads to

$$P(k,\omega) = \chi(k,\omega) E(k,w) \tag{8-17}$$

where $\chi(k, \omega)$ is, in general, a complex analytic function of the angular frequency ω. The complex function $\chi(k, \omega)$ unifies the two concepts of a low-frequency polarizability χ' and a low-frequency conductivity σ of mobile charges to a single complex quantity

$$\lim_{\omega \to 0} \chi(\omega) = \chi'(\omega) + 4\pi i \frac{\sigma(\omega)}{\omega} \tag{8-18}$$

At high frequencies the separation of the two concepts breaks down, since above the frequencies of optical phonon modes in the IR the bound charges are unable to follow the electric field, whereas below the phonon modes the charges can follow this motion.[14] The usual form in which the susceptibility enters the equations for optical purposes is via the dielectric function

$$\varepsilon(k,\omega) = 1 + \chi(k,\omega) \tag{8-19}$$

Inserting the dielectric function into the material equation (8-15) gives

13 这是一个无磁性、匀质、各向同性、完美的绝缘介质，更是一种完美的线性光学材料。

14 在高频率区域，这两个物理量（极化率和导电性）的分开变得不可能。因为当电磁场频率高于红外区域的光学声子振动频率时，束缚电荷跟不上电场变化，而频率低于这类声子时，电荷是可以跟上电场变化的。

15 通过忽略介电函数为二阶张量的本质，我们将讨论限制在各向同性的理想光学材料上。

$$D(k,\omega)=\varepsilon_0\varepsilon(k,\omega)E(k,w) \tag{8-20}$$

Here, we restrict ourselves to ideal optically isotropic materials by neglecting the nature of the dielectric function as a second-rank tensor.[15] With the same steps as in Equations (8-10, 8-11), a wave equation can be derived which has the following form in Fourier space:

$$\left[k^2-\varepsilon(k,\omega)\frac{\omega^2}{c^2}\right]E_0=0 \tag{8-21}$$

The expression in brackets in Equation (8-21) defines the dispersion relation for an optically linear, homogeneous, isotropic material.

Ⅳ. General Form of the Dielectric Function

For most optical materials the dielectric function has a form in which a transparent frequency (or wavelength) window is bounded at the high energy site by electron—hole excitations (dominating the UV edge) and at the low energy site by IR absorptions given by optical phonon modes (lattice vibrations). The general form of the dielectric function is given by the Kramers-Heisenberg equation [Ch. Kittel: *Introduction to Solid State Physics* (Oldenbourg, Munich 1988)]

$$\varepsilon(k,\omega)=1+\sum_j\frac{\alpha_{k,j}}{\omega^2-\omega_{k,j}^2-i\omega\eta_{k,j}} \tag{8-22}$$

here $\alpha_{k,j}$ is the amplitude, $\omega_{k,j}$ the frequency and $\eta_{k,j}$ the damping of the particular excitation j. A schematic view of the dielectric function is plotted in Figure 8.3. Here we use a model for a transparent homogeneous, isotropic solid (such as glass) with one generic absorption at low energies (ω_{IR} in the infrared, IR) and another one at large phonon energies (ω_{UV} in the ultraviolet spectral range, UV). In the following this model solid is used to discuss optical material properties.

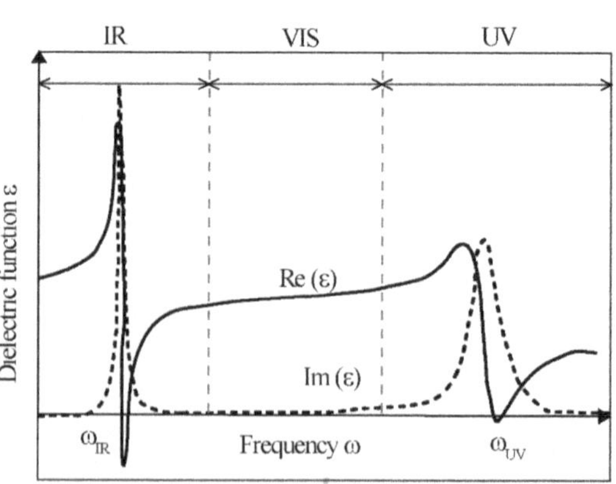

Figure 8.3 Dielectric function $\varepsilon(\omega)$ for the "model optical solid" with one generic absorption in the infrared ω_{IR} and a second one in the ultraviolet ω_{UV}. The dielectric function is plotted on a logarithmic energy scale. The solid line is the real part and the dashed line is the imaginary part of $\varepsilon(\omega)$.

Ⅴ. Refractive Index

The refractive index n is the most widely used physical quantity in optical design. It is the square root of the dielectric function. The dynamic refractive index is generally a complex quantity

$$\tilde{n}(\omega)=n(\omega)+i\kappa(\omega) \tag{8-23}$$

and must fulfill the Kramers–Kronig relations. In practical use, the wavelength dependence is often exploited

$$\tilde{n}(k,\lambda)=\sqrt{\varepsilon(k,2\pi c/\lambda)} \tag{8-24}$$

If one further restricts to wavelengths which is far away from absorption ($\omega^2-\omega^2_{k,j} \ll \omega\eta_{k,j}$), the Sellmeier formula which is widely used for characterizing optical materials, is obtained

$$n(\lambda)^2 \approx 1+\sum_j \frac{B_j\lambda^2}{\lambda^2-\lambda_j^2} \tag{8-25}$$

Normally B_j and λ_j are just fitting constants to describe the dispersion of the refractive index over a certain wavelength range. They are, however, connected to the microscopic fundamental absorption behavior of the material. Sometimes also $n(\lambda)$ and not $n(\lambda)^2$ is approximated with a Sellmeier formula. Since $n(\lambda)$ as well as $n(\lambda)^2$ are complex differential (analytic) functions both formula give refractive indices and dispersions with the same accuracy. However, care has to be taken, which quantity is expressed when using a Sellmeier formula.

Ⅵ. Refraction and Reflection

We now derive the laws of refraction and reflection for the ideal transparent medium just described.[16] They are obtained by solving Maxwell's equations at the (infinite) boundary between two materials of different refractive indices n_1 and n_2 (see Figure8.4). As boundary conditions one obtains that the normal component of the electric displacement (and magnetic induction) and the tangential component of the electric (and magnetic) field have to be continuous at the interface

$$D_1^n=D_2^n, E_1^t=E_2^t \tag{8-26}$$

Further, a phase shift of an incoming wave occurs upon reflection

$$\theta_r=\pi-\theta_i \tag{8-27}$$

where θ_r, θ_i are the phases of the reflected and incident wave, respectively. If we solve Maxwell's equations for an incoming plane wave (applying the boundary conditions stated above), Snell's law of refraction is obtained

$$n_1\sin\alpha_1=n_2\sin\alpha_2 \tag{8-28}$$

16 现在，我们推导出刚刚描述过的透明理想介质的折射和反射规律。

together with that of reflection

$$\alpha_r = \alpha_1 \tag{8-29}$$

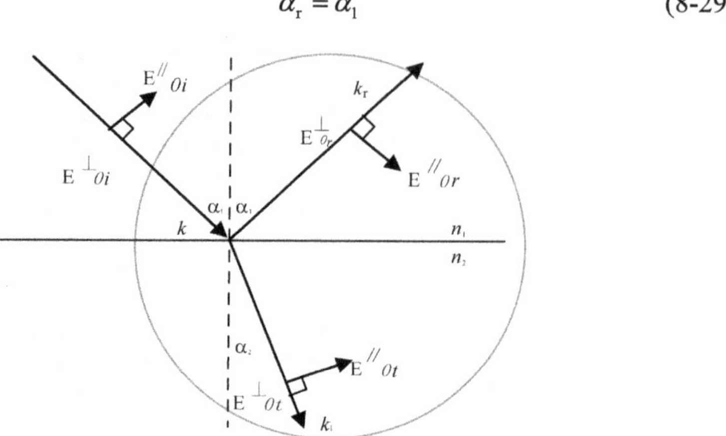

Figure 8.4 The polarization directions of the E field for reflection and refraction at an interface between two optical materials of different refractive indices are shown. A circle indicates that the vector is perpendicular to the plane shown.

Now, the electric field E is decomposed into its components which are defined relative to the plane outlined by the three beams of incoming, transmitted and reflected light. This decomposition is shown in Figure 8.4. The coefficients for reflection and transmission are defined as

$$\begin{aligned} r_{\parallel} &= \frac{E_{0r}^{\parallel}}{E_{0i}^{\parallel}};\ t_{\parallel} = \frac{E_{0t}^{\parallel}}{E_{0i}^{\parallel}} \\ r_{\perp} &= \frac{E_{0r}^{\perp}}{E_{0i}^{\perp}};\ t_{\perp} = \frac{E_{0t}^{\perp}}{E_{0i}^{\perp}} \end{aligned} \tag{8-30}$$

The Fresnel formula for these coefficients can be derived as

$$\begin{aligned} r_{\perp} &= \frac{n_1\cos(\alpha_1) - n_2\cos(\alpha_2)}{n_1\cos(\alpha_1) + n_2\cos(\alpha_2)} = -\frac{\sin(\alpha_1 - \alpha_2)}{\sin(\alpha_1 + \alpha_2)} \\ r_{\parallel} &= \frac{n_2\cos(\alpha_1) - n_1\cos(\alpha_2)}{n_1\cos(\alpha_2) + n_2\cos(\alpha_1)} = -\frac{\tan(\alpha_1 - \alpha_2)}{\tan(\alpha_1 + \alpha_2)} \\ t_{\perp} &= \frac{2n_1\cos(\alpha_1)}{n_1\cos(\alpha_1) + n_2\cos(\alpha_2)} = -\frac{2\sin(\alpha_2)\cos(\alpha_1)}{\sin(\alpha_1 + \alpha_2)} \\ t_{\parallel} &= \frac{2n_1\cos(\alpha_1)}{n_1\cos(\alpha_2) + n_2\cos(\alpha_1)} = -\frac{2\sin(\alpha_2)\cos(\alpha_1)}{\sin(\alpha_1 + \alpha_2)\cos(\alpha_1 - \alpha_2)} \end{aligned} \tag{8-31}$$

Here the usual convention has been used that the coefficients of reflectivity obtain an additional minus sign in order to indicate back traveling of light. The quantities that are measured in an experiment are intensities. The relationship between the intensities defines the reflectivity and transmissivity of a material:

$$R_{\perp} := |r_{\perp}|^2,\ R_{\parallel} := |r_{\parallel}|^2 \qquad (8\text{-}32)$$
$$T_{\perp} := |t_{\perp}|^2,\ T_{\parallel} := |t_{\parallel}|^2$$

The angular-dependent coefficients of reflection from (8-31) are displayed in Figure 8.5. In Figure 8.5a the case of light propagating from an optically thin medium with refractive index n_1 to an optically thicker medium with refractive index $n_2 > n_1$ is plotted. At the so called Brewster angle α_B the reflected light is completely polarized. α_B is given by the condition $\alpha_1 + \alpha_2 = \pi/2$. Therefore, the Brewster angle α_B results as a solution of

$$\alpha_1 = \frac{\pi}{2} - \arccos\left(\frac{n_2}{n_1}\cos\alpha_1\right) \qquad (8\text{-}33)$$

which gives $\alpha_B = \arctan(n_2/n_1)$. In Figure 8.5b the case of light propagating from an optically thick to an optically thinner medium is plotted. Here an additional special angle occurs—the angle of total reflection, α_T. All light approaching the surface at an angle larger than α_T is totally reflected. At $\alpha_1 = \alpha_T$ the angle for refraction in the medium with refractive index n_2 is $\alpha_2 = \pi/2$. For α_T it follows that

$$\alpha_T = \arcsin\left(\frac{n_2}{n_1}\right) \qquad (8\text{-}34)$$

Evaluating (8-31) for the special case of incident light as $\lim_{\alpha\to 0}$ allows one to calculate the reflectivity for normal incidence

$$R_{\text{norm}} = \left|\frac{n_1 - n_2}{n_1 + n_2}\right|^2 \qquad (8\text{-}35)$$

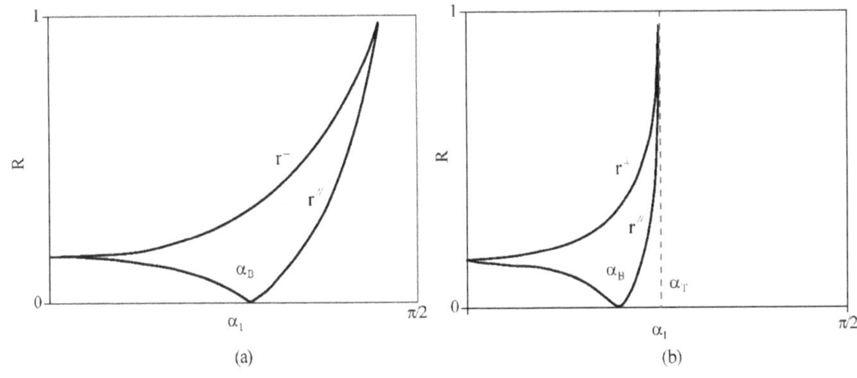

Figure 8.5 The reflection coefficients are plotted as a function of incident scattering angle for light propagating from a medium of (a) smaller refractive index into a medium of larger index and (b) larger refractive index into a medium of smaller index. Here, total reflection occurs at an angle α_T and α_B is the Brewster angle.

8.2.2.2 Microscale Interaction of Light with Solid

Ⅰ. Absorption in Glass

Absorption in optical medium is characterized by a decrease in

17 光学介质的吸收特征在于，不考虑由表面的反射损失或者内部散射时，光通过样品后透射光强度减弱。

transmitted light intensity through the sample that is not accounted for by reflection losses at the surface or scattering by inclusions.[17] Absorption is not uniform across all wavelengths of interest (UV−VIS−IR ≈ 200-2000 nm) and can be characterized by absorption bands. The absorption bands are due to both intrinsic and extrinsic effects.

The quantity used to discuss absorption as a function of wavelength in optical medium is the transmittance (T), which is the ratio of the transmitted light intensity (I) to the initial light intensity (I_0) after passing through a sample plate of thickness L

$$T = I / I_0 \tag{8-36}$$

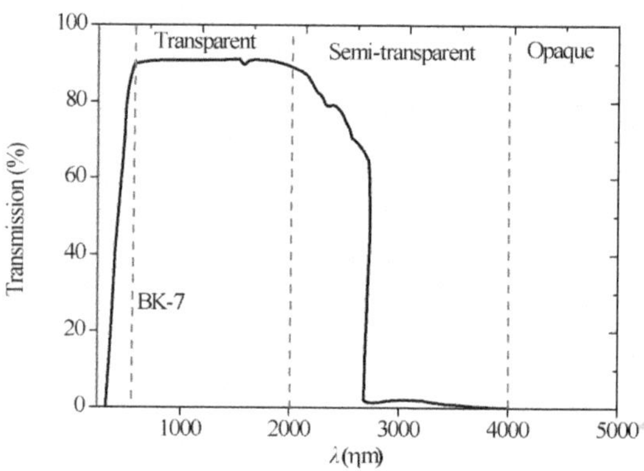

Figure 8.6 Transmission through 10 mm thick BK-7 optical glass showing regions of transparency (visible), semi- transparency (near IR) and opacity (far IR).

18 玻璃，作为典型的简单固态光学材料，从近紫外到近红外范围都是透明的。

Glasses, as typical candidates for the simple optical solid model, are transparent from the near UV to the near IR.[18] The regions of transparency and semi-transparency in the visible and near IR are shown for a typical optical glass (BK-7) in Figure 8.6. The intrinsic absorption is dominated by electronic transitions in the UV−VIS range, while in the IR it is dominated by molecular vibrations. There are multiple extrinsic mechanisms, which occur with the inclusion of ionic complexes, insulators (crystals and phase separation), semiconductors, and conductors.

(1) Intrinsic Absorption

The UV−VIS absorption is dominated by electronic transitions. The electronic transitions that occur in the UV−VIS range involve a transition from the highest occupied molecular orbital (HOMO) to the lowest unoccupied molecular orbital (LUMO). The difference in energy (ΔE) between the LUMO (E_{LUMO}) and HOMO(E_{HOMO}) energy levels corresponds to the energy of the corresponding absorption maxima, which can be related to the wavelength, and frequency of the absorption

$$\Delta E = E_{\text{LUMO}} - E_{\text{HOMO}} = h\nu = hc / \lambda \tag{8-37}$$

The bonds present dominate the energy gap between the HOMO and LUMO, and thus the energy of the transition.[19] For example, the UV absorption is very similar between crystalline and fused silica, indicating that the short-range structure dominates the electronic transitions. It is in the long-range structure of silica that fused and crystalline differ due to the random connections of the [SiO_4] tetrahedral (i.e., one Si atom attached with four O atoms to from tetrahedral) in fused silica.

19 化学键的存在决定了HOMO和LUMO之间的能隙，从而决定着跃迁的能量。

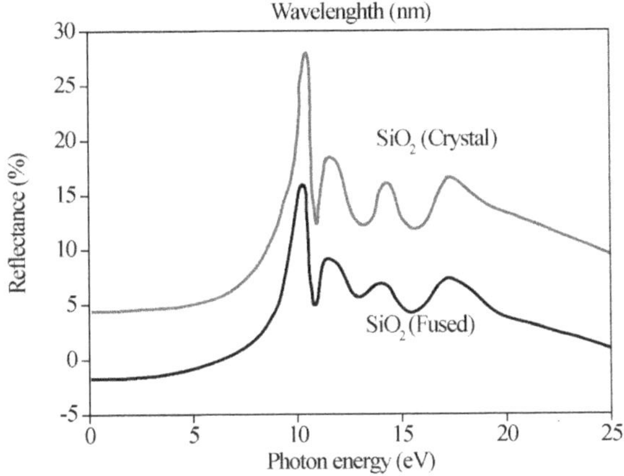

Figure 8.7 Intrinsic absorption of glassy and crystalline SiO_2 [H. Bach, N. Neurorth (Eds.): *The Properties of Optical Glass* (Springer, Berlin, Heidelberg 1998)]. The values for fused silica have been lowered by 5% for improved visibility.

The absorption band present in Figure 8.7 represents the intrinsic UV absorption in silica. The increase in peak width is due to the distribution of tetrahedral in the fused silica. The lowest energy peak is associated with a Wannier exciton, but the next three peaks are associated with electronic transitions between the oxygen 2p electrons with the silicon 3d electrons.

The intrinsic absorption of glass in the IR is due to molecular vibrations, which begin at the multi-phonon edge.[20] These vibrations can be understood using a harmonic oscillator model.

20 玻璃在红外区域的固有吸收是由分子振动引起的，该振动始于多声子边缘。

(2) Ionic Absorption

Absorption in glass due to ionic impurities involves transition-metal (TM) and rare-earth (RE) ions that act as the central atom (CA) in ligand complexes absorbing light in different wavelength bands. The absorption due to TM and RE ions results from electronic transitions within the d or f orbital of the TM or RE elements. The energy of the absorption band is dependent upon the coordination environment of the CA (coordination number and geometry), the chemical identity of the CA (Cr, Co, Cu ...) and the ligand (O versus F). The basic composition of the glasses affects the nature of the ligand bonds available for the CA. This change affects the energy of the absorption bands and thus the resulting color of the glass. This effect is taken advantage of when designing filter glass, and can be used to explain the differences in color of the same TM or RE in different glass host. The absorption results from a splitting of the d orbital for TM elements and the 4f orbital for the RE elements in the presence of a ligand field. The absorption cross section for TM elements is approximate 100 times that of the absorption cross section for RE elements. This results in distinct coloration at the ppm concentration level for TM elements.

Figure 8.8 shows the variation in absorption due to various TM elements in BK7 glass at the 1 ppm dopant level. The intensity of cobalt absorption is clearly shown. These intraband transitions due to electronic transitions within the ionic species d or f orbital are formally Laporte-forbidden transitions by selection rules, so the intensity is low. Charge-transfer transitions involve an electron transition between the CA and the coordinating ligand (interband). These interband transitions are formally allowed, and thus have approximate 100–1000 times greater intensity than the intraband electronic transitions.

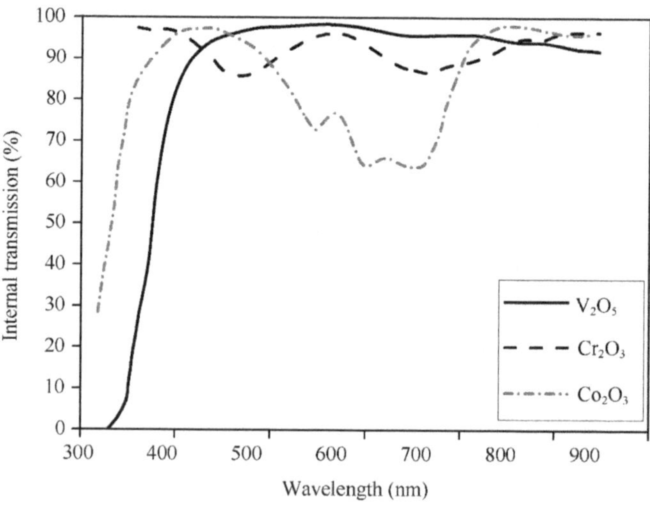

Figure 8.8 Absorption due to 1 ppm of transition-metal elements in BK7 glass in 100mm thick samples.

(3) Absorption by Semiconductor Particles

Semiconductor particles in glass are typically too small (1–10 nm) to scatter visible light, however they absorb light over a continuum of wavelengths corresponding to energies greater than the band gap (Eg) of the semiconductor particles.[21] The band gap of the semiconductor particles is controlled by the size of the particle and the chemical composition.

21 玻璃中的半导体颗粒，通常太小（1 - 10 nm），不足以引起可见光的散射；但是在大于半导体颗粒带隙（Eg）的一个连续波长范围内，它们是吸收光的。

Typically, semiconductor glasses are melted with Zn, Cd, S, Se and Te raw materials in the batch and upon casting the glasses cool colorless. Secondary heat treatment (striking) results in the crystallization of various semiconductor crystal phases in the glass, or a mixture there of: ZnS, ZnSe, ZnTe, CdS, CdSe, and CdTe. The size and distribution of the semiconductor particles can be controlled by the heat treatment and thus so can the optical properties of the resulting glasses. With the proper heat treatment, one glass can be "struck" into multiple glasses with absorptions leading to red, yellow and orange coloration. When the band gap of the particles is large enough, the absorption edge is shifted into the UV and the glass appears colorless.[22] The opposite can also occur when the band gap energy is so low that it absorbs all visible light, and the sample appears black. Semiconductor-doped glasses are often used as low-pass filter glasses because of their sharp absorption cutoff.

22 当粒子的带隙足够大时，吸收边界转移到紫外区，玻璃呈现出无色。

(4) Absorption by Conducting Particles

The coloration of glasses due to small conducting particles in the glass is a combination of both absorption and scattering effects. For small metallic particles, the scattering is governed specifically by Mie's theory but can also be treated in general with the Rayleigh scattering theory.[23] Mie and Rayleigh scattering are covered next section.

23 对于小的金属颗粒，这种散射尤其受控于 Mie 散射理论，但也可以用瑞利散射理论进行解释。

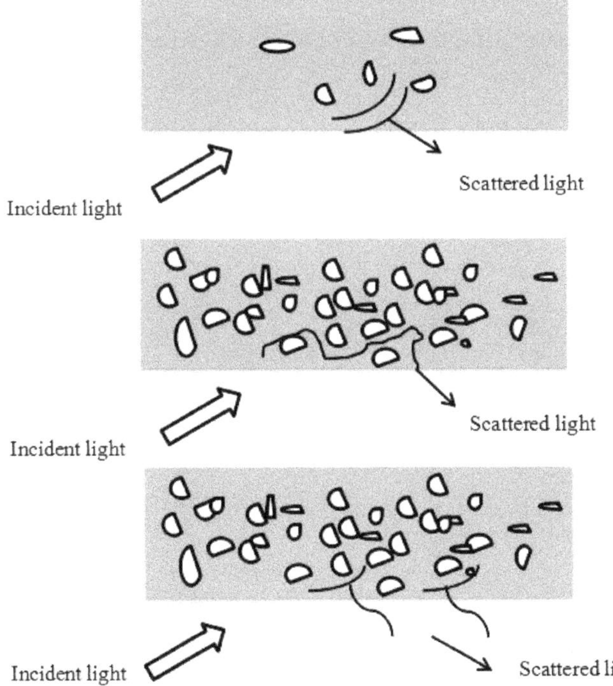

Figure 8.9 Illustration of the subdivision of volume scattering: single scattering (top), multiple scattering (middle), and coherent scattering (bottom).

Ⅱ. Scattering

Scattering is defined here as energy absorption of incident light followed by re-emission of part of this light at the same frequency.[24] Thus inelastic effects such as Raman scattering or Brillouin scattering are not included.

24 在这里，散射被定义为入射光的能量被吸收后，一部分的这种相同频率光的再发射。

The origins of light scattering are surface irregularities. From this point, volume scattering is scattering at the surface of particles in a certain volume. Surface scattering is scattering due to surface irregularities. Obviously scattering is strongly related to diffraction, as seen from the definition of scattering. Indeed, diffraction is scattering by a flat particle.

(1) Volume Scattering

Volume scattering is subdivided into single scatter, multiple scattering, coherent scattering, etc (Figure 8.9). Volume scattering can be regarded as the sum of single scattering events as long as the density of scattering particles is not too high.[25]

25 只要散射粒子的密度不是太高，体积散射可以视为一个个单一散射事件的总和。

- **Single Scattering (Mie Scattering)**

A very important geometry of single scatters is the sphere.[26] In 1908, Mie derived an analytic theory that completely describes the scattering from a sphere embedded in a non-conducting medium. Many basic characteristics of scattered power can be obtained by studying scattering by a sphere. The calculated scattering cross section normalized to the geometric cross section of a glass sphere (= a^2 where a is the sphere radius) can be seen in Figure 8.10.

26 单一散射的一个非常重要的几何模型是球体。

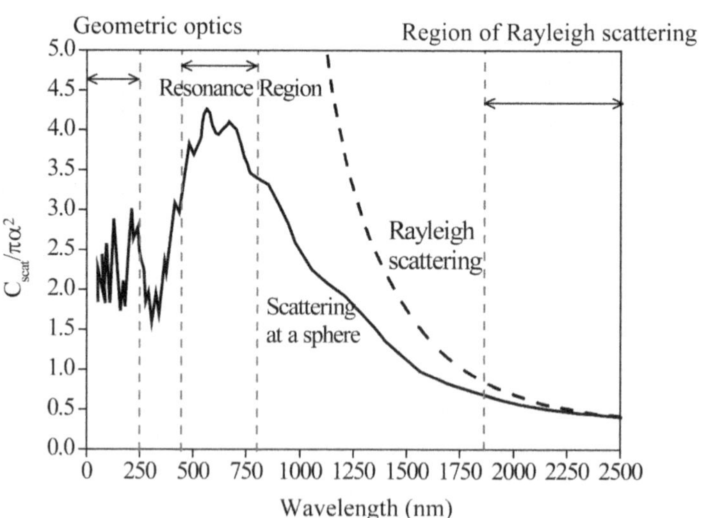

Figure 8.10 Scattering cross section per sphere area versus wavelength for a glass sphere (n_{sphere}= 1.50) of radius a = 400 nm in air (n_{medium}= 1.0).

Figure 8.10 demonstrates the different wavelength dependence of the scattering cross section and thus of the scattered power. Depending on the geometric sphere size (characterized by the diameter $2a$) relative to the wavelength, three important regions can be identified.

- $\lambda << 2a/n_{medium}$, region of geometric optics. Surprisingly the term $C_{scat}/\pi a^2 = 2$. This effect is called the extinction paradox and the factor 2 is due to diffraction effects.
- $\lambda \approx 2a/n_{medium}$, resonance effect, the term $C_{scat}/\pi a^2$ has its maximum. Therefore, the highest scattered power can be measured if the geometric size of an object is of the order of the wavelength.
- $\lambda >> 2a/n_{medium}$, Rayleigh scattering. The scattered power is proportional to $1/\lambda^4$. This effect is responsible for the blue sky, where small particles inside the atmosphere scatter blue light (≈ 400 nm) more than red light (≈ 800 nm).
- **Multiple Scattering**

Multiple scattering is the weighted superposition of many single scatters without interference effects.[27] Multiple scattering with interference effects is called coherent scattering and will discussed later.

Multiple scattering can only be described numerically and must be used whenever the packing fraction η is larger than about 0.3. Based on the knowledge of the scattering function (described by the differential scattering cross section) of a single scatter, the overall scattering function is calculated by tracing rays throughout the volume.

- **Coherent Scattering**

Coherent scattering i.e., the phenomenon in which a series of independent single-particle scattering events is replaced by collective light manipulation from an ensemble of different scattering centers is the focus of this section.[28] When dealing with materials containing scatters separated by distances greater than the coherence length (the distance necessary for propagating waves to lose their coherence), the scattering events can be treated as independent occurrences even under multiple scattering conditions. Such scattering is called incoherent, and the resulting intensity of such radiation is simply a summation of the intensity contributions of all the independent scattering centers.

However, when the distances between scatters are on the order of or less than the coherence length, coherent scattering effects must be considered. In this case, wave package interference takes place as the package is capable of interacting with more than one scattering center at a

27 在不考虑干涉效应时，多重散射是许多单一散射的加权叠加。

28 相干散射，即一系列独立的单粒子散射被含不同散射中心的集体光操控所取代，这是本节的重点。

time; several scatters distort the photon package simultaneously. Thus, a relationship exists between the phases of the light signals arising from the different scatters. The events are no longer separate; they are correlated, and the resulting intensity of transmitted light is no longer a simple sum. Thus, in coherent scattering, wave packages experience a combined interaction with several scattering centers that affects their transport through the medium. In photon group interference (a quantum effect), the wave package comes apart upon scattering but the different sinusoidal components meet again and interfere. This interference affects the intensity of light transmitted through the sample; thus, coherent scattering studies can yield valuable structural information.

(2) Surface Scattering

29 表面散射的发生是由于表面的不规则性。

Surface scattering is due to surface irregularities.[29] The Rayleigh criterion assesses when scattering is important and must be included. How roughness is related to the scattered light can be estimated by the so called total integrated scatter (TIS). TIS is an integrated (over all directions) value, so the angular dependence of scattering is not covered. The angular dependence of surface scattering can be estimated by assuming a sinusoidal surface roughness.

Perfectly smooth surfaces only reflect light specularly. As stated in previous section, due to surface irregularities light is also scattered in directions other than the specular one. This is illustrated in Figure 8.11.

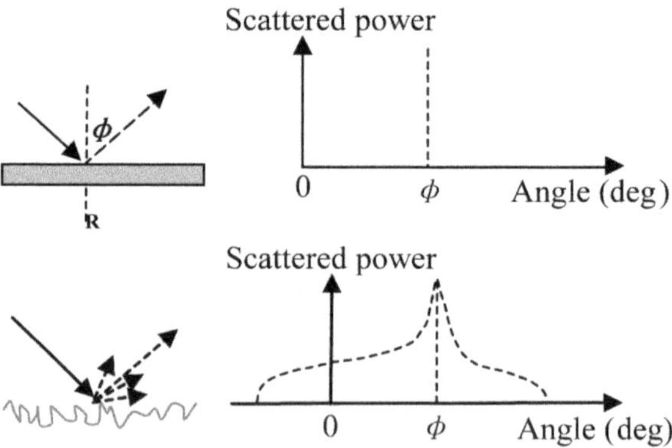

Figure 8.11 Specular light reflection at a perfectly smooth surface (top) and light scattering at a rough surface (bottom). Dotted line = scattered (reflected) light, solid line = incident light.

30 表面越粗糙，远离镜面反射方向的散射光越多。

As can be seen from Figure 8.11 in addition to the specular reflected light, scattered light is also generated due to the rough surface. This stray light is scattered over a wide angle around the specular direction. The rougher the surface the more light is scattered away from the specular direction.[30]

But when is a surface smooth? This is answered by the so called Rayleigh criterion.

As can be seen from Figure 8.12, a plane wave incident on a rough surface experiences a phase difference between the two outer "rays". If this phase difference Ψ is smaller than $\pi/2$, the surface is called smooth (i.e. Rayleigh criterion)

$$\Delta h < \frac{\lambda}{8\cos\alpha} \tag{8-38}$$

where Δh is the RMS roughness (standard deviation).

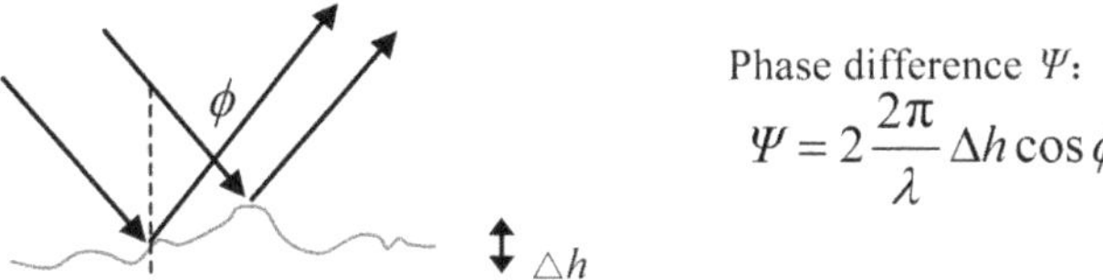

Figure 8.12 A plane wave incident on a rough surface and its phase difference due to different optical paths.

The way in which surface roughness is related to scattered light can be seen in the so called TIS.[31] The TIS is the ratio of scattered power P_{scat} (without specular reflected light) in one hemisphere to the specularly reflected power P_{ref}. If the surface roughness is smaller than the wavelength, then TIS is given by

31 表面粗糙度与散射光的关联，可以在被称为总积分散射（TIS）中导出。

$$\text{TIS} = \frac{P_{scat}}{P_{ref}} = \left[\frac{4\pi\cos\alpha_R}{\lambda}\right]^2 (\Delta h)^2 \tag{8-39}$$

where α_R is the angle of incidence (reflectance) and Δh is the RMS roughness of the surface (Figure 8.13).

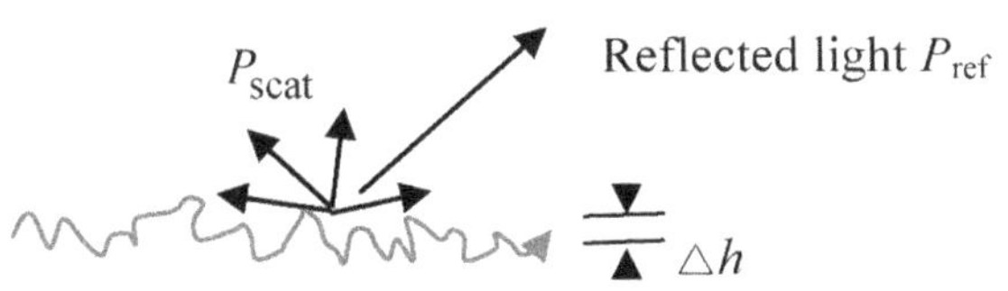

Figure 8.13 Surface roughness and TIS.

From Equation (8-39) it can be seen that the scattered power is proportional to the square of the surface roughness and depends as $1/\lambda^2$ on the wavelength. The surface roughness vs wavelength for various TIS values is plotted in Figure 8.14. Consider a specific glass surface and incident light at 500 nm. For this wavelength, only a thousandth part of the reflected light should be allowed to be scattered (TIS = 10^{-3}). For this wavelength and TIS value the RMS roughness is just 1.3nm. This value is between normal and super-polished glass surface. Hence, normal polishing is not sufficient here. Therefore, a more expensive super-polishing must be used.

32 总积分散射也可以被用于估算表面粗糙度。

The TIS can also be used for estimating the surface roughness.[32] By measuring the TIS at a determined wavelength the surface roughness can be calculated with the help of Equation (8-39).

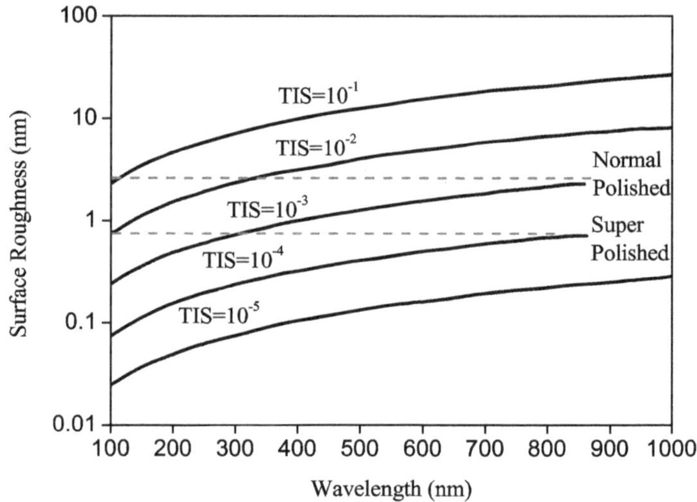

Figure 8.14 Surface roughness plotted against wavelength for various TIS values. Typical roughness values for super and normal polished glass are also plotted. Note the logarithmic scale of the roughness axis.

Ⅲ. Atomic/Electronic Interactions

33 在固体材料内发生的光学现象，涉及电磁辐射和原子、离子、和/或电子之间的相互作用。

The optical phenomena that occur within solid materials involve interactions between the electromagnetic radiation and atoms, ions, and/or electrons.[33] Two of the most important of these interactions are electronic polarization and electron energy transitions.

(1) Electronic Polarization

34 电磁波的一个组成部分，仅仅是一个快速波动的电场。

One component of an electromagnetic wave is simply a rapidly fluctuating electric field.[34] For the visible range of frequencies, this electric field interacts with the electron cloud surrounding each atom within its path in such a way as to induce electronic polarization or to shift the electron cloud relative to the nucleus of the atom with each change in direction of electric field component, as demonstrated in Figure 8.15. Two consequences of this polarization are as follows: some of the radiation energy may be absorbed, and light waves are decreased in velocity as they pass through the medium; the second consequence is manifested as refraction.[35]

35 这一极化有两个结果：部分辐射能量可以被吸收，导致速度的降低；另一个结果表现为折射。

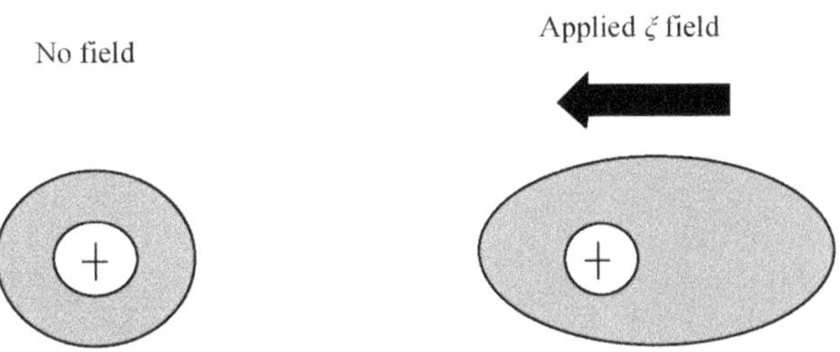

Figure 8.15 Electronic polarization that results from the distortion of an atomic electron cloud by an electric field.

(2) Electron transitions

The absorption and emission of electromagnetic radiation may involve electron transitions from one energy state to another.[36] The microscopic processes between these absorption and emission levels can have different origins, for example several levels of 4f electron systems of a rare earth atom. For the sake of this discussion, consider an isolated atom, the electron energy diagram for which is represented in Figure 8.16. An electron may be excited from an occupied state at energy E_2 to a vacant and higher-lying one, denoted E_4, by the absorption of a photon of energy. The change in energy experienced by the electron, ΔE, depends on the radiation frequency as follows

$$\Delta E = h\nu \tag{8-40}$$

where, again, h is Planck's constant. At this point, it is important to understand several concepts: First, because the energy states for the atom are discrete, only specific ΔE exist between the energy levels; thus, only photons of frequencies corresponding to the possible ΔE for the atom can be absorbed by electron transitions. Furthermore, all of a photon's energy is absorbed in each excitation event.[37]

36 电磁辐射的吸收和发射可能涉及从一个能量状态到另一个能量状态的电子跃迁。

37 进一步，在每一个激发事件中，光子的能量被全部吸收。

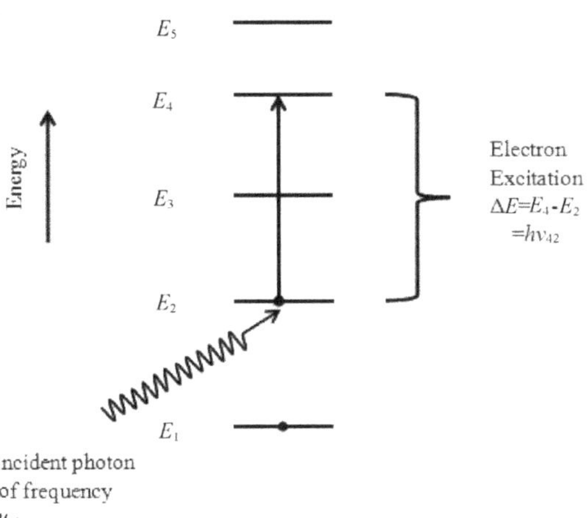

Figure 8.16 For an isolated atom, a schematic illustration of photon absorption by the excitation of an electron from one energy state to another. The energy of the photon($h\nu_{42}$) must be exactly equal to the difference in energy between the two states ($E_4 - E_2$).

38 第二个重要的概念是，一个受激的电子不能无限地停留在激发态；一段短时间后，通过电磁辐射的再发射或者发热，它跃迁或衰变到基态或未激发的能级。

39 考虑到金属的电子能带结构，较高能量的带只是部分地充满电子。金属是不透明的，因为在可见光范围内的入射辐照可以将电子激发到费米能级以上的空置能级（由于入射光被吸收，金属不透明）。

A second important concept is that a stimulated electron cannot remain in an excited state indefinitely; after a short time, it falls or decays back into its ground state, or unexcited level, with reemission of electromagnetic radiation or dissipating heat.[38] Several decay paths are possible, and these are discussed later with material optical property. In any case, there must be a conservation of energy for absorption and emission electron transitions.

8.3 Optical Property of Materials

8.3.1 Metals

Considering the electron energy band schemes for metals, a high-energy band is only partially filled with electrons. Metals are opaque because the incident radiation having frequencies within the visible range excites electrons into unoccupied energy states above the Fermi energy, as demonstrated in Figure 8.17a.[39] As a consequence, the incident radiation is absorbed, in accordance with Equation (8-40). Total absorption is within a very thin outer layer, usually less than 100 nm; thus only metallic films thinner than 100 nm are capable of transmitting visible light.

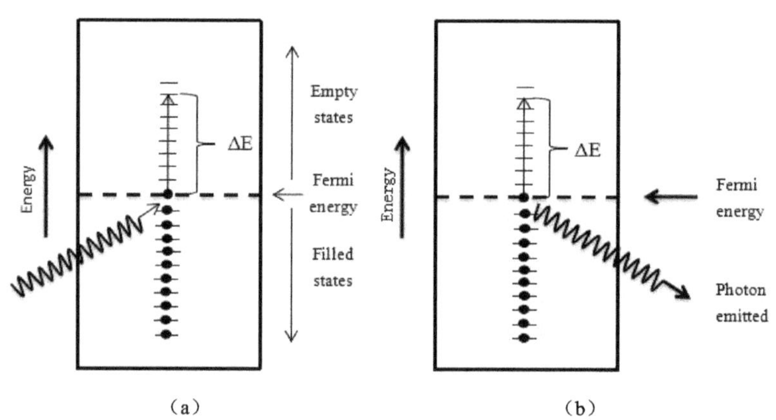

Figure 8.17 (a) Schematic representation of the mechanism of photon absorption for metallic materials in which an electron is excited into a higher-energy unoccupied state. The change in energy of the electron E is equal to the energy of the photon. (b) Reemission of a photon of light by the direct transition of an electron from a high to a low energy state.

All frequencies of visible light are absorbed by metals because of the continuously available empty electron states, which permit electron transitions as in Figure 8.17a. In fact, metals are opaque to all electromagnetic radiation on the low end of the frequency spectrum, from radio waves, through infrared and the visible, and into about the middle of the ultraviolet radiation. Metals are transparent to high-frequency radiation.

Most of the absorbed radiation is reemitted from the surface in the form of visible light of the same wavelength, which appears as reflected light; an electron transition accompanying re-radiation is shown in Figure 8.17b. The reflectivity for most metals is between 0.90 and 0.95, some small fraction of the energy from electron decay processes is dissipated as heat.

Because metals are opaque and highly reflective, the perceived color is determined by the wavelength distribution of the radiation that is reflected and not absorbed.[40] A bright silvery appearance when exposed to white light indicates that the metal is highly reflective over the entire range of the visible spectrum. In other words, for the reflected beam, the composition of these reemitted photons, in terms of frequency and number, is approximately the same as for the incident beam. Aluminium and silver are two metals that exhibit this reflective behavior. Copper and gold appear red-orange and yellow, respectively, because some of the energy associated with light photons having short wavelengths is not reemitted as visible light.

40 因为金属是不透明且高反射的，我们感知到的颜色是由不被吸收的反射光的波长分布决定的。

8.3.2 Nonmetals

By virtue of their electron energy band structures, nonmetallic materials may be transparent to visible light.[41] Therefore, in addition to reflection and absorption, refraction and transmission phenomena must also be considered.

41 缘于非金属的电子能带结构，它们可以对可见光透明。

(1) Absorption

Nonmetallic materials may be opaque or transparent to visible light, if transparent, they often appear colored. In principle, light radiation is absorbed in this group of materials by two basic mechanisms that also influence the transmission characteristics of these nonmetals. One of these is electronic polarization. Absorption by electronic polarization is important only at light frequencies in the vicinity of the relaxation frequency of the constituent atoms. The other mechanism involves valence band-conduction band electron transitions, which depend on the electron energy band structure of the material.

Absorption of a photon of light may occur by the promotion or excitation of an electron from the nearly filled valence band, across the band gap, and into an empty state within the conduction band,[42] as demonstrated in Figure 8.18a, a free electron in the conduction band and a hole in the valence band are created. Again, the energy of excitation E is related to the

42 光子的吸收可以通过将电子从几乎充满的价带，穿过带隙，跃迁或者激发到导带内空置的态。

absorbed photon frequency through Equation (8-41). These excitations with the accompanying absorption can take place only if the photon energy is greater than that of the band gap E_g, that is, if

$$h\nu > E_g \tag{8-41}$$

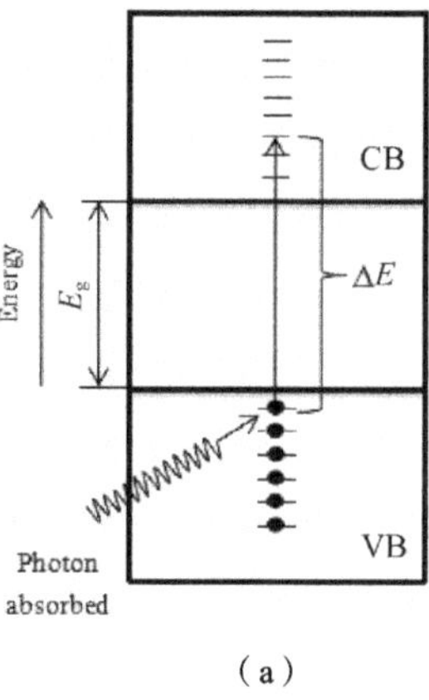

(a)

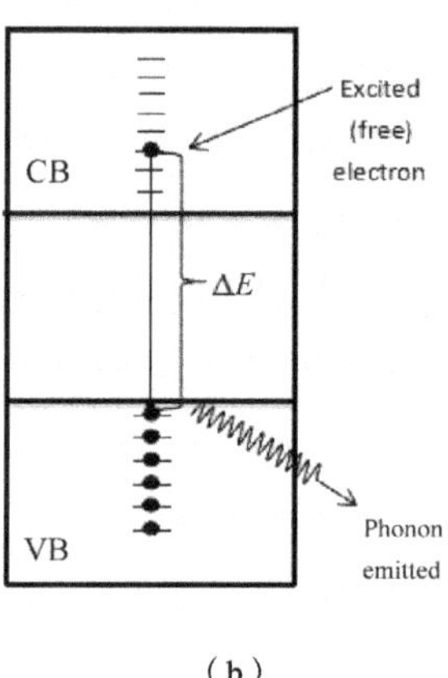

(b)

Figure 8.18 (a) Mechanism of photon absorption for nonmetallic materials in which an electron is excited across the band gap, leaving behind a hole in the valence band. The energy of the photon absorbed is E, which is necessarily greater than the band gap energy E_g. (b) Emission of a photon of light by a direct electron transition across the band gap.

The minimum wavelength for visible light, λ(min), is about 0.4 μm, thus the maximum band gap energy E_g(max)for which absorption of visible light can take place is 3.1 eV. In other words, no visible light is absorbed by nonmetallic materials having band gap energies greater than about 3.1eV, these materials, if of high purity, appear transparent and colorless. However, the maximum wavelength for visible light, λ (max), is about 0.7 μm, computation of the minimum band gap energy E_g (min) for which there is absorption of visible light gives 1.8 eV.

This result means that all visible light is absorbed by valence band to conduction band electron transitions for semiconducting materials that have band gap energies less than about 1.8 eV, thus, these materials are opaque.[43] Only a portion of the visible spectrum is absorbed by materials having band gap energies between 1.8 and 3.1 eV, consequently, these materials appear colored.

43 这一结果意味着，通过带隙小于约 1.8 eV 的半导体材料中价带到导带的电子跃迁，所有可见光被吸收。因此，这些材料是不透明的。

Every nonmetallic material becomes opaque at some wavelength, which depends on the magnitude of its E_g. For example, diamond, having a band gap of 6.6 eV, is opaque to radiation having wavelengths less than about 0.22 μm.

Interactions with light radiation can also occur in dielectric solids having wide band gaps, involving other than valence band−conduction band electron transitions. If impurities or other electrically active defects are present, electron levels within the band gap may be introduced, such as the donor and acceptor levels, except that they lie closer to the center of the band gap.[44] Light radiation of specific wavelengths maybe emitted as a result of electron transitions involving these levels within the band gap. For example, consider Figure 8.19a, which shows the valence band–conduction band electron excitation for a material that has one such impurity level. Again, the electromagnetic energy that is absorbed by this electron excitation must be dissipated in some manner, several mechanisms are possible. For one, this dissipation may occur via direct electron and hole recombination according to the reaction

$$\text{electron} + \text{hole} \rightarrow \text{energy}\ (\Delta E) \quad (8\text{-}42)$$

which is represented schematically in Figure 8.19b. In addition, multiple-step electron transitions may occur that involve impurity levels lying within the band gap. One possibility, as indicated in Figure 8.19b, is the emission of two photons: one is emitted as the electron drops from a state in the conduction band to the impurity level, the other as it decays back into the valence band. Alternatively, one of the transitions may involve the generation of a phonon (Figure 8.19c), in which the associated energy is dissipated in the form of heat.

44 如果杂质或其他电活性缺陷存在，带隙内可以引入电子能级，例如排除更接近于带隙的中心的情形，将产生给体或者受体。

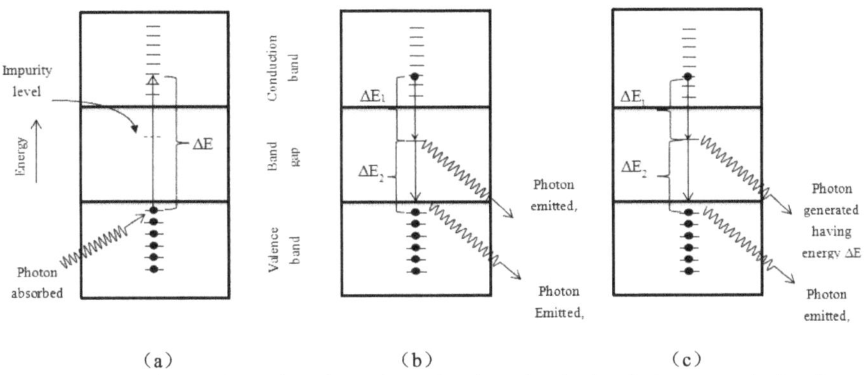

Figure 8.19 (a) Photon absorption via a valence band–conduction band electron excitation for a material that has an impurity level that lies within the band gap. (b) Emission of two photons involving electron decay first into an impurity state and finally to the ground state. (c) Generation of both a phonon and a photon as an excited electron falls first into an impurity level and finally back to its ground state.

The intensity of the net absorbed radiation is dependent on the character of the medium and the path length within.[45] The intensity of transmitted or unabsorbed radiation I_T continuously decreases with the distance x that the light traverses:

$$I_T = I_0 e^{-\beta x} \quad (8\text{-}43)$$

45 净吸收强度取决于介质特性和光的内部路径长度。

where I_0 is the intensity of the non-reflected incident radiation and β the absorption coefficient (in mm^{-1}), is characteristic of the particular material. β varies with the wavelength of the incident radiation. The distance parameter x is measured from the incident surface into the material. Materials with large β values are considered highly absorptive.

(2) Reflection

When light radiation passes from one medium into another having a different index of refraction, some of the light is reflected at the interface between the two media, even if both are transparent.[46] The reflectivity R represents the fraction of the incident light that is reflected at the interface, or

46 当光照射从一种介质进入另一种不同折射率的介质时，即使二者都是透明的，一些光将在两个介质之间的界面被反射。

$$R = \frac{I_R}{I_0} \tag{8-44}$$

where I_0 and I_R are the intensities of the incident and reflected beams, respectively. If the light is normal (or perpendicular) to the interface, then

$$R = \left(\frac{n_2 - n_1}{n_2 + n_1}\right)^2 \tag{8-45}$$

where n_1 and n_2 are the indices of refraction of the two media. If the incident light is not normal to the interface, R depends on the angle of incidence. When light is transmitted from a vacuum or air into a solid s, then

$$R = \left(\frac{n_s - 1}{n_s + 1}\right)^2 \tag{8-46}$$

because the index of refraction of air is very nearly unity. Thus, the higher the index of refraction of the solid is, the greater the reflectivity.[47] For typical silicate glasses, the reflectivity is approximately 0.05. Just as the index of refraction of a solid depends on the wavelength of the incident light, so does the reflectivity vary with wavelength. Reflection losses for lenses and other optical instruments can be minimized significantly by coating the reflecting surface with very thin layers of dielectric materials.

47 因此，固体的折射率越高，反射率就越大。

(3) Refraction

Light that is transmitted into the interior of transparent materials experiences a decrease in velocity, and, as a result, is bent at the interface. This phenomenon is termed as refraction. The index of refraction n of a material is defined as the ratio of the velocity in a vacuum c to the velocity in the medium v, or

$$n = \frac{c}{v} \tag{8-47}$$

The magnitude of n (or the degree of bending) depends on the wavelength of the light. This effect is graphically demonstrated by the familiar dispersion or separation of a beam of white light into its component colors by a glass prism. Each color is deflected by a different amount as it passes into and out of the prism, which results in the separation of the colors. Not only does the index of refraction affect the optical path of light, but also, as explained shortly, it influences the fraction of incident light reflected at the surface.[48] Just as Equation (8-1) defines the magnitude of c, an equivalent expression gives the velocity of light v in a medium as

$$v = \frac{1}{\sqrt{\varepsilon\mu}} \tag{8-48}$$

where ε and μ are, respectively, the permittivity and permeability of the particular substance. From Equation (8-47), we have

$$n = \frac{c}{v} = \frac{\sqrt{\varepsilon_0\mu_0}}{\sqrt{\varepsilon\mu}} = \sqrt{\varepsilon_\gamma\mu_\gamma} \tag{8-49}$$

where ε_γ and μ_γ are the dielectric constant and the relative magnetic permeability, respectively. Because most substances are only slightly magnetic, $\mu_\gamma \approx 1$, and

$$n \approx \sqrt{\varepsilon_\gamma} \tag{8-50}$$

Thus, for transparent materials, there is a relation between the index of refraction and the dielectric constant.[49] As already mentioned, the phenomenon of refraction is related to electronic polarization at the relatively high frequencies for visible light, thus, the electronic component of the dielectric constant may be determined from index of refraction measurements using Equation (8-50).

Because the retardation of electromagnetic radiation in a medium result from electronic polarization, the size of the constituent atoms or ions has considerable influence on the magnitude of this effect—generally, the larger an atom or ion, the greater the electronic polarization, the slower the velocity, and the greater the index of refraction. The index of refraction for a typical Sodalime glass is approximately 1.5. Additions of large barium and lead ions (as BaO and PbO) to a glass increases n significantly. For example, highly leaded glasses containing 90% PbO have an index of refraction of approximately 2.1.

For crystalline ceramics with cubic crystal structures and for glasses, the index of refraction is independent of crystallographic direction (i.e., it is isotropic). Non cubic crystals, however, have an anisotropic n—that is, the index is greatest along the directions that have the highest density of ions. Table8.1 gives refractive indices for several glasses, transparent ceramics,

48 折射率不仅影响光的路径，简单解释的话，它还影响入射光在表面上的反射比例。

49 因此，对于透明材料来说，折射率和介电常数之间有着一定的关系。

and polymers. Average values are provided for the crystalline ceramics in which n is anisotropic.

Table 8.1 Refractive indices for several glasses, transparent ceramics, and polymers.

	Average Index of Refraction
Glass and Ceramic	
Silica glass	1.458
Borosilicate (Pyrex) glass	1.47
Soda-lime glass	1.51
Quartz (SiO_2)	1.55
Dense optical flint glass	1.65
Spinel ($MgAl_2O_4$)	1.74
Periclase (MgO)	1.74
Corundum (Al_2O_3)	1.76
Polymers	
Polytetrafluoroethylene	1.35
Poly(methyl methacrylate)	1.49
Polypropylene	1.49
Polyethylene	1.51
Polystyrene	1.60

(4) Transmission

The phenomena of absorption, reflection, and transmission may be applied to the passage of light through a transparent solid[50], as shown in Figure 8.20. For an incident beam of intensity I_0 that impinges on the front surface of a specimen of thickness l and absorption coefficient β, the transmitted intensity at the back face I_T is

50 光的吸收、反射和透射现象可以被应用到光通过一个透明固体这一过程中。

$$I_T = I_0(1-R)^2 e^{-\beta l} \tag{8-51}$$

where R is the reflectance. For this expression, it is assumed that the same medium exists outside both front and back faces.

Figure 8.20 The transmission of light through a transparent medium for which there is reflection at the front and back faces, as well as absorption within the medium.

Thus, the fraction of incident light that is transmitted through a transparent material depends on the losses that are incurred by absorption and reflection.[51] Again, the sum of the reflectivity *R*, absorptivity *A*, and transmissivity *T*, is unity if no scattering is considered. Also, each of the variables *R*, *A*, and *T* depends on light wavelength. This is demonstrated over the visible region of the spectrum for a green glass in Figure 8.21. For example, for light having a wavelength of 0.4 μm, the fractions transmitted, absorbed, and reflected are approximately 0.90, 0.05, and 0.05, respectively. However, at 0.55 μm, the respective fractions shift to about 0.50, 0.48, and 0.02.

51 因此，入射光通过一个透明材料后的透射部分，取决于吸收和反射所产生的损失。

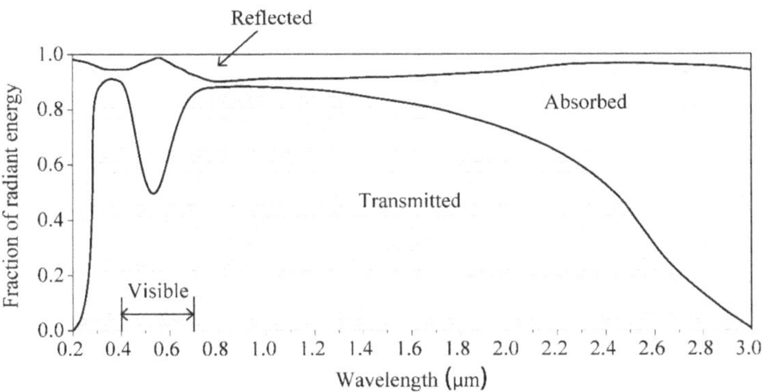

Figure 8.21 The variation with wavelength of the fractions of incident light transmitted, absorbed, and reflected through a green glass.

(5) Color of Nonmetals

Transparent materials appear colored as a consequence of specific wavelength ranges of light that are selectively absorbed; the color discerned is a result of the combination of wavelengths that are transmitted. If absorption is uniform for all visible wavelengths, the material appears colorless; examples include high-purity inorganic glasses and high-purity and single-crystal diamonds and sapphire.

Usually, any selective absorption is by electron excitation.[52] One such situation involves semiconducting materials that have band gaps within the range of photon energies for visible light (1.8 to 3.1 eV). Thus, the fraction of the visible light having energies greater than Eg is selectively absorbed by valence band-conduction band electron transitions. Some of this absorbed radiation is reemitted as the excited electrons drop back into their original, lower-lying energy states. It is not necessary that this reemission occur at the same frequency as that of the absorption.[53] As a result, the color depends on the frequency distribution of both transmitted and reemitted light beams. For example, cadmium sulphide (CdS) has a band gap of about 2.4 eV; hence, it absorbs photons having energies greater than about 2.4 eV,

52 通常情况下，任何选择性的吸收都是缘于电子激发。

53 再发射光的频率没有必要一定与吸收光的频率相同。

which correspond to the blue and violet portions of the visible spectrum; some of this energy is reradiated as light having other wavelengths. Nonabsorbed visible light consists of photons having energies between about 1.8 and 2.4 eV. Cadmium sulphide takes on a yellow-orange color because of the composition of the transmitted beam.

In insulator ceramics, specific impurities also introduce electron levels within the forbidden band gap.[54] Photons having energies less than the band gap may be emitted as a consequence of electron decay processes involving impurity atoms or ions, as demonstrated in Figures 8.19b and 8.19c. Again, the color of the material is a function of the distribution of wavelengths in the transmitted beam. For example, high-purity and single-crystal aluminium oxide or sapphire is colorless. Ruby, which has a brilliant red color, is sapphire to which has been added 0.5% to 2% chromium oxide (Cr_2O_3). The Cr^{3+} ion substitutes for the Al^{3+} ion in the Al_2O_3 crystal structure and introduces impurity levels within the wide energy band gap of the sapphire. Light radiation is absorbed by valence band-conduction band electron transitions, some of which is then reemitted at specific wavelengths as a consequence of electron transitions to and from these impurity levels. The transmittance as a function of wavelength for sapphire and ruby is presented in Figure 8.22. For the sapphire, transmittance is relatively constant with wavelength over the visible spectrum, which accounts for the colorless of this material. However, strong absorption peaks (or minima) occur for the ruby in the blue-violet region (at about 0.4 μm) and the other for yellow-green light (at about 0.6 μm). That unabsorbed or transmitted light mixed with reemitted light imparts to ruby its deep-red color.[55]

54 在绝缘体陶瓷中，特定杂质也在禁带内引入电子能级。

55 那些未被吸收光或者透射光与再发射光的混合，赋予了红宝石深红色。

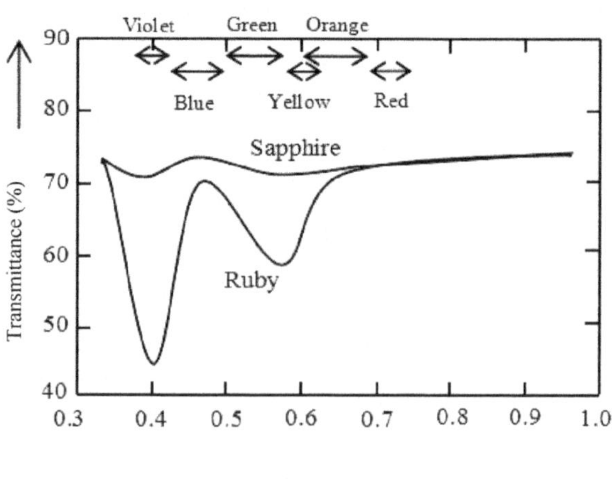

Figure 8.22 Transmission of light radiation as a function of wavelength for sapphire and ruby. The sapphire appears colorless, whereas the ruby has a red tint due to selective absorption over specific wavelength ranges.

Inorganic glasses are colored by incorporating transition or rare earth ions while the glass is in the molten state. Representative color-ion pairs include Cu^{2+}, blue-green; Co^{2+}, blue-violet; Cr^{3+}, green; Mn^{2+}, yellow; and Mn^{3+}, purple. These colored glasses are also used as glazes and decorative coatings on ceramic ware.

(6) Opacity and translucency in insulators

The extent of translucency and opacity for inherently transparent dielectric materials depends to a great degree on their internal reflectance and transmittance characteristics.[56] Many dielectric materials that are intrinsically transparent may be made translucent or even opaque because of interior reflection and refraction. A transmitted light beam is deflected in direction and appears diffuse as a result of multiple scattering events.[57] Opacity results when the scattering is so extensive that virtually none of the incident beam is transmitted undeflected to the back surface.

56 固有透明介电材料的半透明和不透明的程度，在很大程度上取决于其内部的反射和透射特性。

57 透射光束在方向上发生偏转，并出现漫反射行为，缘于多次散射事件。

This internal scattering may result from several different sources. Polycrystalline specimens in which the index of refraction is anisotropic normally appear translucent. Both reflection and refraction occur at grain boundaries, which causes a diversion in the incident beam. This results from a slight difference in index of refraction *n* between adjacent grains that do not have the same crystallographic orientation.

Scattering of light also occurs in two-phase materials in which one phase is finely dispersed within the other. Again, the beam dispersion occurs across phase boundaries when there is a difference in the refractive index for the two phases; the greater this difference, the more efficient the scattering.[58] Glass-ceramics, which may consist of both crystalline and residual glass phases, appear highly transparent if the sizes of the crystallites are smaller than the wavelength of visible light and when the indices of refraction of the two phases are nearly identical (which is possible by adjustment of composition).

58 在细小颗粒分散在另一相的两相材料体系中，也会发生散色。再者，当这两相的折射率不同时，在相界面会发生光束色散，差异越大，色散效率越大。

As a consequence of fabrication or processing, many ceramic pieces contain some residual porosity in the form of finely dispersed pores. These pores also effectively scatter light radiation.

Figure 8.23 demonstrates the difference in optical transmission characteristics of single-crystal, fully dense polycrystalline, and porous (~5% porosity) aluminium oxide specimens. Whereas the single crystal is totally transparent, polycrystalline and porous materials are, respectively, translucent and opaque.

Figure 8.23 The light transmittance of three aluminum oxide specimens. From left to right: single-crystal material (sapphire), which is transparent; a polycrystalline and fully dense (nonporous) material, which is translucent; and a polycrystalline material that contains approximately 5% porosity, which is opaque

For intrinsic polymers (without additives and impurities), the degree of translucency is influenced primarily by the extent of crystallinity.[59] Some scattering of visible light occurs at the boundaries between crystalline and amorphous regions, again as a result of different indices of refraction. For highly crystalline specimens, this degree of scattering is extensive, which leads to translucency and, in some instances, even opacity. Highly amorphous polymers are completely transparent.

59 对于本征聚合物（无添加剂和杂质），其半透明程度主要受结晶程度影响。

8.4 Applications of Optical Materials

8.4.1 Optical Fibers in Communications

The communications field recently experienced a revolution with the development of optical fiber technology; virtually all telecommunications are transmitted via this medium rather than through copper wires.[60] Signal transmission through a metallic wire conductor is electronic (i.e., by electrons), whereas using optically transparent fibers, signal transmission is photonic, meaning that it uses photons of electromagnetic or light radiation. Use of fiber optic systems has improved speed of transmission, information density, and transmission distance, with a reduction in error rate; furthermore, there is no electromagnetic interference with fibers. The bandwidth (i.e., data transfer rate) of optical fibers is amazing; in 1 s, an optical fiber can transmit 15.5 terabits of data over a distance of 7000 km; at this rate it would transmit the entire iTunes catalog from New York to London within minutes. Furthermore, a single fiber is capable of transmitting 250 million phone conversations every second. It would require 30,000 kg of copper to transmit the same amount of information over one

60 随着光纤技术的发展，通信领域最近经历了一场革命，几乎所有的通信信号都是通过这种介质传输，而不是通过铜线传输。

mile as only 0.1 kg of optical fiber.

The present treatment centers on the characteristics of optical fibers; however, it is worthwhile to first briefly discuss the components and operation of the transmission system.[61] A schematic diagram showing these components is presented in Figure 8.24. The information (e.g., a telephone conversation) in electronic form must first be digitized into bits-that is, 1 and 0; this is accomplished in the encoder. It is next necessary to convert this electrical signal into an optical (photonic) one, which takes place in the electrical-to-optical converter (Figure 8.24). This converter is normally a semiconductor laser, which emits monochromatic and coherent light. The wavelength normally lies between 0.78 and 1.6 μm, which is in the infrared region of the electromagnetic spectrum; absorption losses are low within this range of wavelengths. The output from this laser converter is in the form of pulses of light; a binary 1 is represented by a high-power pulse (Figure 8.25a), whereas a 0 corresponds to a low-power pulse (or the absence of one), (Figure 8.25b). These photonic pulse signals are then fed into and carried through the fiber-optic cable (sometimes called a waveguide) to the receiving end. For long transmissions, repeaters may be required; these are devices that amplify and regenerate the signal. Finally, at the receiving end the photonic signal is reconverted to an electronic one and then decoded (undigitized).

61 虽然目前的处理方法集中于光纤特性，但是在这之前简要讨论一下传输系统的组件和操作也是值得的。

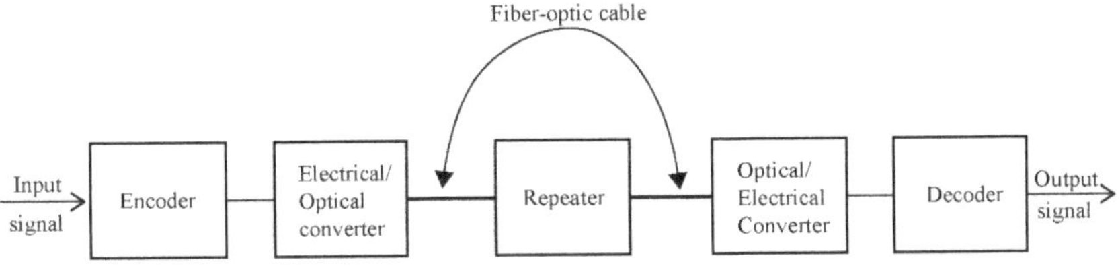

Figure 8.24 Schematic diagram showing the components of an optical-fiber communications system.

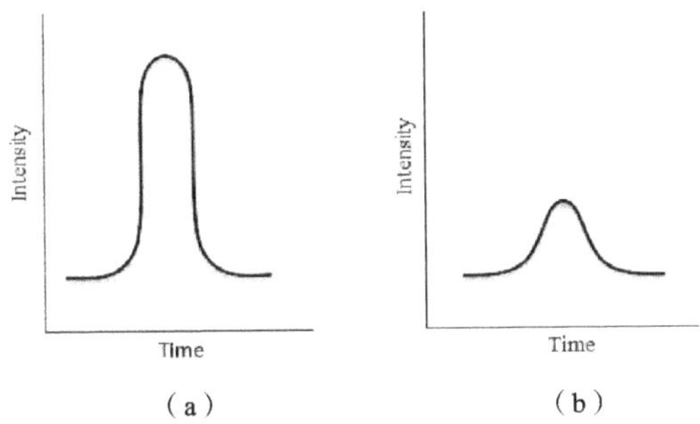

Figure 8.25 Digital encoding scheme for optical communications. (a) A high-power pulse of photons corresponds to a 1 in the binary format. (b) A low-power photon pulse represents a 0.

62 它必须在很长的距离内传导这些光脉冲，并且没有明显的信号功率损耗（即衰减）和脉冲失真。

63 信号从纤芯传导，而周围包层将光线限制在内部核心；外部涂层保护核心和包层免受磨损和外部压力的损害。

The heart of this communication system is the optical fiber. It must guide these light pulses over long distances without significant signal power loss (i.e., attenuation) and pulse distortion.[62] Fiber components are the core, cladding, and coating; these are represented in the cross-section profile shown in Figure 8.26. The signal passes through the core, whereas the surrounding cladding constrains the light rays to travel within the core; the outer coating protects core and cladding from damage that might result from abrasion and external pressures.[63]

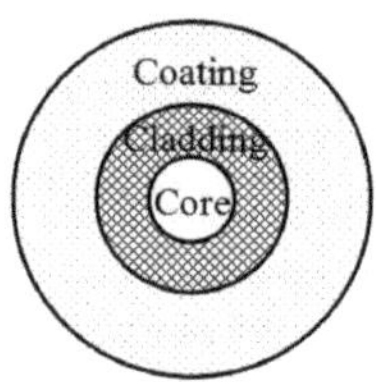

Figure 8.26 Schematic cross section of an optical fiber.

High-purity silica glass is used as the fiber material; fiber diameters normally range between about 5 to 100 μm. The fibers are relatively flaw-free and, thus, remarkably strong; during production the continuous fibers are tested to ensure that they meet minimum strength standards.

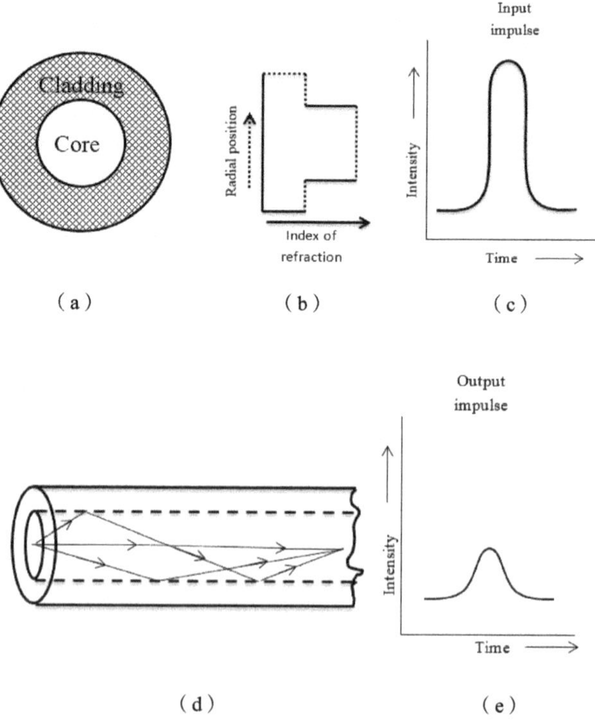

Figure 8.27 Step-index optical fiber design. (a) Fiber cross section. (b) Fiber radial index of refraction profile. (c) Input light pulse. (d) Internal reflection of light rays. (e) Output light pulse.

Containment of the light to within the fiber core is made possible by total internal reflection-that is, any light rays traveling at oblique angles to the fiber axis are reflected back into the core. Internal reflection is accomplished by varying the index of refraction of the core and cladding glass materials. In this regard, two design types are used. With one type (termed step-index), the index of refraction of the cladding is slightly lower than that of the core. The index profile and the manner of internal reflection are shown in Figures 8.27b and 8.27d. For this design, the output pulse is broader than the input one (Figures 8.27c and 8.27e), a phenomenon that is undesirable because it limits the rate of transmission. Pulse broadening results in various light rays, although injected at approximately the same instant, arrive at the output at different times; they traverse different trajectories and, thus, have a variety of path lengths.[64]

64 脉冲展宽导致了不同的光线，虽然这些光线在大致相同的瞬间输入，但是它们在不同的时间到达输出端；它们经过不同的轨迹，因此，就有了不同的路径长度（所以到达输出端的时间不同）。

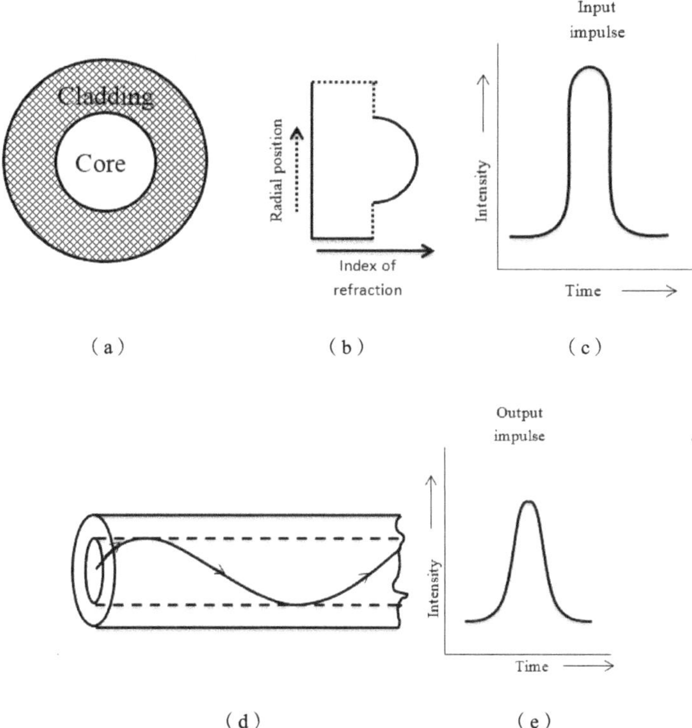

Figure 8.28 Graded-index optical-fiber design. (a) Fiber cross section. (b) Fiber radial index of refraction profile. (c) Input light pulse. (d) Internal reflection of a light ray. (e) Output light pulse.

Pulse broadening is largely avoided by use of the graded-index design. Here, impurities such as boron oxide (B_2O_3) or germanium dioxide (GeO_2) are added to the silica glass such that the index of refraction is made to vary parabolically across the cross section (Figure 8.28b). Thus, the velocity of light within the core varies with radial position, being greater at the

65 所以，沿纤芯的外层周边传过较长路径的光线，在低折射率材料中传播较快，与通过纤芯中心部分的没有偏离光线几乎是同时到达输出端。

periphery than at the center. Consequently, light rays that traverse longer path lengths through the outer periphery of the core travel faster in this lower-index material and arrive at the output at approximately the same time as undeviated rays that pass through the center portion of the core.[65]

Exceptionally pure and high-quality fibers are fabricated using advanced and sophisticated processing techniques, which are not discussed here. Impurities and other defects that absorb, scatter, and thus attenuate the light beam must be eliminated. The presence of copper, iron, and vanadium is especially detrimental; their concentrations are reduced to on the order of several parts per billion. Likewise, water and hydroxyl contaminant contents are extremely low. Uniformity of fiber cross-sectional dimensions and core roundness is critical; tolerances of these parameters to within 1 μm over 1 km of length are possible. In addition, bubbles within the glass and surface defects have been virtually eliminated. The attenuation of light in this glass material is imperceptibly small. For example, the power loss through a 16-km thickness of optical-fiber glass is equivalent to the power loss through a 25-mm thickness of ordinary window glass!

8.4.2 Plastic Optics

8.4.2.1 Introduction

Injection molded precision plastic optics in high volumes were first produced during the 1960's. After the development of sophisticated measuring and manufacturing methods in the late sixties also precise aspheric surfaces were as easy to make as spheric contours. Today plastic optics is a widely used low-cost option compared to glass with even more degrees of freedom for optical and component design. Polymer optical systems at present are used in sensor applications, visual systems, cameras (mobile phones, video-conferencing cameras), scanners, security systems and so on.

Of course the properties of glass materials are very different from that of plastic materials. But glass and plastic optics both offer unique advantages and help to solve various engineering problems.

Generally speaking, glass materials are harder and more durable than plastic materials. Glass materials are also more stable (temperature and humidity) than plastic. The variety of optical glass available from well-known suppliers comprises hundreds of different materials. Compared with this the choice for plastic materials is limited only to about 10 different materials (and even less optical parameter variations). The large selection of glass materials allows the designer to choose materials with desirable optical properties to gain better optical performance. However, plastic optics offers other design freedoms that are not achievable with glass optics.[66]

The manufacturing technologies for glass and plastic optics are totally different. Glass lenses are made by a grinding and polishing process whereas precision plastic lenses are in the whole made by injection-molding or compression molding. Because of the materials characteristics and the manufacturing process plastic optics have some unique advantages:

- **High Production Numbers at Low Costs**

Injection molding is ideal for high volume production with low unit costs.[67] Moderate raw material costs and multi cavity moulds (up to 32 cavities) allow large production volumes at a reasonable unit price. In spite of normally considerable tooling costs the cost breakeven compared to glass design versions can be relatively low.

- **Lightweight and Hardiness**

For a given volume glass is much heavier than plastic (by a factor of 2.3 to 4.9). Plastic materials are relatively shatter and impact resistant. These features are important, e.g., for head-mounted systems.

- **Design Potentials**

Injection molding makes it economical to produce sophisticated optical shapes such as a spheres, diffractive optical elements or even freeform surface structures.[68] From the design point of view, the more sophisticated surface shapes reduce costs or obtain better performance or even performance unrealizable by optics in glass (Fresnel structures, lens-arrays, diffractive optical elements etc.).

- **Optical Systems and Component Assembly**

For typical optical systems, the optical components (mirrors, lenses, prisms etc.) must be fixed in a mounting. With plastic optics, it is possible to mold mounting elements, posts or alignment notches integrally with the optical component.[69] This can reduce part and assembly costs considerably. Technologies adapted to plastic materials like ultrasonic and laser welding, gluing and integrated snap-in structures allow fast and cost efficient automated and manually operated assembling solutions. Two component injection molding is the choice for integrating optical elements and mounts in one injection cycle.

8.4.2.2 Materials for Plastic Optics

As already mentioned, the choice for plastic optical material is limited only to about 10 different types of material. Optical properties (Abbé value, refractive index, Figure 8.29) as well as mechanical, thermal and humidity

66 玻璃材料的巨大选择范围允许设计师选择具有理想光学性质的材料，以获得更好的光学性能。然而，塑胶光学提供了其他使用玻璃光学材料不可实现的设计自由。

67 注塑是低成本大规模量产的理想制备方法。

68 注塑可以经济廉价地生产复杂的光学形状，如球形、衍射光学元件或者自由曲面结构。

69 塑胶光学器件制造时可以与有效元件部分一起同时设计制作出支撑结构，安装对准凹槽等部件。

boundary conditions are decisive for the material choice. Different to glass during plastic injection molding the process affects not only the geometry but also the inner properties like refractive index and birefringence.[70]

70 不同于玻璃，在塑胶注入的过程中不仅影响外在几何形状，也影响材料内在属性，如折射率和双折射。

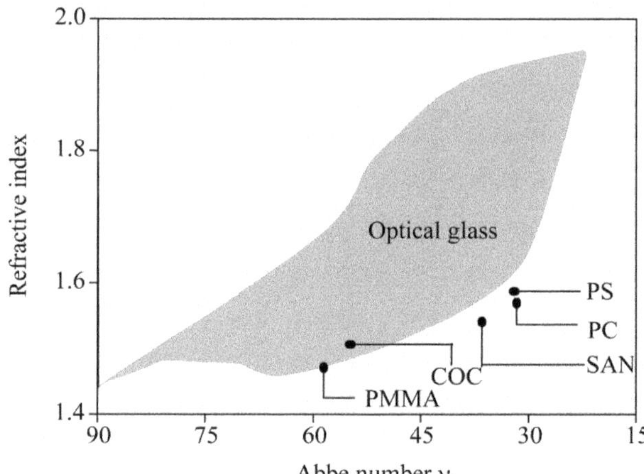

Figure 8.29 Abbe diagram of optical materials

Ⅰ. Optical Properties

Fundamental optical properties are defined by optical transmission, refractive index and dispersion. For plastic optics birefringence is an important parameter too. Although the total number of plastic materials has increased in recent years, the range of refractive index and dispersion characteristic is limited to almost two major groups i.e., crown-like materials as acrylic (PMMA), polyolefin (COC, COP) and Flint-like materials as polystyrene, polycarbonate and SAN. This strongly limited variety of plastic optical materials restricts the optical design that can be made in plastic considerably.

Ⅱ. Physical Properties

Important physical properties are weight, impact and abrasion resistance and thermal properties like temperature resistance and thermal expansion (Table 8.2).

It is important to be taken into consideration of both mechanical and thermal properties during the initial optical system design. The thermal expansion of optical plastics is approximately ten times higher than that of glass materials. In an optical system this effect has to be compensated by optical design and mounting.[71]

71 光学塑胶材料的热膨胀比玻璃高约十倍。在一个光学系统中，这种效果必须通过光学设计和安装方式进行补偿。

Typically, most of optical plastics can withstand temperatures up to 90℃. The maximum service temperature of polycarbonate and polyolefine materials reaches 120°C. The specific gravity of plastic optical materials ranges from 1 to 1.3. Polycarbonate has the highest impact resistance of all optical plastics and is used for windshields and crash helmets. Acrylic has the best abrasion resistance.

Ⅲ. Most Common Materials

Some of the most common optical polymers are:

- **PMMA (Polymethyl Methacrylate)**

Acrylic is the most commonly used optical plastic. It is moderately priced, easy to mould, scratch resistant and not very water absorptive. Its transparency is greater than that of most optical glasses. Additives to acrylic (as well as to several other plastics) considerably improve its ultraviolet transmittance and stability. So acrylic is used in almost 80% of all plastic optical applications.

- **Polystyrene**

Polystyrene is a low-cost material with excellent molding properties. Styrene has a higher index and a lower numerical dispersion value than other plastics. It is often used as the flint element in color corrected plastic optical systems. Compared with acrylic, styrene has lower transmission in the UV portion of the spectrum. It does not have the UV radiation resistance nor the scratch resistance of Acrylic. Because its surface is less durable, styrene is more typically used in nonexposed areas of a lens system.

- **NAS Copolymer**

NAS is a copolymer of certain fractions of polystyrene and acrylic (typically 70/30), this allows to adjust refractive index. It is a low-cost material with excellent molding properties.

- **PC (Polycarbonate)**

Polycarbonate is very similar to styrene in terms of such optical properties as transmission, refractive index and dispersion. Polycarbonate, however, has a much broader operating temperature up to 120℃. For this reason, it is used as the flint material for systems that have to withstand severe thermal conditions. Another advantage is the high impact resistance of polycarbonate. Safety glasses and systems requiring durability often consist of polycarbonate.

- **Cyclic Olefin Polymer and Copolymer (COP/COC)**

Cyclic olefin (co-)polymer provides a high temperature alternative to acrylic. Its refractive index and transmittance are similar but the heat distortion temperature is about 30℃ higher than for acrylic. Its water absorption capability is significant smaller.

Table 8.2 gives summarization for the optical properties of polymers.

Table 8.2 Plastic optical material properties

Material Properties	Characteristics	Acrylic (PMMA)	Poly-styrene (PS)	Poly-carbonate (PC)	Styrol-Acryl-nitril (SAN)	Cyclo-olefine (ZEONEX)	Polyether-sulfone (PES)	Acrylnitril-Butadien-Styrolcopoly-mere (ABS)	Optical glass (BK7)
Optical	Spectral passing band (nm)	390-1600	400-1600	360-1600	395-1600	300-1600			
	Refractive index at 587nm and 20℃	1.4918	1.5905	1.5855	1.5674	1.5261	1.6600	1.538	1.517
	Abbé value $(n_D - 1)/(n_F - n_C)$	57.2	30.7	30	34.8	56	19.4		64.4
	Transmittance (%) thickness 3.2 mm	92	88	90	88	92	80	85	
	Haze (%) thickness 3.2 mm	1.3	1.5	1.7	1.5	1.5			
Physical	Specific gravity (g/cm^3)	1.18	1.06	1.25	1.07	1.02	1.37	1.05	2.53
	Max. service temperature (℃)	80	90	120	95	125	200	90	400
	Linear expansion coefficient (1/K)	6.8×10^{-5}	8.0×10^{-5}	6.6×10^{-5}	7.0×10^{-5}	7.0×10^{-5}	5.5×10^{-5}	8.5×10^{-5}	1.1×10^{-5}
	Abrasion resistance (1-10)	10	4	2		6			
	Izod impact strength (kJ/m^2)	2.0	2.0			2.4	7.0	25.0	
Environ-mental	$dn/dT\ (10^{-6})$	-105							
	Sensitvity to humidity	high	low	low	medium	low	high	medium	
	Water absorbtion (weight%) 23℃, ISO 61	0.60	0.10	0.15	0.30	0.01	0.70	0.45	0
Manufac-turability	Processability	Excellent	Good	Poor	Excellent	Good			
	Birefrigence	Low	High	High		Low			
Chemical	Resitance to alcohol	Limited	Good	Limited			Good		
Costs	Approx. material costs (EUR/kg)	3.3	2.5	4.4	4.4	27.1	21.0	3.5	25

8.4.2.3 Manufacturing Process

Ⅰ. Optical and System Design

On the one hand optic design for plastic optics in principle uses the same mathematical algorithms to optic design for glass. On the other hand, designing plastic optics requires a profound understanding of the material properties and the manufacturing processes. Knowledge of production technologies, material characteristics and assembling methods together with design expertise are needed to fully exhaust what precision plastic optics can be.[72]

Simply substituting the indices of refraction and re-optimizing the design will not succeed. Expert design assistance is essential at this stage. Designing plastic optics with modern design tools allows great design freedom. The advantage of combining integral mounting structures with the optical surfaces to create mounting flanges, alignment and snap features provides the ability to automated assembly. Aspheric, cylindric or toroidal surfaces are as easy to realize spherical ones. Microstructures such as diffractive optical elements can be integrated too.

Ⅱ. Prototyping

After designing a plastic optical system, lens prototypes can be made by diamond point turning in various plastic materials. The best surface finishes can be obtained with PMMA. Materials such as polycarbonate, Polyimid or Zeonex do not yield very smooth surface finishes.

A major problem is the availability of semi-finished plastic blocks in various materials.[73] For PMMA a wide variety of bars or plates are available. For other materials or colored options semi-finished items have to be made by injection molding.

Because of the high manufacturing costs diamond point turning is only recommended for making a limited number of prototypes to verify functionality of the optical design and to perform first tests. In this stage optical systems are normally assembled from single elements. Housing parts often are made from aluminium. The resulting surface quality and system performance cannot be a validation of the manufacturability by injection molding. To get reliable knowledge about this, making a molded prototype from a single-cavity prototype mould is recommended.

Ⅲ. Injection Mould

A high quality injection mould is obviously essential for precise plastic optic parts.[74] The parts never can be better than the tool—but good tooling,

72 精密的塑胶光学需要彻底地掌握生产技术，材料特性和组装方法，以及设计的专业经验。

73 一个主要的问题是获得各种材料的半成品塑料块的可能性。

74 高质量的注塑模具对精密塑料光学零件的加工有着重要的意义。

however, does not guarantee good parts. A strong understanding of the whole manufacturing process is the key to producing precision plastic components.

Thermoplastic material shrinks during cooling down in the mould. This geometric effect has to be compensated for in the injection mold. Exact shrink rates can be calculated and the tool can be modified. For manufacturing optical plastic parts by injection molding the optical surfaces (plano or as pherical shapes, diffractive, conical, lenticular and cylindrical surfaces) are generated as separate inserts in the tool.

Aspheric inserts are manufactured by two steps. First a best-fit curve is generated on a stainless steel substrate. The substrate is then subjected to a nickel plating process (electroless nickel) that deposits a thin layer of nickel (up to 500μm). In the second step single-point diamond turning produces the final aspheric or diffractive curve in the nickel. Because the hardness of a nickel-plated insert is less than that of a steel spherical insert, it will be more susceptible to scratch defects.

Ⅳ. Pre–Production

A pre-production stage is recommended to check the manufacturing process. Typically, this is done by using a single-cavity prototype mould. This mold can be used to find optimal molding conditions. An optical mechanical design can be verified with real molded components and design revision is possible to affordable conditions.[75] Often the prototype moulds is used to start limited production since production tooling may take much more time.

75 光学机械设计可以通过真正的模压品来验证，且通过并反馈设计是可以达到满意的条件的。

Ⅴ. Series Production

For series production of high volumes multi-cavity production molds are required. Depending on quality, volume, throughput and cost, the production tool may have 2, 4, 8 up to 32 cavities. The production molds function for at least several hundred of thousands of injection cycles.

8.4.2.4 Coating on Plastic Optical Materials

Because of their limited temperature and UV resistance plastic lenses must not be coated in an elevated temperature and radiation environment. During the deposition of thin films onto plastic, the coating chamber temperature is significantly lower than that for glass optics.[76] This requires deposition techniques such as ion-assisted deposition to apply antireflective, conductive, mirror and beam-splitter coatings.

76 在塑胶上沉积薄膜时的真空室温度，明显低于在光学玻璃镀膜的温度。

Today multi-layer dielectric coatings are routinely deposited on plastic components. Typical broadband antireflection coatings reduce reflection to about 0.5% per surface across the entire visible spectrum. Narrowband, multilayer antireflection coatings can achieve surface reflectance less than 0.2%. Multi-layer dielectric coatings can be modified easily to scratch resistant designs for front lenses and windows.

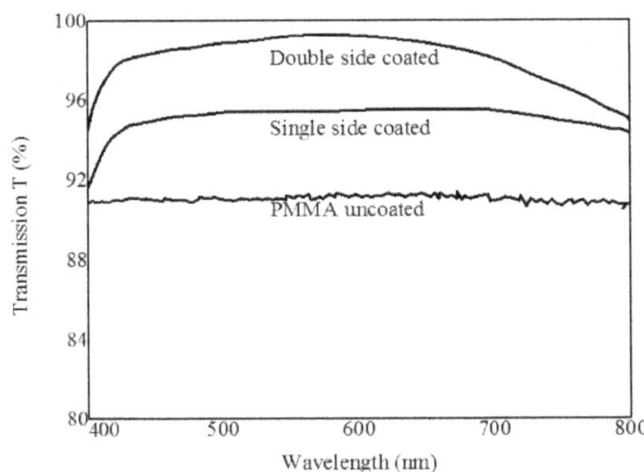

Figure 8.30 Transmission of coated plastic optics

Several front and back surface reflector coatings are available for plastic substrates. Standard coating metals include aluminium, silver and gold. Aluminium coatings provide surface reflectance greater than 88% across the visible spectrum, and gold coatings greater than 95% for the near infrared region.

8.4.2.5 New Developments

In the last years several new developments in the field of plastic optics took place. The permanent improvement process in tooling technologies in combination with diamond turning, milling and grinding led to new design and tolerance levels. New materials like the cycloolefins led to better material performance and coating stability.

The fabrication process for diffractive elements (done primarily by single-point diamond turning) has improved considerably. Micro and nanostructures like moth-eyed surfaces can be achieved by injection mold and compression mold and allow functionalized surfaces.[77] Even calibration structures for scanning probe microscopy applications can be replicated by injection mold up to 25 nm structures.

Plastic optic technology permanently expands its traditional limits and fields of application. On the one hand plastic optic is still limited by material properties but on the other hand design and assembling freedoms allow new approaches.[78]

77 微型和纳米结构，如复眼透镜的表面可以通过模压来实现，并且允许对表面功能化。

78 一方面，塑料光学仍然受到材料性能的限制，但另一方面，设计和组装的自由允许新的方法。

8.4.3 Luminescent Materials (Inorganic Only)

Luminescent materials, also called phosphors, are mostly solid inorganic materials consisting of a host lattice, usually intentionally doped with impurities. The impurity concentrations generally are low in view of the fact that at higher concentrations the efficiency of the luminescence process usually decreases (concentration quenching). In addition, most of the phosphors have a white body color. Especially for fluorescent lamps, this is an essential feature to prevent absorption of visible light by the phosphors used. The absorption of energy, which is used to excite the luminescence, takes place by either the host lattice or by intentionally doped impurities. In most cases, the emission takes place on the impurity ions, which, when they also generate the desired emission, are called activator ions. When the activator ions show too weak an absorption, a second kind of impurities can be added (sensitizers), which absorb the energy and subsequently transfer the energy to the activators. This process involves transport of energy through the luminescent materials. Quite frequently, the emission color can be adjusted by choosing the proper impurity ion, without changing the host lattice in which the impurity ions are incorporated. On the other hand, quite a few activator ions show emission spectra with emission at spectral positions which are hardly influenced by their chemical environment. This is especially true for many of the rare-earth ions.

The luminescent materials can be grouped as center luminescence which the emission results from optical transitions between host lattice band state or from a transition between two centers[79], such as the emission of Mn^{4+} in $Mg_4GeO_{5.5}F$: Mn; charge transfer luminescence which the optical transition takes place between different kinds of orbitals or between electronic states of different ions, such as a very well-known example, $CaWO_4$ used for decades for the detection of X-rays, which shows luminescence originating from the $(WO_4)^{2-}$ group; donor acceptor pair luminescence which is found in some semi-conducting materials doped with both donors and acceptors; long afterglow phosphors which the optical excitation energy is stored in the lattice by trapping of photo excited charge carriers.

79 这样的发光材料可以归类为中心发光，即发光来自于主体材料的能带之间的光学跃迁，或者来自于两个发光中心之间的光学跃迁。

本章小结

1. 内容概要

在本章中，光学性质即材料在电磁辐射下尤其是在可见光辐照下的反应，将会被讨论。首先，介绍与光相关的一些基础概念；然后，分别在宏观现象和微观理论两方面，阐述光与材料之间的相互作用；之后，讨论金属和非金属材料的光学行为，如吸收，散射，反射和透射等，着重于光的吸收和散射；最后，介绍两种典型的材料，光学纤维和塑胶光学，对它们的应用也有所阐述。

2. 基本概念

光、光波、电磁、介电函数、折射率、折射、反射、吸收、散射、电子极化、电子跃迁、光学纤维、塑胶光学。

3. 主要公式

（1）真空中光速：$c=\dfrac{1}{\sqrt{\varepsilon_0\mu_0}}$

（2）介质中光速：$v=\dfrac{1}{\sqrt{\varepsilon\mu}}=\dfrac{c}{\sqrt{\varepsilon_{\rm r}\mu_{\rm r}}}$

（3）光子能量：$E=hv=\dfrac{hc}{\lambda}$

（4）介电函数：$\varepsilon(k,\omega)=1+\sum_{\rm j}\dfrac{\alpha_{\rm k,j}}{\omega^2-\omega_{\rm k,j}^2-i\omega\eta_{\rm k,j}}$

（5）折射率经验公式：$n(\lambda)^2\approx1+\sum_{\rm j}\dfrac{B_{\rm j}\lambda^2}{\lambda^2-\lambda_{\rm j}^2}$

（6）菲涅尔公式：$n_1\sin\alpha_1=n_2\sin\alpha_2$

（7）全反射角公式：$\alpha_{\rm T}=\arcsin\left(\dfrac{n_2}{n_1}\right)$

（8）垂直入射反射率公式：$R_{\rm norm}=\left|\dfrac{n_1-n_2}{n_1+n_2}\right|^2$

（9）总积分散射（*TIS*）：$TIS=\dfrac{P_{\rm scat}}{P_{\rm ref}}=\left[\dfrac{4\pi\cos\alpha_{\rm R}}{\lambda}\right]^2(\Delta h)^2$

（10）折射率：$n=\dfrac{c}{v}=\dfrac{\sqrt{\varepsilon_0\mu_0}}{\varepsilon\mu}=\sqrt{\varepsilon_{\rm r}\mu_{\rm r}}\approx\sqrt{\varepsilon_{\rm r}}$

Vocabulary

absorption/absorbance	系数/吸光度
chromatic dispersion	色散
collision	碰撞
concave lens	凹透镜
convex lens	凸透镜
dielectric constant	介电常数
diffuse reflection	漫反射
dispersion coefficient	色散系数
double refraction	双折射
electromagnetic wave	电磁波
electron polarization	电子极化
energy flux rate	能流速率
extraordinary beam	非常光
extraordinary index of refraction	非常光折射率
far infrared	远红外
incident light	入射光
irradiate	照射
magnetic permeability	磁导率
medium	介质
middle infrared	中红外
near infrared	近红外
non-polarized light	非偏振光
normal line	法线
opaque	不透明的
optical axis	光轴
plane of incidence light	光线入射面
polarization	偏振
polarized light	偏振光
prism	三棱镜
prismatic decomposition of light	三棱镜分光

propagation	传播
reflection/reflectance	反射/反射度
refraction	折射
scattering	散射
semitransparent	半透明
transmission/transmittance	透射/透射比
transparent/transparency	透明的/透明度
ultraviolet	紫外线的
visible light	可见光

Problems

1. When talking about optical properties, please give at least three processes/parameters.

2. Please state the mechanism of light refraction by material.

3. Please give three types of interactions between light and material, and briefly point out the micro-mechanism.

4. Why metals are generally opaque to visible light?

5. True or false questions.

(1) The degree of light reflection depends on its wavelength.

(2) The degree of light absorption depends on both the wavelength of light and the electronic structure of material.

(3) The degree of light refraction is independent of light wavelength but only dependent on material.

(4) The higher the material refractive index, the higher the reflectance.

(5) The ratio of the absorbed light is independent to the intensity of the incident light, but is proportional to the length of the medium.

(6) Generally speaking, the absorbed light by metal tends to be released as emission with wavelength different to the absorption.

(7) In metals, the small part of the absorbed photons that does not involve radioactive decay converts to heat eventually.

(8) A non-metal material can be either transparent or opaque depending on its absorption property and its stacking modes.

(9) The refractive index of materials is generally greater than 1.

(10) The color of a metal is determined by the wavelengths of reflection photons.

(11) The color of a non-metal material is dependent on the wavelengths of absorption.

(12) Adding heavy atoms to a non-metal material can increase refractive index.

(13) Besides material polarization, refractive index of a non-metal material is dependent on crystal structure and stress.

(14) Electrons at excited state have high energy, and they tend to decay to ground state only via light emission.

6. Please state the decay processes of the excited electrons by optical excitation in metals.

7. What is the determining factor for the color of metal? Explain.

8. Explain the meaning of the terms: fluorescence, phosphorescence, photoluminescence, electroluminescence and laser.

9. Please explain reflection phenomena in metal and non-metal materials. please state the relationship between reflection and scattering.

10. Please give the reason for the direction change in refraction process.

11. According the visible light wavelength of 400 to 700 nm, please state the transparency of a material based on its band-gap.

12. Please define refractive index.

13. Please define chromatic dispersion.

Chapter 9 Thermal Properties

In this chapter, thermal properties of materials will be discussed. In the first part, the heat capacity is interpreted in different models. Then, the differences among the heat capacity of various real materials are introduced. In the second part, thermal conductivity is illustrated in terms of definition and mechanism, followed by the discussion of the thermal conduction in real materials. Specifically, the relationship between thermal and electrical conductivity in metals is exhibited. Finally, thermal expansion and its connection with heat capacity are present.

9.1 Heat Capacity

9.1.1 Definition

To determine the change of temperature in materials during heating, we have mole heat capacity and specific heat capacity. The mole heat capacity, also termed as heat capacity, is the energy required to raise the temperature of one mole of a material by one degree. The specific heat (specific heat capacity) is defined as the energy needed to increase the temperature of one gram of a material by one degree.[1] Thus

1 摩尔热容也被称为热容，是指将一摩尔物质升高一度所需的能量。比热（比热容）的定义是将一克物质升高一度所需的能量。

$$specific\ heat = \frac{heat\ capacity}{molar\ weight} \tag{9-1}$$

The heat capacity can be expressed either at constant pressure, C_P, or at a constant volume, C_V. These are expressed as:

$$C_P = \left(\frac{\partial H}{\partial T}\right)\Big|_P = \left(\frac{\partial (U+PV)}{\partial T}\right)\Big|_P \tag{9-2}$$

$$C_V = \left(\frac{\partial U}{\partial T}\right)\Big|_V \tag{9-3}$$

in which H is the enthalpy, U is the internal energy, P is the pressure, T is the temperature, and V is the volume. The relationship between C_P and C_V can be deduced as following:

Performing the indicated differentiation to Equation (9-2)

$$C_P = \left(\frac{\partial U}{\partial T}\right)\Big|_P + P\left(\frac{\partial V}{\partial T}\right)\Big|_P + V\left(\frac{\partial P}{\partial T}\right)\Big|_P \tag{9-4}$$

and noting that $\partial P/\partial T = 0$ at constant pressure, one get:

$$C_P = \left(\frac{\partial U}{\partial T}\right)\Big|_P + P\left(\frac{\partial V}{\partial T}\right)\Big|_P \tag{9-5}$$

The quantity ∂U can be expressed as

$$\partial U = \left(\frac{\partial U}{\partial V}\right)\Big|_T \partial V + \left(\frac{\partial U}{\partial T}\right)\Big|_V \partial T \tag{9-6}$$

and when divided by ∂T this becomes

$$\frac{\partial U}{\partial T} = \left(\frac{\partial U}{\partial V}\right)\Big|_T \left(\frac{\partial V}{\partial T}\right)\Big|_P + \left(\frac{\partial U}{\partial T}\right)\Big|_V \tag{9-7}$$

Thus, C_P (Equation 9-5) becomes

$$C_P = \left(\frac{\partial U}{\partial V}\right)\Big|_T \left(\frac{\partial V}{\partial T}\right)\Big|_P + \left(\frac{\partial U}{\partial T}\right)\Big|_V + P\left(\frac{\partial V}{\partial T}\right)\Big|_P \tag{9-8}$$

When this is used with Equation (9-3)

$$C_\mathrm{p} - C_\mathrm{V} = \left(\frac{\partial U}{\partial V}\right)\Big|_\mathrm{T}\left(\frac{\partial V}{\partial T}\right)\Big|_\mathrm{P} + \left(\frac{\partial U}{\partial T}\right)\Big|_\mathrm{V} + P\left(\frac{\partial V}{\partial T}\right)\Big|_\mathrm{P} - \left(\frac{\partial U}{\partial T}\right)\Big|_\mathrm{V}$$
$$= \left(\frac{\partial U}{\partial V}\right)\Big|_\mathrm{T}\left(\frac{\partial V}{\partial T}\right)\Big|_\mathrm{P} + P\left(\frac{\partial V}{\partial T}\right)\Big|_\mathrm{P} = \left[\left(\frac{\partial U}{\partial T}\right)\Big|_\mathrm{T} + P\right]\left(\frac{\partial V}{\partial T}\right)\Big|_\mathrm{P} \quad (9\text{-}9)$$

For most solids, especially metals at normal pressures, $(\partial V/\partial T)_\mathrm{P}$ is relatively small, $\Delta V/\Delta T$ being on the order of 10^{-4} to 10^{-5} $\mathrm{cm^3 \cdot K^{-1}}$. Thus, for many purposes it can be assumed that this quantity is negligible and that

$$C_\mathrm{P} \cong C_\mathrm{V} \quad (9\text{-}10)$$

This approximation will be employed for solids where applicable and convenient. It gives a maximum error of < 0.5 $\mathrm{cal \cdot mol^{-1} \cdot K^{-1}}$.

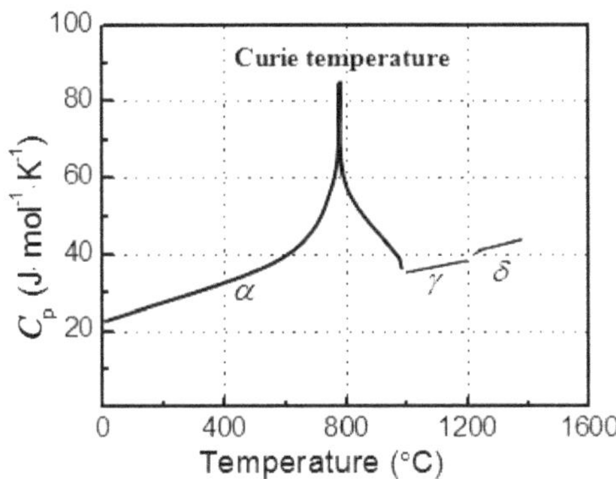

Figure 9.1 The effect of temperature on the heat capacity of iron. Both the change in crystal structure and the change from ferromagnetic to paramagnetic behavior are indicated.

Figure 9.1 shows the temperature dependence of the iron heat capacity. The most important factor affecting heat capacity is the lattice vibrations or phonons; however, other factors affect the heat capacity. One striking feature occurring in ferromagnetic materials such as iron is that an abnormally high heat capacity is observed in iron at the Curie temperature, where the normally aligned magnetic moments of the iron atoms are randomized and the iron becomes paramagnetic.[2] Heat capacity also depends on the crystal structure, as shown in Figure 9.1 for iron.

2 铁磁性材料，比如铁，一个显著的特点是在居里温度下拥有异常高的热容。在个这温度时，原本有序排列的铁原子磁矩被随机化（无序化）了，同时铁转变为顺磁性的。

9.1.2 Classical Model (Dulong-Petit Law)

In classical theory, when interpreting the pressure of ideal gas, equipartition principle was introduced as following: In equilibrium condition of an ideal gas, an average kinetic energy of each degree of freedom is

$$E_\mathrm{k}(x) = \frac{1}{2}m\overline{v}_\mathrm{x}^2 = \frac{1}{2}kT \quad (9\text{-}11)$$

where, $\overline{v}_\mathrm{x}$ is the average velocity of a gas molecule in x-direction, T is absolute temperature, k is the Boltzmann constant. Since there are three degrees of freedom, the average kinetic energy of a gas molecule is

$$E_k = \frac{3}{2}kT \qquad (9\text{-}12)$$

The classical theory assumed that the internal energy of a solid could be considered to reside in the ion cores of a solid. A solid was thought to consist of an assembly of ion cores that behaved as simple harmonic oscillators, each vibrating about an equilibrium position, in thermal equilibrium at a given temperature.[3]

3 固体被看作由做简谐振动的离子实聚集而成，每一个谐振子都在平衡位置，即在一定温度下在热平衡位置附近振动。

In 19th century, Dulong-Petit introduced the equipartition principle to solid, which contains the following main points:

(1) The internal energy of a solid could be considered to reside in the ion cores of a solid.

(2) A solid was thought to consist of an assembly of ion cores that behaved as simple harmonic oscillators, each vibrating about an equilibrium position, with mutually conversional potential and kinetic energy. And for average,

$$E_k = E_p = \frac{1}{2}kT \qquad (9\text{-}13)$$

(3) The oscillating ions were considered to have three degrees of freedom that correspond to their energies of translation parallel to the three cartesian coordinates. The average internal energy

$$U = 3\times(E_k + E_p) = 3\times\left(\frac{kT}{2}+\frac{kT}{2}\right) = 3kT \qquad (9\text{-}14)$$

For one mole the internal energy is

$$U = 3N_A kT = 3RT \qquad (9\text{-}15)$$

where N_A is Avogadro's number (6.02×10^{23} atoms per mol). From Equation (9-3), the heat capacity is

$$C_V = \left(\frac{\partial U}{\partial T}\right)\Big|_V = 3N_A k = 3R \qquad (9\text{-}16)$$

It will be recalled that $R = N_A k$, is the gas constant (= 8.31 $J\cdot mol^{-1}\cdot K^{-1}$). This provides a value of C_V for an element equal to approximately 24.9 $J\cdot mol^{-1}\cdot K^{-1}$, a value that agrees with experiments at high temperature. However, this fails to provide any temperature dependence of the heat capacity, that generally exits (Figure 9.2).[4]

4 这表明单质材料的 C_V 大约等于 24.9 $J\cdot mol^{-1}\cdot K^{-1}$。在高温下这个值与实验结果相符。然而，它却不能给出任何热容对温度的依赖关系，而这通常是存在的（图 9.2）。

The Dulong-Petit law neglects any contribution of the valence electrons to the internal energy of the solid. In addition, similar to the ideal gases, the thermal equilibrium of the ion cores is treated as such that the energy distribution is continuous, and makes use of the Maxwell-Boltzmann statistics and equipartition of energy.[5]

5 杜隆—珀替定律忽视了价电子对固体内能的贡献。而且，与理想气体相似，处于热平衡离子实的能量被视为连续分布的，并利用了麦斯威尔—波尔兹曼统计和能量均分原理。

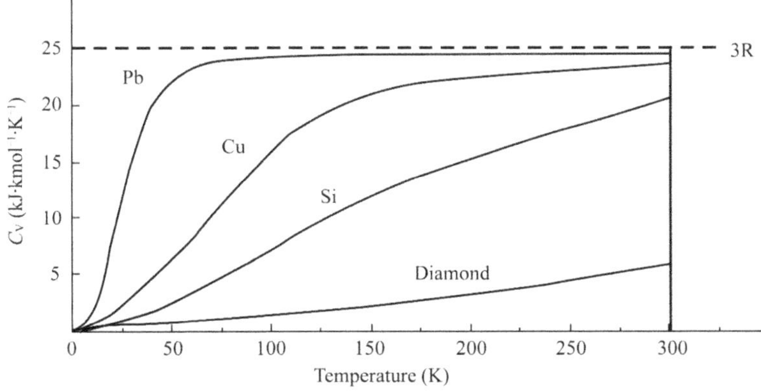

Figure 9.2 The molar heat capacities at constant volume of some solids as a function of temperature.

Implicit in this oversimplified treatment is the idea that each ion in the solid is treated as an independent oscillator. It is apparent that the oscillations of a given ion will vary with temperature and will affect those of its neighbors. Models of heat capacity that are more representative of the actual, more complex conditions must be used to account for such ion-ion interactions.[6]

6 这一简化处理隐含的是固体中每一离子都被视为一个独立的谐振子。显而易见，一个给定离子的振动将因温度而变化，并影响邻近的其他离子。为了表示这种离子—离子之间的相互作用，必须采用考虑了更复杂条件的、更能够代表实际情况的热容模型。

9.1.3 Einstein Model

In 1906, Einstein developed a model that partially overcomes the failure of the classical approach to describe the heat capacity of a solid as a function of temperature. Instead of using oscillation with continuous energy to describe ion cores in solid, he expressed energy discretely in terms of the frequency of the oscillating ion cores. Such a single quantized oscillation is known as a phonon. By means of the Planck idea of discrete energies, with angular frequency of ω, the vibrational energy should be:

$$E_{\mathrm{i}} = \left(n + \frac{1}{2}\right)\hbar\omega_{\mathrm{i}} \tag{9-17}$$

where n is the quantum number, $\frac{1}{2}\hbar\omega_{\mathrm{i}}$ is the zero-point energy of the system, $\hbar = \frac{h}{2\pi}$, is Dirac constant.

Before further discussion, we will say something about the zero-point energy of a system. The molar zero-point energy of a solid (U_0) is given by summing all of the ions, as

$$U_0 = \int_0^{\omega_c} \frac{1}{2}\hbar\omega N(\omega)\mathrm{d}\omega = \int_0^{\nu_c} \frac{1}{2}h\upsilon N(\nu)\mathrm{d}\nu \tag{9-18}$$

Finally:

$$U_0 = \frac{9}{2}\frac{N_{\mathrm{A}}h}{\nu_{\mathrm{c}}^3}\int_0^{\nu_c} \nu^3\mathrm{d}\nu = \frac{9}{8}N_{\mathrm{A}}h\nu_{\mathrm{c}} \tag{9-19}$$

where v_c is the maximum frequency of oscillators; v, frequency; N_A, Avogadro constant.

Introducing Debye temperature:

$$U_0 = \frac{9}{8} N_A k \Theta_D = \frac{9}{8} R \Theta_D \tag{9-20}$$

Because energy is a relative quantity, the lowest possible internal energy of a solid is taken to represent that condition in which both all of the ions and all of the electrons that constitute the solid reside in a unique lowest possible energy state at absolute zero.[7] For certain thermodynamic purposes such a condition is assigned as being the zero of energy. This is not at variance with thermodynamics, because energy measurements are relative. When the zero-point energy is used as a reference for energy measurements, the differences or changes in energy are actually measured, and the zero-point energy cancels out. In the same way, the entropy of a substance is arbitrarily designated as being equal to zero for this set of conditions at 0 K. This means zero-point energy has no contribution to heat capacity, and can be omitted in the development of heat capacity equation.[8] Thus energy of phonon is:

7 由于能量是一个相对的量，一个固体可能的最低内能，用来表示这样的情况：在绝对零度下，组成固体的所有离子和电子都处于一个最低可能并唯一的能量状态。

8 类似地，在 0K 的这一系列条件下，一个物质的熵可以任意指定为 0。这意味着零点能对热容没有作用，并且在热容方程中可以被忽略。

$$E_i = n\hbar\omega_i \tag{9-21}$$

According to Boltzmann statistics, the number of oscillators with energy E_i is proportional to:

$$f(E_i) \propto e^{-\frac{n\hbar\omega_i}{kT}} \tag{9-22}$$

The average energy of one such oscillator is found by obtaining the average from the energies of all of the ions in the solid. This average energy is given by

$$\overline{E}_i = \frac{\sum_{n=0}^{\infty} n\hbar\omega_i e^{-n\hbar\omega_i/kT}}{\sum_{n=0}^{\infty} e^{-n\hbar\omega_i/kT}} \approx \frac{\hbar\omega_i}{e^{\hbar\omega_i/kT}-1} \tag{9-23}$$

An expression is thus obtained for the average energy of an oscillator; this is quite different to the previous propose of oscillation with equal average kinetic and potential energies under equipartition principle.

Based on the classical picture of a solid heat capacity, Einstein extended it by applying this average energy (Equation 9-23). As in the classical case, it was assumed that the internal energy of a solid was associated only with the ions. The energy of the electrons was not taken into account. The solid thus was treated as an assembly of independent simple-harmonic oscillators in thermal equilibrium; i.e., each ion of the solid independently oscillated with the same, given angular frequency, ω_E. As noted for the classical case, this is too great an oversimplification because each oscillating ion at least would be expected to affect the oscillations of its nearest neighbors so that their frequencies could not possibly be all the same.[9] The inclusion of the quantum-mechanic approach did constitute a major advance for the new physics.

For 1 mol atoms containing solid, the N_{total} atoms can be regarded as $3N_{\text{total}}$ linear harmonic oscillators, i.e., each atom corresponds to three harmonic oscillators, which vibrate in the three perpendicular directions, x, y and z, with a single basic angular frequency ω_{E} around their average positions independently of all their neighbours.[10] Therefore, the internal energy U of 1 mol of solid is given by the sum of the average energies of all the atoms in the solid:

$$U=\frac{3N_{\text{A}}\hbar\omega_{\text{E}}}{e^{\hbar\omega_{\text{E}}/kT}-1}=3N_{\text{A}}\frac{k\Theta_{\text{E}}}{e^{\Theta_{\text{E}}/T}-1} \tag{9-24}$$

Define:

$$\Theta_{\text{E}}=\frac{\hbar\omega_{\text{E}}}{k} \tag{9-25}$$

as Einstein temperature. Its common values are in the range of 100 – 300 K. The mole heat capacity of the crystal vibration is:

$$C_{\text{mV}}=\frac{\partial U}{\partial T}\Big|_{\text{V}}=3N_{\text{A}}k\left(\frac{\Theta_{\text{E}}}{T}\right)^2\frac{e^{\Theta_{\text{E}}/T}}{(e^{\Theta_{\text{E}}/T}-1)^2} \tag{9-26}$$

From Equation (9-26), it can be concluded that:

(a) At high temperature, $T>>\Theta_{\text{E}}$,

$$e^{\frac{\Theta_{\text{E}}}{T}}-1=\left[1+\frac{\Theta_{\text{E}}}{T}+\frac{1}{2!}\left(\frac{\Theta_{\text{E}}}{T}\right)^2+\frac{1}{3!}\left(\frac{\Theta_{\text{E}}}{T}\right)^3+\cdots\right]-1\approx\left[1+\frac{\Theta_{\text{E}}}{T}\right]-1$$
$$=\frac{\Theta_{\text{E}}}{T}$$

$$e^{\frac{\Theta_{\text{E}}}{T}}\approx e^0=1$$

Thus,

$$C_{\text{mV}}=3N_{\text{A}}k\left(\frac{\Theta_{\text{E}}}{T}\right)^2\frac{e^{\Theta_{\text{E}}/T}}{(e^{\Theta_{\text{E}}/T}-1)^2}\approx 3N_{\text{A}}k\left(\frac{\Theta_{\text{E}}}{T}\right)^2\frac{1}{\left(\frac{\Theta_{\text{E}}}{T}\right)^2}=3N_{\text{A}}k$$

This means that at high temperature, Einstein model equals to the Dulong-Petit law.

(b) At low temperature, $e^{\Theta_{\text{E}}/T}>>1$

$$C_{\text{mV}}=3N_{\text{A}}k\left(\frac{\Theta_{\text{E}}}{T}\right)^2\frac{e^{\Theta_{\text{E}}/T}}{(e^{\Theta_{\text{E}}/T}-1)^2}=3N_{\text{A}}k\left(\frac{\Theta_{\text{E}}}{T}\right)^2e^{-\Theta_{\text{E}}/T} \tag{9-27}$$

As the temperature approaches zero, the negative exponential term approaches zero, and causes C_{mV} to approach zero in an exponential fashion. Thus, the Einstein model fits the observed conditions at the limits. Figure 9.3

9 就像在经典模型时注意到情况一样，该模型太过于简化，因为每个振动离子至少会影响最邻近的振动离子，因此它们的频率不可能都一样。

10 对于含有一摩尔原子的固体来说，N_{total} 个原子可以看作是 $3N_{\text{total}}$ 个线性谐振子，即：每个原子对应三个相互垂直方向 x, y, z 的谐振子，并且拥有一个在其平均位置附近、独立于他们周围谐振子的单一基础角频率 ω_{E}。

shows how this fits the experimental data over a range of temperatures. It is apparent that the Einstein relationship gives a poor fit in the lower range of temperatures because of its exponential rather than the observed T^3 behaviour; this deviation decreases as the temperature increases.[11] However, the degree of agreement is surprisingly good considering the simplicity of the model.

11 显然，在较低温度范围内爱因斯坦关系式与实际不符合，因为在爱因斯坦关系中热容对温度是指数依赖关系，而不是实际的三次方关系。这一偏差随着温度的上升而减小。

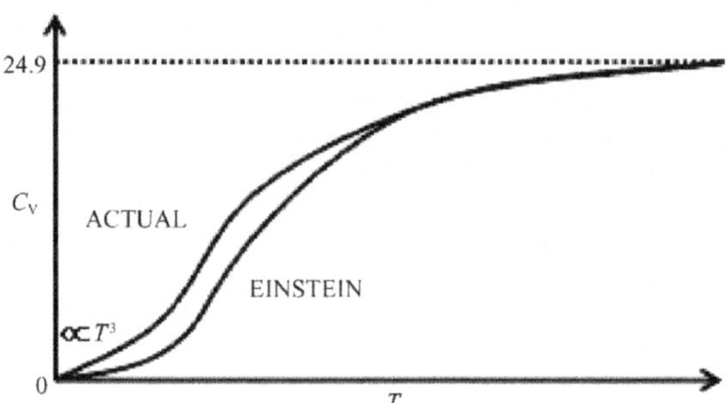

Figure 9.3 Comparison of the Einstein results for heat capacity with experimental behavior.

The reasons for inconsistence between the Einstein model and the experiment can be summarized as:

(a) Einstein assumed that all the atoms in the crystal were vibrating with the same frequency and that the vibrations of each atom or harmonic oscillator were independent of all the others.[12]

12 爱因斯坦假设，晶体中所有原子都是以相同的频率振动，并且每个原子或谐振子的振动都是独立于其他所有原子或谐振子。

(b) Einstein assumed, in analogy with Planck's photon theory of blackbody radiation, that there is no upper limit of the oscillators or the phonon energies. (in $n\hbar\omega_i$, $n \to \infty$)

9.1.4 Debye Model

Ⅰ. Main Points

Debye included all phonons varying with values in the whole crystal in his calculations, and introduced interaction between them. In addition, he realized that there must be a maximum phonon energy $h\nu_{max}$. The condition implied that the range of vibrational frequencies must be cut off at a maximum frequency, called the Debye frequency, ν_D. The upper limit of the phonon frequencies is equal to the Debye frequency ν_D.[13]

13 德拜在计算中包含整个晶体中所有不同值的声子，并且引入声子之间的相互作用。此外，他意识到有一个最大能量的声子 $h\nu_{max}$。这一情况意味着，振动频率的范围一定是在一个最大频率处截止，这一最大频率被称为德拜频率 ν_D。声子频率的上限等于德拜频率。

For crystal with one mole atoms possessing 3-dimension vibration:

$$\int_0^{\omega_m} \rho(\omega)\mathrm{d}\omega = 3N_A \tag{9-28}$$

where ω is the angular frequency for vibration (0 to ω_m); $\rho(\omega)$, the probability of vibration with energy of $\hbar\omega$.

For elastic wave in crystal, it has been proved that the frequency distribution has the following form:

$$\rho(\omega)=\frac{3V}{2\pi^2}\frac{\omega^2}{v_p^3} \tag{9-29}$$

where V is volume of the crystal; v_p, the velocity of the phonon wave. In the obtaining of expression (9-29), interactions between the oscillations are considered.

Put $\rho(\omega)$ into the above integral Equation (9-28), the maximum frequency for vibration can be got:

$$\omega_m^3=\frac{6N_A\pi^2v_p^3}{V} \tag{9-30}$$

Thus the average internal energy of one mole crystal:

$$U=\int_0^{\omega_m}\frac{\hbar\omega}{e^{\hbar\omega/kT}-1}\rho(\omega)d\omega=\frac{3V}{2\pi^2v_p^3}\int_0^{\omega_m}\frac{\hbar\omega^3}{e^{\hbar\omega/kT}-1}d\omega \tag{9-31}$$

Define $\chi=\frac{\hbar\omega}{kT}$, and Debye temperature as $\Theta_D=\frac{\hbar\omega_m}{k}$, then:

$$\chi_m=\frac{\hbar\omega_m}{kT}=\frac{\Theta_D}{T} \tag{9-32}$$

Combining $\omega_m^3=6N_A\pi^2v_p^3/V$ and Equation (9-31), we have:

$$U=\frac{3V}{2\pi^2v_p^3}\int_0^{\omega_m}\frac{\hbar\omega^3}{e^{\hbar\omega/kT}-1}d\omega=9N_AkT\left(\frac{T}{\Theta_D}\right)^3\int_0^{\Theta_D/T}\frac{x^3}{e^x-1}dx \tag{9-33}$$

Thus:

$$C_V=\frac{\partial U}{\partial T}|_v=9N_Ak\left(\frac{T}{\Theta_D}\right)^3\int_0^{\Theta_D/T}\frac{x^4e^x}{(e^x-1)^2}dx \tag{9-34}$$

Examining Equation (9-34), we get:

(a) At high temperature, $T>>0$, $\Theta_D/T<<1$, so

$$\left(\frac{T}{\Theta_D}\right)^3\int_0^{\Theta_D/T}\frac{x^4e^x}{(e^x-1)^2}dx\approx\frac{1}{3}$$

Therefore, $C_V\approx3N_Ak=3R$, this is the Dulong-Petit law.

(b) At very low temperature, $T\to0$, $\Theta_D/T\to\infty$, so:

$$\int_0^{\infty}\frac{x^4e^x}{(e^x-1)^2}dx\approx\frac{4}{15}\pi^4 \tag{9-35}$$

$$C_V=9N_Ak\left(\frac{T}{\Theta_D}\right)^3\int_0^{\Theta_D/T}\frac{x^4e^x}{(e^x-1)^2}dx=9N_Ak\left(\frac{T}{\Theta_D}\right)^3\frac{4}{15}\pi^4$$

$$=\frac{12}{5}\pi^4R\left(\frac{T}{\Theta_D}\right)^3=bT^3 \tag{9-36}$$

This is the Debye third power law. It predicts that the heat capacity of solid should approach zero as T^3 approaches zero. It, therefore, agrees with the observed experimental behavior in the very low temperature range of up

14 因此，在直到的极低温度范围内，德拜模型与很多材料的实验情况相一致。德拜模型也消除了爱因斯坦模型中 C_V 指数接近于绝对零度这一内在固有的难点。

to about $\Theta_D/10$ for many materials. It also removes the difficulty inherent in the Einstein model, in which C_V approaches zero exponentially.[14] Many metals agree well with Debye model both in high and low temperature, as shown in Figure 9.4. Table 9.1 gives Debye temperature and heat capacity of some element materials.

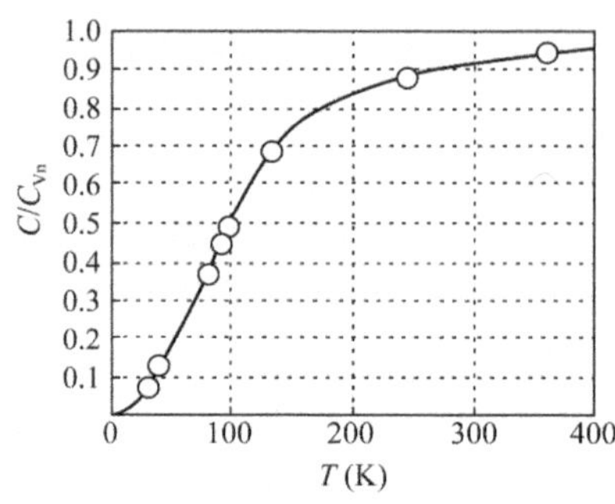

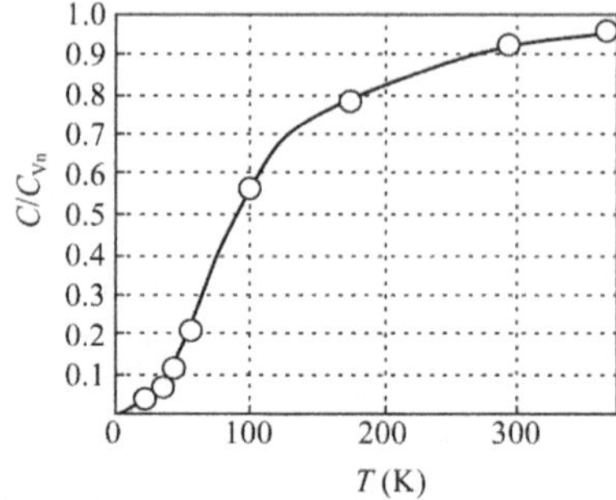

Figure 9.4 The heat capacity of Al (left) and Cu (right) at different temperature (empty symbol), compared with Debye model (solid line).

Table 9.1 Debye temperatures and heat capacities of some elements.

Element	Θ_D(K)[a]	C_P^b	Element	Θ_D(K)[a]	C_P^b	Element	Θ_D(K)[a]	C_P^b
Li	400	4.95	Mn	400	6.71	Sn(gray)	260	—
Be	1000	5.21	Fe	420	5.30	Sn(white)	170	6.30
B	1250	4.62	Co	385	5.93	Sb	200	6.03
Diamond	1860	3.18	Ni	375	6.16	La	132	6.65
Ne	63	4.97	Cu	315	5.86	Pr	74	6.45
Na	150	6.71	Zn	234	6.07	Gd	152	7.03
Mg	318	5.88	Ga	240	6.24	Ta	225	6.43
Al	394	5.82	Ge	360	6.22	W	310	5.97
Si	625	5.91	As	285	5.89	Pt	230	6.19
Ar	85	4.97	Zr	250	6.92	Au	170	6.03
K	100	7.12	Mo	380	5.67	Hg	100	6.50
Ca	230	6.28	Pd	275	6.21	Tl	96	6.29
Ti	400	6.00	Ag	215	5.70	Pb	88	6.12
V	390	5.67	Cd	120	6.19	Bi	120	6.10
Cr	460	5.24	In	129	6.50	Th	100	6.29

Ⅱ. Insufficiency of Debye Model:

Similar to previously discussed models, Debye model is not perfect:

(a) At low temperature, Debye model cannot completely satisfy the reality, especially at very low T, e.g. heat capacity in superconductor.

(b) Although Debye model consists with experimental results in a quite wide temperature range for atomic crystals (Al, Ag, C) and for simple ionic crystals (KCl, Al_2O_3). For complicate structures, such as polycrystalline/ multiphase inorganic materials, due to the various coupling among high frequency vibration, Debye model is generally not suitable.[15]

It must be recalled that the Debye model assumes an ideal solid. Impurities and imperfections exist in even the best single-crystal specimens. In polycrystalline materials, the deviation from the ideal is compounded by grain boundaries that further complicate it.[16] Additionally, if the solid is a metal, the role of the valence electrons or conduction electrons have been neglected. This factor must be considered, particularly at low temperatures.

Ⅲ. Debye Temperature

The Debye temperature is an important point of reference because it is the temperature above which the vibrational energy of the ion is large enough to give average ionic displacements, $\bar{x}$:[17]

$$\frac{1}{2}K\,\bar{x}^2 = mean\ potential\ energy = \frac{1}{2}kT \tag{9-37}$$

where K is the "spring constant". So the classical mechanics may be employed. Below this temperature quantum mechanics must be used. It can be shown that Θ_D is the temperature at which C_V is at 96% of its asymptotic value. This explains the reasonable findings of Dulong-Petit, whose data were taken at or near Θ_D.

Debye temperature is related to the velocity of phonon in solid, i.e. acoustic velocity (c_s). Using $\nu_c = c_s/\lambda_c$, and substituting this into the expression for Θ_D gives

$$\Theta_D = \frac{hc_s}{k\lambda_c} \tag{9-38}$$

It is apparent that Θ_D depends primarily upon the velocity of sound in the solid. This property is not a constant, but is dependent upon such factors as the temperature of the solid, its lattice spacing, or its density, and the way in which the phonons, or quanta of lattice vibrational energy, are transmitted and/or scattered by the lattice.[18] Thus, the Debye temperature is not exactly a constant. It also varies slightly depending upon the technique used in its experimental determination as well as upon the range of temperatures over which it is determined. Equation (9-38) provides one way to determine Θ_D; it also may be calculated from measurements of electrical resistivity.

15 尽管德拜模型在相对宽的温度范围内对于原子晶体（Al, Ag, C）和简单的离子晶体（KCl, Al_2O_3）与实验结果一致，但是对于复杂结构，如多晶或多相无机材料，由于在高频振动中存在多种耦合作用，德拜模型通常是不适合的。

16 在多晶材料中，对照理想模型，又混入了晶界的偏离，使得体系更加复杂化。

17 德拜温度是一个重要的参数，因为高于这个温度时，离子振动能量变得足够大，导致平均离子位移 $\bar{x}$：

18 德拜温度不是一个常量，而是依赖于固体的温度、晶格间距、密度，也依赖于声子或者说一个晶格振动能量的量子透过晶格和/或被晶格散射的方式。

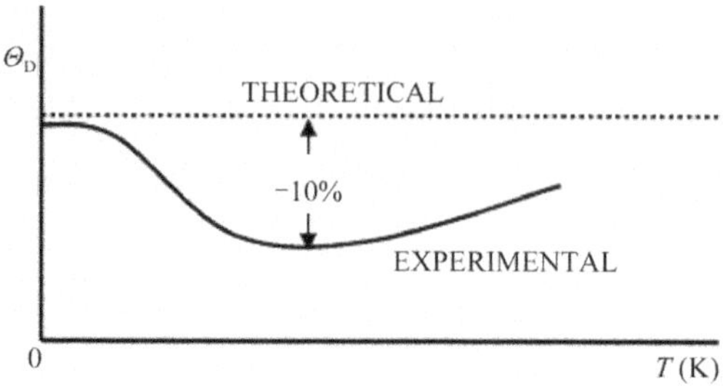

Figure 9.5 Schematic variation of the Debye temperature of a metal as a function of temperature; $T \ll \Theta_D$.

Typical temperature variations of the Debye temperature are shown schematically in Figure 9.5. The treatment given here is applicable only to elemental solids containing no free electrons and to the ion cores of metallic lattices, since the electron effects are not included. The theory treats Θ_D as being constant with temperature, as shown in the figure as a broken line. The experimentally determined values for this parameter for copper show a variation of about 10%. Other substances show somewhat different ranges of variations, but most appear to have the same general sigmoidal type of variation of Θ_D with T, as shown in the figure.[19]

19 这一处理方法仅适用于没有自由电子的单质固体和金属晶格离子实，因为电子效应没有被考虑。这一理论把德拜温度视为不随温度变化的常量，如图中的虚线所示。铜 Θ_D 的实验值与理论值相差约10%。其他物质或多或少有不同程度的差异，但是大多数都有着与图中所示相似的S型 Θ_D-T 变化曲线，如图中所示。

Θ_D can be roughly estimated by lattice vibration frequency:

$$\Theta_D = \frac{\hbar \omega_m}{k} = 0.76 \times 10^{-11} \omega_m \tag{9-39}$$

It is reasonable to suppose that at the temperature of melting point T_m, the lattice can be destroyed, at which time the vibration of lattice reaches its maximum. Hence, T_m is related to ω_m:

$$\nu_m = 2.8 \times 10^{12} \sqrt{\frac{T_m}{MV_a^{2/3}}} \tag{9-40}$$

and

$$\Theta_D = 137.7 \sqrt{\frac{T_m}{MV_a^{2/3}}} \tag{9-41}$$

where M is relative atomic mass; V_a, atomic volume. This is understandable, since the binding force (related to oscillation ω_m) for material with high melting point is also high thus its Θ_D is high.

9.1.5 Comparison of Models

Between Einstein and Debye models, the same general expression was employed to obtain a relationship for the internal energy for the solid. This is given by

$$U = \int \text{average energy of an oscillator} \times \text{density of states} \tag{9-42}$$

Each model employed the same expression for the average energy of an oscillator:

$$\bar{E} = \frac{hv}{e^{hv/k_B T} - 1} \tag{9-43}$$

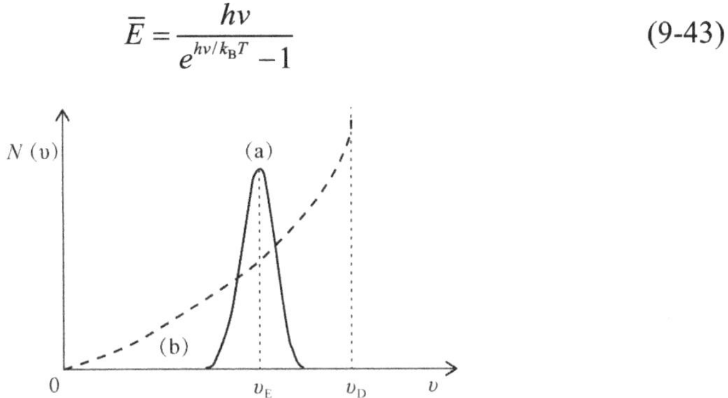

Figure 9.6 Densities of states for (a) the Einstein model, (b) the Debye model.

Thus, the theories may be compared by examining the differences in the ways the densities of states are determined. The Einstein model employed only one frequency, ω_E, for all independently oscillating ions of the solid, while the original Debye approach assumed that the oscillating ions could be considered as constituting quantized elastic waves in the solid. The resultant densities of states of these two models are shown graphically in Figure 9.6. Clearly, the density of states implicit in the Einstein model is based on the probable distribution of the frequencies about a value of ν_E, and the distribution of Debye frequencies is based on continuous oscillations between zero and ν_D.[20]

20 爱因斯坦模型只使用一个频率 ω_E 来表示固体中所有独立振动的离子。德拜模型假设振动离子在固体中是一系列量子化的弹性波。基于这 2 个模型的态密度在图 9.6 中展示。可以清楚的看到，爱因斯坦模型中隐含的态密度是基于频率 ν_E 的最可几分布。德拜频率的分布是基于 0 和 ν_D 之间的连续振动。

Figure 9.7 shows the comparison between two models and experiment results. It can be seen that Einstein model is inconsistent with experimental data in the lower temperature range though at higher temperature it gives more satisfied results. In contrast, the Debye approximation provides the best model for the behavior of the heat capacity of lattices, especially at very low temperatures. It is frequently employed along with the electron contribution for the determination of properties of metals at very low temperatures.[21]

21 从图中可见，爱因斯坦模型在较低温度范围内与实验数据不符，虽然高温时结果较为满意。相比之下，德拜近似给出了晶格热容最好的模型，尤其是在很低的温度下。德拜模型加上电子贡献，经常被应用于来决定金属在极低温度下的性质。

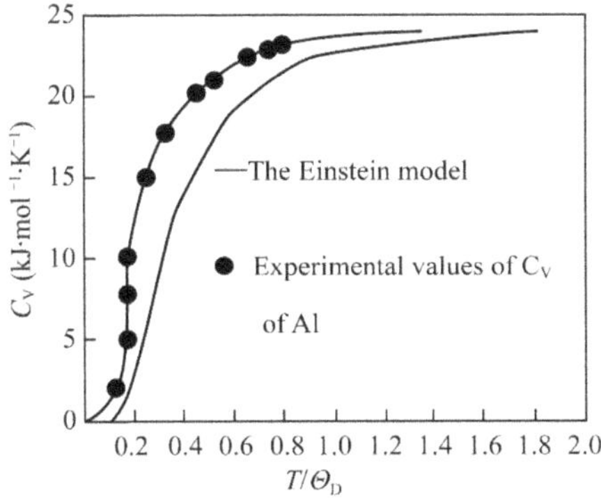

Figure 9.7 Comparison between two models and experimental results.

9.1.6 Heat Capacity in Real Materials

Ⅰ. Metal and alloy

In metals and alloys, besides lattice vibration, free electrons in metal also contributes to heat capacity. The average energy of a free electron in metal:

$$\bar{E} = \frac{3}{5}E_{\mathrm{F}}^{0}\left[1+\frac{5}{12}\pi^{2}\left(\frac{kT}{E_{\mathrm{F}}^{0}}\right)^{2}\right] \tag{9-44}$$

For one mole (N_A) electron, their contribution to heat capacity is:

$$C_{\mathrm{mV}}^{\mathrm{e}} = N_{\mathrm{A}}Z\left(\frac{\partial \bar{E}}{\partial T}\right) = N_{\mathrm{A}}Z\frac{\pi^{2}k^{2}}{2E_{\mathrm{F}}^{0}}T = \frac{\pi^{2}ZkR}{2E_{\mathrm{F}}^{0}}T \tag{9-45}$$

Z is the number of valence electrons

Example: the density of Cu is 8.9×10^{3} kg/m^{3}, atomic number 63, the number of free electrons in unit volume (Z) can be calculated. Then,

$$C_{\mathrm{mV}}^{\mathrm{e}} = 0.64\times10^{-4}T \tag{9-46}$$

At room temperature, compared to atomic mole capacity ($3R$), the electron contribution can be omitted. Therefore, for real metals, the value of heat capacity can be divided into three regions as shown in Figure 9.8: (a) Region I (0-5 K): Phonons are frozen, electrons are the main contribution to heat capacity, $C \propto T$; (b) Region II (relatively high but below Debye temperature): In this region, the main factor for heat capacity is lattice vibration, the Debye model is well satisfied, $C \propto T^{3}$; (c) At Debye temperature, $C = 3R$, (d) Region III: in this region, the lattice vibration and the electron both contribute to heat capacity above Debye Temperature, therefore, $C > 3R$.[22]

22 （a）范围 I（0～5 K）：声子是被冻结的，电子对于热容起到主要作用，$C \propto T$；（b）范围 II（相对高但低于德拜温度）：在这个范围内，热容的主要影响因素是晶格振动，德拜模型很好地满足这一关系，$C \propto T^{3}$；（c）在德拜温度，$C = 3R$；（d）范围 III：在这个范围，温度高于德拜温度，晶格振动和电子都对热容有贡献，因此 $C > 3R$。

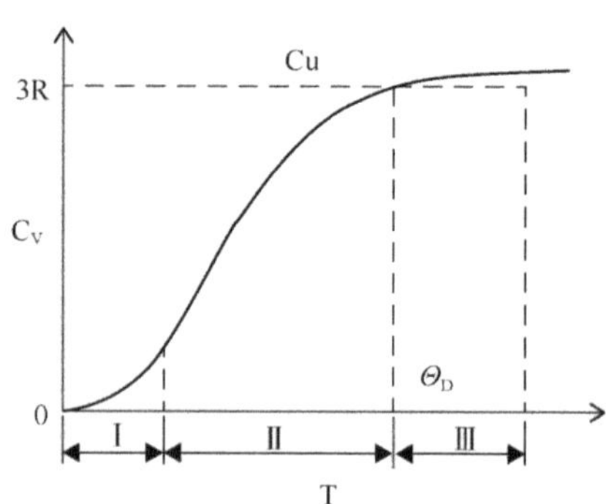

Figure 9.8 Heat capacity for copper.

The situation of heat capacity for alloy is similar to pure metals, and the heat capacity of alloy is a linear combination of the component heat capacity, because the vibration of atom in alloy is almost the same as in element metal.[23]

23 合金热容的情况与纯金属相似，其值等于各组份热容的线性加和，因为合金中原子的振动大体上与单质金属中的类似。

$$C = pC_{1} + qC_{2} \tag{9-47}$$

This equation is Neumann-Kopp law, where p and q are the percentage for M_1 and M_2; C_1 and C_2 are atomic heat capacity for M_1 and M_2. For multiple components,

$$C = \sum g_i C_i \tag{9-48}$$

where g_i and C_i are the percentage and heat capacity of the i component.

Ⅱ. Ceramic

There are hardly free electrons in ceramics, and thus, ceramics are more conformed with Debye model (Figure 9.9). And due to the different values of Θ_D in different ceramic, the temperatures reaching $3R$ are different. Above Θ_D, heat capacity of ceramics almost keeps unchanged with a few exceptions (e.g. MgO).[24]

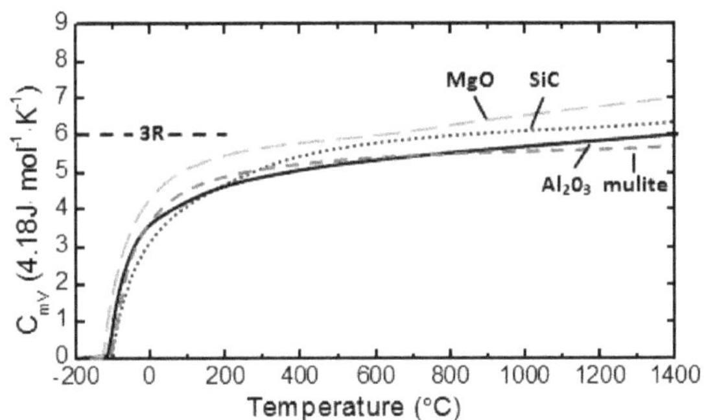

Figure 9.9 Heat capacity of ceramics.

24 陶瓷中几乎没有自由电子，因此陶瓷材料的热容与德拜模型更一致（见图 9.9）。由于不同陶瓷Θ_D的不同，它们到达 $3R$ 的温度是不同的。当温度高于Θ_D时，陶瓷的热容几乎保持不变，只有少数例外（如 MgO）。

9.2 Thermal Conductivity

9.2.1 Thermal Conductivity and Thermal Diffusivity

Thermal conduction refers to the transfer of heat from hot side to cold side. Thermal conductivity is a measure of the rate at which heat is transferred through a material. The treatment of thermal conductivity is similar to that of diffusion coefficient.[25]

Under steady state, i.e., the temperature of every point of a material does not change with time, along the x direction, the amount of heat passing the cross-section Δs is proportional to temperature gradient of $\mathrm{d}T/\mathrm{d}x$, the area Δs, and the time Δt[26](Figure 9.10):

$$Q = -k_t \frac{\mathrm{d}T}{\mathrm{d}x} \Delta S \Delta t \tag{9-49}$$

The proportionality factor k_t is defined as thermal conductivity. The minus sign means that under positive temperature gradient, along x-direction, it is an exothermal process.

25 热导是指热量从热端到冷端的转移。热导率是度量材料中热量转移速率的参数。热导率的处理方法与扩散系数的处理方法类似。

26 在稳态下，也就是说材料中每一点的温度都不随时间变化，沿 x 轴方向，通过横截面 Δs 的热量与温度梯度 $\mathrm{d}T/\mathrm{d}x$、横截面积 Δs、以及时间 Δt 成正比。

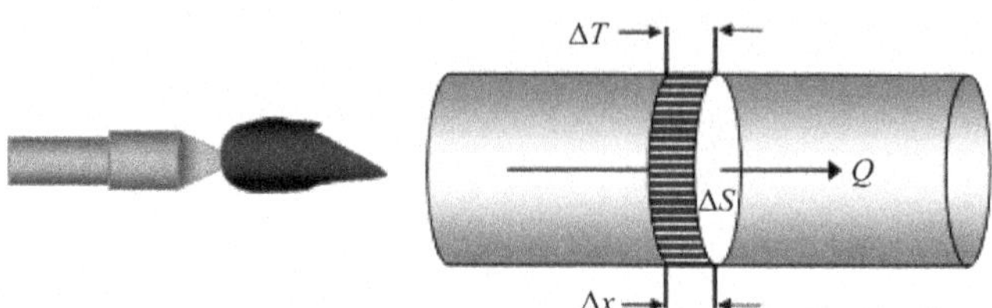

Figure 9.10 Heat transfer from hot side to cold side.

Defining heat flux density (J) as heat being transferred per unit area and time, and combining with Equation (9-49):

$$J = \frac{Q}{\Delta S \Delta t} = -k_t \frac{dT}{dx} \tag{9-50}$$

The heat flux is proportional to temperature gradient, and the thermal conductivity is their proportionality factor.

For general scientific use, thermal conductance is the quantity of heat that passes in unit time through a plate of particular area and thickness when its opposite faces differ in temperature by one Kelvin:[27]

$$H_t = k_t \Delta s \Delta x \tag{9-51}$$

where H_t is the thermal conductance, Δs is area, and Δx is thickness. The unit for conductance is in $W \cdot K^{-1}$ (equivalent to: $W \cdot ℃^{-1}$). The thermal conductance of that particular construction is the inverse of the thermal resistance. Thermal conductivity and conductance are analogous to electrical conductivity ($A \cdot m^{-1} \cdot V^{-1}$) and electrical conductance ($A \cdot V^{-1}$).

27 对于通常的科学用途，热导是指在单位时间内，由一个特定面积和厚度的平面到另外一个温差为 1 K 的平面传导的热量。

Thermal conductivity reflects the degree of thermal conduction under steady state. For unsteady state, the thermal conducting ability is determined by thermal diffusivity (α), which combines time and space:

$$\frac{\partial T}{\partial t} = \alpha \frac{\partial^2 T}{\partial x^2} \tag{9-52}$$

Under the same heating or cooling condition (dT/dt), a large α means temperature difference within a material is small, i.e. the speed of thermal conduction is fast. Therefore, the temperature is evenly distributed in material and the thermal strain is also small.[28] α can be deduced as:

$$\alpha = \frac{k_t}{dC_P} \tag{9-53}$$

where k_t is thermal conductivity; d, material density; C_P, heat capacity at constant pressure.

28 在相同的加热或冷却条件下（dT/dt），α 值越大意味着材料中温度变化小，也就是说热传导快。因此，在材料中温度是均匀分布的，并且热应力也比较小。

Table 9.2 Thermal conductivity k_t in $J \cdot cm^{-1} \cdot s^{-1} \cdot K^{-1}$ at room temperature.

Al	Cu	Na	Ag	NaCl	KCl	Cr-Al-alloy
2.26	3.94	1.38	4.19	0.071	0.071	0.019

Thermal conductivity (k_t), similar to the diffusion coefficient, is a microstructure sensitive property. The magnitude of k_t determines whether a material is a good or a poor thermal conductor. Usually excellent electrical conductors, i.e. metals, are also good thermal conductors, and vice versa; electrical insulators usually also have a low thermal conductivity[29] (Table 9.2). However, the thermal conductivity depends on temperature (Figure 9.11). At low temperatures some insulators, such as diamond, also have excellent thermal conductivity. The thermal conductivity is always degraded by the introduction of lattice defects, e.g., even isotopes (Figure 9.11b) in an otherwise ideal crystal can lower thermal conductivity.

29 与扩散系数相似，热导率是一种微观结构敏感的性质。热导率的大小决定了材料是否是良好的导热体。通常极佳的导电体，如金属，也是很好的导热体，反之亦然；绝缘体的热导率通常比较低。

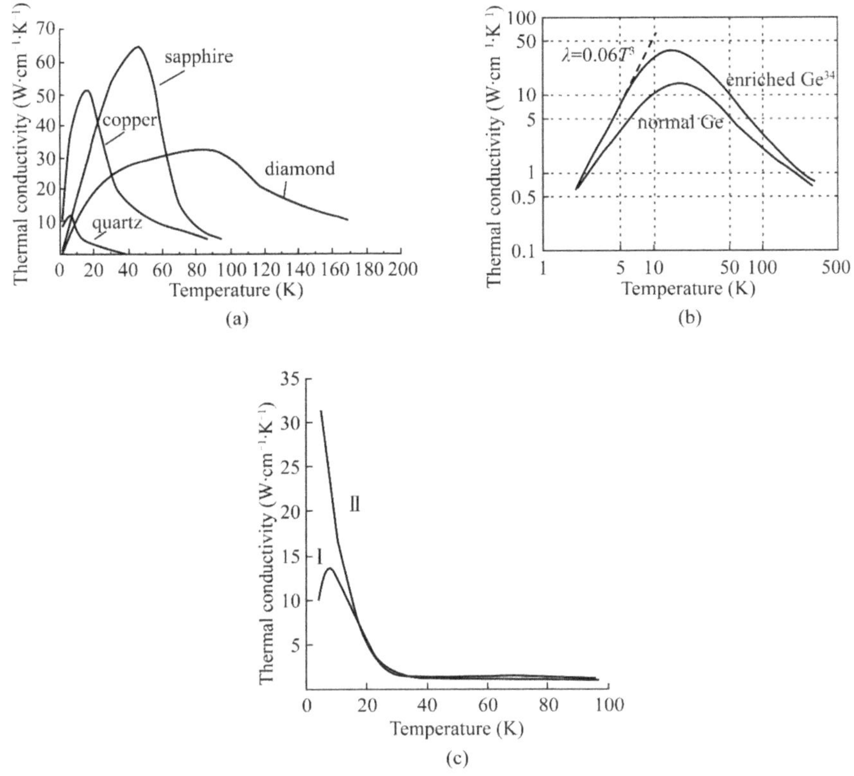

Figure 9.11 (a) Thermal conductivity of copper, quartz, synthetic sapphire and diamond; (b) the influence of isotopic composition on thermal conductivity in germanium. Regardless of its maximum value, thermal conductivity is proportional to T^3 at low temperature. (c) Thermal conductivity of a high purity sodium crystal (II) and impure sodium (I).

9.2.2 Mechanism for Heat Conduction

Thermal energy is transferred by two important mechanisms: transfer of lattice vibrations (i.e. phonons) and free electrons. In insulators there are few freely moving charges (electrons), therefore, the thermal conductivity proceeds almost exclusively via phonons through the crystal lattice. In metals, the number of free electrons is very high, therefore the thermal conductivity is mainly via the free electrons. In addition, at very high temperature, photons can also take part in thermal conduction.[30]

30 在绝缘体中几乎没有自由移动的电荷（电子），因此导热性能完全是通过晶格中声子进行的。在金属中，自由电子的数量很多，因此导热性能主要通过自由电子完成。此外，在非常高的温度下，光子也在热导中起到作用。

Ⅰ. Thermal Conduction Via Lattice Vibration

The thermal conductivity through the crystal lattice proceeds by exchange of high energy vibrations among atoms. This can be visualized most easily, if the lattice vibrations are considered as quasi-particles, or so-called phonons. More specifically, a phonon is the energy quantum of a vibration. The model and terminology are analogous to electromagnetic radiation. An electromagnetic energy quantum (for instance a light wave) is called a photon, which can also be identified with a particle to simplify understanding. In analogy with electromagnetic waves, elastic waves of atoms in a solid also occur in discrete quanta, the unit of which is referred to as a phonon. The transmission of a lattice vibration can be visualized in this model by a collision of one phonon with another phonon, by which energy is exchanged, namely heat.[31]

31 与电磁波类似，固体中原子的弹性波也以分立的量子存在，其单位被称为声子。在这种模型中晶格振动的传播可以通过一个声子与另一个声子的碰撞来形象地描述。通过碰撞，能量即热量被交换了。

Considering two phonons with distance of λ_p, i.e. the free length of phonon, meaning the average distance between phonon collisions, the temperature difference (ΔT) can be expressed as,

$$\Delta T = -\lambda_p \frac{dT}{dx} \tag{9-54}$$

At the two sides of λ_p the energy transferred by the phonon with higher energy:

$$Q_p = C_V^P \Delta T = -C_V^P \lambda_p \frac{dT}{dx} \tag{9-55}$$

C_V^P is the contribution of a phonon to heat capacity at constant volume. Then the heat flux in volume of V:

$$\begin{aligned} J &= \frac{Q}{\Delta S \cdot \Delta t} = \frac{C_V^P \cdot V \cdot \Delta T}{\Delta S \cdot \Delta t} = \frac{C_V^P \cdot \Delta s \cdot \lambda_p \cdot \Delta T}{\Delta S \cdot \Delta t} = C_V^P \cdot v_{px} \cdot \Delta T \\ &= -C_V^P \cdot v_{px} \cdot \lambda_p \frac{dT}{dx} \end{aligned} \tag{9-56}$$

where v_{px} is the average phonon velocity along the x-direction.

Considering the relaxation time (τ_p) between two phonon collisions: $\lambda_p = \tau_p v_{px}$

$$J = -C_V^p v_{px}^2 \tau_p \frac{dT}{dx} \tag{9-57}$$

Based on energy equipartition principle:

$$\overline{v}_{px}^2 = \frac{1}{3} v_p^2 \tag{9-58}$$

$\overline{v}_{px}^2$ is the average value for v_{px}^2, and $\overline{v}_p$ is the average phonon velocity.

$$J = -\frac{1}{3}C_{\mathrm{v}}^{\mathrm{P}}\overline{v}_{\mathrm{p}}^{2}\tau_{\mathrm{p}}\frac{\mathrm{d}T}{\mathrm{d}x} = -\frac{1}{3}C_{\mathrm{v}}^{\mathrm{P}}\overline{v}_{\mathrm{p}}\lambda_{\mathrm{p}}\frac{\mathrm{d}T}{\mathrm{d}x} \tag{9-59}$$

Compare with $J = -k_{\mathrm{t}}\frac{\mathrm{d}T}{\mathrm{d}x}$, yielding:

$$k_{t}^{P} = \frac{1}{3}C_{\mathrm{v}}^{\mathrm{P}}\overline{v}_{\mathrm{p}}\lambda_{\mathrm{p}} \tag{9-60}$$

The specific heat of the phonons $C_{\mathrm{V}}^{\mathrm{P}}$ is the specific heat of the lattice, $\overline{v}_{p}$ is the velocity of propagation of elastic waves, which is the temperature independent speed of sound. At high temperatures $C_{\mathrm{V}}^{\mathrm{P}}$ is constant and λ_{p} approximately proportional to $1/T$. At low temperatures λ_{p} becomes eventually as large as the specimen dimensions, in which case the temperature dependence of the thermal conductivity is given by the specific heat that decreases proportionally to T^{3}.[32] The maximum of $k_{\mathrm{t}}^{\mathrm{P}}$ is caused by the competing effects of a decreasing specific heat at lower temperatures and increasing mean free-path. Perturbations of the perfect crystal, like point defects, dislocations or impurities, even isotopes cause phonon scattering and, therefore, reduce the mean free-path.[33]

Ⅱ. Thermal Conduction Via Electron

Valence electrons gain energy, move toward the colder areas of the material, and transfer their energy to other atoms. Similar to phonon, the thermal conductivity by electron can be expressed as:

$$k_{\mathrm{t}}^{\mathrm{e}} = \frac{1}{3}C_{\mathrm{V}}^{\mathrm{e}}v_{\mathrm{e}}\lambda_{\mathrm{e}} \tag{9-61}$$

Symbols in the right hand of Equation (9-61) are heat capacity from electron, electron velocity, electron free mean path, respectively. Taking

$$C_{\mathrm{V}}^{\mathrm{e}} = \frac{\pi^{2}k^{2}n_{\mathrm{e}}}{2E_{\mathrm{F}}^{0}}T\ ,\quad E_{\mathrm{F}}^{0} = \frac{1}{2}mv_{\mathrm{e}}^{2}\ \text{and}\quad \lambda_{\mathrm{e}} = \tau_{\mathrm{e}}v_{\mathrm{e}}$$

yielding:

$$k_{\mathrm{t}}^{\mathrm{e}} = \frac{\pi^{2}n_{\mathrm{e}}k^{2}T\tau_{\mathrm{e}}}{3m} \tag{9-62}$$

where n_{e}, τ_{e}, and m are electron density, average relaxation time, and electron mass, respectively.

The specific heat of electrons changes in proportion to the temperature. The speed of electrons is given by the Fermi-energy, respectively the Fermi-velocity, both of which are independent of temperature since $E_{\mathrm{F}} = 1/2mv^{2}$ is a material constant. The mean free-path, λ, is determined by the scattering of electrons by phonons and lattice defects.[34]

Ⅲ. Thermal Conduction by Photon

At high T, the vibration variation of molecule, atom and electron will

32 在高温下，$C_{\mathrm{V}}^{\mathrm{P}}$是常数，$\lambda_{\mathrm{p}}$近似地与$1/T$成比例。在低温下，$\lambda_{\mathrm{p}}$最终变得和样品尺寸一样大，在这种情况下，热导率与温度的关系由与T^{3}成反比的比热决定。

33 k_{t}的最大值是由较低温下比热的降低与平均自由程的增加共同决定。完美晶体中的扰动，如点缺陷、位错、或者杂质，甚至是同位素，都能造成声子散射，并因此减少声子的平均自由程。

34 电子比热的变化与温度成正比。电子速度由费米能级及相应的费米速率决定，这两者都与温度无关，因为$E_{\mathrm{F}} = 1/2mv^{2}$是一种材料常数。平均自由程$\lambda$是由声子和晶格缺陷导致的电子散射决定的。

emit electromagnetic wave. Among them, the radiation with wavelength between 400 to 40000 nm possesses rather strong thermal effect. According to the black body radiation:

$$E_{\mathrm{T}} = 4\sigma_0 n^3 T^4 / v \tag{9-63}$$

where $\sigma_0 = 5.67\times10^{-8}\ \mathrm{W\cdot m^{-2}\cdot K^{-4}}$ is Stefan-Boltzmann constant, n is refractive index, v is light velocity. The radiative heat capacity is:

$$C_{\mathrm{r}} = \frac{\partial E_{\mathrm{T}}}{\partial T} = 16\sigma_0 n^3 T^3 / v \tag{9-64}$$

In material, $v_{\mathrm{r}} = v/n$, and the thermal conductivity of photon is:

$$k_{\mathrm{t}}^{\mathrm{r}} = \frac{1}{3} C_{\mathrm{r}} v_{\mathrm{r}} \lambda_{\mathrm{r}} = \frac{16}{3}\sigma_0 n^2 T^3 \lambda_{\mathrm{r}} \tag{9-65}$$

where λ_{r} is the mean free path of photon.

Ⅳ. Total Effect

As can be seen from Equation (9-60), (9-62) and (9-65), the thermal conductivity of solids varies widely as a function of temperature. It is apparent from prior discussions that each of the three factors assumes a different degree of importance in various ranges of temperature.

Considering Equation (9-60) with Θ_{D} as the lower limit, at such temperatures the number of phonons, n, is very high and may be considered as approaching an upper limit. The mean free path of the phonons is expected to vary inversely as their number. Thus, λ_{p} is expected to be very small. As n approaches its upper limit, λ_{p} approaches a corresponding lower limit.[35] In addition, as n becomes larger, the greater is the probability of a phonon interacting with other phonons and of being scattered. This is known as an “umklapp” (flip-over) process. Such interactions can result in complex coupling or anharmonic effects, which effectively diminish the transfer of energy. Under such conditions the absorbed thermal energy is transmitted very inefficiently.[36]

35 考虑公式9-60，将 Θ_{D} 作为温度下限时，此温度下声子的数量 n 非常大，可以看作接近于上限。声子平均自由程将随着声子数量的增加而减少。因此，平均自由程 λ_{p} 被认为非常小。当 n 趋近于上限时，λ_{p} 趋于相应的下限。

36 此外，随着 n 的增加，一个声子与其他声子相互作用和被散射的概率增大。这就是所谓的翻转过程。这种相互作用会造成复杂的耦合或非谐振效应，这会很有效地降低能量的传递。在这种情况下，被吸收热能的传递效率很低。

In the case of metals, this very low efficiency of energy transmission by the ions of the lattice at high temperatures emphasizes the importance of the role of the electrons in the transport of thermal energy (Equation 9-62). Due to the much smaller size of an electron compared to a phonon, electrons are not scattered by phonons to the extent that phonons are scattered by the phonon gas.[37] Metals are good thermal conductors because of the relative freedom from scattering of their valence electrons. For $T > \Theta_{\mathrm{D}}$ the thermal energy transport afforded by the electrons is about 50 times that of the phonons. Therefore, in the range of temperatures above Θ_{D}, the ionic or lattice contribution to k_{t} is low because the effective phonon λ_{p} is small.

37 由于电子比声子的尺寸小很多，电子被声子气体的散射不会达到声子被声子气体散射的程度。

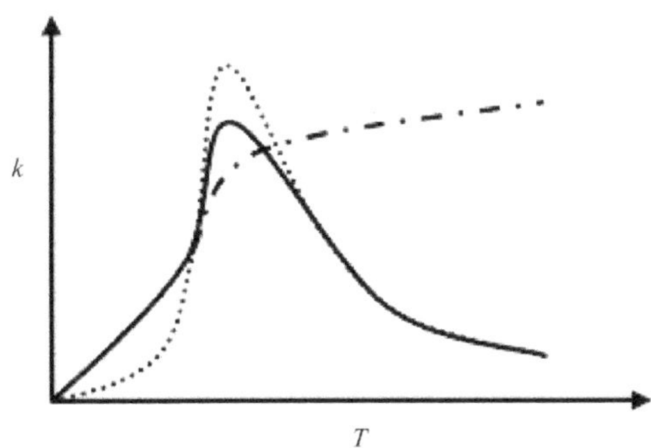

Figure 9.12 Theoretic thermal conductivity. Solid line: the theoretical thermal conductivity of an ionic lattice as a function of temperature; Dot line: phonon effect; Dot-dash curve: electron effect.

Therefore, for an crystalline insulator with electron impurities, the thermal conductivity can be described as such (Figure 9.12): (1) at very low temperature, due to the frozen of phonon, the main factor for thermal conduction originates from electron; (2) with the increase of temperature, the number of phonon increases, the effect from phonon for thermal conduction dominates; (3) if further increase temperature, after coexisting effects from phonon and electron, when the phonon mean free path is negligible, the effect of phonon thermal conduction is low, leading to significant decrease of thermal conduction.[38]

38 （1）在非常低的温度下，由于声子的冻结，热导的主要因素是电子；（2）随着温度的升高，声子的数量增加，声子对热导的影响起到了主导作用；（3）如果进一步升高温度，在声子和电子的共同效应后，并且声子平均自由程可以忽略时，声子对热导的影响变小，致使热导率显著降低。

9.2.3 Thermal Conduction in Real Materials

Ⅰ. Thermal Conductivity in Pure Metals

There are free electrons in metals. Therefore, both phonons (Equation 9-60) and electrons (Equation 9-62) take part in thermal conduction. When the temperature of the material increases, two competing factors affect thermal conductivity. Higher temperatures are expected to increase the energy of the electrons, and increase the contribution from lattice vibrations; these effects increase the thermal conductivity. At the same time, the increased lattice vibrations scatter the electrons, reducing their mobility, and therefore tend to decrease the thermal conductivity.[39] The combined effect of these factors leads to very different behaviors for different metals. Furthermore, thermal conductivity in metals also depends on crystal structure defects, microstructure, and processing. Thus, cold-worked metals, solid-solution-strengthened metals, and two-phase alloys might display lower conductivities compared with their defect-free counterparts.[40]

39 温度升高电子能量增加，并且晶格振动对热导的贡献增加，这些影响提高热导率。同时，提高晶格振动使电子发生散射，降低电子迁移率，因此使热导率降低。

40 此外，金属的热导率也依赖于晶体结构缺陷、微观结构和加工过程。因此，冷加工金属、固溶体强化的金属和两相合金可能表现出相对于无缺陷金属较低的热导率。

In metals, taken the following approximation:

$$C_V^e / C_V^P \approx 0.01$$
$$v_p \approx 5 \times 10^3 \mathrm{m/s}$$
$$v_e \approx 10^6 \mathrm{m/s}$$
$$\overline{\lambda}_p \approx 10^{-9} \mathrm{m}$$
$$\overline{\lambda}_e \approx 10^{-8} \mathrm{m}$$

the contribution of free electrons and phonons to thermal conduction can be estimated as:

$$\frac{k_t^e}{k_t^P} = \frac{C_V^e v_e \overline{\lambda_e}}{C_V^P v_p \overline{\lambda_p}} \approx 20 \tag{9-66}$$

This means the electron contribution is larger compared to that from phonon in terms of thermal conductivity in metals.

In metals, the free electrons carry both thermal conductivity, k_t, and electrical conductivity, σ, so both quantities are correlated. The correlation between k_t and σ is given by the Wiedemann-Franz-Lorenz law:

$$\frac{k_t}{\sigma} = LT \tag{9-67}$$

where L is the Lorenz number, T is absolute temperature. If only electron conduction, both for thermal and electrical, is considered, we have

$$k_t^e = \frac{\pi^2 n_e k^2 T \tau_e}{3m} \text{ and } \sigma = \frac{e^2 n_e \tau_e}{m}$$

Thus,

$$L = \frac{k_t^e}{\sigma T} = \frac{\pi^2 n_e k^2 T \tau_e / 3m}{(e^2 n_e \tau_e / m)T} = \frac{\pi^2 k^2}{3e^2} \approx 2.45 \times 10^{-8}\ V^2 \cdot K^{-2} \tag{9-68}$$

For metals with high electrical conductivity and above Θ_D, $k_t/\sigma T$ is almost constant (Figure 9.13), because this value is almost equal to L, the Lorenz number.

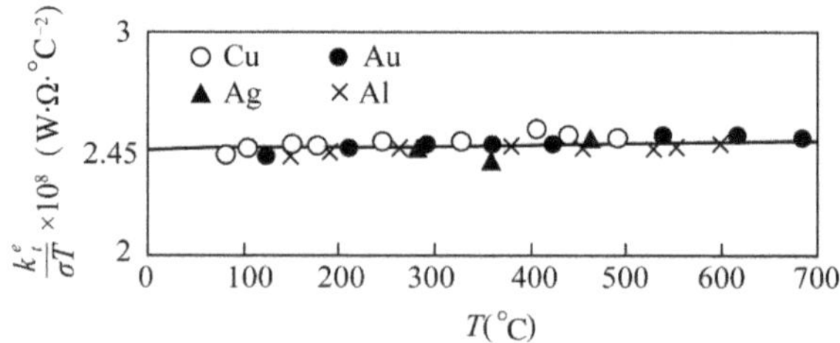

Figure 9.13 The Wiedemann-Franz temperature dependent curves for the electrical and thermal conductions in some metals.

For metals with low electrical conductivity and at relatively low temperature, $k_t/\sigma T$ is a variable (Figure 9.14).

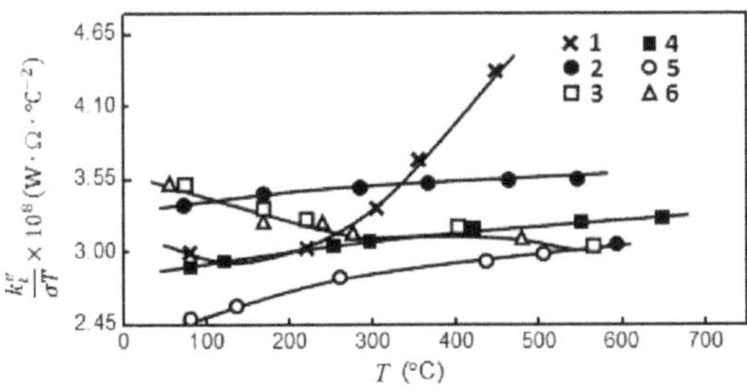

Figure 9.14 Wiedemann-Franz temperature-dependent curves for electrical and thermal conduction in metals or alloys with relatively low electrical conductivity under relatively low temperature. 1: pure Fe, 2: casting Ti (96.9%), 3: Ti, 4: Pt, 5: Ni (99.9%), 6: Zr (99.9%).

This is attributed to the arising contribution from the thermal conduction by phonon.

$$\frac{k_t}{\sigma T}=\frac{k_t^e}{\sigma T}+\frac{k_t^P}{\sigma T}=L+\frac{k_t^P}{\sigma T} \tag{9-69}$$

when σT is very large, $\frac{k_t^P}{\sigma T}\to 0$, Wiedemann-Franz-Lorenz law is valid.

Ⅱ. Thermal Conductivity of Metal Alloy

In two components alloy, the second components can be treated as impurity. Increasing impurity concentration, the electron mean free path decreases, leading to the reduction of thermal conduction. At 50% impurity, thermal conductivity has the lowest value[41] (Figure 9.15).

41 在两相合金中，第二种组分可以被视为杂质。增加杂质的浓度，电子平均自由程降低，导致热导的降低。含有 50% 杂质的合金，热导率达到最低值。

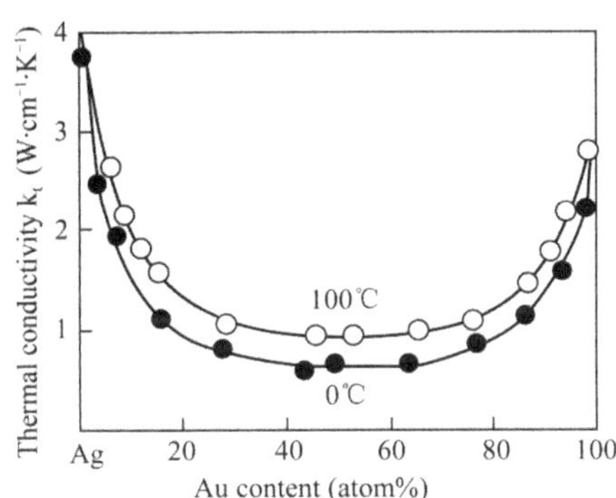

Figure 9.15 Thermal conductivity of Ag-Au alloy with different concentration at temperature of 0°C and 100°C.

Ⅲ. Thermal Conductivity of Ceramic

In ceramics, the energy gap is too large for many electrons to be excited into the conduction band except even at very high temperatures. Thus, the transfer of heat in ceramics occurs primarily by lattice vibrations (i.e. phonons). But at high temperature, photons should be included. Since the electronic contribution is absent, the thermal conductivity of most ceramics is much lower than that of metals.[42] Equation for thermal

42 因陶瓷材料能隙太大,除非在很高的温度下,许多电子很难被激发到导带。因此,在陶瓷材料中热量的传递主要靠晶格振动（即声子），但是在高温下,需要同时考虑光子的热导。因为电子的作用是被忽略的,所有大多数陶瓷材料的热导率比金属热导率低得多。

43 在低温条件下，声子平均自由程达到最大值，同时声子的热容与 T^3 成正比。因此，在低温条件下，陶瓷材料的热导率与 T^3 成正比。升高温度，热容增加缓慢，但是声子平均自由程随温度升高而降低。因此，热导率达到最大值。进一步升高温度，热容达到 3*R*。因为在这一范围内，平均自由程与温度成反比，所以热导率随着温度升高而降低。进一步升高温度，由于光子的参与，热导率再次随着温度升高而增加。

44 晶体陶瓷及存在大量晶型沉淀的玻璃陶瓷的结构越有序性，导致了声子散射的降低。与玻璃相比，这些材料有较高的热导率。

45 在透明材料中，低温下声子的热传导起主导作用。因此，k_t随着温度增加而增加。进一步升高温度，光子热传导出现，并能够在较高的温度下迅速地增加。在不透明材料中，光子对热导没有作用。在升高到一定温度后，热导率将停止增加。

conduction in ceramics is:

$$k_t = k_t^P + k_t^r = \frac{1}{3} C_V^P v_p \overline{\lambda_p} + \frac{16}{3} \sigma_0 n^2 T^3 \lambda_r \tag{9-70}$$

where, the phonon propagation velocity, v_p, can be regarded as constant, i.e., not varying with temperature. At low temperature, the mean free path of phonons reaches maximum, concomitantly, the phonon heat capacity is proportional to T^3. Therefore, at low temperature, thermal conductivity of ceramics is proportional to T^3. Increasing temperature, heat capacity increases slowly, but the mean free path of phonon decreases with temperature. Thus, the thermal conductivity approaches its maximum. Further increasing temperature, heat capacity reaches 3*R*. Since the mean free path is reciprocal to temperature, at this range, the thermal conductivity decreases with temperature. Increasing temperature furthermore, due to the photon involvement for thermal conducting, the thermal conductivity increases with temperature again.[43]

The more ordered structures of the crystalline ceramics, as well as glass-ceramics that contain large amounts of crystalline precipitates, cause less scattering of phonons. Compared with glasses, these materials have a higher thermal conductivity.[44] As the temperature increases, however, scattering becomes more pronounced, and the thermal conductivity decreases (Figure 9.16). At still higher temperatures, heat transfer by radiation becomes significant, and the conductivity may increase. The thermal conductivity of polycrystalline ceramic materials is typically lower than that of single crystals.

In amorphous ceramics, at high temperature, the thermal conduction largely depends on their transparency (Figure 9.17). In transparent material, at low temperature, phonon thermal conduction dominates. As a result, k_t increases with temperature. Further increase of temperature, photon thermal conduction appears and can increase sharply at even higher temperature. In opaque materials, there is no photon contribution. After certain temperature, the thermal conductivity will stop increasing.[45]

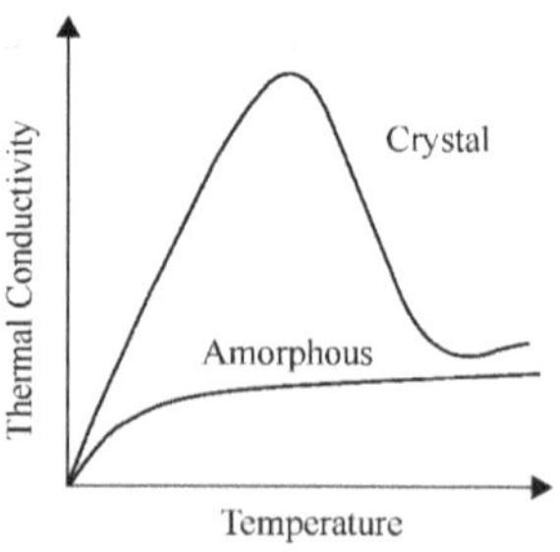

Figure 9.16 Thermal conductivity of ceramics in the states of crystal and amorphous.

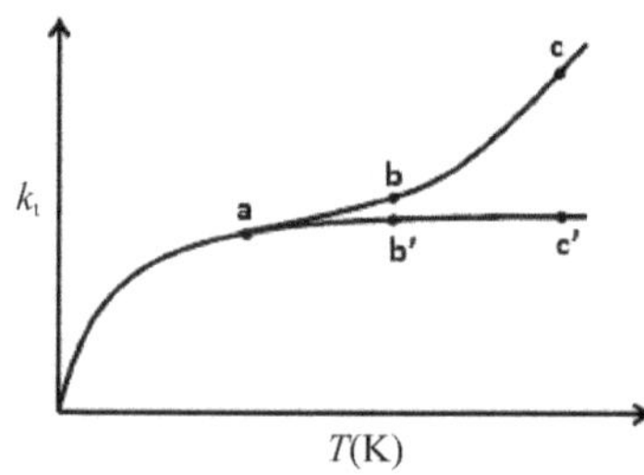

Figure 9.17 Thermal conduction in transparency (abc) and opaque (ab'c') amorphous ceramics.

Another important factor that influences thermal conductivity is porosity. For example, the best insulating brick, contains a large porosity fraction. Effective sintering reduces porosity (and therefore increases thermal conductivity).[46]

Some ceramics such as diamond have very high thermal conductivities. Materials with a close-packed structure and high modulus of elasticity produce high energy phonons that encourage high thermal conductivities. It has been shown that many ceramics with a diamond-like crystal structure (e.g., AlN, SiC, BeO, BP, GaN, Si, AlP) have high thermal conductivities.[47] Although SiC and AlN are good thermal conductors, they are also electrical insulators; therefore, these materials are good candidates for use in electronic packaging substrates where heat dissipation is needed.

Similar to alloy, the thermal conductivity of ceramic composite is lower than ceramic with one component (see Figure 9.18).

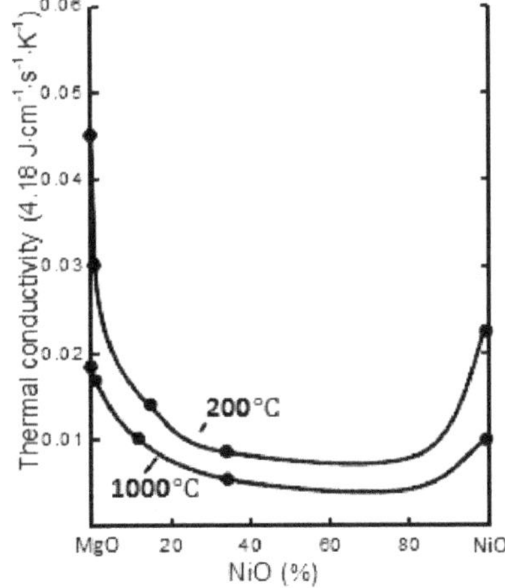

Figure 9.18 The conductivity of MgO-NiO solid solution with different NiO concentration.

Ⅳ. Thermal Conductivity of Polymer

In organic materials, there are no free electrons, and interaction between molecules is weak. Thermal conduction mainly originates from phonons, i.e. collision among molecules. The electrical and thermal conductivities of organic materials are all relatively low, and they are typically considered to be electrical and thermal insulators. At low temperature region, thermal conductivity increases with temperature, reaching maximum value at glass transition temperature (T_g). Above T_g, molecular package becomes loose, thermal conduction decreases. Increasing the degree of polymerization, increasing the crystallinity, minimizing branching, and providing extensive cross-linking all produce a more rigid structure and provide for higher thermal conductivity.[48]

46 另外一种影响热导率的重要因素是孔隙结构。例如，最佳的隔热砖含有大量的孔隙部分。有效的烧结能够减小孔隙（因此提高热导率）。

47 拥有密堆积结构和高弹性模量的材料，能产生有利于高热导率的高能声子。许多类金刚石晶体结构的陶瓷材料表现出高热导率（如AlN, SiC, BeO, BP, GaN, Si, AlP）。

48 在低温区域，热导率随着温度的增加而增加，并在玻璃转化温度（T_g）时达到最大值。高于玻璃转化温度时，分子的堆积变得松散，所以热导率降低。提高聚合度，提升结晶度，减少分支，以及提供大量的交联，都能够产生比较刚性的结构从而得到较高的热导率。

The thermal conductivity of many engineered polymers depends upon the volume fractions of different phases and their connectivity. Silver-filled epoxies are used in many heat-transfer applications related to microelectronics. Unusually good thermal insulation is obtained by using polymer foams, often produced from polystyrene or polyurethane.[49]

Ⅴ. Thermal Conductivity in Semiconductor

In semiconductors, heat is conducted by both phonons and electrons. At low temperatures, phonons are the principal carriers of energy, but at higher temperatures, electrons are excited through the small energy gap into the conduction band, and thermal conductivity increases significantly.[50]

Ⅵ. Comparison of the Thermal Conductivity in Different Materials

The thermal conductivities of some typical materials are shown in Figure 9.19. The high conductivities of aluminum and copper are to be expected because of the transport properties of their nearly free valence electrons. The behavior of graphite is explained by the trigonal sp^2 hybridized bonding; the electron delocalization within the conjugated basal plane as well as the weak bond between its basal planes is readily available for conduction.[51] The comparably high properties of the single Al_2O_3 crystal result from phonon conduction arising from its unusually high purity and crystalline perfection.

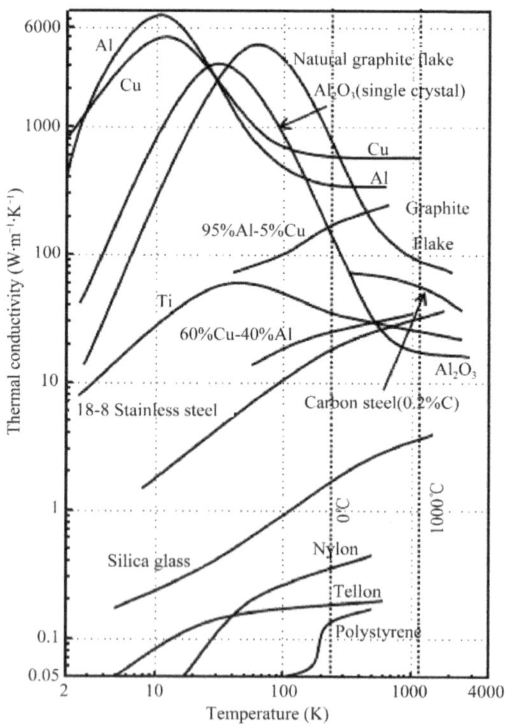

Figure 9.19 Thermal conductivity of some typical materials.

49 许多工业聚合物的热导率由不同相的体积分数和它们的链接性决定。填充银的环氧树脂被用于许多与微电子相关的热传递应用中。良好的绝热体通常通过使用聚合物泡沫获得，这些聚合物泡沫通常由聚苯乙烯和聚氨酯制得。

50 在半导体中，热量是通过声子和电子共同传导的。在低温条件下，声子是主要的能量载体；但在较高的温度下，电子被激发，由较小的能隙进入导带，从而使得热导率显著提高。

51 石墨的性质通过平面三角形的 sp^2 杂化键来阐述；电子在共轭基面内的离域，以及基面之间的弱作用力，有利于传导。

The effects of impurities, in the materials shown in the figure, are largely the result of intentionally added alloying elements. These are shown by the properties of 95%Al-5%Cu and 60%Cu-40%Ni when compared to those of the respective "pure" elements. The comparison of properties of the 18-8 stainless steel with those of the 0.2%C steel also shows a significant contrast that results from large differences in alloying elements.[52]

52 如图所示材料的杂质效应在很大程度上是通过有意加入的合金元素 95%铝-5%铜合金和 60%铜-40%镍合金的性能，与其相应的纯金属的性能相比较，来展示。18-8 不锈钢与含 0.2%碳钢的比较，也能够说明由于合金元素的较大差异而导致的显著不同。

The organic polymeric materials show the lowest thermal conductivities because of the strong covalent bonding involved in these materials and their virtually amorphous molecular structures; the energy transfer, consequently, is very low and occurs almost entirely by phonons in these virtually noncrystalline solids.[53] Data for some typical ceramic and organic materials are given in Table 9.3.

53 由于有机聚合材料中强的共价键和几乎无定形的分子结构，其热导率最低。因此，能量传递效率很低并且几乎完全是靠声子在这些几乎非晶态固体中的传递。

Table 9.3 Thermal conductivities of some selected ceramic and organic materials

Material	37.8 ℃ (100°F)	93.3 ℃ (200°F)	148.9 ℃ (300°F)
Single crystals			
Silicon carbide	0.21	0.21	0.20
Periclase	0.11	0.09	0.08
Spinel $MgO\text{-}Al_2O_3$	0.03	0.03	0.02
Quartz (c axis)	0.03	0.02	0.02
Quartz (basal plane)	0.01	0.01	0.01
Fluorite	0.02	0.02	0.01
Polycrystalline materials			
BeO (pure, hot pressed)	0.52	0.43	0.38
MgO (spec, pure)	0.09	0.08	0.07
ThO_2 (hot pressed)	0.04	0.03	0.03
PbO	0.007	0.005	0.004
Organic materials	k_t near room temperature		
Polyethylene	0.002		
Rubber	0.007		
Urethane foam	0.0001		

9.3 Thermal Expansion

9.3.1 Length and Volume Thermal Expansions

In the absence of change in structure, most solid materials expand when temperature increases and shrinks with decreasing temperature. The change in the dimensions of the material per unit length is given by the linear coefficient of thermal expansion:

$$\alpha_l = \frac{\partial l}{l\,\partial T} \tag{9-71}$$

where l is the length of specimen at temperature T, α_l is the linear thermal expansion coefficient. Table 9.4 shows the linear coefficient of thermal expansion for some materials.

Linear expansion is closely related to the volume thermal expansion of solids. In analogy with the one-dimensional case we have

$$\alpha_V = \frac{\partial V}{V\,\partial T} \tag{9-72}$$

Table 9.4 The linear coefficient of thermal expansion at room temperature for selected materials

Material	Linear Coefficient of Thermal Expansion $10^{-8}\cdot ℃^{-1}$	Material	Linear Coefficient of Thermal Expansion $10^{-6}\cdot ℃^{-1}$
Al	25.0	Epoxy	55.0
Cu	16.6	6.6-nylon	80.0
Fe	12.0	6.6-nylon-33% glass fiber	20.0
Ni	13.0	Polyethylene	100.0
Pb	29.0	Polyethylene-30% glass fiber	48.0
Si	3.0	Polystyrene	70.0
W	4.5	Al_2O_3	6.7
1020 steel	12.0	Fused silica	0.55
3003 aluminum alloy	23.2	Partially stabilized ZrO_2	10.6
Gray iron	12.0	SiC	4.3
Invar (Fe-36%Ni)	1.54	Si_3N_4	3.3
Stainless steel	17.3	Soda-lime glass	9.0
Yellow brass	18.9		

where V is the volume of specimen at temperature T, α_V is the volume thermal expansion coefficient. The relationship between α_l and α_V depends on the structure of the crystalline solid. For isotropic materials, $a_v = 3a_l$; and for anisotropic materials, $\alpha_v = \alpha_{lx} + \alpha_{ly} + \alpha_{lz}$. Table 9.5 gives some of these values.

Table 9.5 Relationship between linear (length) and volume expansion coefficients in crystals

Type of crystal structure	Relationship between α_l and α_v
Cubic structure	$\alpha_v \approx 3\alpha_l$
Hexagonal, trigonal and tetragonal structures	$\alpha_v \approx \alpha_{lx} + 2\ \alpha_{ly}$
Rhomic, mono- and triclinic structures	$\alpha_v \approx \alpha_{lx} + \alpha_{ly} + \alpha_{lz}$

9.3.2 Theory for Thermal Expansion

The coefficient of thermal expansion of a material is related to the atomic bonding. It is well known that thermal expansion of solids is caused by a change in the interatomic distances in the crystal lattice and to a minor extent vacancy formation at higher temperatures below the melting point of the solid.[54] A very reasonable explanation of length expansion is the harmonicity of the vibrations of the atoms around their equilibrium positions in the lattice, which corresponds to increasing distances between the atoms with increase in temperature.

54 众所周知，固体热膨胀是由于晶体晶格内原子间距离的变化，在很小程度上也缘于，低于固体熔点的较高温度下的空位形成。

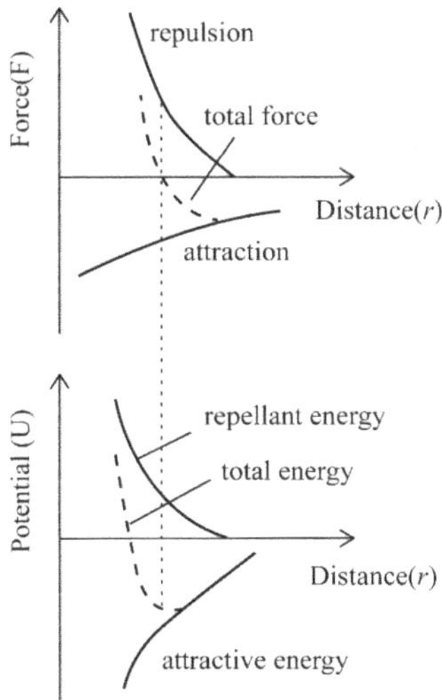

Figure 9.20 The relationship between atomic interaction/potential and atomic distance.

As shown in Figure 9.20, at r_0, the potential energy is the lowest, and the attraction force is balanced by the repulsion force (their sum is zero). When atom deviates from its equilibrium position(r_0), for the same value of distance increase or decrease, the change of repulsion force/energy is far much larger than the attraction force/energy. This means that with the increase of temperature, the atom deviation toward distance-enlargement should be larger than atom deviation toward distance reduction direction, leading to atomic expansion.[55]

55 这意味着，随着温度的升高，原子在距离上的偏移，其正方向上的增加变化量应大于其负方向上的减小变化量，从而导致原子膨胀。

Considering two atoms with equilibrium distance of r_0, increasing distance of δ, the potential energy change to:

$$E(r) = E(r_0 + \delta) \tag{9-73}$$

Expand Equation (9-73) to Taylor series:

$$E(r) = E(r_0) + \left(\frac{\partial E}{\partial r}\right)_{r_0} \delta + \frac{1}{2!}\left(\frac{\partial^2 E}{\partial r^2}\right)_{r_0} \delta^2 + \frac{1}{3!}\left(\frac{\partial^3 E}{\partial r^3}\right)_{r_0} \delta^3 + \cdots \tag{9-74}$$

Because $\left(\frac{\partial E}{\partial r}\right)_{r_0} = 0$, and define $\beta = \left(\frac{\partial^2 E}{\partial r^2}\right)_{r_0}$, $\beta' = -\frac{1}{2}\left(\frac{\partial^2 E}{\partial r^3}\right)_{r_0}$ we have:

$$E(r) \approx E(r_0) + \frac{1}{2}\beta\delta^2 - \frac{1}{3}\beta'\delta^3 + \cdots \tag{9-75}$$

Only considering the first two items, the force between atoms can be expressed as:

$$F(r) = -\frac{\mathrm{d}E}{\mathrm{d}r} = -\beta\delta \tag{9-76}$$

This means that an increase of distance linearly increases the attraction force, and vice visa. Therefore, although the higher temperature leads to larger deviation from equilibrium position, since the changes in the positive (attraction) and negative directions (repulsion) are the same, the equilibrium position does not change, hence no thermal expansion.[56] This does not agree with experimental results. Considering the first three items in Equation (9-75), we get

56 因此，尽管较高的温度会导致较大的距平衡位置的偏离，因为正向（相吸）和反向的变化（排斥）是相同的，所以平衡位置不会发生改变，因此没有热膨胀。

$$F(r) = -\frac{\mathrm{d}E}{\mathrm{d}r} = -\beta\delta + \beta'\delta^2 \tag{9-77}$$

This means that the change of force with different r is no longer linear, and hence the equilibrium position changes. As shown in Figure 9.21, when temperature increases, the oscillation amplitude of atom toward x direction is larger than toward $-x$ direction, leading to the deviation of equilibrium position from a_0 point and hence thermal expansion occurs.[57]

57 如图 9.21 所示，当温度升高，原子沿 x 正方向的振动振幅要大于沿 x 轴负方向的振动振幅，从而导致了 a_0 平衡位置的偏移，因此发生热膨胀。

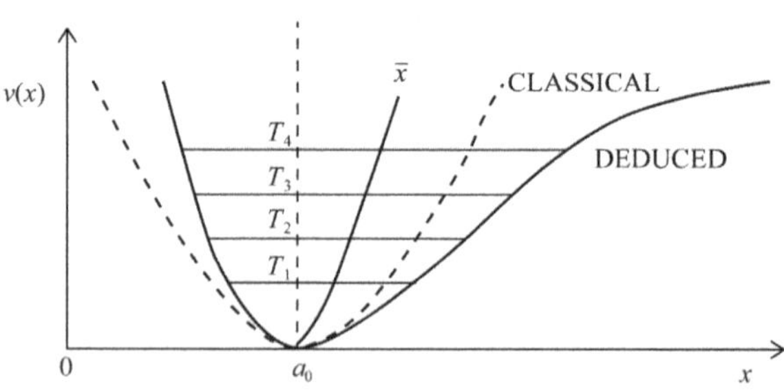

Figure 9.21 Calculation of the average amplitudes of an oscillating particle as function of temperature ($T_4 > T_3 > T_2 > T_1 > 0$ K) under different models.

Using Bolzmann statistics, the average distance change can be expressed as,

$$\bar{\delta} = \frac{\beta' kT}{\beta^2} \tag{9-78}$$

The linear thermal expansion coefficient is:

$$\alpha_1 = \frac{\mathrm{d}\bar{\delta}}{r_0 \mathrm{d}T} = \frac{\beta' k}{r_0 \beta^2} \tag{9-79}$$

Figure 9.22 shows the increase of the average distance with increasing temperature in some metals and alloy.

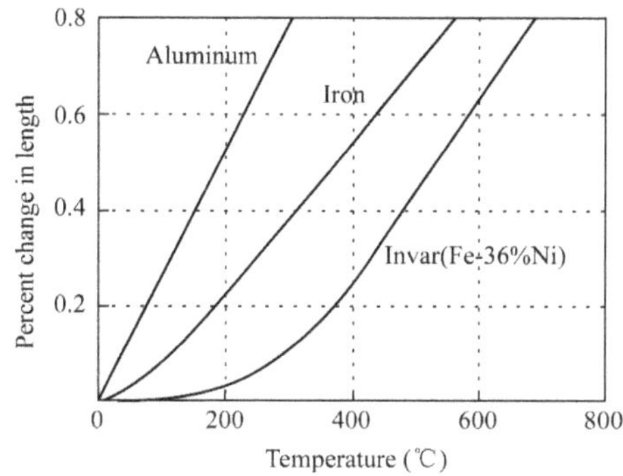

Figure 9.22 The expansion of some metals and alloys with temperature.

9.3.3 Thermal Expansion and Heat Capacity

Thermal expansion is the dimension/volume increase caused by lattice vibration, and heat capacity is the energy increase caused by lattice vibration. Gruneisen rule can express their relationship.[58]

$$\alpha_1 = \frac{\gamma}{KV} C_{\mathrm{V}} \tag{9-80}$$

here γ is Gruneisen parameter, referring to nonlinear lattice vibration, generally with values of 1.5 - 2.5; K is bulk modulus; V is volume. Figure 9.23 shows the temperature dependency of heat capacity and thermal expansion in Al. It can be seen that thermal expansion and heat capacity have similar changing tendency, demonstrating their similar origination.[59]

58 热膨胀是由晶格振动造成的尺度或体积的增加，热容则是由晶格振动造成的能量的增加。格律乃森规能够表达这两者之间的关系。

59 图 9.23 展示了在铝金属中，温度与热容和热膨胀的关系。可以看出，热膨胀和热容有相似的变化趋势，证实了两者相似的来源。

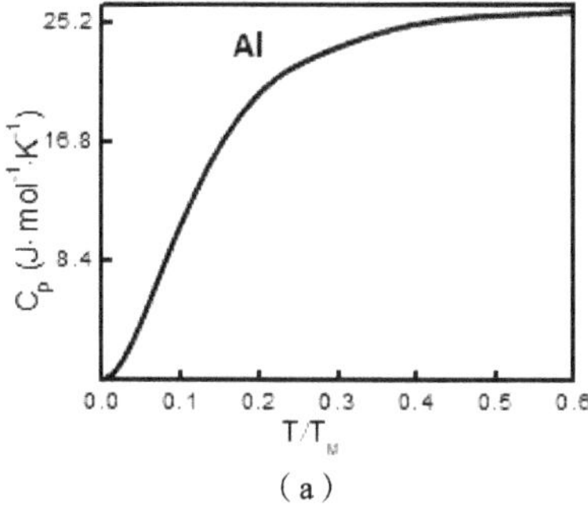

(a)

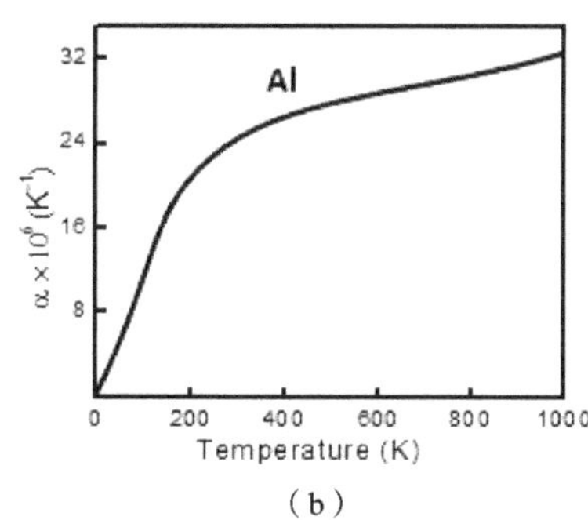

(b)

Figure 9.23 Temperature dependency of (a) heat capacity and (b) thermal expansion in Al.

9.3.4 Thermal Expansion and Melting Point

Gruneisen proposed that the volume will expand 6% when pure metal change from 0 K to T_m. Because when volume expands 6%, the bonding between atoms is so weak that the solid state cannot maintain, so

$$\frac{V_{T_m} - V_0}{V_0} \approx 6\% \tag{9-81}$$

The lower the melting point, the larger the thermal expansion coefficient. There is an empirical equation:

$$\alpha_1 T_m = b \tag{9-82}$$

where b is constant in a material.

Figure 9.24 shows the experimental results of the linear coefficient of thermal expansion versus the melting temperature in metals. As can be seen that higher melting point metals tend to expand to a lesser degree. Introducing Debye temperature of Equation (9-41), we get

$$\alpha_1 = \frac{A}{MV_a^{2/3}\Theta_D^2} \tag{9-83}$$

where M is relative atomic mass, V atomic volume, A constant.

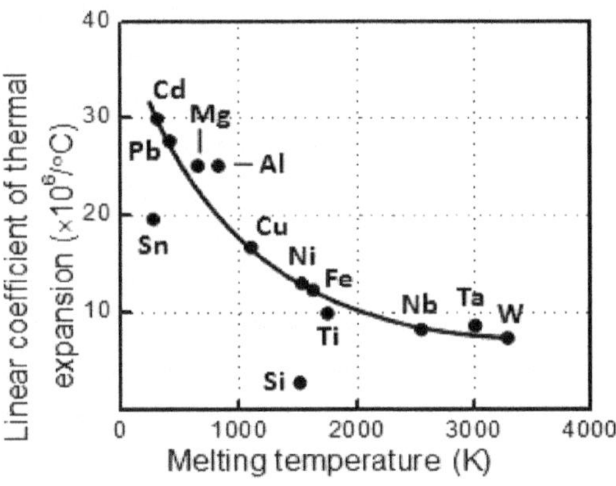

Figure 9.24 The linear coefficient of thermal expansion versus the melting temperature in metals.

9.4 Thermal Stability and Thermal Shock

(1) Thermal Stability

Thermal stability is the quality of a substance to resist irreversible change in its chemical or physical structure, often by resisting decomposition or polymerisation, at a relative higher temperature.[60] Thermostable materials may be used industrially as fire retardants, most of which are ceramic and refractory metal alloys etc. Thermal stability is also one of the very important

60 热稳定性指的是材料抵抗不可逆的物理或者化学变化的特性，通常以在相对较高温度下，材料抵抗分解或者聚合来表示。

material parcameters for optoelectronic applications.

(2) Thermal Shock

Thermal shock describes the way in which some materials are prone to damage if they are exposed to a sudden change in temperature or exist a temperature gradient.[61] Glass and certain other materials are vulnerable to this process, in part because they do not conduct thermal energy very well. This is readily observed when a hot glass is exposed to ice water — the result is a cracked, broken, or even shattered glass.

The damage is a reaction to a rapid and extreme temperature fluctuation, but the process is somewhat more complicated than this.[62] The shock is the result of a thermal gradient, which refers to the fact that temperature change occurs in an uneven fashion. Temperature change causes expansion of the molecular structure of an object, due to weakening of the bonds that hold the atoms/molecules in formation. The existence of the thermal gradient means this expansion occurs unevenly.[63]

In the example of the hot glass, this means that the rapid temperature change causes some parts of the glass to quickly become much hotter than other parts. This, in turn, causes uneven expansion, which puts stress on the molecular structure. If the stress becomes great enough, the strength of the material is overcome and the glass breaks.

Improving the shock resistance of glass and ceramics can be achieved by improving the strength of the material or reducing its tendency to uneven expansion.[64] One example of success in this area is Pyrex®, the brand name of a type of glass that is most well-known to consumers as cookware, but which is also used to manufacture laboratory glassware. The type of glass traditionally used to make Pyrex® is called borosilicate glass, due to the addition of boron, which prevents shock by reducing the tendency of the glass to expand.

Reinforced carbon-fiber is extremely resistant to thermal shock, due to graphite's extremely high thermal conductivity and low expansion coefficient, the high strength of carbon fiber, and a reasonable ability to deflect cracks within the structure.[65]

61 热冲击描述的是当材料暴露在突然的温度变化或者存在温度梯度时，倾向于被破坏的方式。

62 （热冲击）损害是对快速且极限温度波动的反应，但其过程在某种程度上比该反应复杂得多。

63 由于温度变化引起原子/分子结合键的削弱，导致物体的原子/分子结构的膨胀。热梯度的存在意味着不均匀的膨胀。

64 可以通过提高材料强度或者降低非均匀膨胀的倾向来改善玻璃和陶瓷对热冲击的抵抗。

65 增强碳纤维有非常好的热冲击抵抗力，是源于石墨非常高的热导和低热膨胀系数、碳纤维的高强度以及在结构内抵制裂痕的能力。

本章小结

1. 内容概要

本章讨论了材料的热学性质，包括热容、热导和热膨胀。首先，本章介绍了不同的热容模型，并比较了它们之间的差异，介绍了真实材料的热容性质；其次，从定义和机理角度介绍了热导，对不同材料的热导进行了解释，对材料的热导和电导的关系也有所陈述；最后，引入热膨胀的概念，并简明地指出其与热容之间的联系。

2. 基本概念

热容、爱因斯坦温度、德拜温度、热导、热导率、导温系数、热膨胀、体积弹性模量、自由度均分原则、热流密度、线性热膨胀系数、表面热传导系数、比热、热应变、热应力、体积膨胀系数、零点能。

3. 主要公式

（1）热容：$specific\ heat = \dfrac{heat\ capacity}{molar\ weight}$

（2）杜隆—珀替定律：$C_{\mathrm{V}} = \left(\dfrac{\partial U}{\partial T}\right)|_{\mathrm{V}} = 3N_{\mathrm{A}}k = 3R$

（3）爱因斯坦模型:

$$C_{\mathrm{mV}} = 3N_{\mathrm{A}}k\left(\frac{\Theta_{\mathrm{E}}}{T}\right)^2 \frac{e^{\Theta_{\mathrm{E}}/T}}{(e^{\Theta_{\mathrm{E}}/T}-1)^2} \approx 3N_{\mathrm{A}}k\left(\frac{\Theta_{\mathrm{E}}}{T}\right)^2 e^{-\Theta_{\mathrm{E}}/T}(@T \ll \Theta_{\mathrm{E}})$$

（4）德拜模型:

$$C_{\mathrm{V}} = 9N_{\mathrm{A}}k\left(\frac{T}{\Theta_{\mathrm{D}}}\right)^3 \int_0^{\frac{\Theta_{\mathrm{D}}}{T}} \frac{x^4 e^x}{(e^x-1)^2}\mathrm{d}x \approx 9N_{\mathrm{A}}k\left(\frac{T}{\Theta_{\mathrm{D}}}\right)^3 \frac{4}{15}\pi^4 = bT^3(@T \ll \Theta_{\mathrm{D}})$$

（5）德拜温度：$\Theta_{\mathrm{D}} = \dfrac{\hbar\omega_{\mathrm{m}}}{k} = 0.76\times10^{-11}\omega_{\mathrm{m}}$

（6）德拜温度与熔点关系：$\Theta_{\mathrm{D}} = 133.7\sqrt{\dfrac{T_{\mathrm{m}}}{MV_{\mathrm{a}}^{2/3}}}$

（7）自由电子热容：$C_{\mathrm{mV}}^{\mathrm{e}} = N_{\mathrm{A}}Z\left(\dfrac{\partial \bar{E}}{\partial T}\right) = N_{\mathrm{A}}Z\dfrac{\pi^2k^2}{2E_{\mathrm{F}}^0}T = \dfrac{\pi^2 ZkR}{2E_{\mathrm{F}}^0}T$

（8）二组分热容：$C = pC_1 + qC_2$

（9）多组分热容：$C = \sum g_{\mathrm{i}}C_{\mathrm{i}}$

（10）热扩散率（导温系数）：$\alpha = \dfrac{k_{\mathrm{t}}}{dC_{\mathrm{P}}}$

（11）声子热导率（导热系数）：$k_{\mathrm{t}}^{\mathrm{P}} = \dfrac{1}{3}C_{\mathrm{V}}^{\mathrm{P}}\overline{v}_{\mathrm{p}}\lambda_{\mathrm{p}}$

（12）自由电子热导率（导热系数）：$k_{\mathrm{t}}^{\mathrm{e}} = \dfrac{1}{3}C_{\mathrm{V}}^{\mathrm{e}}v_{\mathrm{e}}\lambda_{\mathrm{e}} = \dfrac{\pi^2 n_{\mathrm{e}}k^2T\tau_{\mathrm{e}}}{3m}$

（13）光子热导率（导热系数）：$k_t^r = \frac{1}{3}C_r v_r \lambda_r = \frac{16}{3}\sigma_0 n^2 T^3 \lambda_r$

（14）魏德曼-费兰兹-洛伦兹定律：$\frac{k_t}{\sigma} = LT$

（15）洛伦兹常量：$L = \frac{k_t^e}{\sigma T} = \frac{\frac{\pi^2 n_e k^2 T \tau_e}{3m}}{(e^2 n_e \tau_e / m)T} = \frac{\pi^2 k^2}{3e^2} \approx 2.45 \times 10^{-8}\ \mathrm{V^2 \cdot K^{-2}}$

（16）陶瓷热导率：$k_t = k_t^P + k_t^r = \frac{1}{3}C_V^P v_p \overline{\lambda_p} + \frac{16}{3}\sigma_0 n^2 T^3 \lambda_r$

（17）线性热膨胀系数：$\alpha_l = \frac{d\overline{\delta}}{r_0 dT} = \frac{\beta' k}{r_0 \beta^2}$

（18）格律乃森规则：$\alpha_l = \frac{\gamma}{KV}C_V = \frac{A}{MV_a^{2/3}\Theta_D^2}$

Vocabulary

atomic distance	原子距离
atomic number	原子序数
attraction force	吸引力
bulk modulus	体积弹性模量
constant pressure	恒压
constant volume	定容
deterioration	退化
equilibrium position	平衡位置
equipartition principle	自由度均分原则
exothermal	放热的
expand	膨胀
gas constant	气体常数
heat capacity	热容
heat flux	热流，热流密度
internal energy	内能
kinetic energy	动能
linear coefficient of thermal expansion	线性热膨胀系数
multiply	乘
phonon	声子
potential energy	势能
quantized	量子化的
repulsion force	排斥力

shrink	缩小，收缩
specific heat	比热
steady state	稳态
surface thermal transfer coefficient	表面热传递系数
temperature gradient	温度梯度
thermal	热的
thermal conductivity	热导率，导热系数
thermal diffusivity	导温系数
thermal expansion	热膨胀
thermal shock	热冲击
thermal strain	热应变
thermal stress	热应力
unsteady state	非稳态
volume coefficient of thermal expansion	体热膨胀系数
zero point energy	零点能

Problems

1. What is the meaning of the gas constant *R*?

2. Contrast the classical, Einstein and Debye models for the internal energies of solids. What is the significance of the differences between them?

3. Explain the basis for the failure of the Einstein model of the heat capacity of solids at low temperature.

4. Explain the partial success of the classical theory for heat capacity.

5. Please give definition of heat capacity and specific heat capacity.

6. According to the following figure for metal heat capacity, please explain the different temperature dependence in each region.

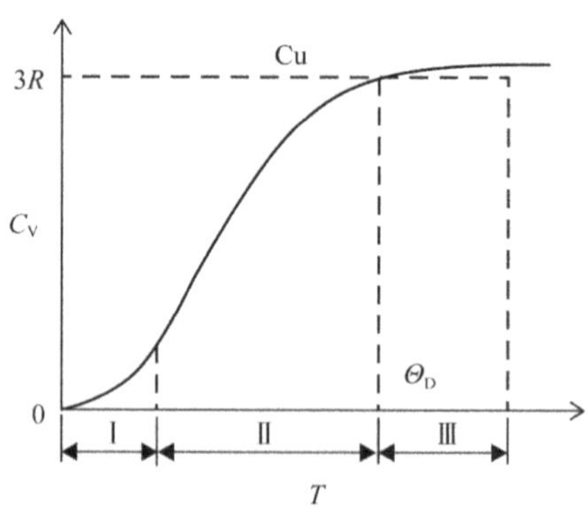

7. Please explain Debye temperature.

8. True or false questions

(1) Heat capacity is independent of spin arrangement.

(2) The heat capacity of one material is dependent on crystal structure.

(3) The factors that contribute to heat capacity include atomic vibration and electron vibration.

(4) Compared to metal, the capacity of ceramics shows more agreement with Debye model because there are hardly free electrons there.

(5) In copper, the dominant factor for thermal conduction is phonon.

(6) For metals with relatively low electrical conductivity and at relatively low temperature, both electron and phonon are the main contribution for thermal conduction.

(7) The thermal conductivity of materials is independent of temperature.

(8) The thermal conduction by photon takes effect only at high temperature.

(9) In a good conductor and above Debye temperature, the ratio of thermal and electrical conductivities is almost proportional to absolute temperature.

(10) Generally speaking, a thermal conductivity of a metal alloy is lower than corresponding metal.

(11) The thermal conductivity of a ceramic with crystalline structure is generally lower than that in amorphous state.

(12) Compared to opaque amorphous ceramic, the sharper increasing of thermal conduction at high temperature in a transparency amorphous ceramic is due to photon thermal conduction.

(13) In an isotropic material, the volume expansion coefficient is three times the linear expansion coefficient.

(14) Generally speaking, the thermal expansion coefficient is reciprocal to melting point.

9. Please characterize a phonon (energy, momentum, statistics, formation and annihilation).

10. Compare the thermal conducting factors of nonconducting solids with metals.

11. Please points out the three factors that are responsible for thermal conduction.

12. Please state the origination for thermal expansion.

13. Explain the conditions under which a nonmetallic solid would be a better conductor of heat than a metal. How might these ideas be applied to a cryogenic insulation system?

参考文献

1. 李志林. 材料物理[M]. 北京：化学工业出版社，2009.

2. FREDRIKSSON H, AKERLIND U. Physics of Functional Materials [M]. West Sussex, England: John Wiley & Sons, Ltd, 2008.

3. ASKELAND D R, FULAY P P, WRIGHT W J. The Science and Engineering of Materials, Six Ed[M]. Stamford, USA:Cengage Learning, 2011.

4. SHACKELFORD J F. Introduction to Materials Science for Engineers, Seventh Ed [M].New Jersey, USA: Pearson Prentice Hall, 2009.

5. SOLYMAR L, WALSH D. Electrical Properties of Materials, Seventh Ed[M]. Oxford, UK: Oxford University Press, 2004.

6. POLLOCK D D. Physical Properties of Materials for Engineers, Second Ed [M]. Florida, USA: CRC Press, 1993.

www.ingramcontent.com/pod-product-compliance
Ingram Content Group UK Ltd.
Pitfield, Milton Keynes, MK11 3LW, UK
UKHW062005290726
14090UKWH00022B/1410